Reproductive Toxicology

TARGET ORGAN TOXICOLOGY SERIES

Series Editors

A. Wallace Hayes, John A. Thomas, and Donald E. Gardner

Reproductive Toxicology

Third Edition

Edited by

Robert W. Kapp, Jr.
BioTox
Monroe Township, New Jersey, U.S.A.

Rochelle W. Tyl
RTI International
Research Triangle Park, North Carolina, U.S.A.

CRC Press
Taylor & Francis Group
Boca Raton London New York

CRC Press is an imprint of the
Taylor & Francis Group, an **informa** business

CRC Press
Taylor & Francis Group
6000 Broken Sound Parkway NW, Suite 300
Boca Raton, FL 33487-2742

First issued in paperback 2019

© 2010 by Taylor & Francis Group, LLC
CRC Press is an imprint of Taylor & Francis Group, an Informa business

No claim to original U.S. Government works

ISBN-13: 978-1-4200-7343-0 (hbk)
ISBN-13: 978-0-367-38360-2 (pbk)

This book contains information obtained from authentic and highly regarded sources. While all reasonable efforts have been made to publish reliable data and information, neither the author[s] nor the publisher can accept any legal responsibility or liability for any errors or omissions that may be made. The publishers wish to make clear that any views or opinions expressed in this book by individual editors, authors or contributors are personal to them and do not necessarily reflect the views/opinions of the publishers. The information or guidance contained in this book is intended for use by medical, scientific or health-care professionals and is provided strictly as a supplement to the medical or other professional's own judgement, their knowledge of the patient's medical history, relevant manufacturer's instructions and the appropriate best practice guidelines. Because of the rapid advances in medical science, any information or advice on dosages, procedures or diagnoses should be independently verified. The reader is strongly urged to consult the relevant national drug formulary and the drug companies' and device or material manufacturers' printed instructions, and their websites, before administering or utilizing any of the drugs, devices or materials mentioned in this book. This book does not indicate whether a particular treatment is appropriate or suitable for a particular individual. Ultimately it is the sole responsibility of the medical professional to make his or her own professional judgements, so as to advise and treat patients appropriately. The authors and publishers have also attempted to trace the copyright holders of all material reproduced in this publication and apologize to copyright holders if permission to publish in this form has not been obtained. If any copyright material has not been acknowledged please write and let us know so we may rectify in any future reprint.

Except as permitted under U.S. Copyright Law, no part of this book may be reprinted, reproduced, transmitted, or utilized in any form by any electronic, mechanical, or other means, now known or hereafter invented, including photocopying, microfilming, and recording, or in any information storage or retrieval system, without written permission from the publishers.

For permission to photocopy or use material electronically from this work, please access www.copyright.com (http://www.copyright.com/) or contact the Copyright Clearance Center, Inc. (CCC), 222 Rosewood Drive, Danvers, MA 01923, 978-750-8400. CCC is a not-for-profit organization that provides licenses and registration for a variety of users. For organizations that have been granted a photocopy license by the CCC, a separate system of payment has been arranged.

Trademark Notice: Product or corporate names may be trademarks or registered trademarks, and are used only for identification and explanation without intent to infringe.

A CIP record for this book is available from the British Library.

Library of Congress Cataloging-in-Publication Data available on application

Typeset by MPS Limited, a Macmillan Company

Visit the Taylor & Francis Web site at
http://www.taylorandfrancis.com

and the CRC Press Web site at
http://www.crcpress.com

To my wife, Beverly, for her encouragement and understanding during this project; to my children Bob, Sheri, and Bonney—all successful in their chosen fields; and to the memory of my father, Robert W. Kapp, Sr., MD (1919–2008), an accomplished surgeon who was my inspiration and mentor and, sadly, passed away during the preparation of this manuscript. His life was dedicated to healing and helping the infirmed in desperate situations. His zest for life was amazing and is appropriately captured in the following:

> *You have a choice every day regarding the attitude you will embrace for that day. You cannot change what has already happened to you or control how people will act toward you or control the inevitable. The only thing you can control is your attitude in the present . . . which is more important than money, than circumstances, than failures, than successes, than possessions, than what other people think or say or do . . . life is 10% what happens to you, and 90% how you react to it. Attitude will make or break a company . . . a church . . . a home . . . attitude is virtually everything. If you control your attitude . . . your life will follow in concert with it . . .*

Modified from Charles R. Swindoll
Robert W. Kapp, Jr.

To my husband, Tom, my inspiration and strongest supporter; to my children, Dr Jenifer Wolkowski and Mr Jeffrey Tyl, my greatest achievements; and to my many teachers, professors, students, collaborators, staff, and friends whom I have had the honor and privilege of learning from and working with over these many years.

Rochelle W. Tyl

Preface

This third edition of *Reproductive Toxicology* is a collaborative effort by the authors and editors to capture our rapidly growing understanding of reproductive and developmental biology and toxicology from the level of molecules to gametes to adult organisms. This explosion of information is made possible by integrative approaches (e.g., systems biology, in silico initiatives), new areas and technologies (e.g., genomics/genetics, epigenetics, proteomics, and metabolomics), new methodologies and applications, etc. We are beginning to understand the regulatory processes involved in the exquisite, orchestrated processes of male and female development in terms of reproductive structure and functions and the coordinated development of the major organ systems. The chapters cover the normal and abnormal development and function of the reproductive systems of males and females, the processes involved in pregnancy, fertilization, and development of the central nervous, cardiovascular, renal, respiratory, immune, and the endocrine systems. We also focus on the consequences of exposures to toxic substances during vulnerable life stages on the development, structures, and functions of these biological systems. These chapters also include epigenetics, -omics, and metallic environmental factors in reproduction and development.

We have included overviews of the national and international governmental regulatory agencies, which provide oversight (testing guidelines, technical guidance, etc.) for pharmaceuticals by the Food and Drug Administration (FDA) and the International Conference on Harmonization (ICH, United States, Japan, and European Union), and for pesticides and commodity chemicals by the Environmental Protection Agency (EPA), Office of Prevention, Pesticides and Toxic Substances (OPPTS), and the Organization for Economic Cooperation and Development (OECD), now including over 26 countries (including the United States, Japan, European Union, etc.). We also included an overview of Investigative New Drug (IND) submissions to the FDA.

We fervently hope (and expect) that the expertise, insight, and excitement in the content of these 20 chapters on our accelerated acquisition of knowledge and clear understanding of reproductive toxicology (which the authors have conveyed so well) will inspire our readers, as they have us, the editors of this book.

Robert W. Kapp, Jr.
Rochelle W. Tyl

Acknowledgments

The previous editions of *Reproductive Toxicology* in the Target Organ Toxicology Series (Robert L. Dixon edited the first edition in 1985 and Raphael J. Witorsch edited the second in 1995) have set the bar high, with insightful and in-depth reviews of critical current topics in reproductive toxicology. The scientific advancements in reproductive toxicology over the past 15 years have been extraordinary. With this third edition, we have striven to maintain the quality of these previous editions by keeping the topics contemporary and providing current developments in the field. This task was daunting because of the explosive growth of the science and could not have been completed without the many contributing authors. We greatly appreciate their excellent work and are thankful that, in spite of their many other responsibilities, they were able to participate in the project. We are grateful to our many colleagues, friends, and students who have provided critical evaluation and meaningful suggestions during the course of this edition. We further thank the staff of Informa for their patient and helpful assistance throughout the production of this volume.

Lastly, we are deeply indebted to the series editors, Drs John A. Thomas, A. Wallace Hayes, and Donald E. Gardner who provided much needed guidance and encouragement throughout the project. We acknowledge and appreciate the opportunity to create this third edition *of Reproductive Toxicology.*

Contents

Contributors

Mallikarjuna Basavarajappa   Department of Veterinary Biosciences, University of Illinois at Urbana-Champaign, Urbana-Champaign, Illinois, U.S.A.

Chad Blystone   Reproductive Toxicology Branch, US Environmental Protection Agency, Research Triangle Park, North Carolina, U.S.A.

Jason Burgess   Discovery Sciences Division, RTI International, Research Triangle Park, North Carolina, U.S.A.

Leigh Ann Burns-Naas   Drug Safety Research and Development, Pfizer Global Research and Development, San Diego, California, U.S.A.

Gregg D. Cappon   Pfizer Global Research and Development, Groton, Connecticut, U.S.A.

Edward W. Carney   Toxicology, Dow Chemical Company, Midland, Michigan, U.S.A.

Mark Collinge   Drug Safety Research and Development, Pfizer Global Research and Development, Groton, Connecticut, U.S.A.

Ralph L. Cooper   Endocrine Toxicology Branch, National Health and Environmental Effects Research Laboratory, US Environmental Protection Agency, Research Triangle Park, North Carolina, U.S.A.

John M. DeSesso   Center for Toxicology and Mechanistic Biology, Exponent, Alexandria, Virginia and Department of Biochemistry, Molecular & Cellular Biology, Georgetown University School of Medicine, Washington, DC, U.S.A.

Kimberly Ehman   RTI International, Research Triangle Park, North Carolina (Now at Toxicology Regulatory Services [TRS], Inc., Charlottesville, Virginia), U.S.A.

Timothy Fennell   Discovery Sciences Division, RTI International, Research Triangle Park, North Carolina, U.S.A.

Jodi A. Flaws   Department of Veterinary Biosciences, University of Illinois at Urbana-Champaign, Urbana-Champaign, Illinois, U.S.A.

Paul Foster   National Institute of Environmental Health Sciences, Research Triangle Park, North Carolina, U.S.A.

Johnathan R. Furr   Reproductive Toxicology Branch, US Environmental Protection Agency, Research Triangle Park, North Carolina, U.S.A.

Jerome M. Goldman Endocrine Toxicology Branch, National Health and Environmental Effects Research Laboratory, US Environmental Protection Agency, Research Triangle Park, North Carolina, U.S.A.

Mari S. Golub Department of Environmental Toxicology, University of California Davis, Davis, California, U.S.A.

L. Earl Gray, Jr. Reproductive Toxicology Branch, US Environmental Protection Agency, Research Triangle Park, North Carolina, U.S.A.

Rupesh Gupta Department of Veterinary Biosciences, University of Illinois at Urbana-Champaign, Urbana-Champaign, Illinois, U.S.A.

Andrew K. Hotchkiss Reproductive Toxicology Branch, US Environmental Protection Agency, Research Triangle Park, North Carolina, U.S.A.

Kembra L. Howdeshell Reproductive Toxicology Branch, US Environmental Protection Agency, Research Triangle Park, North Carolina, U.S.A.

Mark E. Hurtt Pfizer Global Research and Development, Groton, Connecticut, U.S.A.

Kamin J. Johnson Nemours Biomedical Research, Alfred I. duPont Hospital for Children, Wilmington, Delaware, U.S.A.

Robert W. Kapp, Jr. BioTox, Monroe Township, New Jersey, U.S.A.

Thomas T. Kawabata Drug Safety Research and Development, Pfizer Global Research and Development, Groton, Connecticut, U.S.A.

Harriet Karimi Kinyamu Laboratory of Molecular Carcinogenesis, National Institute of Environmental Health Sciences, Research Triangle Park, North Carolina, U.S.A.

Christy R. Lambright Reproductive Toxicology Branch, US Environmental Protection Agency, Research Triangle Park, North Carolina, U.S.A.

Susan L. Makris National Center for Environmental Assessment, US Environmental Protection Agency, Washington, DC, U.S.A.

M. Sue Marty Toxicology, Dow Chemical Company, Midland, Michigan, U.S.A.

Harihara M. Mehendale Department of Toxicology, University of Louisiana at Monroe, Monroe, Louisiana, U.S.A.

Retha Newbold Laboratory of Molecular Toxicology and National Toxicology Program, National Institute of Environmental Health Sciences, Research Triangle Park, North Carolina, U.S.A.

Robert M. Parker Princeton Research Center, Huntingdon Life Sciences, East Millstone, New Jersey, U.S.A.

John Parrish   Department of Animal Sciences, University of Wisconsin at Madison, U.S.A.

Merle G. Paule   National Center for Toxicological Research, Jefferson, Arkansas, U.S.A.

Tessie Paulose   Department of Veterinary Biosciences, University of Illinois at Urbana-Champaign, Urbana-Champaign, Illinois, U.S.A.

Jackye Peretz   Department of Veterinary Biosciences, University of Illinois at Urbana-Champaign, Urbana-Champaign, Illinois, U.S.A.

Reza J. Rasoulpour   Toxicology, Dow Chemical Company, Midland, Michigan, U.S.A.

Cynthia V. Rider   Reproductive Toxicology Branch, US Environmental Protection Agency, Research Triangle Park and Duke University, Durham, North Carolina, U.S.A.

Kartik Shankar   Arkansas Children's Nutrition Center, University of Arkansas for Medical Sciences, Little Rock, Arkansas, U.S.A.

William Slikker, Jr.   National Center for Toxicological Research, Jefferson, Arkansas, U.S.A.

Rodney Snyder   Discovery Sciences Division, RTI International, Research Triangle Park, North Carolina, U.S.A.

Susan Sumner   Discovery Sciences Division, RTI International, Research Triangle Park, North Carolina, U.S.A.

Rochelle W. Tyl   Discovery Sciences Division, RTI International, Research Triangle Park, North Carolina, U.S.A.

Arthi G. Venkat   Georgetown University School of Medicine, Washington, DC, U.S.A.

Vickie S. Wilson   Reproductive Toxicology Branch, US Environmental Protection Agency, Research Triangle Park, North Carolina, U.S.A.

Raymond G. York   RG York and Associates, Manlius, New York, U.S.A.

Ayelet Ziv-Gal   Department of Veterinary Biosciences, University of Illinois at Urbana-Champaign, Urbana-Champaign, Illinois, U.S.A.

1 Biology and physiology of fertilization

John Parrish

INTRODUCTION

Fertilization is the sequence of molecular and cellular events that culminates in the union of the male and female haploid genome (Fig. 1). This review summarizes the current understandings of fertilization-related events that occur both to the male and female gamete. Interaction of the gametes in the female reproductive tract is also included as this is often not discussed related to how dramatically it can either support or interfere with fertilization. As this review is a current summary of information, the reader is directed to more detailed reviews on the spermatozoon, fertilization-related events, and gamete interactions with the oviduct, if needed (1–3).

In mammals, fertilization occurs within the confines of the oviduct. Preceding the act of fertilization, the haploid gametes must be formed. In males, spermatogenesis in the testis results in the production of a haploid sperm that then undergoes a series of modifications in the epididymis that ensure the sperm can arrive in the oviduct at the correct time following deposition in the female. In the female, the primary oocyte resumes meiosis following the luteinizing hormone (LH) surge and then becomes arrested as a secondary oocyte. Ovulation ensues and the union of the sperm and oocyte in the oviduct results in the final resumption of meiosis by the oocyte. The fusion of the paternal and maternal pronuclei culminates the reestablishment of the diploid state and concludes the events of fertilization discussed in this review.

SPERMATOGENESIS AND EPIDIDYMAL MATURATION

In mammals, sperm complete spermatogenesis in the testis and then undergo further maturation in the epididymis. This portion of the review is not intended to be an extensive coverage of the process of spermatogenesis or epididymal maturation (4,5), it only covers the development of structures in spermatozoa essential for fertilization.

During spermatogenesis, the essential structure of the spermatozoan is formed. Critical to fertilization is the formation of the acrosome from the Golgi, segmentation of the plasma membrane, development of the locomotion apparatus in the flagella, and reduction of the DNA to the haploid state, along with nuclear condensation via the addition of protamines and packaging of the DNA in toroids to be attached to the nuclear matrix. The completion of these and other steps lead to a spermatozoan leaving the testis with the potential for fertilization as clearly indicated by the ability to fertilize an oocyte through the use of intracytoplasmic sperm injection. However, sperm leaving the testis cannot on their own fertilize an oocyte. These sperm remain only vibratory in motion, cannot undergo sustained movement, and so could not reach the site of fertilization in the female oviduct even if ejaculated. Additional maturation thus ensues in the epididymis (6). It is here that the DNA becomes further condensed via the addition of disulfide bonds. This appears essential to prevent damage from reactive oxygen species to the DNA once sperm get into the female reproductive tract or even during storage in the cauda epididymis. Modifications also occur to the plasma membrane of the sperm to prevent premature fusion of the outer acrosome membrane with the inner plasma membrane. During the normal process of fertilization, the zona pellucida of the oocyte induces this fusion after sperm have undergone capacitation.

1

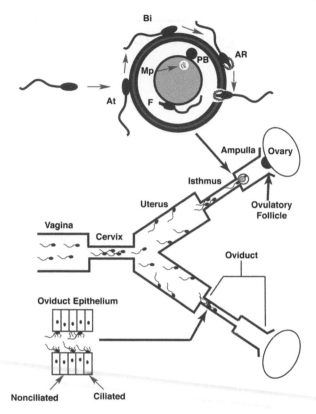

FIGURE 1 Physiological events of mammalian fertilization. The diagram utilizes a bovine female reproductive tract to generically illustrate events in mammals. Sperm are deposited in the female during mating and then make their way up the reproductive tract interacting with various epithelial cells of the female until the sperm reservoir is established in the isthmus of the oviduct. As capacitation is completed and ovulation occurs, sperm move up the oviduct to interact with the oocyte. The oocyte is surrounded by the zona pellucida and is arrested in metaphase II of meiosis as indicated by the presence of a metaphase plate (Mp) and the first polar body (PB). The interaction of sperm with the oocyte involves first attachment (At), then binding (Bi), and finally induction of the acrosome reaction (AR). Enzymatic digestion of the zona pellucida occurs to allow sperm to enter the perivitelline space. Fusion (F) of the sperm and oocyte plasma membrane occurs and is followed by several molecular events activating the oocyte.

The modifications in the epididymis appear to delay capacitation from occurring prematurely. Similar events may also occur due to components of seminal fluid contributed by accessory sex glands during the ejaculatory process.

OOGENESIS

In the female mammal, mitotic divisions of oogenesis, entry into meiosis, and meiotic arrest as a primary oocyte, all occur prior to birth in the fetal stage with only a few exceptions. The ability to ovulate a mature oocyte only develops in a female post puberty when the correct endocrine profile is present. A primary oocyte, however, undergoes a period of growth during which the zona pellucida, cortical granules, and calcium stores develop (2). The role of these will become clear later in this review. In addition, there is synthesis of mRNA that is to be stored until fertilization and early embryo development prior to the turn-on of the embryonic genome. The correct storage profile of the mRNA is the principal basis of cytoplasmic maturation. What triggers the follicle and oocyte to initiate this period of growth is unknown. If, however, the growing oocyte and follicle are at the correct stage of development and a female has an LH surge, then meiosis will resume. Meiosis then continues with the first division of meiosis and extrusion of the first polar body. The oocyte becomes the secondary oocyte and progresses through the cell cycle arresting at metaphase. At this point, the cell cycle regulatory proteins hold the oocyte in the metaphase arrest, pending the calcium wave at oocyte activation by the sperm (2). During this time, the cortical granules also will align themselves in the cortical region of the oocyte, and the cumulus cells surrounding the oocyte will go through expansion. Final maturation of the oocyte occurs as indicated by the ability to respond to the sperm with a calcium wave of activation and the ability to complete the final phases of meiosis, and is termed nuclear maturation. In vivo, the mature oocyte is ovulated, picked up by the cilia of the oviductal fimbria, transported into the ostium (mouth of the oviduct), and then moves down the oviduct to the ampullary-isthmic junction via muscular contractions. The oocyte then stops at this location and awaits the fertilizing sperm.

When the surface of the ovulated egg is examined, we find that most of the oocyte is covered with microvilli (7). However, the region over the meiotic spindle is devoid of these microvilli (8). This will become important later as sperm binding to the oocyte membrane occurs only in regions of microvilli. Thus, sperm cannot bind just anywhere on the surface of an oocyte.

SPERM IN FEMALE TRACT

Sperm enter the reproductive tract of the female at the time of mating. The exact timing of mating is controlled in most species by the expression of estrus, a discrete period in which the female will allow the male to mount. Estrus is triggered by increased estrogens from the preovulatory follicle. The increased estrogens also likely change the physiology of the female reproductive tract to be conducive to sperm survival and to provide the correct coordinated transport to the site of fertilization as the oocyte is ovulated and picked up by the fimbria.

As sperm enter the reproductive tract during mating, they are either deposited in the anterior vagina, cervix, or pass through as a bulk ejaculate directly into the uterus. Three phases of sperm movement or transport within the female then follow. First, sperm move rapidly up through the uterus and oviduct in a process known as rapid transport (3). This occurs in a matter of a few minutes (10–15) and was once thought to be the primary mechanism by which the fertilizing sperm reached the site of fertilization. Many of these sperm are now known to be nonmotile and unlikely to participate in fertilization. They may be cleared from the oviduct by continuation of this transport phenomenon into the peritoneal cavity followed by removal via phagocytosis. A second phase of transport, first described by Harper (9) and then by Overstreet and Cooper (10) in the rabbit, followed by work in the pig and sheep (11), involves the movement over 6 to12 hours of sperm up to the lower isthmus of the oviduct just beyond the uterotubal junction. This type of transport has been referred to as sustained transport and is thought to be present in most, if not all, mammals. Once sperm arrive in the oviduct, they have been found to bind to uterine oviductal cells and retain viability for extended lengths of time, and form what is known as the oviductal sperm reservoir (12,13). In addition, the binding of sperm in the lower isthmus prevents excess sperm from reaching the site of fertilization at the ampullary-isthmic junction (14).

Much is now known about how sperm interact with the oviductal epithelium. In most mammals, there are no specialized structures in the lower isthmus of the oviduct. Sperm interaction with mouse oviductal cells in vivo was first noted by Suarez (15).

In vitro observations of binding of sperm to oviduct epithelial cells have been observed in many species (12) and are not restricted to the isthmus; ampullary cells might bind to sperm even better. The location of the sperm reservoir is likely then the first site at which viable sperm interact with the oviduct cells. The specific binding is clearly carbohydrate mediated with a variety of mono- and oligosaccharides shown to inhibit binding (12,16–18). A variety of in vitro assays demonstrated the importance of fucose in bovine sperm recognition of bovine oviduct epithelial cells, especially if in a Lewis-a trisaccharide configuration (19). In comparison to the bovine, porcine sperm appear to bind to porcine oviduct epithelial cells via an oligomannose high-affinity site as well as a low-affinity galactose residue in glycan chains (20).

Utilizing an affinity purification approach with Lewis-a fucose, the protein PDC109 on the surface of ejaculated bovine sperm was found responsible for binding to oviductal cells (14). Interestingly, the protein is of seminal plasma origin, previously called BSP-A1/A2. When ejaculated bovine sperm are exposed to oviduct epithelial cells in vitro, they do not bind as well as sperm exposed to seminal plasma or PDC109 (21). This effect of seminal plasma has also been noted in the porcine (22), although AQN-1 may be the seminal plasma protein responsible for sperm binding in pigs (20). The PDC109 belongs to a family of seminal plasma proteins, several of which have been shown to be able to replace PDC109 in inducing bovine epididymal sperm binding to oviduct cells (23). There are also several proteins in the porcine seminal plasma, and it may be that these proteins have similar properties in the mechanisms by which they interact with sperm and the oviduct cells of the respective species. Similar proteins may be expressed in epididymal secretions in other species such as the hamster as these epididymal sperm can bind to hamster oviductal cells in vivo (16).

When the bovine sperm receptor on oviduct cells was investigated, several members of the annexin family of proteins were identified by utilizing sperm proteins as traps (24). It is interesting that there are not only several sperm proteins potentially involved in oviduct binding but perhaps several oviduct receptors of sperm. The complexity of molecules involved in the formation of the isthmic sperm reservoir suggests it is critical in mammalian fertilization.

The next question is what is the purpose of sperm binding. The isthmic cells of the oviduct clearly stop functionally normal sperm and appear to hold them in a viable yet uncapacitated state for hours (14,20). Some coordination must then occur to allow sperm to be released from these isthmic cells as ovulation takes place. While numerous theories have been advanced for this integration of events, no evidence supporting any point of view has been experimentally obtained. Whatever the mechanism, sperm are released and then rapidly move up the oviduct, likely under their own power, to the ampullar-isthmic junction where they encounter the recently ovulated oocyte. This final phase of sperm transport might then be called oviductal transport. As the mammalian sperm approaches the oocyte, there has been evidence of chemotaxis on the millimeter scale of distances that lead the sperm into interaction with the oocyte, but exact mechanism(s) has not been resolved (25,26).

CAPACITATION

While ejaculated mammalian sperm have all the needed equipment to fertilize an oocyte, surprisingly they require a period of time either in vivo in the female repro-ductive tract or in vitro under various species-specific conditions to be able to penetrate the zona pellucida around the oocyte (27). The processes that occur to the sperm, which convey this ability to penetrate an oocyte, have been termed capacitation. Capacitation then culminates in the ability of the zona pellucida to induce an acrosome reaction and the sperm to express hyperactivation. Through experiments in a variety of species, a common pathway of capacitation is beginning to emerge (2,28). Proteins from the male epididymis or seminal secretions are known to bind to sperm during epididymal maturation or ejaculation. For example, in the bovine, these may be the BSP-A1/A2 (29), while in the porcine the important ones may be AQN-1, AQN-3, and AWN (30). These proteins have been termed spermadhesion proteins and for the most part inhibit the capacitation process until their removal (31–33). Once sperm enter the female reproductive tract, these proteins may either be stripped off or interact with reproductive tract secretions to promote initial portions of capacitation. The initial changes in sperm are increases in intracellular calcium, pH, and cholesterol efflux. The increase in calcium also appears to load the acrosome with calcium during capacitation (34,35). Although involvement of membrane channels or extracellular cholesterol acceptors have been shown, no clear, all encompassing pathway by which these increases occur has been identified. Preventing any of these, however, inhibits capacitation and demonstrates their essential nature. In addition to these specific changes, bicarbonate is present in the female reproductive tract and it, along with the other noted changes in sperm, stimulates a sperm-specific soluble adenylate cyclase. This, in turn, raises cyclic adenosine monophosphate (cAMP) and protein kinase A (PKA) activity in sperm. Cross talk with tyrosine kinase(s) then results in tyrosine phosphorylation of a variety of proteins. It is this phosphorylation event that is similar in all species studied (36–38). Although numerous, the tyrosine phosphorylation targets have recently been identified in the human (39,40), hamster (41), boar (42) and mouse (43). The phosphorylated proteins fall into several groups associated with glycolysis, motility, voltage-activated ionic channels, and response to oxygen free radicals. The results of the tyrosine phosphorylation and other changes then lead to the ability of sperm to undergo a zona pellucida–induced acrosome reaction. The increase in cAMP and PKA activity also appears to be tied to F-actin polymerization in bull, human, murine, and ram spermatozoa (44), which is essential to the acrosome reaction.

At some point in capacitation there is a change in the membrane protein and lipid organization (45). The ability of the proteins and lipids to diffuse in the membrane is not random and begins to change during capacitation. A portion of the initial lipid and protein distribution as well as changes that occur later may be due to only minimal cytosol, and resulting availability of the membrane to interact with various cytoskeletal elements that are regionally distributed to both areas within the head as well as tail of the sperm. As noted above, changes to the cytoskeleton protein, actin, occur during capacitation and so may physically promote some of these membrane changes. The concept of lipid rafts has now become central to explaining how these changes in lipids are affecting the capacitated state of sperm. As cholesterol is removed from the sperm membrane during capacitation (46,47), there is a clustering of the lipid rafts, in particular in the apical region of the plasma membrane over the acrosome. This is the region that will later interact with the zona pellucida of the oocyte and trigger the acrosome reaction. The process of capacitation produces not only the intracellular changes to PKA and tyrosine kinase phosphorylation but also changes to membrane architecture.

Another main feature of capacitation is the initiation of what is termed hyperactivation of motility. The amplitude of the sperm tail movement increases with a whiplash motion that results in a figure 8–type pattern to sperm movement when in culture medium. In the confines of the reproductive tract or among cumulus cells, this is likely to result in increased force generated from this type of motility and its ability then to drive a sperm forward. A repetitive pulse of calcium is necessary for normal motility and may be enhanced during hyperactivation. Recent discovery in mice of several cation channels of sperm (CatSper) ion channels have demonstrated they are essential for the hyperactivation via influx of calcium (48). Interestingly, calcium entering sperm via CatSper channels and release from intracellular stores may initiate glycolysis and production of ATP needed for motility during hyperactivation.

ACROSOME REACTION, ZONA PELLUCIDA BINDING, AND SPERM FUSION

One of the most important organelles of sperm is the acrosome for which its formation and role in fertilization has recently been reviewed (49). The acrosome is formed in the late pachytene spermatocyte through mid spermiogenesis. It is derived from a series of vesicles produced from the Golgi, which coalesce in the perinuclear region and eventually form a single vesicle, the acrosome, in each spermatid. The acrosome of a mature sperm has an outer membrane that is just below the plasma membrane of the anterior sperm head and an inner membrane that is in tight association with the nuclear membrane. During the acrosome reaction, exocytosis of the acrosome vesicle occurs with fusion of the outer acrosomal membrane with the plasma membrane. The contents of the acrosomal matrix, a series of proteases, glycosidases, and some binding proteins are then released. Many of these enzymes exist in the acrosomal matrix as inactive precursors of their active form. It has been believed that a central function of the acrosome was the autocatalytic cleavage of proacrosin to its active form, acrosin, during the acrosome reaction. Acrosin is a protease and esterase that was considered to be zona pellucida lysin, which allowed sperm penetration through the zona pellucida of the oocyte. This was brought into question by retained fertility of transgenic mice that are null for the proacrosin gene. The proacrosin null mice, however, do have decreased ability to penetrate the zona pellucida of the oocyte so, in practice, acrosin may still play the role of the zona lysin.

As capacitated sperm are released from the oviductal sperm reservoir, it is likely that sperm move on their own in periods of hyperactivated and nonhyperactivated movement until they come in contact with the oocyte at the ampullary-isthmic junction (3). In some species such as mice and hamsters, the oocytes are surrounded by cumulus cells as the sperm approach. Yet, in others, such as the bovine, the cumulus cells, although present at ovulation, are rapidly removed prior to the time the oocyte reaches the ampulla-isthmic junction. Most indications are that only acrosome-intact sperm interact successfully with either the cumulus mass or the zona pellucida. With regards to the cumulus mass, it is made of the cumulus cells separated by the glycosamino-glycan, hyaluronan (hyaluronic acid). Penetration through the cumulus has for a long time believed to be due to hyaluronidase from sperm (50). While sperm do contain hyaluronidase, most of the originally identified enzyme was found within the acrosome, questioning how an acrosome-intact sperm could penetrate through the zona without first undergoing the acrosome reaction to release the active hyaluronidase. Later, one form was found on the surface of sperm as a glycosylphosphatidylinositol (GPI)-anchored protein, PH-20, but knockouts for this protein did not interfere with fertilization. Additional forms of hyaluronan hydrolyzing proteins have now been identified with some in and on sperm. One or several of these are now thought responsible for sperm penetration through the cumulus mass in species, which encounter this structure prior to fertilization.

Once sperm reach the zona pellucida, they undergo a series of steps defined by first a loose attachment followed by a more tenacious binding event and induction of the acrosome reaction (2). The zona pellucida is made up of three proteins referred to as ZP1, ZP2, or ZP3, the naming based on electrophoretic mobilities of zona proteins isolated in the mouse (51). An alternative nomenclature, ZPA, ZPB, ZPC, is based on the length of the cDNAs, with ZPA as the longest (52). Interestingly, the zona pellucida proteins are related to egg coats of nonmammalian species that are separated by hundreds of millions of years in evolution (53). Egg coats in many species are capable of inducing the acrosome reaction within the same species but generally not in other species.

The sperm binds to the zona in what was originally thought to be a strictly species-specific fashion but is now recognized to be more species restrictive (2). Only a capacitated sperm then undergoes the acrosome reaction and penetrates the zona pellucida. Several events during capacitation may regulate this phenomenon. First capacitated sperm hyperpolarize and the most hyperpolarized ones are those that undergo the acrosome reaction (54). The same is true about calcium accumulation within the acrosome, with the highest acrosomal calcium found in sperm that undergo a zona pellucida–induced acrosome reaction (34,35). Lastly, as already discussed, capacitation is associated with changes in lipid rafts that aggregate in the apical region of the sperm head where the sperm binds to the zona (45). The acrosome-reacting ability of the zona resides within the ZP3. The binding of the capacitated sperm to ZP3 produces a G-protein activation, intracellular alkalinization, and transient Ca^{2+} entry (54). There is generation of inositol 1,4,5-trisphosphate (IP_3) at this time, which likely activates intracellular calcium release from the acrosome or other stores in the neck of sperm (34,55). A phosphoinositide 3 (PI3) kinase in combination with enkurin appears to open a canonical transient receptor potential channel (TRPC), allowing a sustained influx of Ca^{2+} that triggers downstream events of the fusion process (54). In addition to effects on the acrosome reaction, calcium release from intracellular sperm stores such as the neck region may occur in response to the zona, cumulus cells, progesterone, and/or nitric oxide in or near the cumulus cells and regulate flagellar activity (55).

Several molecules have been identified on sperm which interact with the zona pellucida, specifically ZP3 to induce the acrosome reaction in mammalian sperm (56). One of the first to be identified was sperm protein 56 (sp56) in the mouse and the guinea pig homologue, AM67. While originally these proteins were thought to be surface proteins on acrosome intact sperm, they were eventually identified as components of the acrosomal matrix. So how then could it be involved in induction of the acrosome reaction? Research into this question has unraveled how proteins in sperm may undergo trafficking to different region of the sperm. In mice transgenic for expression of the green fluorescent protein (GFP) in the acrosome, it was demonstrated that sp56 was not present on uncapacitated sperm, but as capacitation progressed, it was found on the sperm surface where it could bind to anti-sp56 antibodies. At this point in time there was no loss of GFP from the sperm, and through the light microscope could not be seen to be undergoing the acrosome reaction. Sperm then underwent the acrosome reaction with increased binding to the anti-sp56 and only minimal loss of GFP. Later, substantial GFP was lost from sperm as they completed the acrosome reaction and could no longer bind anti-sp56. Results of these and supporting experiments suggest that, in the latter steps of capacitation, small pores must form between the acrosome and plasma membrane allowing mixing of sperm acrosomal proteins onto the surface of sperm to facilitate zona binding and complete induction of the acrosome reaction via the stimulus-induced response previously discussed.

Another sperm protein identified to bind to ZP3 and able to induce the acrosome reaction in mouse sperm is β1-4 galactosyltransferase (GalT) (57). The role of sperm GalT came into question when GalT-null mice were found to bind to ovulated zona pellucida but not soluble ZP3. It was demonstrated that a ZP3-independent mechanism was involved and a novel ligand of ovulated but not ovarian oocytes was present. Second, a homologue of a pig zona-binding protein has been identified in the mouse and named SED1 that is known to be involved with a variety of cell-cell interactions. The protein SED1 binds to both ZP3 and, to some extent, Zp2. According to Shur (57), it is hypothesized that several ZP3-indendent and ZP3-dependent molecules initiate the weak initial binding of capacitated sperm to the zona pellucida. This then allows a more tenacious binding with a binding molecule such as sperm GalT that results in the induction of the acrosome reaction. What is the exact mechanism of sperm binding to the zona remains unclear with the conflicting sp56 and GalT not resolved. As indicated by Shur (57), the binding may be in a multimeric complex with many redundancies built in to ensure this important event occurs.

Once a sperm undergoes the acrosome reaction on the surface of the zona pellucida, enzymatic digestion is presumed to allow sperm to penetrate this protein coat. An alternative mechanism, for example, physical force, by which this occurs have also been put forward by Bedford (6). Regardless, a motile and acrosome-reacted sperm penetrates the zona pellucida reaching the perivitelline space. The acrosome reaction appears essential for the sperm then to bind and fuse with the oocyte's plasma membrane (27). In addition, motility, at least initially, also is important. The acrosome reaction appears to reorganize the sperm plasma membrane with its specific modifications to the equatorial region that allow this part of the sperm to then bind to the oocyte. As discussed previously, the sperm interacts with microvilli on the surface of the oocyte but does not interact with the nonmicrovilli area above the meiotic spindle and first polar body (8). On interaction with the microvilli of the oocyte, fusion begins to occur in the equatorial segment and extends to the posterior head. The motility of the sperm ceases as fusion begins and the sperm is taken up by the oocyte in what is described as a phagocytotic-like event (27).

The fusion of the sperm with the oocyte membrane appears to be modified by proteins on the sperm and an integrin in the oocyte (58). On the sperm side, a series of proteins were identified that belongs to a class of proteins known as ADAMs, which contain a disintegrin domain and a metalloprotease domain. These molecules were identified in sperm as ADAM1 and ADAM2, initially called fertilin α and fertilin β. Another molecule ADAM3, first called cyritestin, has also been identified. Members of this family have been found in many tissues and may function as cell adhesion molecules or are important for proteolysis. In addition to ADAMs, a protein of epididymal origin and found on sperm, cysteine-rich secretory protein 1 (CRISP1), has also been identified as an inhibitor of sperm-oocyte fusion during in vitro fertilization, but it remains unclear how this is mediated. A second protein, CRISP2, of testicular origin found on sperm, has also recently been implicated in the fusion event (59). On the oocyte side there are many potential integrins that could bind to sperm ADAMs. Integrins are made of an α and a β chain, but there are known to be up to 18 different α chains and 8 β chains. Thus, a large potential number of different integrins is possible in oocytes. It is speculated that the important one in oocytes and involved in sperm-oocyte fusion may be $\alpha_9\beta_1$, although several others have also been found in oocytes. In addition, the fusion properties may be mediated by a fusion protein called CD9, which is a member of the tetraspanin protein family and known to be involved in cell fusion in other cell types (60). The current view on sperm-oocyte adhesion and fusion are that on the sperm side, fertilin α, fertilin β and cyritestin interact to mediate the adhesion process with an integrin on the oocyte side. Through as yet an undefined mechanism, CRISP1 is involved in either adhesion or fusion. While the exact mechanism of fusion is not clear, there may be involvement of CD9 or a similar type of fusion protein.

OOCYTE ACTIVATION

It has been known for many years that the sperm must activate the oocyte to trigger resumption of meiosis and extrusion of the cortical granules via a calcium-dependent mechanism (61). The cortical granules are released into the perivitelline space and modify the zona pellucida such that it cannot induce additional capacitated sperm to undergo an acrosome reaction or at least blocks penetration of the zona pellucida by previously acrosome-reacted sperm. The calcium involved in the activation of the oocyte is released from internal stores and propagated as repetitive calcium waves or oscillations across the oocyte from the site of sperm interaction with the oocyte. The mechanism of the calcium increase was demonstrated to be via IP_3 production in the oocyte followed by calcium release from the endoplasmic reticulum. How the initial generation of IP_3 was stimulated by sperm remained unclear for many years. There is now consensus that sperm release a soluble and sperm-specific phospholipase C zeta (PLCζ) (62). The PLCζ has now been found in eight mammalian species (63). During fusion of the sperm and oocyte, PLCζ, which is present in the sperm head, is released into the oocyte. It then binds to and cleaves PIP_2 in the oocyte. The location of phosphatidylinositol 4,5-biphosphate (PIP_2) in the oocyte plasma membrane or in other organelles is not known. Importantly, PLCζ can work at the free calcium levels found in the oocyte, 100 to 800 nM, and so can generate IP_3 at resting calcium concentrations, and has even more activity during each calcium pulse of the subsequent oscillations. The IP_3 appears to generate the calcium release after interaction with the IP_3 receptor on the intracellular calcium stores within the oocyte. As calcium is released, increasing amounts of IP_3 are produced by PLCζ and the calcium wave increases and calcium-induced calcium release propagates the wave across the oocyte. Calcium levels decrease as the stores reload with calcium and some calcium is pumped out of the oocyte. The action of PLCζ will then reinitiate the wave as intracellular oocyte stores become refilled. The interval of the calcium oscillation is between a few minutes in rodents up to 30 or 40 minutes in species such as the bovine. The exact role and need for the calcium oscillations are also controversial. It is clear that a calcium increase is needed and leads to release of cortical granules as well as destruction of cyclin B and inactivation of maturation-promoting factor (MPF) also called CDK1 (cell division kinase 1), which is at the core of the anaphase-promoting complex. A more detailed explanation of the effects of the calcium signal on the cell cycle regulatory machinery in the oocyte can be found in Florman and Ducibella (2). The need for repetitive oscillations is not clear other than these events occur more completely when oscillations are present in contrast to a single calcium pulse. An alternative role of the oscillations may involve metabolism of egg mitochondria, which are stimulated and produce more ATP with each calcium pulse. Reduced ATP does lead to poorer embryonic development potential (64). While a single calcium spike can cause resumption of the meiosis and progression of the cell cycle, embryonic development is clearly better with calcium oscillations (65). The calcium oscillations appear to stop with formation of the pronuclei and the initiation of S-phase. The reason for this appears to be the translocation of PLCζ to the pronuclei due to a nuclear-targeting portion of the protein.

SPERM NUCLEAR DECONDENSATION AND PRONUCLEAR FORMATION

After the fertilizing sperm triggers resumption of meiosis via the calcium oscillations, the sperm nucleus enters the oocyte's cytoplasm, decondenses in response to oocyte glutathione, which reduces the disulfide bonds existing among and between adjacent protamine molecules (66–69). The protamines are then removed from the DNA and replaced by histones (70) with the paternal pronuclear envelope then forming around the decondensed sperm nucleus. There is a 10 to 100 times greater volume in the paternal pronucleus than in the original sperm nucleus. At the same time, the maternal chromatin goes through the second division of meiosis, the second polar body is extruded, and then condenses followed by decondensation and formation of the maternal pronucleus in synchrony with the paternal chromatin. There appears to be a window of opportunity for the sperm to decondense after its activation signal triggers the oocyte to resume meiosis. If the sperm fails to complete decondensation as the oocyte is going from metaphase through telophase and extrusion of the second polar body, the maternal chromatin will enter interphase but sperm chromatin remodeling will be arrested and fertilization will fail (68,69).

An additional role of the calcium activation event is the release of the oocyte cortical granules. Within 15 minutes of fertilization, release of the cortical granules into the perivitelline space is complete (2). Modification of the zona pellucida by cortical granule release is the major mechanism used by mammals to prevent polyspermy, although an additional block at the level of the oocyte plasma membrane may be present and apparently absent in aged oocytes (8,71). Following fertilization, there are clearly changes to the zona pellucida that prevent induction of the acrosome reaction and entry of more sperm into the perivitelline space, but it has been difficult to exactly relate this to specific enzymatic components found within cortical granules (2).

CONCLUSIONS

Fertilization involves many events that are initiated both in the male and female reproductive systems with completion of meiosis. The gametes must then meet each other in a coordinated sequence of events that not only involve ovulation in the female but ejaculation and semen deposition by the male into the female along with sperm transport to the site of fertilization in the oviduct. As described in this review, many novel proteins are involved in sperm interaction with the female reproductive tract and the oocyte to promote sperm survival, capacitation, hyperactivation, the acrosome reaction, sperm fusion with the oocyte, activation of the oocyte via calcium waves, and prevention of polyspermy. Most, if not all, of these events are subject to disruption that decreases the efficiency of fertilization and eventual embryo development. The process of fertilization is therefore a very sensitive area for effects of toxicological agents.

REFERENCES
1. Eddy EM. The spermatozoon. In: Neill JD, ed. Physiology of Reproduction. San Diego, CA: Elsevier, 2006:3–54.
2. Florman HM, Ducibella T. Fertilization in mammals. In: Neill JD, ed. Physiology of Reproduction. San Diego, CA: Elsevier, 2006:55–112.
3. Suarez SS. Gamete and zygote transport. In: Neill JD, ed. Physiology of Reproduction. San Diego CA: Elsevier, 2006:113–145.
4. Kerr JB, Loveland KL, O'Bryan MK, et al. Cytology of the testis and intrinsic control mechanisms. In: Neill JD, ed. Physiology of Reproduction. San Diego, CA: Elsevier, 2006:827–948.
5. Robaire B, Hinton BT, Orgebin-Crist M. The epididymis. In: Neill JD, ed. Physiology of Reproduction. San Diego, CA: Elsevier, 2006:1071–1148.
6. Bedford JM. Puzzles of mammalian fertilization—and beyond. Int J Dev Biol 2008; 52:415–426.
7. Longo FJ, Chen DY. Development of cortical polarity in mouse eggs: involvement of the meiotic apparatus. Dev Biol 1985; 107:382–394.
8. Dalo DT, McCaffery JM, Evans JP. Ultrastructural analysis of egg membrane abnormalities in post-ovulatory aged eggs. Int J Dev Biol 2008; 52:535–544.
9. Harper MJ. Relationship between sperm transport and penetration of eggs in the rabbit oviduct. Biol Reprod 1973; 8:441–450.
10. Overstreet JW, Cooper GW. Sperm transport in the reproductive tract of the female rabbit. II. The sustained phase of transport. Biol Reprod 1978; 19:115–132.
11. Hunter RH, Nichol R. Transport of spermatozoa in the sheep oviduct: preovulatory sequestering of cells in the caudal isthmus. J Exp Zool 1983; 228:121–128.
12. Suarez SS. Formation of a reservoir of sperm in the oviduct. Reprod Domest Anim 2002; 37:140–143.

13. Töpfer-Petersen E, Wagner A, Friedrich J, et al. Function of the mammalian oviductal sperm reservoir. J Exp Zool 2002; 292:210–215.
14. Suarez SS. Regulation of sperm storage and movement in the mammalian oviduct. Int. J Dev Biol 2008; 52:455–462.
15. Suarez SS. Sperm transport and motility in the mouse oviduct: observations in situ. Biol Reprod 1987; 36:203–210.
16. Demott RP, Lefebvre R, Suarez SS. Carbohydrates mediate the adherence of hamster sperm to oviductal epithelium. Biol Reprod 1995; 52:1395–1403.
17. Lefebvre R, Lo MC, Suarez SS. Bovine sperm binding to oviductal epithelium involves fucose recognition. Biol Reprod 1997; 56:1198–1204.
18. Suarez SS. Carbohydrate-mediated formation of the oviductal sperm reservoir in mammals. Cells Tissues Organs 2001; 168:105–112.
19. Suarez SS, Revah I, Lo M, et al. Bull sperm binding to oviductal epithelium is mediated by a Ca2+-dependent lectin on sperm that recognizes Lewis-a trisaccharide. Biol Reprod 1998; 59:39–44.
20. Töpfer-Petersen E, Ekhlasi-Hundrieser M, Tsolova M. Glycobiology of fertilization in the pig. Int J Dev Biol 2008; 52:717–736.
21. Gwathmey TM, Ignotz GG, Suarez SS. PDC-109 (BSP-A1/A2) promotes bull sperm binding to oviductal epithelium in vitro and may be involved in forming the oviductal sperm reservoir. Biol Reprod 2003; 69:809–815.
22. Petrunkina A, Gelharr R, Drommer W, et al. Selective sperm binding to pig oviductal epithelium in vitro. Reproduction 2001; 121:889–896.
23. Gwathmey TM, Ignotz GG, Mueller JL, et al. Bovine seminal plasma proteins PDC-109, BSP-A3, and BSP-30-kDa share functional roles in storing sperm in the oviduct. Biol Reprod 2006; 75:501–507.
24. Ignotz GG, Cho MY, Suarez SS. Annexins are candidate oviductal receptors for bovine sperm surface proteins and thus may serve to hold bovine sperm in the oviductal reservoir. Biol Reprod 2007; 77(6):906–913.
25. Eisenbach M. Chemotaxis. London: Imperial College Press, 2004.
26. Gakamsky A, Schechtman E, Caplan SR, et al. Analysis of chemotaxis when the fraction of responsive cells is small—application to mammalian sperm guidance. Int J Dev Biol 2008; 52:481–487.
27. Yanagimachi R. Mammalian fertilization. In: Knobil E, Neill JD, eds. The Physiology of Reproduction. New York: Raven Press, Ltd. 1994:189–317.
28. Vadnais ML, Galantino-Homer HL, Althouse GC. Current concepts of molecular events during bovine and porcine spermatozoa capacitation. Arch Androl 2007; 53(3):109–123.
29. Manjunath P, Therien I. Role of seminal plasma phospholipid binding proteins in sperm membrane lipid modification that occurs during capacitation. J Reprod Immunol 2002; 53:109–119.
30. Calvete J, Ensslin M, Mburu J, et al. Monoclonal antibodies against boar sperm zona pellucida-binding protein AWN-1. Characterization of a continuous antigenic determinant and immunolocalization of AWN epitopes in inseminated sows. Biol Reprod 1997; 57:735–742.
31. Dukelow W, Chernoff H, Williams W. Properties of decapacitation factor and presence in various species. J Reprod Fertil 1967; 14:393–399.
32. Fraser L, Harrison RA, Herod J. Characterization of a decapacitation factor associated with epididymal mouse spermatozoa. J Reprod Fertil 1990; 89:135–148.
33. Oliphant G, Reynolds A, Thomas T. Sperm surface components involved in the control of the acrosome reaction. Am J Anat 1985; 174:269–283.
34. Parrish JJ, Susko-Parrish JL, Graham JK. In vitro capacitation of bovine spermatozoa: role of intracellular calcium. Theriogenology 1999; 51:461–472.
35. Parrish JJ, Susko-Parrish JL. Calcium increases in the anterior head of bovine sperm during capacitation. Biol Reprod 2001; 64(suppl 1):112.
36. Visconti P, Bailey J, Moore G, et al. Capacitation of mouse spermatozoa. I. Correlation between the capacitation state and protein tyrosine phosphorylation. Development 1995; 121:1129–1137.
37. Leclerc P, De Lamirande E, Gagnon C. Regulation of protein tyrosine phosphorylation and human sperm capacitation by reactive oxygen derivatives. Free Radic Biol Med 1996; 22:643–656.

38. Galantino-Homer HL, Visconti PE, Kopf GS. Regulation of protein tyrosine phosphorylation during bovine sperm capacitation by a cyclic adenosine 3,50-monophosphate-dependent pathway. Biol Reprod 1997; 56:707–719.
39. Naaby-Hansen S, Flickinger CJ, Herr JC. Two-dimensional gel electrophoretic analysis of vectorially labeled surface proteins of human spermatozoa. Biol Reprod 1997; 56:771–787.
40. Ficarro S, Chertihin O, Westbrook VA, et al. Phosphoproteome analysis of capacitated human sperm. Evidence of tyrosine phosphorylation of a kinase-anchoring protein 3 and valosin-containing protein/p97 during capacitation. J Biol Chem 2003; 278:11579–11589.
41. Kumar V, Rangaraj N, Shivaji S. Activity of pyruvate dehydrogenase a (PDHA) in hamster spermatozoa correlates positively with hyperactivation and is associated with sperm capacitation. Biol Reprod 2006; 75:767–777.
42. Bailey JL, Tardif S, Dube C, et al. Use of phosphoproteomics to study tyrosine kinase activity in capacitating boar sperm kinase activity and capacitation. Theriogenology 2005; 63:599–614.
43. Arcelay E, Salicioni AM, Wertheimer E, et al. Identification of proteins undergoing tyrosine phosphorylation during mouse sperm capacitation. Int J Dev Biol 2008; 52: 463–472.
44. Cohen G, Rubinstein S, Gur Y, et al. Crosstalk between protein kinase A and C regulates phospholipase D and F-actin formation during sperm capacitation. Dev Biol 2004; 267:230–241.
45. Gadella BM, Tsai P, Boerke A, et al. Sperm head membrane reorganisation during capacitation. Int J Dev Biol 2008; 52:473–480.
46. Cross NL. Role of cholesterol in sperm capacitation. Biol Reprod 1998; 59:7–11.
47. Flesch FM, Brouwers JF, Nievelstein PF, et al. Bicarbonate stimulated phospholipid scrambling induces cholesterol redistribution and enables cholesterol depletion in the sperm plasma membrane. J Cell Sci 2001; 114:3543–3555.
48. Navarro B, Kirichok Y, Chung J, et al. Ion channels that control fertility in mammalian spermatozoa. Int J Dev Biol 2008; 52:607–613.
49. Buffone MG, Foster JA, Gerton GL. The role of the acrosomal matrix in fertilization. Int J Dev Biol 2008; 52:511–522.
50. Kim E, Yamashita M, Kimura M, et al. Sperm penetration through cumulus mass and zona pellucida. Int J Dev Biol 2008; 52:677–682.
51. Wassarman PM, Litscher ES. Mammalian fertilization: the egg's multifunctional zona pellucida. Int J Dev Biol 2008; 52:665–676.
52. Hedrick JL. Anuran and pig egg zona pellucida glycoproteins in fertilization and early development. Int J Dev Biol 2008; 52:683–701.
53. Litscher ES, Wassarman PM. Egg extracellular coat proteins: from fish to mammals. Histol Histopathol 2007; 22:337–347.
54. Florman HM, Jungnickel MK, Sutton KA. Regulating the acrosome reaction. Int J Dev Biol 2008; 52:503–510.
55. Bedu-Addo K, Costello S, Harper C, et al. Mobilisation of stored calcium in the neck region of human sperm—a mechanism for regulation of flagellar activity. Int J Dev Biol 2008; 52:615–626.
56. Wassarman PM. Mammalian fertilization: the strange case of sperm protein 56. BioEssays 2009; 31:153–158.
57. Shur BD. Reassessing the role of protein-carbohydrate complementarity during sperm-egg interactions in the mouse. Int J Dev Biol 2008; 52:703–715.
58. Evans JP. The molecular basis of sperm-oocyte membrane interactions during mammalian fertilization. Hum Reprod Update 2002; 18(4):297–311.
59. Cohen DJ, Busso D, Da Ros V, et al. Participation of cysteine-rich secretory proteins (CRISP) in mammalian sperm-egg interaction. Int J Dev Biol 2008; 52:737–742.
60. Stein KK, Primakoff P, Myles D. Sperm-egg fusion: events at the plasma membrane. J Cell Sci 2004; 117:6269–6274.
61. Swann K, Yu Y. The dynamics of calcium oscillations that activate mammalian eggs. Int J Dev Biol 2008; 52:585–594.
62. Saunders CM, Larman MG, Parrington J, et al. PLCζ: a sperm-specific trigger of Ca^{2+} oscillations in eggs and embryo development. Development 2002; 129:3533–3544.
63. Swann K, Saunders CM, Rogers N, et al. PLCζ(zeta): a sperm protein that triggers Ca^{2+} oscillations and egg activation in mammals. Semin Cell Dev Biol 2006; 17:264–273.

64. Van Blerkom J, Davis PW, Lee J. ATP content of human oocytes and developmental potential and outcome after in-vitro fertilization and embryo transfer. Hum Reprod 1995; 10:415–424.
65. Ozil JP, Banrezes B, Toth S, et al. Ca^{2+} oscillatory pattern in fertilized mouse eggs affects gene expression and development to term. Dev Biol 2006; 300:534–544.
66. Gall WE, Ohsumi Y. Decondensation of sperm nuclei in vitro. Exp Cell Res 1976; 102:349–358.
67. Huret JL. Variability of the chromatin decondensation ability test on human sperm. Arch Androl 1983; 11(1):1–7.
68. Perreault SD, Barbee RR, Slott BL. Importance of glutathione in the acquisition and maintenance of sperm nuclear decondensing activity in maturing hamster oocytes. Dev Biol 1988; 125(1):181–186.
69. Perreault SD. Regulation of sperm nuclear reactivation during fertilization. In: Bavister BD, Cummings J, Roldan ERS, eds. Fertilization in Mammals. Norwell, MA: Serono Symposia, 1990:285–296.
70. Perreault SD. Chromatin remodeling in mammalian zygotes. Mutat Res 1992; 296(1–2): 43–55.
71. Jaffe LA, Gould M. Polyspermy-preventing mechanisms. In: Metz CB, Monroy A. Biology of Fertilization. New York: Academic Press, 1985:223–250.

2

Normal development of the male reproductive system

Reza J. Rasoulpour, M. Sue Marty, Kamin J. Johnson, and Edward W. Carney

INTRODUCTION AND SCOPE

Excepting some rare and interesting conditions, mammalian offspring generally come in two basic varieties, male and female. The initial gender-determining event occurs at conception as established by the inherited chromosomal constitution. However, the establishment of "genetic sex" at the time of conception is only the first in a complex, highly orchestrated series of events, all of which are necessary to manifest normal male, or female, reproductive function. Other key processes include the determination of "gonadal sex," whereby the bipotential gonad of a genetic male develops into a testis. While the testis is charged with the all-important task of producing sperm, a network of ducts (e.g., ductus deferens) and secretory organs (e.g., prostate) are also necessary to deliver sperm to the female at the right time, in adequate concentration, and in a condition supportive of subsequent motility and fertilization capacity. The development of these male-specific organs and characteristics is referred to as "phenotypic sex." To ensure that all of these branches of the male reproductive system function in a coordinated fashion, overall control is managed by the neuroendocrine system, particularly the hypothalamus, pituitary gland, and interstitial cells of the testis.

In this chapter, we outline the embryology, molecular biology, and neuroendocrine regulatory processes involved in the development of genetic, gonadal, and phenotypic sex. These events represent a continuum that begins at conception but may extend over weeks or even years until puberty is achieved. Also to be reviewed is the attainment of basic reproductive functions, with a particular emphasis on those that are commonly evaluated in reproductive toxicology assessments. Although beyond the scope of this chapter owing mainly to its complexity, one cannot divorce development of the brain from reproductive function, as sexually dimorphic development of certain brain regions is necessary for regulation of mating as well as other male-specific traits (e.g., refusing to ask for directions when lost).

EMBRYOLOGY—INDIFFERENT STAGE

The construction of the male reproductive system begins with what amounts to a biological coin toss, with those embryos inheriting a Y chromosome from the father destined to become males and those fertilized by an X-bearing sperm set on a female path. Surprisingly, the outcome of this coin toss is not revealed by any overt changes in embryo morphology until very late in organogenesis, with the first signs of male structures not evident until embryonic day (E) 12 in the mouse or the seventh week of gestation in humans (1). During this time, traditionally referred to as the indifferent stage, it is virtually impossible to visually distinguish male and female embryos.

Nonetheless, underlying biochemical and molecular changes are occurring during this period of embryogenesis, with three major developmental lineages involved in forming a complete male reproductive system (Fig. 1). The first of these is the germ cell lineage that will differentiate into spermatogonial cells (and eventually sperm) and is signaled by the appearance of primordial germ cells (PGCs) at around the time of

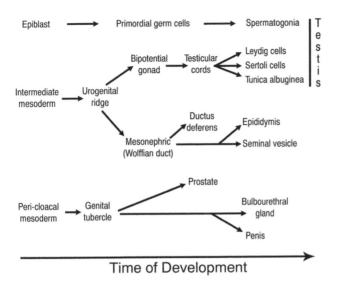

FIGURE 1   Major lineages in male reproductive system development.

primitive streak formation (E7 in the mouse). These PGCs are detectable by alkaline phosphatase staining and are first found as a cluster of about 50 such cells in the caudal wall of the yolk sac adjacent to the allantois (reviewed in Ref. 2). PGCs are thought to be derived from the epiblast of the early embryo and are the only cells of the post-implantation embryo that are totipotent. Just after primitive streak formation, the PGCs begin a lengthy migration facilitated by a combination of their own ameboid movement, morphogenetic forces, chemotactic factors, and extracellular matrix signals (reviewed in Ref. 1). This journey takes them toward the base of the allantois, through the hindgut endoderm, and ultimately to the urogenital ridge (described later), which serves as a common anlage to both the renal and reproductive systems [Fig. 2 (1)]. Concomitant with migration, the PGCs actively proliferate under the influence of *steel* factor, fibroblast growth factors (FGFs), and leukemia inhibitory factor, such that by E13.5 in the mouse, approximately 25,000 PGCs populate this region.

All other components of the male reproductive tract originate from embryonic mesoderm (Fig. 1), which can be subdivided into intermediate mesoderm located in the midsection of the embryo and the pericloacal mesoderm, found at the caudal end of the embryo. Intermediate mesoderm derivatives include all testis cell types except for the germ cells, the ductal system (i.e., rete testis, epididymis, ductus deferens), and the seminal vesicles. Initially the intermediate mesoderm appears as a bilateral pair of cylindrical thickenings running lateral to the dorsal aortae along the posterior abdominal wall. It subsequently develops into a primitive renal system comprised of the mesonephros and the mesonephric ducts, also known as the Wolffian ducts. Next, the mesonephros forms an additional pair of thickenings sandwiched between it and the midline, resulting in a ridge between the two organs called the urogenital ridge. Even prior to the arrival of PGCs, the epithelium of the urogenital ridge (also called the gonadal ridge) begins to proliferate and penetrate its underlying mesenchyme, resulting in the formation of the primitive sex cords. This structure is now identifiable as a gonad, but it is still not yet possible to determine if it will become a testis or an ovary. Hence, it is referred to as the bipotential gonad. Associated with the mesonephros are the

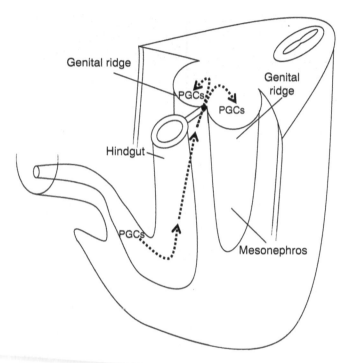

FIGURE 2   Migration of PGCs. *Source:* From Ref. 1

mesonephric and paramesophrenic ducts (also called Müllerian ducts), which eventually give rise to the male and female duct systems, respectively.

The final lineage derives from pericloacal mesoderm, defined as cells that migrated through the primitive streak around the cloacal membrane. Subsequent elevation of these cells leads to formation of a pair of cloacal folds. At the anterior end of the cloaca, these folds fuse and swell to form the genital tubercle, a common progenitor of the male and female external genitalia. The pericloacal mesoderm and genital tubercle also contribute to the urogenital sinus, an internal canal, from which evaginations of epithelium lead to development of the prostate and bulbourethral glands (3).

These initial development processes constitute the basic events of reproductive system development that set the stage for subsequent expression of the male and female phenotypes. Next we discuss how these "indifferent" embryological structures transform into individual male reproductive organs, beginning with the indifferent gonad and PGCs combining to form the testis, followed by development of the ductal system as derived from the Wolffian ducts, and then the external genitalia, prostate, and coagulating glands originating from the genital tubercle.

TESTIS
Adult Testis Structure
The adult testis is divided into a spermatogenic compartment, termed the seminiferous cords in utero and seminiferous tubules in postnatal life, and the interstitium. These two interwoven compartments are encapsulated by a fibrous membrane, the tunica albuginea. The seminiferous cords and tubules are elongated cylindrical structures comprised of an epithelium containing germ cells embedded within the somatic Sertoli cells,

in an arrangement much like fingers stuck into a balloon. Firmly attached to, and surrounding, the seminiferous epithelium are flattened myoid-like cells called peritubular myoid cells. Residing within the interstitium are androgen-secreting Leydig cells, vascular cells, and other mesenchymal cells. As described later, differentiation of Sertoli cells drives all aspects of testis development.

Development of the Bipotential Gonad

Between approximately E10.5 and 11.5 in both female and male mice, the urogenital ridge enlarges to become the bipotential gonad. The genes driving the expansion include *Cbx2* (4), *Emx2* (5), *Lhx9* (6), *Nr5a1* (7), *Pod1* (8), and *Wt1* (9). Enlargement of the bipotential gonad is thought to expand the population of cells destined to become Sertoli cell precursors and provide the correct tissue milieu for subsequent expression of *Sry*, the sex-determining region-Y-chromosome, in these cells (10). In fact, the presence of *Sry* alone is sufficient for male gonadal sex determination, as demonstrated by transgenic mouse experiments where XX mice expressing Sry developed as males (11).

Sex Determination

Development of a phenotypically male reproductive system begins at approximately mouse E11.5, when expression of *Sry* in sufficient numbers of pre-Sertoli cells, during a critical six-hour time window, tips the balance toward testis development (12,13). Proteins regulating *Sry* expression include Wt1 (14), insulin receptor family members (15), and a Gata4/Fog2 complex (16). Within hours of *Sry* expression, Sry and Nr5a1 bind to a *Sox9* enhancer element in pre-Sertoli cells to increase *Sox9* transcription (17). In the absence of *Sox9* expression, a genetically male gonad will become phenotypically female (18). Sox9 maintains its own expression in the testis by setting up a feed-forward loop consisting of Fgf9, the Fgf9 receptor Fgfr2, and prostaglandin D2 synthase (Ptgds) that recruits additional progenitor cells into the Sertoli cell lineage (19–21).

Around E12 in the mouse, pre-Sertoli cells produce factors driving formation of the testicular vasculature and seminiferous cords. Unlike the ovary, an extensive vascular system is assembled in the fetal testis requiring migration of endothelial cells from the adjacent mesonephros (22,23). Initially, mesonephric endothelial cells traverse the gonad and aggregate to form the prominent coelomic vessel (24). This migration utilizes the adhesion protein VE-cadherin expressed on endothelial cells as well as Pdgfra (22,25) and partitions the gonad into areas where Sertoli, germ, and peritubular myoid cells will aggregate to form seminiferous cords (22). The factors contributing to seminiferous cord formation are unclear, but germ cells are not required (26). Progenitors of peritubular myoid cells had been thought to have emigrated from the mesonephros, but this has been discounted recently (22,23); instead, peritubular myoid cells probably arise from an interstitial cell population, perhaps via the action of the Sertoli cell–derived paracrine factor Dhh (27).

Leydig Cells

Located in the interstitium surrounding the seminiferous tubules are the Leydig cells, which are primarily involved in secretion of androgens and other steroids that are required for male sexual development (28). These fetal Leydig cells are not precursors to adult Leydig cells but instead are a morphologically and functionally distinct cell type (29). During fetal life, the cellular origin of progenitor Leydig cells is unclear, with data suggesting that they either come from the coelomic epithelium or from the mesonephros (25,30). At any rate, their expansion may be controlled by Notch signaling (31), while gene-deletion experiments indicate that their differentiation is regulated by the *Dhh* gene (32). Studies with *Arx* knockout mice also suggest that *Arx* expressed in peritubular myoid cells, endothelial cells, and other undefined interstitial cells is necessary for fetal Leydig cell differentiation (33). By E12.5 in mice, Leydig cell differentiation has been completed, as defined by expression of the steroidogenic genes *Cyp11a1*, *Cyp17a1*, and *Star*. Thereafter, additional precursor cells acquire the Leydig cell phenotype, and production of hormones (Insl3 and androgen) crucial for testis descent and differentiation of male reproductive organs occurs. In humans, hCG and LH stimulation is required for fetal Leydig cell steroidogenesis, but the crucial factor(s) maintaining rodent fetal testis steroidogenesis in vivo remains unknown (34). Interestingly, genetic ablation of the mouse pituitary dramatically reduces fetal testosterone production (35), which suggests that this crucial factor(s) in fetal testis steroidogenesis may be a pituitary-derived hormone(s).

Progenitors of adult Leydig cells first appear by postnatal day (PND) 10 in the rat testis as spindle-shaped cells with 3-beta-hydroxysteroid dehydrogenase (3β-HSD or Hsd3b1), LH receptor, and androgen receptor expression (reviewed in Ref. 36). A series of studies in the rat determined that adult Leydig cells arise from mesenchymal precursors recruited on PND 14–28, followed by cell division to populate the developing testis (37).

Postnatal proliferation and differentiation of adult Leydig cells is primarily induced by LH signaling (for review, see Ref. 38), while proliferation and differentiation of fetal Leydig cells is largely independent of pituitary gonadotropin control (34,39). In fact, hypogonadal (*hpg*) mice, which are gonadotropin deficient, undergo normal development of fetal Leydig and Sertoli cells, but are deficient in mature adult Leydig cells during postnatal testis development (40). In addition to stimulation by LH, the presence of androgen (41) and follicle stimulating hormone (FSH) (42) also appear critical for Leydig cell function. While postnatal Leydig cells express androgen receptor, only Sertoli cells express the FSH receptor, suggesting that FSH stimulation of Leydig cells occurs through secondary signaling via Sertoli cells (for review, see Ref. 43).

Steroidogenesis in adult Leydig cells is activated by LH binding to the LH receptor and begins with cholesterol (obtained intracellularly through lipid droplets or plasma membrane) transferred to the outlet leaflet of the mitochondria in a protein kinase A–dependent mechanism. Cholesterol transfer from the outer to inner mitochondrial membrane is regulated primarily by steroidogenic acute regulatory protein (Star) and secondarily by peripheral benzodiazepine receptor. Within the mitochondrial inner membrane, cholesterol is converted to pregnenolone by the P450 side-chain cleavage (Cyp11a1 or P450scc) enzyme (Fig. 3). Pregnenolone is then transferred to the smooth endoplasmic reticulum where it is converted to testosterone and dihydrotestosterone (DHT) through the function of enzymes such as Hsd3b1 (3β-HSD), Cyp17a1 (17α-hydroxylase/17,20 lyase), Hsd17b (17β-HSD), Cyp19a1 (P450aro or aromatase), and 5α-reductase (for reviews, see Refs. 38,44,45).

Sertoli Cells

The Sertoli cell plasma membrane has many convolutions and processes, which act in a fluid nature to encompass various germ cell types (46). Sertoli cells provide complete support for germ cells throughout spermatogenesis by maintaining and compartmentalizing the seminiferous epithelium, secreting fluid and proteins, phagocytosing apoptotic germ cells, actively translocating germ cell populations through the seminiferous epithelium, and mediating para- and endocrine signals to germ cells (47). While highlights of these functions will be discussed, a full description of Sertoli cells can be found elsewhere (46).

As mentioned previously, there are a finite number of germ cells that can be supported by a given Sertoli cell. In rats, Sertoli cells have a discrete proliferative

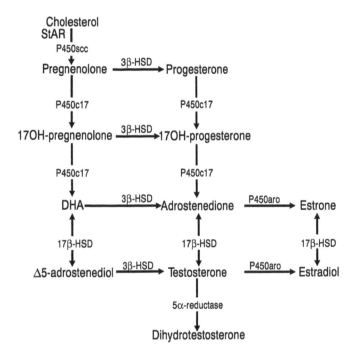

FIGURE 3 The process of steroidogenesis.

window that begins on E16 (E12-15 in mice) and begins to decline by birth. This proliferative window continues, albeit at a slower rate, until PND 15 (48), reaching a final population of approximately one million cells per testis (49). After this point, the cells become quiescent and undergo differentiation. So dependent are germ cells on functional Sertoli cells that the success of Sertoli cell proliferation and differentiation during this perinatal period directly correlates to lifetime spermatogenic potential. Extension of the Sertoli cell proliferation window through neonatal exposure to the neonatal goitrogen 6-propyl-2-thiouracil, which results in increased FSH production, greatly increases adult testis weight that correlates with an increase in both Sertoli cells and germ cells (spermatocytes and spermatids) (50). Conversely, exposure to a gonadotropin-releasing hormone antagonist from PND 1 to 15 results in fewer Sertoli cells, decreased testis weight, and infertility (51).

In humans, the prepubescent testis largely consists of numerous immature Sertoli cells associated with relatively few undifferentiated germ cells (52). A wave of Sertoli cell proliferation begins after birth and lasts until six months of age (53).

Hypothalamic-Pituitary-Gonadal Axis

The hypothalamic-pituitary-gonadal axis involves gonadotropin-releasing hormone secretion from the hypothalamus-stimulating LH and FSH release from the anterior pituitary. LH and FSH act on Leydig and Sertoli cells, respectively, to stimulate spermatogenesis (36). Testosterone from Leydig cells and inhibin B from Sertoli cells, negatively feedback to the anterior pituitary and hypothalamus to inhibit LH and FSH production, respectively (47). During normal pre- and postnatal development of the human male reproductive system, concerted circulating hormone surges occur that stimulate sex differentiation and spermatogenesis. In the rat, LH is not required for fetal testosterone production, and fetal androgen only appears to modulate fetal Sertoli cell number. In prenatal development, a surge of anti-Müllerian hormone (AMH) followed by testosterone is necessary for masculinization of the bipotent gonad. Postnatally, the production of FSH, by PND 20 in rat, followed by LH and testosterone are all required for induction of puberty and population of the testis with the first wave of spermatogenesis (34,39).

WOLFFIAN DUCT DERIVATIVES
Wolffian Duct Development
As mentioned earlier, embryos initially contain both Wolffian ducts (mesonephric ducts) and Müllerian ducts (paramesonephric ducts), which are the anlagen for the male and female reproductive tracts, respectively. The critical period for Wolffian duct development, differentiation, and subsequent patterning into epididymides, ductus deferens, seminal vesicles, and ejaculatory ducts are approximately E15–20 in the rat, E15.5–19.5 in the mouse (54,55), and gestation weeks 6 to 13 in humans (for reviews see Refs. 56–58).

In the absence of any sex-specific differentiation factors, development will default to a phenotypic sex of female, whereas formation of a male requires active male-specific signals. Among the most important of these is AMH, which is secreted from the fetal testis and induces regression of the Müllerian ducts in males. Timing for AMH secretion is critical as the Müllerian ducts quickly lose their responsiveness to this hormone, resulting in feminization (59,60).

A number of other genes, namely *Pax2/Pax8*, *Emx2*, and *Lim1*, are critical in early Wolffian duct growth and development. In mice with null mutations in either *Pax2* or *Emx2*, the Wolffian duct elongates partially, but then regresses, resulting in agenesis of the epididymis, ductus deferens, and seminal vesicles (5,61). Although *Pax8* null mice develop normally, *Pax2/Pax8* double knockout mice completely lack Wolffian duct development (62). Likewise, *Lim1* null mice lack Wolffian ducts (63), and Lim1 may be a downstream mediator of Pax genes (64).

Shortly after the Müllerian duct begins to regress, the fetal Leydig cells begin producing testosterone, which is secreted directly into the Wolffian duct resulting in high local concentrations (57,65). This secretion of androgen triggers a series of gene changes that occur differentially along the length of the duct, which lead to coiling, evagination, and cell proliferation in selected regions, ultimately forming the epididymis in its anterior-most reaches, seminal vesicles at the caudal end, and ductus deferens in-between. By E19.5 in rat, Wolffian duct patterning has already been established and no longer requires high levels of androgen (55).

Following an oft-repeated script in developmental biology, mesenchymal-epithelial interactions and differential expression of patterning genes drive the transformation of Wolffian duct into epididymis, ductus deferens, and seminal vesicles. For example, culture of seminal vesicle mesenchyme adjacent to anterior Wolffian duct epithelium (that normally would become epididymis) switched epithelial fate to become seminal vesicle (66). *Inhba* expressed within anterior Wolffian duct appears to regulate coiling and epithelial proliferation. Evidence for this comes from *Inhba* null mice, in which the Wolffian duct remains as a straight tube and epithelial cell proliferation is reduced (67). Other factors involved in patterning of the Wolffian duct are the homeobox genes, particularly *Hoxa10*, which are differentially expressed along its length (68).

Epididymis
At the anterior end of the mesonephric duct, as the mesonephros is regressing, some residual tubules from the mesonephros join with ducts emanating from the testis. The latter ducts, called the efferent ductules and rete testis, provide a conduit for the exit of

spermatazoa from the testis and into the Wolffian duct derivatives. Just below this connection point, the Wolffian duct elongates and becomes highly convoluted, forming the early epididymis. Subsequent development of the epididymis can be divided into three phases: an undifferentiated phase, a period of differentiation, and a period of expansion. In the rat, these respective phases occur from birth to PND 15, PND 16–44, and >PND 44 (69). In phase 1, the efferent ducts and the epididymis elongate and become more convoluted shortly after birth. During this remodeling, Lrg4 appears to regulate proliferation of the epithelium and mesenchyme, as well as their interaction, which contributes to growth and cellular organization of the ducts and epididymis (70). Phase 2 is induced by both testosterone and the arrival of germ cells and their bathing fluid from the rete testis (71). In phase 3, spermatozoa appear in the epididymis and expand the lumen, corresponding with increases in epididymal weight (72).

Seminal Vesicles

In humans, formation of the seminal vesicles occurs during the fourth month of gestation and largely involves androgen-induced branching (73). The seminal vesicles and ductus deferens are arranged in a similar manner to the adult by the sixth month of gestation. During this period, the seminal vesicles develop a larger lumen than the ductus deferens. The adult form of the seminal vesicles is largely attained by the seventh month of gestation, except that the mucous membrane around the lumen does not arrange itself into folds until full term (74).

With the attainment of a primarily adult form at seven months of gestation, secretory activity begins. Seminal vesicle secretions slowly increase during the remainder of gestation and continue for a considerable time after birth (i.e., detected at 17 months), but subsequently decline [e.g., not been detected at four years of age (75)]. During this relatively quiescent period, the seminal vesicles, ampulla of the ductus deferens, and the ejaculatory ducts develop slowly until puberty when the glands form saclike structures that contribute approximately 70% of seminal fluid (~50% of ejaculate) (76). Seminal vesicle secretions, which arise from epithelium due to androgen stimulation, are hypothesized to facilitate sperm motility and semen coagulation, stabilize sperm chromatin, contain antioxidants, and have immunosuppressive properties in the female reproductive tract (77). The seminal vesicles also reabsorb fluids and dissolved substances and degrade damaged sperm (73).

In rats, the seminal vesicles begin to form on E19.5 (78). In newborns, the seminal vesicle epithelium shows no signs of secretion and has poorly developed rough endoplasmic reticulum (RER) and Golgi complex (79). By PND 10, the basic pattern for seminal vesicle formation is present. During PND 2–15, slow development of the seminal vesicle lumen occurs and the seminal vesicles increase markedly in size between PND 11 and 24. As animals approach and progress through puberty, the seminal vesicles continue to proliferate and differentiate in conjunction with increasing testosterone levels. The RER and Golgi complex continue development until the adult ultrastructure is established around PND 40–60 (79). Secretory properties and adult appearance are attained between PND 40 and 50 (80). Testosterone is needed to maintain seminal vesicle epithelial cells and their secretory function.

Ductus Deferens

The segment of Wolffian duct lying between the highly coiled epididymis, formed at the anterior end of the Wolffian duct, and the seminal vesicles at the caudal end becomes the ductus deferens (also called the vas deferens). This segment remains as a relatively straight or slightly curved tube and obtains a thick muscular wall that functions to transport sperm during the process of ejaculation (1).

PERICLOACAL MESODERM DERIVATIVES

As described earlier, the pericloacal mesoderm, located at the caudal end of the embryo, contributes to the urogenital sinus and ultimately the internally located prostate gland, as well as the genital tubercle that gives rise to the external genitalia. The development and regulation of these structures are described below.

Prostate Gland

On the surface, the adult human prostate appears as a single gland, but actually consists of four parts: (*i*) the anterior nonglandular fibromuscular stroma, which makes up one-third of the gland concentrated on the anterior surface; (*ii*) the central zone, which comprises approximately 25% of the gland and contains a group of ducts (81); (*iii*) the peripheral zone, which surrounds the central zone and comprises approximately 75% of the gland; and (*iv*) the preprostatic segment (periurethral portion), which includes the urethral segment and has only limited ducts (82). Most carcinomas arise from the peripheral zone (82), whereas benign prostatic hyperplasia is primarily localized to the preprostatic segment.

The human prostate forms during gestation weeks 10 to 13 as an evagination of the urogenital sinus, with its morphological development completed in utero (83). Initially, the prostate appears as a few widely spaced tubules supported by stromal cells (75). As gestation progresses, prostatic tubules increase in number and density. As seen before, the mesenchyme plays an important role in inducing development and maturation of the epithelial layer (82). Maternal and placental estrogens interact with estrogen receptors in the fetal prostate (Esr1 or ERα in the stroma and Esr2 or ERβ in the epithelium), resulting in squamous metaplasia of the prostatic tubular epithelium in the middle and lateral prostate lobes. These metaplastic cells are shed at birth when estrogen exposure decreases, resulting in the release of prostatic epithelial cells into the lumen (75). This desquamation leads to lumen formation and the saclike structure. The anterior and posterior lobes exhibit little or no metaplastic changes (75). While estrogen stimulates squamous metaplasia, androgens stimulate prostatic secretion; thus, androgen-stimulated secretion initially occurs from the periphery of the lateral and anterior prostatic lobes at approximately gestation week 14 and increases throughout the remainder of the fetal period until the last month of gestation (75). Because squamous metaplasia and secretion are localized to different areas of the prostate based on the local hormonal milieu, the central zone begins secretion after the peripheral zone, when squamous metaplasia has been completed in these areas (75).

Prostatic secretion is present at the time of birth. For approximately one month after birth, the histology of the prostate is relatively stable. Secretions can be steadily detected at this stage but become more variable thereafter (75,84). By three months of age, the metaplastic changes that occurred during the fetal period regress, leaving predominantly empty tubules with some debris from desquamated cells. The tubule epithelium also regresses (84).

During puberty, androgen stimulates further maturation of the prostatic secretory cells (epithelium) (82). Epithelial cells provide secretions that empty through ducts into the urethra to form a major component of the seminal plasma of the ejaculate. The prostate is stimulated to grow and maintained in size and secretory function by androgens, of which DHT is the most important. As maturation occurs, more elaborate glandular structures form distally in the prostate and acid phosphatase, the major prostatic secretory protein, increases (85). The noncellular stroma and connective tissue provide the extracellular matrix, which play important roles in development and control of cellular functions (86), including contraction to expel secretions (82).

Unlike the human prostate, the rat prostate consists of three distinct regions—ventral, lateral, and dorsal lobes (87). In immature and young rats, separation of the prostatic lobes is easier than in adult rats as there is no distinct border between the dorsal and lateral lobes in adults (84). Interestingly, the rat ventral prostate does not have an embryological correlate in man (78).

In rats, the critical period of androgen action is reported as E15.5–19.5 (54,78). Exposure to antiandrogens during fetal development can alter androgen signaling and subsequent prostate development in the rat (88).

Much of rat prostate development occurs postnatally. At birth, the prostate is rudimentary, but it undergoes marked morphological differentiation during PND 1–15 (84,87). The prostatic lobes form between PND 1 and 7 with prostatic tubular lumen forming between PND 7 and 14. The prostatic mesenchyme appears to direct the transformation of epithelial cells during this period (89,90) as the mesenchyme expresses both AR and ER at birth, whereas the epithelium does not express either (91). In contrast, the mesenchyme maintains high functional AR expression throughout postnatal development, whereas the epithelial cells begin to express AR at approximately PND 10 during the time of lumen formation in the ducts (91). The mesenchyme is hypothesized to stimulate AR expression and/or activation in developing epithelial cells via diffusible factors (92). It has been hypothesized that epithelial AR are involved in secretory function of the prostate, whereas stromal AR influence prostate growth (92).

Secretory granules are evident in the developing prostate between PND 14 and 21. Prostatic binding protein is a major secretory protein of the prostate in rats (vs. acid phosphatase in human males) (82). This protein may regulate secretory cells as it has been shown to inhibit nuclear uptake of the AR complex into ventral prostate nuclei (93). Paralleling an increase in androgen levels, the prostate attains its adult appearance between PND 28 and 35. In the fully developed rat ventral prostate, gland size remains constant with epithelial cell proliferation only occurring to restore subnormal glandular cells numbers (82,94).

Some postnatal prostate development can occur in the absence of androgens; however, testosterone is needed for normal morphogenesis, growth, and secretory activity of the prostate (90). Several other hormones also may influence prostate growth and development (90), and there is considerable heterogeneity among the lobes with respect to development patterns (78) and hormone responsiveness (90). For example, neonatal estrogen exposure can influence prostate morphology, size, and function in adult rats in a lobe-specific manner, including lobe-specific effects on differentiation and AR expression and responsiveness (84,90,95,96).

Penis

In humans, the external genitalia begin to masculinize at approximately nine weeks of gestation (97), although others (83) indicate weeks 11–13. The anogenital distance (AGD) increases and the labioscrotal swellings fuse with the swellings on each side forming half the scrotum (98). The edges of the urethral fold fuse to form the penis and the external urethral orifice moves toward the glans (98). The mesenchymal tissue in the phallus develops into the corpora cavernosa and the corpus spongiosum of the penis. The external genitalia show signs of masculinization by gestation week 14 (98), but do not increase in size until week 20 continuing until term. Neonatal testosterone concentrations have been correlated with penile development (99).

DHT converted from testosterone by 5α-reductase is involved in formation of the external genitalia as human males deficient in 5α-reductase exhibit undervirilization of external genitalia, although Wolffian duct development is normal (57). DHT has a higher affinity for the androgen receptor and stabilizes the hormone-receptor complex, making it more efficient than testosterone for development of distal structures.

In the rat, the penis appears to have a slightly different window for androgen stimulation than the other parts of the male reproductive system. Rats also unique in that they have a bony component called the os penis (100). Penis length and ossification of the penile os bone occur under androgen regulation during the critical window of development (E15.5-19.5), whereas penis weight is increased by androgen stimulation on E19.5-21.5 and perhaps postnatally as well (54).

Anogenital Distance

AGD, which is defined as the distance between the genital tubercle and the anus, is typically 2.5-fold greater for males than females and is used as an external indicator of gender. Rhesus monkey fetuses are reported to exhibit a similar sex difference in AGD (101,102). Mean AGD for PND 0 Sprague-Dawley rats have been reported as 3.51 mm (3.27–3.83) for males and 1.42 mm (1.29–1.51) for females. Absolute AGD is correlated with body size and age (103).

The relationship between AGD in laboratory animals and adverse effects in humans is not clear. Studies have reported human AGD measurements (104,105); however, there are no reports of decreased AGD in pseudohermaphroditic men, who lack 5α-reductase. Furthermore, finasteride, a 5α-reductase inhibitor, failed to alter the AGD of male monkeys at doses causing hypospadias (102).

FUNCTIONAL DEVELOPMENT
Overview

The process of mammalian spermatogenesis is a delicate balancing act in which totipotent germ line stem cells undergo successive rounds of mitotic proliferation, differentiation, meiosis, and chromatin repackaging to produce haploid elongate spermatids. Spermatogenesis can be split into three sections with corresponding suffixes associated with the germ cells, with spermatogonia as the diploid early germ cells undergoing successive rounds of mitosis, spermatocytes as germ cells in meiosis, and the haploid spermatids that undergo differentiation (spermiogenesis) and repackaging of chromatin. The limiting factor in the amplification process of spermatogenesis is the supportive somatic cells of the seminiferous tubule, the Sertoli cells. Sertoli cells exist in constant contact with multiple layers of germ cells throughout spermatogenesis like fingers stuck in a balloon. Not surprisingly, a single Sertoli cell can support only a finite number of germ cells through spermatogenesis (106,107). Therefore, adult testis size is determined by Sertoli cell numbers and consequently germ cell numbers.

Germ Cell Differentiation

Although germ cells are not required for testis formation, gonocytes exhibit a well-characterized sexually dimorphic entry into meiosis (see Ref. 108 for a review). In the E13.5 mouse ovary, retinoic acid produced by the mesonephros drives expression of meiotic genes (such as *Stra8*) in fetal germ cells (109,110). Germ cell responsiveness to the extrinsic retinoic acid signal requires intrinsic expression of *Dazl* (111) and is mediated by binding to nuclear retinoic acid receptors (109). Unlike the female, somatic cells in the fetal testis around the time of cord formation begin to express high levels of retinoic acid–metabolizing protein Cyp26b1 (109,112), and this effectively reduces testicular retinoic acid levels below those required to initiate germ cell meiosis. Without entry into meiosis, fetal testis germ cells continue to divide until late gestation, a point at which germ cell intercellular bridges are observed in humans and rodents (113–115).

Spermatogenesis

Spermatogonial stem cells within the seminiferous tubule are perpetually renewed to replenish the tremendous number of spermatids that are shed into the tubule lumen toward the epididymis. This proliferation is tightly regulated through endocrine (e.g., FSH and LH), paracrine (e.g., SCF/c-kit or GDNF/Ret/GFRα1,2), and autocrine (e.g., TGF-β1) mechanisms (116–119). The timing of stages is vital to the maintenance of the ratio of cell proliferation versus cell loss (via shedding or apoptosis) and is controlled by germ cells (120). Germ cell intercellular bridges are created between all non-stem germ cells via incomplete cytokinesis. Mathamatically, a single stem germ cell could lead to 4096 spermatids; however, because of a finite amount of support from Sertoli cells, germ cell apoptosis ensures that fewer spermatids are created from one stem cell division (Fig. 4). A majority of these 12 divisions (from stem cell to spermatid) occur during spermatogonial mitotic amplification (121). Type A spermatogonia divide eight times to yield up to 256 Intermediate (In) spermatogonia. These In spermatogonia then divide to form type B spermatogonia, which are the most differentiated germ cells to still reside in the basal compartment (below Sertoli cell tight junctions) of the seminiferous tubule. These type B spermatogonia then divide to become preleptotene spermatocytes as the germ cells enter the lengthy stages of meiosis (122,123).

In male mice, the meiotic germ cells, named spermatocytes, undergo meiosis I for 14.5 days going from preleptotene to leptotene to zygotene fairly rapidly and then entering the long phase of pachytene meiotic phase (47). A long meiosis I allows for crossing over events to occur, and starting at leptotene spermatocytes, the synaptone-mal complex is formed. After a short meiosis II, germ cells become haploid and undergo spermiogenesis from round spermatids to elongate spermatids. This period of differentiation involves condensation of DNA and replacement of histones with transition proteins, and eventually protamines (for reviews, see Refs. 124,125).

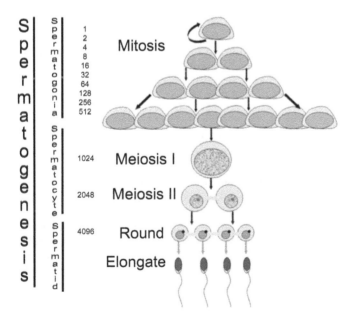

FIGURE 4 Stem cell division.

Stages of Spermatogenesis

Stages of spermatogenesis were assigned by early investigators to more clearly characterize the spatial associations of different spermatogenic cells that coexist within a single cross section of a seminiferous tubule (for more details, see Ref. 47). Roman numerals from I–XII in mouse, I–XIV in rat, and I–VI in humans identify the different stages. The strict timing and regulation of spermatogenesis, as mentioned earlier, form the biological basis for these distinct stages. For example, a stage VII seminiferous tubule in the mouse is identified by layers of preleptotene and pachytene spermatocytes, round spermatids at step 7 with a well-defined acrosome, and step 16 elongate spermatids that are near release (47). All of the germ cells in a given stage transition to the germ cell cohort of the next stage at the same time. For example, imagine a miniature human sitting next to a rodent seminiferous tubule whose germ cell cohorts were in stage VIII. If this miniature human did not move for a day, they would witness the elongate spermatids shedding into the lumen, preleptotene spermatocytes becoming leptotene, pachytene spermatocytes continuing meiosis, and round spermatids differentiating into step 9. These germ cells now define stage IX, and our minihuman is now witnessing that stage.

Germ Cell Proliferation

Spermatogonia are the mitotic germ cells that are largely responsible for the huge germ cell amplification occurring during spermatogenesis. Spermatogonial stem cells are termed type A_{is} (isolated) and exist in distinctive stem cell niches along the seminiferous tubule basement membrane (126). Type A_{is} spermatogonia undergo self-renewing to maintain stem cells and the daughters divide to eventually form A_{pr} (paired) spermatogonia. There is a highly proliferative phase where A_{pr} divides into A_{al} (aligned) and these cells undergo three successive rounds of mitosis (127). Again, these cells all have connecting intercellular bridges and undergo division and differentiation as syncitia in the original cohort created by A_{pr} spermatogonia (128,129). The final A_{al} population then undergoes differentiation into a series of type A spermatogonia named type A_1 through A_4. In-between each numbered type A differentiation step is a round of mitosis to further amplify the germ cell cohort. Type A_4 spermatogonia then undergo division and differentiation to become type In (intermediate) and then type B where there is a final spermatogonial division to become a preleptotene spermatocyte as meiosis I begins (47). In total, there are 10 mitotic events that occur in spermatogonia, followed by meiosis I and II in spermatocytes for a maximum of 12 divisions from A_{is} to round spermatid.

Germ Cell Culling

During spermatogenesis, germ cell apoptosis normally occurs in several spermatogenic stages, such that 25% to 75% of the expected germ cell yield is lost (122,130–133). The expression of pro- and anti-apoptotic genes during spermatogenesis suggests that apoptosis is regulated in specific germ cells and during certain stages of spermatogenesis. The balance of proliferation and culling of germ cells during spermatogenesis is critical to testicular homeostasis, such that inhibition of pro-apoptotic factors (e.g., Bax) can lead to overpopulation of the testis and massive cell death. This is thought to occur because of the limited supportive capacity of Sertoli cells (estimated to be 45 germ cells per Sertoli cell). In rats, type A3 and A4 spermatogonia are constantly undergoing apoptosis as cohorts in a synchronized manner (134). It is still not clear how Sertoli cells "sense" an overwhelming population of germ cells. A number of genes in the apoptotic machinery control this basal level of apoptosis such as Bcl-2 family members, SCF/c-kit, cyclin D_2, and others (for review, see Ref. 135). In mutant mice with deleted pro-apoptotic genes, testes are large initially, due to a lack of culling, but upon adulthood, these testes quickly deteriorate and become atrophic. An example of this phenomenon is the Bax null mouse whose testes have larger populations of type A_2-A_4 spermatogonia, measured at PND 15. Later in postnatal testis development, there is a massive overcrowding of the seminiferous epithelium and subsequent cell death resulting in testicular atrophy (107). In cases where a gene knockout, or overexpression, leads to larger testes (e.g., p53 mutant mice), the modest increase in testis size is often due to an expansion of the Sertoli cell proliferative window.

The tight association between Sertoli and germ cells is typical of the numerous autocrine, paracrine, and endocrine signals that must all work in concert for appropriate spermatogenesis to occur. An example of Sertoli-germ cell paracine signaling related to apoptosis is the Fas system. Fas ligand (FasL) is expressed on Sertoli cells (136–138) and Fas receptor (Fas) on germ cells (137–139). This pathway is essential for the transmission of death signals resulting from different types of injury. For example, mice deficient in FasL (*gld*) are resistant to germ cell apoptosis after toxicant exposure. In germ cells Bclw, an antiapoptotic Bcl-2 family member, is essential for germ cell survival as shown by knockout experiments (140,141). Germ cells in general are typically sensitive to testicular injury even if the proximal target is Sertoli or Leydig cells due to endocrine and paracrine signaling and physical support. Therefore, the endpoint and outcome of most types of testicular injury is germ cell apoptosis and subsequent atrophy.

Blood-Testis Barrier

Freeze-fracture preparations of the human testis show junctional particles at five years of age and plaques of communicating junctions at eight years of age. At puberty, a continuous belt of junctional particles can be seen that composes the blood-testis barrier (142).

In the rat between PND 14 and 19 when Sertoli cell divisions are complete and the first early spermatocytes move to the luminal side of the barrier, Sertoli cell tight junctions form. The blood-testis barrier shows decreased permeability between rat PND 15 and 35 (143,144). At this prepubescent stage, the blood-testis barrier is not fully mature as the adult barrier excludes smaller molecules (143).

Levator Ani Muscle Development

The levator ani/bulbocavernosus muscle plays a role in penile reflexes during erection and ejaculation in the rat. The levator ani/bulbocavernosus muscle weight is an end point in the Hershberger assay to screen chemicals for androgenic/antiandrogenic activity.

Levator ani muscle fiber number is sexually dimorphic due to testosterone-stimulated formation of muscle fibers during the prenatal/perinatal period in the male rat, which results in a dramatic increase in muscle units in males by PND 6 (145,146). By PND 7, testosterone stimulation no longer increases muscle cell number; however, cross-sectional area of the muscle continues to increase, reaching a two-fold higher diameter than the female muscle by PND 30 (146). This period determines the number of levator ani muscle fibers present in the adult. During puberty, high levels of testosterone between PND 40 and 60 in the male cause a hypertrophic response, marked by an increase in myonuclei number and fiber width. Thyroid hormone levels also have been proposed to play a role in gender differences in levator ani development (147).

Testicular Descent

Testicular descent occurs via a similar process in most species, although there are some timing and topographical differences [e.g., gubernacular outgrowth, mesenchymal regression, and development of the cremaster muscle (148)]. In humans, the abdominal cavity enlarges at 10 to 15 weeks of gestation, and the testis is anchored close to the future inguinal region by enlargement of the gubernaculum at 8 to 15 weeks of gestation and regression of the cranial ligament (149). Between 25 and 35 weeks, the gubernaculum expands beyond the external inguinal ring, migrates across the pubic region, and descends to the scrotum as this path is opened up by the processus vaginalis (149,150).

Testosterone has been shown to play a role in the inguinoscrotal phase of testicular descent, including suspensory ligament regression, and shrinkage of the gubernaculum (151,152). Several other factors have been considered important components of testicular descent including AMH, the cremaster muscle, the genitofemoral nerve, and calcitonin gene-related peptide, but the role of these factors in testicular descent is more controversial (reviewed in Ref. 149).

Cryptorchidism (i.e., failure of testicular descent) is believed to be multifocal in origin, reflecting the complex interplay of hormonal, mechanical, and anatomical interactions required for normal testicular descent. The most common form of cryptochidism likely involves mild or transient prenatal androgen insufficiency secondary to decreased maternal hCG or fetal pituitary stimulation (149). Perturbations may be mild or transient as cryptochidism can occur in the absence of genital anomalies. In older boys, it has been proposed that alterations in postnatal elongation of the spermatic cord or failure of the processus vaginalis to degenerate may lead to alterations in gonadal position (153).

In the rat, transinguinal testicular descent is programmed by androgen stimulation during the critical developmental window (E15.5–19.5) (54). The rat testis is attached to the internal inguinal ring with its caudal pole and the cauda epididymis is located within the canal on E20–21. Transinguinal testicular descent occurs at approximately PND 15 (148) and also requires androgens (154). However, inhibition of androgen signals during postnatal descent does not result in cryptorchidism (100). In rats, the inguinal canal remains wide open in adults, allowing the male rat to lift its testicles into its abdomen (148).

Preputial Separation

In prepubescent rats and mice, the glans penis is covered by and joined to a pouch called the prepuce, but at puberty, these two tissues separate to allow for mating to occur. Separation of the prepuce from the glans penis is used as an external indicator of puberty onset in rats (155), and occurs by the androgen-mediated keratinization of the epithelium joining the two tissues. At PND 30, the glans penis is difficult to expose in the male rat (156), but as androgen production increases, the tip of the penis changes from a V-shape to a W-shape (PND 20-30) to a U-shape when PPS is complete. The mean age at PPS for Sprague-Dawley rats has been reported as 43.6 ± 0.95 (X ± SD) days of age (range 41.8–45.9) (101), although other labs may report slightly higher values. Preputial separation also is heavily influenced by general factors, particularly body weight, body fat, and rate of growth (157).

In humans, the prepuce forms as a fold of skin starting at three months gestation, growing distally to surround the glans penis by about five months gestation. Although preputial separation also involves epithelial keratinization, it is a much more protracted and temporally variable process that begins in late gestation, but is not completed until nine months to three years of age (158–160). Therefore, the timing of preputial separation is not linked with puberty onset in humans.

REFERENCES

1. Sadler TW, Langman J. Langman's Medical Embryology. 9th ed. Philadelphia, Pa: Lippincott Williams & Wilkins, 2004.
2. Kaufman MH, Bard JBL. The Anatomical Basis of Mouse Development. San Diego: Academic Press, 1999.

3. Noden DM, DeLahunta A. The Embryology of Domestic Animals: Developmental Mechanisms and Malformations. Baltimore: Williams & Wilkins, 1985.
4. Katoh-Fukui Y, Tsuchiya R, Shiroishi T, et al. Male-to-Female Sex Reversal in M33 Mutant Mice. Nature 1998; 393(6686):688–692.
5. Miyamoto N, Yoshida M, Kuratani S, et al. Defects of urogenital development in mice lacking Emx2. Development 1997; 124(9):1653–1664.
6. Birk OS, Casiano DE, Wassif CA, et al. The LIM homeobox gene Lhx9 is essential for mouse gonad formation. Nature 2000; 403(6772):909–913.
7. Luo X, Ikeda Y, Parker KL. A cell-specific nuclear receptor is essential for adrenal and gonadal development and sexual differentiation. Cell 1994; 77(4):481–490.
8. Cui S, Ross A, Stallings N, et al. Disrupted gonadogenesis and male-to-female sex reversal in Pod1 knockout mice. Development 2004; 131(16):4095–4105.
9. Kreidberg JA, Sariola H, Loring JM, et al. WT-1 is required for early kidney development. Cell 1993; 74(4):679–691.
10. Sinclair AH, Berta P, Palmer MS, et al. A gene from the human sex-determining region encodes a protein with homology to a conserved DNA-binding motif. Nature 1990; 346 (6281):240–244.
11. Koopman P, Munsterberg A, Capel B, et al. Expression of a candidate sex-determining gene during mouse testis differentiation. Nature 1990; 348(6300):450–452.
12. Hiramatsu R, Matoba S, Kanai-Azuma M, et al. A critical time window of Sry action in gonadal sex determination in mice. Development 2009; 136(1):129–138.
13. Koopman P, Gubbay J, Vivian N, et al. Male development of chromosomally female mice transgenic for Sry. Nature 1991; 351(6322):117–121.
14. Hammes A, Guo JK, Lutsch G, et al. Two splice variants of the Wilms' tumor 1 gene have distinct functions during sex determination and nephron formation. Cell 2001; 106(3):319–329.
15. Nef S, Verma-Kurvari S, Merenmies J, et al. Testis determination requires insulin receptor family function in mice. Nature 2003; 426(6964):291–295.
16. Tevosian SG, Albrecht KH, Crispino JD, et al. Gonadal differentiation, sex determination and normal Sry expression in mice require direct interaction between transcription partners GATA4 and FOG2. Development 2002; 129(19):4627–4634.
17. Sekido R, Lovell-Badge R. Sex determination involves synergistic action of SRY and SF1 on a specific Sox9 enhancer. Nature 2008; 453(7197):930–934.
18. Chaboissier MC, Kobayashi A, Vidal VI, et al. Functional analysis of Sox8 and Sox9 during sex determination in the mouse. Development 2004; 131(9):1891–1901.
19. Kim Y, Bingham N, Sekido R, et al. Fibroblast growth factor receptor 2 regulates proliferation and Sertoli differentiation during male sex determination. Proc Natl Acad Sci U S A 2007; 104(42):16558–16563.
20. Bagheri-Fam S, Sim H, Bernard P, et al. Loss of Fgfr2 leads to partial XY sex reversal. Dev Biol 2008; 314(1):71–83.
21. Colvin JS, Green RP, Schmahl J, et al. Male-to-female sex reversal in mice lacking fibroblast growth factor 9. Cell 2001; 104(6):875–889.
22. Combes AN, Wilhelm D, Davidson T, et al. Endothelial cell migration directs testis cord formation. Dev Biol 2009; 326(1):112–120.
23. Cool J, Carmona FD, Szucsik JC, et al. Peritubular myoid cells are not the migrating population required for testis cord formation in the XY gonad. Sex Dev 2008; 2(3):128–133.
24. Coveney D, Cool J, Oliver T, et al. Four-dimensional analysis of vascularization during primary development of an organ, the gonad. Proc Natl Acad Sci U S A 2008; 105 (20):7212–7217.
25. Brennan J, Tilmann C, Capel B. Pdgfr-alpha mediates testis cord organization and fetal Leydig cell development in the XY gonad. Genes Dev 2003; 17(6):800–810.
26. Merchant H. Rat gonadal and ovarioan organogenesis with and without germ cells. An ultrastructural study. Dev Biol 1975; 44(1):1–21.
27. Pierucci-Alves F, Clark AM, Russell LD. A developmental study of the Desert hedgehog-null mouse testis. Biol Reprod 2001; 65(5):1392–1402.
28. Benton L, Shan LX, Hardy MP. Differentiation of adult Leydig cells. J Steroid Biochem Mol Biol 1995; 53(1-6):61–68.

29. Huhtaniemi I, Pelliniemi LJ. Fetal Leydig cells: cellular origin, morphology, life span, and special functional features. Proc Soc Exp Biol Med 1992; 201(2):125–140.
30. Jeays-Ward K, Hoyle C, Brennan J, et al. Endothelial and steroidogenic cell migration are regulated by WNT4 in the developing mammalian gonad. Development 2003; 130 (16):3663–3670.
31. Tang H, Brennan J, Karl J, et al. Notch signaling maintains Leydig progenitor cells in the mouse testis. Development 2008; 135(22):3745–3753.
32. Yao HH, Whoriskey W, Capel B. Desert Hedgehog/Patched 1 signaling specifies fetal Leydig cell fate in testis organogenesis. Genes Dev 2002; 16(11):1433–1440.
33. Kitamura K, Yanazawa M, Sugiyama N, et al. Mutation of ARX causes abnormal development of forebrain and testes in mice and X-linked lissencephaly with abnormal genitalia in humans. Nat Genet 2002; 32(3):359–369.
34. O'Shaughnessy PJ, Morris ID, Huhtaniemi I, et al. Role of androgen and gonadotrophins in the development and function of the Sertoli cells and Leydig cells: Data from mutant and genetically modified mice. Mol Cell Endocrinol 2009; 306(1–2):2–8.
35. Pakarinen P, Kimura S, El-Gehani F, et al. Pituitary hormones are not required for sexual differentiation of male mice: phenotype of the T/ebp/Nkx2.1 null mutant mice. Endocrinology 2002; 143(11):4477–4482.
36. Payne AH, Hardy MP. The Leydig Cell in Health and Disease. Totowa, NJ: Humana Press, 2007.
37. Hardy MP, Zirkin BR, Ewing LL. Kinetic studies on the development of the adult population of Leydig cells in testes of the pubertal rat. Endocrinology 1989; 124(2): 762–770.
38. Habert R, Lejeune H, Saez JM. Origin, differentiation and regulation of fetal and adult Leydig cells. Mol Cell Endocrinol 2001; 179(1-2):47–74.
39. O'Shaughnessy PJ, Baker P, Sohnius U, et al. Fetal development of Leydig cell activity in the mouse is independent of pituitary gonadotroph function. Endocrinology 1998; 139(3):1141–1146.
40. Baker PJ, O'Shaughnessy PJ. Role of gonadotrophins in regulating numbers of Leydig and Sertoli cells during fetal and postnatal development in mice. Reproduction 2001; 122(2):227–234.
41. O'Shaughnessy PJ, Johnston H, Willerton L, et al. Failure of normal adult Leydig cell development in androgen-receptor-deficient mice. J Cell Sci 2002; 115(pt 17): 3491–3496.
42. Kerr JB, Sharpe RM. Follicle-stimulating hormone induction of Leydig cell maturation. Endocrinology 1985; 116(6):2592–2604.
43. Heckert LL, Griswold MD. The expression of the follicle-stimulating hormone receptor in spermatogenesis. Recent Prog Horm Res 2002; 57:129–148.
44. Haider SG. Cell biology of Leydig cells in the testis. Int Rev Cytol 2004; 233:181–241.
45. Haider SG. Leydig cell steroidogenesis: unmasking the functional importance of mitochondria. Endocrinology 2007; 148(6):2581–2582.
46. Russell LD, Griswold MD. The Sertoli cell. 1st ed. Clearwater, FL: Cache River Press, 1993.
47. Russell LD. Histological and Histopathological Evaluation of the Testis. 1st ed. Clearwater, Fl: Cache River Press, 1990.
48. Wang ZX, Wreford NG, De Kretser DM. Determination of Sertoli cell numbers in the developing rat testis by stereological methods. Int J Androl 1989; 12(1):58–64.
49. Orth JM. Proliferation of Sertoli cells in fetal and postnatal rats: a quantitative auto-radiographic study. Anat Rec 1982; 203(4):485–492.
50. Hess RA, Cooke PS, Bunick D, et al. Adult testicular enlargement induced by neonatal hypothyroidism is accompanied by increased Sertoli and germ cell numbers. Endocrinology 1993; 132(6):2607–2613.
51. Huhtaniemi IT, Nevo N, Amsterdam A, et al. Effect of postnatal treatment with a gonadotropin-releasing hormone antagonist on sexual maturation of male rats. Biol Reprod 1986; 35(3):501–507.
52. Muller J, Skakkebaek NE. Quantification of germ cells and seminiferous tubules by stereological examination of testicles from 50 boys who suffered from sudden death. Int J Androl 1983; 6(2):143–156.

53. Lemasters GK, Perreault SD, Hales BF, et al. Workshop to identify critical windows of exposure for children's health: reproductive health in children and adolescents work group summary. Environ Health Perspect 2000; 108(suppl 3):505–509.
54. Welsh M, Saunders PT, Fisken M, et al. Identification in rats of a programming window for reproductive tract masculinization, disruption of which leads to hypospadias and cryptorchidism. J Clin Invest 2008; 118(4):1479–1490.
55. Welsh M, Saunders PT, Sharpe RM. The critical time window for androgen-dependent development of the Wolffian duct in the rat. Endocrinology 2007; 148(7):3185–3195.
56. Archambeault DR, Tomaszewski J, Joseph A, et al. Epithelial-mesenchymal crosstalk in Wolffian duct and fetal testis cord development. Genesis 2009; 47(1):40–48.
57. Hannema SE, Hughes IA. Regulation of Wolffian duct development. Horm Res 2007; 67(3):142–151.
58. Joseph A, Yao H, Hinton BT. Development and morphogenesis of the Wolffian/epididymal duct, more twists and turns. Dev Biol 2009; 325(1):6–14. [Epub 2008, Nov 1].
59. Josso N, Picard JY, Tran D. The anti-Mullerian hormone. Birth Defects Orig Artic Ser 1977; 13(2):59–84.
60. Taguchi O, Cunha GR, Lawrence WD, et al. Timing and irreversibility of Mullerian duct inhibition in the embryonic reproductive tract of the human male. Dev Biol 1984; 106(2):394–398.
61. Torres M, Gomez-Pardo E, Dressler GR, et al. Pax-2 controls multiple steps of urogenital development. Development 1995; 121(12):4057–4065.
62. Bouchard M, Souabni A, Mandler M, et al. Nephric lineage specification by Pax2 and Pax8. Genes Dev 2002; 16(22):2958–2970.
63. Kobayashi A, Shawlot W, Kania A, et al. Requirement of Lim1 for female reproductive tract development. Development 2004; 131(3):539–549.
64. Narlis M, Grote D, Gaitan Y, et al. Pax2 and pax8 regulate branching morphogenesis and nephron differentiation in the developing kidney. J Am Soc Nephrol 2007; 18 (4):1121–1129.
65. Tong SY, Hutson JM, Watts LM. Does testosterone diffuse down the wolffian duct during sexual differentiation? J Urol 1996; 155(6):2057–2059.
66. Higgins SJ, Young P, Cunha GR. Induction of functional cytodifferentiation in the epithelium of tissue recombinants. II. Instructive induction of Wolffian duct epithelia by neonatal seminal vesicle mesenchyme. Development 1989; 106(2):235–250.
67. Tomaszewski J, Joseph A, Archambeault D, et al. Essential roles of inhibin beta A in mouse epididymal coiling. Proc Natl Acad Sci U S A 2007; 104(27):11322–11327.
68. Benson GV, Lim H, Paria BC, et al. Mechanisms of reduced fertility in Hoxa-10 mutant mice: uterine homeosis and loss of maternal Hoxa-10 expression. Development 1996; 122(9):2687–2696.
69. Sun EL, Flickinger CJ. Development of cell types and of regional differences in the postnatal rat epididymis. Am J Anat 1979; 154(1):27–55.
70. Mendive F, Laurent P, Van Schoore G, et al. Defective postnatal development of the male reproductive tract in LGR4 knockout mice. Dev Biol 2006; 290(2):421–434.
71. Pryor JL, Hughes C, Foster W, et al. Critical windows of exposure for children's health: the reproductive system in animals and humans. Environ Health Perspect 2000; 108 (suppl 3):491–503.
72. Scheer H, Robaire B. Steroid delta 4-5 alpha-reductase and 3 alpha-hydroxysteroid dehydrogenase in the rat epididymis during development. Endocrinology 1980; 107 (4):948–953.
73. Aumuller G, Riva A. Morphology and functions of the human seminal vesicle. Andrologia 1992; 24(4):183–196.
74. Robaire B, Hinton BT, Orgebin-Crist MC. The epididymis: from molecules to clinical practice : a comprehensive survey of the efferent ducts, the epidiymus, and the vas deferens. New York: Kluwer Academic/Plenum Publishers, 2002.
75. Zondek LH, Zondek T. Effect of hormones on the human fetal prostate. Contrib Gynecol Obstet 1979; 5:145–158.
76. Mann T, Lutwak-Mann C. Evaluation of the functional state of male accessory glands by the analysis of seminal plasma. Andrologia 1976; 8(3):237–242.
77. Gonzales GF. Function of seminal vesicles and their role on male fertility. Asian J Androl 2001; 3(4):251–258.

78. Price D. Comparative aspects of development and structure in the prostate. Natl Cancer Inst Monogr 1963; 12:1–27.

79. Mata LR. Dynamics of the seminal vesicle epithelium. Int Rev Cytol 1995; 160: 267–302.

80. Brooks JR, Busch RD, Patanelli DJ, et al. A study of the effects of a new anti-androgen on the hyperplastic dog prostate. Proc Soc Exp Biol Med 1973; 143(3):647–655.

81. McNeal JE. The zonal anatomy of the prostate. Prostate 1981; 2(1):35–49.

82. Aumuller G. Morphologic and endocrine aspects of prostatic function. Prostate 1983; 4(2):195–214.

83. Reyes FI, Winter JS, Faiman C. Studies on human sexual development. I. Fetal gonadal and adrenal sex steroids. J Clin Endocrinol Metabol 1973; 37(1):74–78.

84. Spring-Mills E, Hafez ESE. Male Accessory Sex Glands: Biology and Pathology. New York, NY: Elsevier/North-Holland Biomedical Press, 1980.

85. Shaw LM, Yang N, Brooks JJ, et al. Immunochemical evaluation of the organ specificity of prostatic acid phosphatase. Clin Chem 1981; 27(9):1505–1512.

86. Frick J, Aulitzky W. Physiology of the prostate. Infection 1991; 19(suppl 3):S115–S118.

87. Lasnitzki I, Mizuno T. Antagonistic effects of cyproterone acetate and oestradiol on the development of the fetal rat prostate gland induced by androgens in organ culture. Prostate 1980; 1(2):147–156.

88. Gray LE, Ostby J, Furr J, et al. Effects of environmental antiandrogens on reproductive development in experimental animals. Hum Reprod Update 2001; 7(3):248–264.

89. Jarred RA, McPherson SJ, Bianco JJ, et al. Prostate phenotypes in estrogen-modulated transgenic mice. Trends Endocrinol Metab 2002; 13(4):163–168.

90. Prins GS. Neonatal estrogen exposure induces lobe-specific alterations in adult rat prostate androgen receptor expression. Endocrinology 1992; 130(4):2401–2412.

91. Prins GS, Birch L. The developmental pattern of androgen receptor expression in rat prostate lobes is altered after neonatal exposure to estrogen. Endocrinology 1995; 136 (3):1303–1314.

92. Santti R, Newbold RR, Makela S, et al. Developmental estrogenization and prostatic neoplasia. Prostate 1994; 24(2):67–78.

93. Pousette A, Bjork P, Carlstrom K, et al. Influence of prostatic secretion protein on uptake of androgen-receptor complex in prostatic cell nuclei. Prostate 1981; 2(1): 23–33.

94. Bruchovsky N, Lesser B, Van Doorn E, et al. Hormonal effects on cell proliferation in rat prostate. Vitam Horm 1975; 33:61–102.

95. Prins GS, Huang L, Birch L, et al. The role of estrogens in normal and abnormal development of the prostate gland. Ann N Y Acad Sci 2006; 1089:1–13.

96. Prins GS, Woodham C, Lepinske M, et al. Effects of neonatal estrogen exposure on prostatic secretory genes and their correlation with androgen receptor expression in the separate prostate lobes of the adult rat. Endocrinology 1993; 132(6):2387–2398.

97. Jirasek JE. Morphogenesis of the genital system in the human. Birth Defects Orig Artic Ser 1977; 13(2):13–39.

98. Rey R, Picard JY. Embryology and endocrinology of genital development. Baillieres Clin Endocrinol Metab 1998; 12(1):17–33.

99. Boas M, Boisen KA, Virtanen HE, et al. Postnatal penile length and growth rate correlate to serum testosterone levels: a longitudinal study of 1962 normal boys. Eur J Endocrinol 2006; 154(1):125–129.

100. Inomata T, Eguchi Y, Nakamura T. Development of the external genitalia in rat fetuses. Jikken Dobutsu 1985; 34(4):439–444.

101. Clark RL. Endpoints of reproductive system development. In: Daston G, Kimmel C, eds. An Evaluation and Interpretation of Reproductive Endpoints for Human Risk Assessment. Washington, DC: International Life Sciences INstitute/Health and Environmental Science Institute, 1999:27–62.

102. Prahalada S, Tarantal AF, Harris GS, et al. Effects of finasteride, a type 2 5-alpha reductase inhibitor, on fetal development in the rhesus monkey (Macaca mulatta). Teratology 1997; 55(2):119–131.

103. Wise LD, Vetter CM, Anderson CA, et al. Reversible effects of triamcinolone and lack of effects with aspirin or L-656,224 on external genitalia of male Sprague-Dawley rats exposed in utero. Teratology 1991; 44(5):507–520.

104. Salazar-Martinez E, Romano-Riquer P, Yanez-Marquez E, et al. Anogenital distance in human male and female newborns: a descriptive, cross-sectional study. Environ Health 2004; 3(1):8.
105. Swan SH, Main KM, Liu F, et al. Decrease in anogenital distance among male infants with prenatal phthalate exposure. Environ Health Perspect 2005; 113(8):1056–1061.
106. Orth JM, Gunsalus GL, Lamperti AA. Evidence from Sertoli cell-depleted rats indicates that spermatid number in adults depends on numbers of Sertoli cells produced during perinatal development. Endocrinology 1988; 122(3):787–794.
107. Russell LD, Chiarini-Garcia H, Korsmeyer SJ, et al. Bax-dependent spermatogonia apoptosis is required for testicular development and spermatogenesis. Biol Reprod 2002; 66(4):950–958.
108. Bowles J, Koopman P. Retinoic acid, meiosis and germ cell fate in mammals. Development 2007; 134(19):3401–3411.
109. Bowles J, Knight D, Smith C, et al. Retinoid signaling determines germ cell fate in mice. Science 2006; 312(5773):596–600.
110. Koubova J, Menke DB, Zhou Q, et al. Retinoic acid regulates sex-specific timing of meiotic initiation in mice. Proc Natl Acad Sci U S A 2006; 103(8):2474–2479.
111. Lin Y, Gill ME, Koubova J, et al. Germ cell-intrinsic and -extrinsic factors govern meiotic initiation in mouse embryos. Science 2008; 322(5908):1685–1687.
112. Abu-Abed S, MacLean G, Fraulob V, et al. Differential expression of the retinoic acid-metabolizing enzymes CYP26A1 and CYP26B1 during murine organogenesis. Mech Dev 2002; 110(1-2):173–177.
113. Fukuda T, Hedinger C, Groscurth P. Ultrastructure of developing germ cells in the fetal human testis. Cell Tissue Res 1975; 161(1):55–70.
114. Franchi LL, Mandl AM. The ultrastructure of germ cells in foetal and neonatal male rats. J Embryol Exp Morphol 1964; 12:289–308.
115. Wartenberg H. Comparative cytomorphologic aspects of the male germ cells, especially of the "Gonia". Andrologia 1976; 8(2):117–130.
116. Meng X, de Rooij DG, Westerdahl K, et al. Promotion of seminomatous tumors by targeted overexpression of glial cell line-derived neurotrophic factor in mouse testis. Cancer Res 2001; 61(8):3267–3271.
117. Meng X, Lindahl M, Hyvonen ME, et al. Regulation of cell fate decision of undifferentiated spermatogonia by GDNF. Science 2000; 287(5457):1489–1493.
118. Meng X, Pata I, Pedrono E, et al. Transient disruption of spermatogenesis by deregulated expression of neurturin in testis. Mol Cell Endocrinol 2001; 184(1-2):33–39.
119. Le Roy C, Lejeune H, Chuzel F, et al. Autocrine regulation of Leydig cell differentiated functions by insulin-like growth factor I and transforming growth factor beta. J Steroid Biochem Mol Biol 1999; 69(1-6):379–384.
120. Franca LR, Ogawa T, Avarbock MR, et al. Germ cell genotype controls cell cycle during spermatogenesis in the rat. Biol Reprod 1998; 59(6):1371–1377.
121. Huckins C. Cell cycle properties of differentiating spermatogonia in adult Sprague-Dawley rats. Cell Tissue Kinet 1971; 4(2):139–154.
122. Huckins C. The morphology and kinetics of spermatogonial degeneration in normal adult rats: an analysis using a simplified classification of the germinal epithelium. Anat Rec 1978; 190(4):905–926.
123. Huckins C, Oakberg EF. Morphological and quantitative analysis of spermatogonia in mouse testes using whole mounted seminiferous tubules, I. The normal testes. Anat Rec 1978; 192(4):519–528.
124. Grimes SR Jr. Nuclear proteins in spermatogenesis. Comp Biochem Physiol B 1986; 83 (3):495–500.
125. Govin J, Caron C, Lestrat C, et al. The role of histones in chromatin remodelling during mammalian spermiogenesis. Eur J Biochem 2004; 271(17):3459–3469.
126. Ryu BY, Orwig KE, Avarbock MR, et al. Stem cell and niche development in the postnatal rat testis. Dev Biol 2003; 263(2):253–263.
127. Huckins C. The spermatogonial stem cell population in adult rats. I. Their morphology, proliferation and maturation. Anat Rec 1971; 169(3):533–557.
128. Fawcett DW, Ito S, Slautterback D. The occurrence of intercellular bridges in groups of cells exhibiting synchronous differentiation. J Biophys Biochem Cytol 1959; 5(3):453–460.

129. Weber JE, Russell LD. A study of intercellular bridges during spermatogenesis in the rat. Am J Anat 1987; 180(1):1–24.
130. De Rooij DG, Lok D. Regulation of the density of spermatogonia in the seminiferous epithelium of the Chinese hamster: II. Differentiating spermatogonia. Anat Rec 1987; 217(2):131–136.
131. Allan DJ, Harmon BV, Roberts SA. Spermatogonial apoptosis has three morphologically recognizable phases and shows no circadian rhythm during normal spermatogenesis in the rat. Cell Prolif 1992; 25(3):241–250.
132. Billig H, Furuta I, Rivier C, et al. Apoptosis in testis germ cells: developmental changes in gonadotropin dependence and localization to selective tubule stages. Endocrinology 1995; 136(1):5–12.
133. Bartke A. Apoptosis of male germ cells, a generalized or a cell type-specific phenomenon? Endocrinology 1995; 136(1):3–4.
134. Huckins C. Spermatogonial intercellular bridges in whole-mounted seminiferous tubules from normal and irradiated rodent testes. Am J Anat 1978; 153(1):97–121.
135. de Rooij DG. Proliferation and differentiation of spermatogonial stem cells. Reproduction 2001; 121(3):347–354.
136. Koji T, Hishikawa Y, Ando H, et al. Expression of Fas and Fas ligand in normal and ischemia-reperfusion testes: involvement of the Fas system in the induction of germ cell apoptosis in the damaged mouse testis. Biol Reprod 2001; 64(3):946–954.
137. Lee J, Richburg JH, Younkin SC, et al. The Fas system is a key regulator of germ cell apoptosis in the testis. Endocrinology 1997; 138(5):2081–2088.
138. Xu JP, Li X, Mori E, et al. Expression of Fas-Fas ligand in murine testis. Am J Reprod Immunol 1999; 42(6):381–388.
139. French LE, Hahne M, Viard I, et al. Fas and Fas ligand in embryos and adult mice: ligand expression in several immune-privileged tissues and coexpression in adult tissues characterized by apoptotic cell turnover. J Cell Biol 1996; 133(2):335–343.
140. Ross AJ, Waymire KG, Moss JE, et al. Testicular degeneration in Bclw-deficient mice. Nat Genet 1998; 18(3):251–256.
141. Russell LD, Warren J, Debeljuk L, et al. Spermatogenesis in Bclw-deficient mice. Biol Reprod 2001; 65(1):318–332.
142. Camatini M, Franchi E, DeCurtis I. Differentiation of inter-Sertoli junctions in human testis. Cell Biol Int Rep 5 1981; 5:109.
143. Setchell BP, Voglmayr JK, Waites GM. A blood-testis barrier restricting passage from blood into rete testis fluid but not into lymph. J Physiol 1969; 200(1):73–85.
144. Vitale R, Fawcett DW, Dym M. The normal development of the blood-testis barrier and the effects of clomiphene and estrogen treatment. Anat Rec 1973; 176(3):331–344.
145. Niel L, Willemsen KR, Volante SN, et al. Sexual dimorphism and androgen regulation of satellite cell population in differentiating rat levator ani muscle. Dev Neurobiol 2008; 68(1):115–122.
146. Tobin C, Joubert Y. Testosterone-induced development of the rat levator ani muscle. Dev Biol 1991; 146(1):131–138.
147. d'Albis A, Tobin C, Janmot C, et al. Effect of testosterone and thyroid hormone on the expression of myosin in the sexually dimorphic levator ani muscle of rat. J Biol Chem 1992; 267(14):10052–10054.
148. Wensing CJ. The embryology of testicular descent. Horm Res 1988; 30(4–5):144–152.
149. Hutson JM, Hasthorpe S, Heyns CF. Anatomical and functional aspects of testicular descent and cryptorchidism. Endocr Rev 1997; 18(2):259–280.
150. Wyndham NR. A morphological study of testicular descent. J Anat 1943; 77(pt 2): 179–188. 3.
151. Cain MP, Kramer SA, Tindall DJ, et al. Flutamide-induced cryptorchidism in the rat is associated with altered gubernacular morphology. Urology 1995; 46(4):553–558.
152. Spencer JR, Torrado T, Sanchez RS, et al. Effects of flutamide and finasteride on rat testicular descent. Endocrinology 1991; 129(2):741–748.
153. Clarnette TD, Hutson JM. Is the ascending testis actually 'stationary'? Normal elongation of the spermatic cord is prevented by a fibrous remnant of the processus vaginalis. Pediatr Surg Int 1997; 12(2/3):155–157.
154. Wilhelm D, Koopman P. The makings of maleness: towards an integrated view of male sexual development. Nat Rev 2006; 7(8):620–631.

155. Korenbrot CC, Huhtaniemi IT, Weiner RI. Preputial separation as an external sign of pubertal development in the male rat. Biol Reprod 1977; 17(2):298–303.
156. Maeda K, Ohkura S, Tsukamura H. Physiology of reproduction. In: Krinke GJ, ed. The Laboratory Rat. New York: Academic Press, 2000:145–176.
157. Cameron JL. Metabolic cues for the onset of puberty. Horm Res 1991; 36(3–4):97–103.
158. Ben-Ari J, Merlob P, Mimouni F, et al. Characteristics of the male genitalia in the newborn: penis. J Urol 1985; 134(3):521–522.
159. Gairdner D. The fate of the foreskin, a study of circumcision. Br Med J 1949; 2 (4642):1433–1437, illust.
160. Oster J. Further fate of the foreskin. Incidence of preputial adhesions, phimosis, and smegma among Danish schoolboys. Arch Dis Child 1968; 43(228):200–203.

3   Normal development of the female reproductive system

Jerome M. Goldman and Ralph L. Cooper

INTRODUCTION

Species that possess sex organs employ a wide range of reproductive strategies. There are differences in the elicitation of the endocrine trigger for ovulation, in internal or external oocyte fertilization, in the number of viable offspring at birth, and whether there occur live births or births from hatched eggs. However, the initial development of the reproductive system shows a remarkable degree of homology across species, with phenotypic differences emerging from a rudimentary urogenital system that is essentially indistinguishable between males and females. This bipotential nature of inter- and intraspecies gonadal organ development was touched on over 75 years ago by the biologist Frank Lillie (1), who stated that, "There is no such biological entity as sex. What exists in nature is a dimorphism within species into male and female individuals ...."

The scientific exploration of this dimorphism has commonly employed the mouse as a model organism to delineate the importance and chronological progression of phenotypic events in gestational development, particularly given the large-scale use of the mouse for the creation of gene knockouts via alterations in the known sequence of specific genes to render them inoperative.

THE INDIFFERENT GONAD

Early in development [e.g., gestation day (GD) 10 in the mouse], indifferent gonads are present in mammals that are either chromosomally XX or XY. The development of phenotypic males and females is genetically controlled in most mammals by the presence or absence of the Y chromosome. The path toward development of a male reproductive tract is contingent on the gonadal expression of the Y-linked *Sry* (sex determining region Y) gene. In the absence of this gene [in XX or XY mammals (2)], the gonad will develop as an ovary. In contrast, work with chromosomal XX mice has demonstrated that the introduction of the *Sry* gene will induce the differentiation of a male testis (3). Subsequent investigations have indicated that *Sry* is the only gene on the Y chromosome that is both necessary and sufficient for the initiation of testicular development. The protein product of the *Sry* gene acts as a transcription factor that binds to the promoter of the *Sox9* (SRY-box 9) gene and initiates a cascade of other genes that channels the indifferent gonadal tissue in a male direction (4). In the female, no such equivalents of the Sry and Sox9 transcription factors have been found.

Indifferent gonads arise from a differentiation of the intermediate mesoderm on either side of the embryo soon after gastrulation. They emerge as part of the early development of the urogenital system, comprising gonads, kidneys, and the reproductive and urinary tracts. Within the intermediate mesoderm are three regions, the pronephros, mesonephros, and metanephros, which comprise what has been termed the urogenital ridge. The pronephros is evolutionarily a primitive kidney, but is vestigial in mammals. The mesonephros develops from a mesodermal region called the AGM (aorta-gonad-mesonephros) zone and contributes to the development of the aorta and gonads. Morphologically in mice, the indifferent gonads appear (GD10.5) in a gonadal ridge as thickenings of epithelium ventromedial to the mesonephros (Table 1).

TABLE 1   Chronology of Early Embryonic Development of the Female Reproductive Tract

Occurrence	Rat	Mouse	Human
Pronephros, appearance	day 10	day 8	Wk 6
Mesonephros, appearance	day 11.5	day 9.5	Wk 6–7
Appearance of indifferent gonad	day 12	day 10.5	Wk 6–7
Entry of Wolffian duct into urogenital sinus	day 12	day 11	Wk 7–8
Metanephros, appearance	day 12.8	day 11	Wk 9–10
Müllerian duct, appearance	day 13.5	day 11.5	Wk 9–10
Müllerian duct extends to cloaca	day 15.5	day 13.5	Wk 10–11
Wolffian duct, degeneration in female	day 19–22	day 15.5	Wk 25–26

Source: From Refs. 5–9.

In mammals and birds, the metanephros differentiates into the adult kidney. The indifferent gonads contain three principal types of somatic cells: supporting cells, stromal cells, and steroidogenic cells. Each is able to progress in a male or female direction. Supporting cells will provide support to germ cell development (discussed in sect. "Emergence of the Wolffian and Müllerian Ducts"), while the stromal cells will become involved in the structural organization of the gonad, forming the vasculature and extracellular matrix. The steroidogenic cells will differentiate into either embryonic male Leydig cells or female ovarian theca cells.

The nuclear receptor steroidogenic factor-1 (SF-1) is diffusely expressed in the region of the gonadal ridge. Although in the adult mammal it plays a key role in steroidogenic cell function, in the embryo it is essential for the development of the urogenital mesoderm and embryonic survival of the primary steroidogenic organs. Twenty-five percent of *SF-1* knockout mice are born, indicating that *SF-1* is not essential for prenatal survival, but those that do survive will show agenesis of gonads and adrenal glands (10,11) and die from adrenocortical insufficiency. It has also been reported that a mutation in human gene encoding *SF-1* caused not only adrenal failure but a complete XY sex reversal, with female genitalia and a retention of the uterus, in a genotypic male (12), providing evidence that in humans *SF-1* regulates the regression of Müllerian ducts (see later in text).

The switch between male and female genetic pathways appears to involve the participation of a nuclear receptor protein, Dax1, that in the mouse is expressed in the embryonic urogenital ridge on GD10.5 (13). At the time a separation between male and female gonadal tissue begins about GD12 to GD13, *Dax1* expression also shows a sexual dimorphism, turning off in the testes but remaining in ovaries (13,14). A functional role for the *Dax1* gene at this time has been hypothesized on the basis of experiments with a mouse strain in which the expression of *Sry* is low (15). Transgenic experiments introducing extra copies of *Dax1* in XY animals of this strain showed that expression of the *Dax1* transgenes caused the mice to develop as XY sex-reversed females (16), results suggesting that *Dax1* may antagonize the action of *Sry*. The effect in this case is apparently due to a lack of *Sox9* upregulation (17).

Although evidence of the opposing effects of *Sry* and *Dax1* were demonstrated in this particular strain (*Mus musculus domesticus poschiavinus*), no sex reversal was present in the more common *M. musculus musculus* strain, which indicates that such an effect, if present in opposition to *Sry*, is typically overridden (18). Also, there appears to be some species disparity in the contributions of *Dax1*. In humans, for example, the absence of the protein causes an adrenal hypoplasia and hypogonadotropic hypogonadism. In contrast, XY mice that are *Dax1* deficient have normal corticotropin concentrations but are sterile, while *Dax1*-deficient females are fertile (19). Consequently, a role as an ovary-determining gene is questionable.

DEVELOPMENT OF THE FEMALE GENITAL TRACT
Emergence of the Wolffian and Müllerian Ducts

The formation of two pairs of tubular structures, the Wolffian (mesonephric) and Müllerian (paramesonephric) ducts, will respectively provide the structural foundation for the emergence of the male or female reproductive system (20). The Wolffian ducts first become present as short segments within the pronephros, before developing into a continuous tube along the length of the urogenital ridge. The mesonephros, or Wolffian body, represents the primitive kidney and develops in the mesoderm. It is composed of a number of tubules, which at one end open into the Wolffian duct (Fig. 1). The Müllerian duct arises from a tubular invagination from the surface epithelium of the mesonephros and courses, in a cranial-to-caudal direction, parallel to the Wolffian duct, the presence of which is necessary for Müllerian duct growth (21,22). The prior development of the Wolffian duct also runs in a caudal direction, with both terminating at a fusion with the wall of the cloaca (the common cavity for the urinary and rectal passages).

Regression of the Ducts

The expression of the Y-linked *Sry* gene triggers a cascade of integrative events that result in the development of indifferent gonads in a male direction. One of the resultant effects of this gonadal differentiation is the production of Müllerian inhibiting substance (MIS) by Sertoli cells of the embryonic testes. MIS, a member of the transforming growth factor-β (TGF-β) family, induces regression of the Müllerian ducts and the development of the Wolffian duct into the male epididymis and vas deferens. The effects are mediated by a binding of MIS to the MIS type II receptor (MISRII) present in the Müllerian duct mesenchyme (23), leading to changes in the mesenchyme that result in a subsequent disappearance of the ductal epithelium (24). Without the testicular

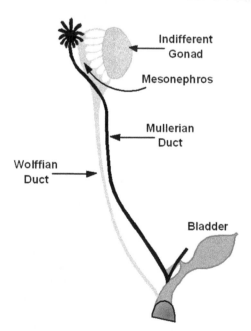

FIGURE 1 Drawing representing the primitive gonad with Wolffian and Müllerian ducts in the sexually undifferentiated embryo.

hormonal production of MIS and testosterone, it is the Wolffian ducts that regress, and the Müllerian ducts develop to form female structures. The proximal region of the Müllerian duct is present at the level of the fetal ovary and develops into the oviduct. The middle region, where Müllerian and Wolffian ducts are closely apposed, differentiates into the uterus, and the cervix and upper part of the vagina are formed from the caudal section. MIS-deficient genetically male mice were found to have female reproductive organs and testes with functional sperm (25). However, the mice were infertile, as the passage of sperm to females at mating was blocked.

OVARIAN FOLLICLE DEVELOPMENT
Primordial Germ Cells
Embryologically, primordial germ cells were first reported in 1954 by Chiquoine (26) as clusters in the genital ridges of an 8.5 day mouse embryo by their high alkaline phosphatase (ALP) activity. Later, work by Ginsburg et al. (27) showed a few ALP-positive cells early on day 7 of gestation, and the numbers continued to increase progressively in embryonic tissue taken throughout the day. By early on day 8, the population of primordial germ cells had increased to between 93 and 193. From GD8.5 to GD12, these cells were seen to migrate into the mesoderm of the mesonephros and then colonize in the gonad (28), a journey that appears to require the involvement of various signaling factors (29–31).

Primordial germ cells in the mammalian embryo are potentially capable of developing into meiotic oocytes or prospermatogonia. In the mouse, the selection of which of the two directions to proceed takes place by GD13.5. In spite of a possible weak, above-mentioned antagonistic influence of *Dax1*, the direction is dictated by the presence or absence of the Y-linked *Sry* gene in gonadal somatic cells, instead of the chromosomal identity of the primordial germ cells. In XX gonads, germ cells obligate to oogenesis one day later than the commitment of XY gonadal germ cells to spermatogenesis. This difference underlies the observation that at GD12.5 coculturing primordial germ cells of XX gonads with male embryonic urogenital ridge tissue can induce a sex reversal in which the germ cells will differentiate into prospermatogonia. In contrast, similar cocultures of XX gonadal germ cells taken at GD13.5 will proceed to differentiate as meiotic oocytes (32).

Within the embryonic mouse ovary, there subsequently appears (GD14.5) groupings of tube-like structures consisting of pregranulosa–germ cell complexes, the ovigerous cords (33,34) within which the pregranulosa cells form the outer wall of the cords. These cords are interconnected by a series of cytoplasmic bridges, resulting in clustered arrangements of germ cells (the oogonia) that are termed germline cysts (35). In the presence of signaling by retinoic acid (36), the oogonia within these cysts will subsequently undergo meiosis to become oocytes within relabeled oocyte nests. The nests will then disperse, and individual cells will become encapsulated within primordial follicles by the pregranulosa somatic cells. During meiosis, the oocytes will proceed through leptotene, zygotene, and pachytene stages until they become arrested in the dictyate stage of prophase I about the time of birth and await later developmental signaling by the gonadotropins after puberty.

In the newborn mouse, few follicles are present, but the number rapidly increases to approximately 40% of the oocyte population by postnatal day 2 and greater than 80% by day 6 (37). The dispersion of the oocyte nests corresponds to a marked amount of developmentally programmed apoptotic activity within the population of mouse germ cells over a two-day window, resulting in a large loss of germ cells that may be a consequence of nest breakdown (37–39).

Assembly of Primordial Follicles

On separation from the nests, individual oocytes become associated with granulosa precursor cells (40), a process that has been found to require the programmed death of random oocytes within the nest. These apoptotic changes appear to be an inductive process for primordial follicle assembly that involves the contribution of tumor necrosis factor-α (TNF-α) (41,42). Endocrine factors also have been observed to play a role in formation of the follicles. In cultured ovaries taken from newborn rats, progesterone acted to inhibit assembly, an effect that could be overcome by the presence of TNF-α (43). Estradiol was also able to suppress follicle assembly, although to a lesser extent. The effect of estradiol was apparently mediated by a probable binding to nuclear estrogen receptors, whereas the influence of progesterone likely involved an association with presumptive surface membrane progesterone receptors (PRs), as the nuclear PR antagonist RU486 was unable to interfere with the inhibition (44). Furthermore, at the time of follicle assembly, microarray techniques have demonstrated the presence of transcripts for progesterone surface receptors, but no detectable levels of the classic nuclear receptors (43). Similar inhibitory effects of estradiol and progesterone on primordial follicle assembly have been reported by Chen et al. (45). The breakdown of oocyte nests was correspondingly suppressed, and the number of primordial follicles that were formed showed an increased percentage of those with multiple oocytes, which the authors believed were created from nests that did not separate. Chen and coauthors speculated that elevated prenatal levels of maternal hormones in the fetus normally serve to retard the dispersion of the nests. When the steroid concentrations fall at birth, the inhibitory signal is removed, causing nests to break apart so that follicle assembly can proceed.

Estradiol and the Appearance of Estrogen Receptors

Early in embryonic rodent development, ovaries formed from the indifferent gonads show no evidence of estradiol synthesis, and only minimal amounts are detectable late in gestation and into the first postnatal week (46–48). This indicates that estrogens are not required for Müllerian duct development in utero, something that also appears to be true for progesterone (49). In contrast, estrogen production by the human embryonic ovary is present, although this fetal estrogen also does not appear to play an essential role in embryonic and fetal development (50).

The use of reverse transcriptase–polymerase chain reactions to investigate the expression of genes in fetal mouse ovaries for various steroidogenic enzymes showed that mRNA expression for CYP11A1, CYP17, and CYP19 was either undetectable from GD13 to GD20 or present in only a small percentage of ovaries examined (51), offering additional support for the lack of basal estradiol production during this time. However, estradiol and estrone production by the rodent fetal ovary late in gestation was reported to be responsive to stimulation and supplementation with steroid precursors. GD19 and GD20 ovaries exposed in culture to precursors progesterone or testosterone formed estrone, although estradiol synthesis was questionable (52). cAMP stimulation of ovaries was observed to induce aromatase activity, again with a predominance of estrone being secreted (53). The stimulated secretion of estrone was seen as early as GD17, with the estrone to estradiol ratio showing a shift in favor of estradiol production between the fetal and postnatal prepubertal stages. Ovarian estrogen secretion was also elevated by follicle-stimulating hormone (FSH), but not until postnatal days 3 and 4, which corresponded to the appearance of FSH receptors (53).

In the indifferent mouse gonad at GD11 to GD13, there is no detectable immunoreactivity for estrogen receptor α (ERα) subtype (54,55), although Nielsen and et al. (55) reported the presence of a few ERα-positive cells in the mouse, surrounding the Müllerian and Wolffian ducts. At GD13.5, ERα immunoreactivity was observed in all Müllerian duct epithelial cells. In mice at 16 to 18 days of gestation, Müllerian vaginal and uterine epithelial tissues were present and appeared morphologically uniform. At this time, the vaginal epithelium had begun to show evidence of ERα and one cytokeratin marker (K14), but the PR was not detectable (49). By postnatal days 1 to 3, vaginal epithelial cells were strongly positive for ERα and K14, although vaginal PR was still absent. ERα in uterine epithelial tissue lagged somewhat behind vaginal tissue, only exhibiting some weakly positive signs by postnatal days 5 to 7. In contrast, PRs were present in uterine epithelium from postnatal day 3 onward.

Estrogen receptor knockout animals for both the α (ERαKO) and β (ERβKO) forms of the receptor have shown minimal effects on the gestational development of any male or female fetal tissue. Development of the female tract in these animals appears normal, although after birth the uterus, fallopian tubes, cervix, and vagina are hypoplastic and lack the typical responsiveness to estradiol (56). The ERαKO females are sterile, and ERβKO females show either infertility or varying degrees of subfertility (57). By adulthood, ovaries in the combined ERαβKO females contained seminiferous-like tubular structures similar to those present in male testes, something that was not observed in separate ERαKO or ERβKO females (58), indicating a process of ovarian redifferentiation had taken place and emphasizing the importance of both forms of the receptor in ovarian maintenance.

Female mice homozygous for a null mutation of the PR were observed to progress normally through embryogenesis, but exhibited wide-ranging abnormalities as adults.

Uterine inflammation and hyperplasia were present, along with problems in mammary development. Females were infertile, and their ovaries contained many unruptured preovulatory follicles (59).

GENETICS OF FEMALE REPRODUCTIVE DEVELOPMENT

With the advent of improved methods to characterize the expression of a large array of genes during the early stages of sexual development, researchers have been able to begin identifying and mapping the chronological participation of individual genes during the generation of a sexually differentiated reproductive tract. Targeted gene knockouts have been extensively employed to characterize the contribution of individual genes to normal development, and the work has resulted in a large catalogue of participants. Homeobox (Hox) genes are homologs of *Drosophila* homeotic genes, which encode transcription factors that control patterning of the embryonic axis. They also generate transcription factors that regulate development by activating or repressing the expression of target genes. Among their roles, members of a subgrouping of these genes are involved in determining positions along the axis of the developing Müllerian ducts that are destined to become the oviduct, uterus, and upper part of the vagina (60).

As the development of a male reproductive tract from the undifferentiated genital ridge must be shifted from a path normally directed toward the female, it is not surprising that the majority of these genes identified thus far are involved in the determination of a testis. Table 2, reproduced from material presented in Wilhelm et al. (20), lists genes separated into sections by their participation in testicular and ovarian development, along with those implicated in early changes within the bipotential gonad. For the purposes of this chapter, focus will be on the expression of those genes relevant to development of the female reproductive tract from the indifferent gonad.

The Wilms' tumor gene (*Wt1*) is important for mesonephros development and is expressed throughout the urogenital ridge (61). It regulates Sry expression and acts as a transcription factor targeting the gene for the nuclear hormone receptor SF-1, which as previously indicated, is an essential factor in both gonadal and adrenal, in addition to kidney, development. As might be expected, inactivation in mice of *Wt1* will result in

TABLE 2   Genes Implicated in Sexual Development in Mammals

Gene	Protein function	Gonad phenotype of null mice	Human syndrome
Indifferent gonad			
Wt1	Transcription factor	Blockage of genital ridge development	Denys-Drash, WAGR, Frasier syndrome
SF-1	Nuclear receptor	Blockage in genital ridge development	Embryonic testicular regression syndrome
Lhx9	Transcription factor	Blockage in genital ridge development	[a]
Emx2	Transcription factor	Blockage in genital ridge development	[a]
M33	Transcription factor	Gonadal dysgenesis	[a]
Testis-determining pathway			
Gata4/ Fog2	Transcription/ cofactor	Reduced Sry levels, XY sex reversal	[a]
Sry	Transcription factor	XY sex reversal	XY sex reversal (LOF), XX sex reversal (GOF)
Sox9	Transcription factor	XY sex reversal	Campomelic dysplasia, XX sex reversal (GOF)
Sox8	Transcription factor	XY sex reversal	[a]
Fgf9	Signaling molecule	XY sex reversal	[a]
Dax1	Nuclear receptor	Impaired testis cord formation/ spermatogenesis	Hypogonadism
Pod1	Transcription factor	XY sex reversal	[a]
Dhh	Signaling molecule	Impaired differentiation of Leydig and surrounding peritubular myoid cells	XY gonadal dysgenesis
Pgdra	Receptor	Reduction in mesonephric cell migration	[a]
Pgds	Enzyme	No phenotype	[a]
Arx	Transcription factor	Abnormal testicular differentiation	X-linked lissencephaly with abnormal genitalia
Atrx	Helicase	ND	ATRX syndrome
Insl3	Signaling factor	Blockage of testicular descent	Cryptorchidism
Lgr8	Receptor	Blockage of testicular descent	Cryptorchidism
Hoxa10	Transcription factor	Blockage of testicular descent	Cryptorchidism
Hoxa11	Transcription factor	Blockage of testicular descent	Cryptorchidism
Amh	Hormone	No Müllerian duct formation	Persistent Müllerian duct syndrome
Misrl1	Receptor	No Müllerian duct formation	Persistent Müllerian duct syndrome
Pax2	Transcription factor	Dysgenesis of mesonephric tubules	[a]
Lim1	Transcription factor	Agenesis of Wolffian and Müllerian ducts	[a]
Dmrt1	Transcription factor	Loss of Sertoli and germ cells	XY female [b]
Ovary-determining pathway			
Wnt4	Signaling molecule	Müllerian duct agenesis, testosterone synthesis, and coelomic vessel formation	XY female (GOF)
FoxL2	Transcription factor	Premature ovarian failure	BPES
Dax1	Nuclear receptor	XY sex reversal (GOF)	XY sex reversal (GOF)

[a]No mutations in human sexual disorders identified to date.
[b]Candidate gene for 9p deletion, XY sex reversal.
Abbreviations: BPES, blepharophimosis-ptosis-epicanthus inversus syndrome; GOF, gain-of-function mutation; LOF, loss-of-function mutation; WAGR, Wilms' tumor–aniridia–genitourinary malformations–mental retardation; ND, not determined.
Source: From Ref. 20, reproduced with permission (Please refer to the original publication for citations for individual genes.)

the absence of both kidneys and gonads (62). *Lhx9* (LIM/homeobox gene 9) is a member of a family of transcription factors and its expression is another essential component in gonadal formation. In mice without a functioning *Lhx9* gene, germ cells will migrate normally, but there is an absence of proliferation in genital ridge somatic cells, and a gonad fails to form (63).

The *Emx2* (empty spiracles homeobox 2) gene is related to the "empty spiracles" gene expressed in the head of developing *Drosophila*. In mammals, it is expressed in pronephros and mesonephros, in the Wolffian and Müllerian ducts, and in the indifferent gonads. Emx2 is critical to the formation of the urogenital system, and homozygous *Emx2* mutations in mice will show a failure of kidneys, ureters, gonads, and genital tracts to develop (64,65). The absence of kidneys causes death soon after birth. The *M33* gene is a homolog of *Drosophila* polycomb genes, and its expression in mammals has also been found to be critical in gonadal formation. Its disruption in mice retards the formation of genital ridges, and there are gonadal growth defects that become evident about the time the *Sry* gene is expressed (66).

Wnt4 is a member of a highly conserved gene class that is involved in multiple aspects of development. The glycoprotein products of this class act as signaling molecules that regulate cell-cell interactions during embryogenesis. The name is a combined designation for the *Drosophila* wingless (*Wg*) gene homolog of the mouse mammary oncogene *int-1*. Several *Wnt* genes are expressed in and around Wolffian and Müllerian ducts and are involved in their development. *Wnt4* expression is especially concentrated in the area surrounding the Müllerian ducts, and in male and female *Wnt4* null mice these ducts are not formed (67). Ovarian dysgenesis is present, along with abnormal development of the adrenals, kidneys, pituitary, and mammary glands (68–70). The ovaries had aggregates that resembled follicles and were positive for anti-Müllerian hormone.

FOXL2 is a member of the *forkhead box* family of transcription factors and is produced in the eye, ovary, and pituitary gland. The name forkhead box is derived from a sequence of amino acids in the DNA binding domain that has a butterfly-like appearance. There is some evidence that expression of the highly conserved *FoxL2* gene may act in an anti-testis capacity and underlie female sex determination in mammals (71,72). In double *Wnt4* and *FoxL2* knockouts, newborn ovaries contained testis-like tubules that expressed *Sox9* and AMH, an indication that granulosa cells had undergone conversion to male Sertoli cells (72). This type of reversal was also a response to a conditional deletion of FoxL2 in sexually mature female mice (73). These data offer a persuasive argument that in mammals the maintenance of an ovarian phenotype depends upon an active repression of the male *Sox9* gene.

The *Dax1* gene, previously discussed in the section "The Indifferent Gonad," is expressed throughout the hypothalamus, pituitary, adrenal, and gonads, implying a critical role for *Dax1* in all of these regions during that time. The name is an abbreviation of the full designation, *dosage sensitive sex reversal, adrenal hypoplasia congenita critical region on the X chromosome, gene 1*. The "dosage-sensitive sex reversal" part of the full name is a consequence of the aforementioned male-to-female reversal seen in response to an insertion in XY mice of two *Dax1* genes on the X chromosome. However, its exclusive role as part of the ovary-determining pathway, as indicated in Table 2, is still open to some questions, and the gene may participate in early testis formation (17).

Another group of genes, members of the Notch family, has been implicated in ovarian development. Its name derives from a strain of *Drosophila* with notches in the wingblades. The genes encode for a number of cell membrane receptors that constitute a highly conserved cell signaling system, and there is now some evidence that Notch expression has a role in the mediation of interactions between germ cells and somatic cells during early follicle development (74).

SEXUAL DIFFERENTIATION OF THE BRAIN

For mammalian tissues, established doctrine regarding sexual differentiation is that chromosomes drive gonadal formation and steroids secreted by the gonads control the phenotypic development of these tissues during sensitive perinatal and subsequent postnatal periods. The impact of such early exposure to these gonadal steroids serves to organize sexually dimorphic brain regions (75,76), physiological and biochemical activity (77), and gender-specific behaviors (78). The bulk of research that has focused on this organizational impact of gonadal hormones during perinatal development has employed the laboratory rat as a model, given that it is arguably the species for which steroid-induced sexual differentiation is best characterized.

The gonadal synthesis of testosterone in the embryonic male precedes ovarian estradiol production in the female, which, as indicated above for the mouse, only begins postnatally. Elevated testosterone production late in gestation is necessary to induce a brain that is phenotypically male, a process that establishes a correspondence between gonadal and brain sexual differentiation. An induction of a male-like phenotypic brain is also true following a treatment of female rats with testosterone during late gestation and up into the first postnatal week (79,80). The sensitive window in females for this exposure closes in the second week, after which a masculinizing treatment with testosterone is ineffective, and the brain becomes phenotypically female (81).

It was determined that the process of brain masculinization was dependent on a conversion of testosterone entering the brain to estradiol by the action of P450 aromatase (CYP19) (82). During the late gestational period, male and female fetuses are exposed to increasing maternal levels of circulating estrogens, so that it is essential for normal sexual differentiation of the brain to shield them, particularly females, from that endocrine milieu. It was discovered that the developing fetal liver and yolk sac produce elevated concentrations of α-fetoprotein, a protein that is potently able to bind and sequester maternal estradiol (83,84), thus preventing the hormone from entering the brain in quantities sufficient to cause masculinization. Experiments have shown that α-fetoprotein knockout female mice are behaviorally defeminized, with a later impairment of steroid-induced surges of luteinizing hormone (LH) normally seen in sexually mature females (85). After birth and the closing of the sensitive window, this sequestration of estradiol by α-fetoprotein is no longer necessary to protect the female against masculinization, and levels of α-fetoprotein then fall. In humans, the masculinizing effect of estradiol has not been demonstrated, and although α-fetoprotein is present during the early neonatal period in infants, its biological role is unclear (86).

As opposed to estradiol, testosterone secreted from the fetal testes is not bound by α-fetoprotein and is permitted to enter the brain to initiate the process of masculinization. The role of testosterone in the process of brain masculinization is mediated by the two forms of the estrogen receptor. In the region of the embryonic mouse brain, ERβ has been detected as early as GD10.5, before a determination of gonadal sex, whereas ERα is expressed somewhat later at GD16.5 (87). Emerging evidence suggests that with the aromatization of testosterone to estradiol, the two receptor subtypes play different roles in the sexual differentiation of the brain, with ERα acting primarily in masculinization and ERβ involved in defeminization of sexual behaviors (88).

Although feminization of the brain occurs in the absence of a masculinizing level of maternal estradiol during the perinatal period, it has been argued that the sexual differentiation of the female brain is an active process mediated by cellular/endocrine mechanisms that act to organize a neural network regulating adult female sex behavior (addressed further in chap. 15) and the appearance of characteristic postpubertal hormonal events critical to ovulation (89–91). Aromatase activity is present in the embryonic and neonatal brain, with peak levels expressed in the hypothalamus and preoptic area during the perinatal sensitive window (92), and Bakker et al. (93) have reported that disruptions of estradiol synthesis in aromatase knockout (ArKO) mice caused a marked impairment in later lordosis behavior.

PUBERTAL DEVELOPMENT AND EMERGENCE OF THE MATURE FEMALE REPRODUCTIVE SYSTEM

In the altricial mammalian female, the emergence of a mature reproductive system occurs over a transitional period of somatic and endocrine changes and represents a culmination of processes that began during infancy. Large numbers of oocytes have already been lost to apoptotic cell death late in gestation and during the weeks after birth. In both rats and humans, early postnatal elevations are present in circulating concentrations of the pituitary gonadotropins, LH and FSH (94,95). The increases in humans correlate with the fall of placental estrogens, and in the rat this increase, particularly in FSH, has been hypothesized to be important for ovarian maturation (96). As postnatal concentrations of rat α-fetoprotein drop, ovarian estradiol production will have a negative feedback effect on pituitary LH and FSH, and gonadotropin levels will decline (97). There subsequently occurs a transition in estradiol feedback, and the rat hypothalamus and pituitary will begin to respond positively, resulting in a stimulatory release of LH (98).

A host of neuroendocrine and other signaling factors is now known to participate in the onset of puberty. In the mature mammal, interactions among the hypothalamus, pituitary, and gonads form the basis for a functional reproductive system. Gonadotropin-releasing hormone (GnRH) secreted from nerve terminals in the hypothalamic median eminence is conveyed in portal vessels to the anterior pituitary, where it stimulates the release of the gonadotropins, LH and FSH. Circulating gonadotropins will then reach the ovaries, where they provide an impetus for follicular maturation (FSH) and ovulation (LH). LH also serves to increase the synthesis of estradiol, which in turn will provide feedback to the hypothalamus and pituitary, regulating GnRH and gonadotropin secretion.

During the third postnatal week in the female rat, a cohort of immature primordial follicles, stimulated by circulating FSH, undergoes a series of maturational changes that will prepare these members of the cohort that have not undergone apoptosis to respond to the first ovulatory surge of LH, the functional expression of a mature reproductive system. The progression in follicle development from a primordial to a preovulatory follicle involves growth in two major cellular regions, the outer theca cell layer and inner granulosa cell layer. In rat, theca cells emerge somewhat later in development than the initial granulosa cell layer, appearing after two to three layers of granulosa cells are formed (99). As the growing follicle matures, an antrum or internal cavity forms within the granulosa cells in a process that is gonadotropin dependent (100), and augmented by estradiol synthesized within the follicle (101). Under LH stimulation, cells in the theca layer synthesize and secrete androgens, which then enter the granulosa cells for conversion to estrogens by the action of aromatase. The estrogens stimulate further follicular growth and enhance LH receptor formation that augments the granulosa cell responsiveness in mature antral follicles to the ovulatory surge of LH (102).

It has been commonly believed that pubertal development in mammals is principally mediated by an increase in the pulsatile activity of GnRH neurons from a more quiescent state during the juvenile period. In the sexually differentiated female brain, this neuronal activation will then trigger the first LH surge that stimulates ovulation. Although many temporal correlates of these changes have been found, the critical initiators of the timing of puberty still remain largely unidentified. The question remains whether the principal trigger is an activation of GnRH neurons or a disinhibition of some type of restraint on these neurons. Various research groups have implicated alterations, for example, in the developmental dynamics of excitatory and inhibitory hypothalamic neurotransmitters (103,104), growth factors (105), and metabolic input (106). Recent studies have revealed that a G-protein-coupled receptor on GnRH neurons, GPR54 (Kiss-1 receptor) and its endogenous ligand kisspeptin (secreted from hypothalamic Kiss-1 neurons), function as essential and powerful regulators of mammalian and nonmammalian GnRH secretion (107–109). Kiss-1 neurons have estrogen receptors, and kisspeptin secretion is responsive to estradiol feedback. Although kisspeptin and its receptor have been identified as key components of events that lead to the emergence in mammals of a mature reproductive system (110–112), the precise mechanism(s) that trigger this signaling during the immediate prepubertal period remains unknown.

CONCLUSIONS

Obviously, considerable progress has been made in characterizing the progression of mammalian reproductive tract development and the importance of an ever-growing number of identified transcriptional and biochemical factors in this complex process. The conserved genetic regulation and limited biological diversity across species during development have permitted us to gain insight into early developmental events from studies confined to a small number of mammalian species. Moreover, this degree of concordance among mammals in the mechanisms of reproductive development has given weight to assessments of risk from early toxicant exposures in test species to the impacts of such exposures in the human population.

ACKNOWLEDGMENTS

The authors appreciate the constructive comments provided by Drs. Rochelle Tyl, John Rogers, and Kembra Howdeshell on an earlier draft of the manuscript.

The material described in this article has been reviewed by the National Health and Environmental Effects Research Laboratory, U.S. Environmental Protection Agency, and approved for publication. Approval does not signify that the contents necessarily reflect the views and policies of the agency, nor does mention of trade names or commercial products constitute endorsement or recommendation for use.

REFERENCES

1. Lillie FR. General biological introduction. In: Allen E, ed. Sex and Internal Secretions. Baltimore: Williams and Wilkins, 1932:6.
2. Lovell-Badge R, Robertson E. XY female mice resulting from a heritable mutation in the primary testis-determining gene, Tdy. Development 1990; 109:635–646.
3. Koopman P, Gubbay J, Vivian N, et al. Male development of chromosomally female mice transgenic for Sry. Nature 1991; 351:117–121.
4. Kobayashi A, Chang H, Chaboissier M-C, et al. Sox9 in testis determination. Ann NY Acad Sci 2005; 1061:9–17.
5. Hoar RM. Comparative female reproductive tract development and morphology. Environ Health Perspect 1978; 24:1–4.
6. Laitinen L, Virtanen I, Saxén L. Changes in the glycosylation pattern during embryonic development of mouse kidney as revealed with lectin conjugates. J Histochem Cytochem 1987; 35:55–65.
7. Kuure S, Vuolteenaho R, Vainio S. Kidney morphogenesis: cellular and molecular regulation. Mech Dev 2000; 92:31–45.
8. Orvis GD, Behringer RR. Cellular mechanisms of Müllerian duct formation in the mouse. Dev Biol 2007; 306:493–504.
9. Welsh M, Sharpe RM, Walker M, et al. New insights into the role of androgens in Wolffian duct stabilization in male and female rodents. Endocrinology 2009; 150: 2472–2480.

10. Luo X, Ikeda Y, Lala D, et al. Steroidogenic factor 1 (SF-1) is essential for endocrine development and function. J Steroid Biochem Mol Biol 1999; 69:13–18.
11. Hanley NA, Ikeda Y, Luo X, et al. Steroidogenic factor 1 (SF-1) is essential for ovarian development and function. Mol Cell Endocrinol 2000; 163:27–32.
12. Achermann JC, Ito M, Ito M, et al. A mutation in the gene encoding steroidogenic factor-1 causes XY sex reversal and adrenal failure in humans. Nat Genet 1999; 22:125–126.
13. Swain A, Zanaria E, Hacker A, et al. Mouse *Dax1* expression is consistent with a role in sex determination as well as in adrenal and hypothalamus function. Nat Genet 1996; 12:404–409.
14. Vilain E, McCabe ERB. Mammalian sex determination: from gonads to brain. Mol Genet Metab 1998; 65:74–84.
15. Swain A, Narvaez V, Burgoyne P, et al. *Dax1* antagonizes Sry action in mammalian sex-determination. Nature 1998; 391:761–767.
16. Ludbrook LM, Harley VR. Sex determination: a 'window' of *DAX1* activity. Trends Endocrinol Metab 2004; 15:116–121.
17. Bouma GJ, Albrecht KH, Washburn LL, et al. Gonadal sex reversal in mutant *Dax1* XY mice: a failure to upregulate *Sox9* in pre-Sertoli cells. Development 2005; 132:3045–3054.
18. Jordan BK, Vilain E. Sry and the genetics of sex determination. Adv Exp Med Biol 2002; 511:1–13.
19. Goodfellow PN, Camerino G. *DAX-1*, an "antitestis" gene. Cell Mol Life Sci 1999; 55:857–863.
20. Wilhelm D, Palmer S, Koopman P. Sex determination and gonadal development in mammals. Physiol Rev 2007; 87:1–28.
21. Carroll TJ, Park JS, Hayashi S, et al. Wnt9b plays a central role in the regulation of mesenchymal to epithelial transitions underlying organogenesis of the mammalian urogenital system. Dev Cell 2005; 9:283–292.
22. Kobayashi A, Kwan KM, Carroll TJ, et al. Distinct and sequential tissue-specific activities of the LIM-class homeobox gene Lim1 for tubular morphogenesis during kidney development. Development 2005; 132:2809–2823.
23. Roberts LM, Hirokawa Y, Nachtigal MW, et al. Paracrine-mediated apoptosis in reproductive tract development. Dev Biol 1999; 208:110–122.
24. Jamin SP, Arango NA, Mishina Y, et al. Genetic studies of MIS signalling in sexual development. Novartis Found Symp 2002; 244:157–164.
25. Behringer RR, Finegold MJ, Cate RL. Müllerian-inhibiting substance function during mammalian sexual development. Cell 1994; 79:415–425.
26. Chiquoine AD. The identification, origin and migration of the primordial germ cells of the mouse embryo. Anat Rec 1954; 118:135–146.
27. Ginsburg M, Snow MH, McLaren A. Primordial germ cells in the mouse embryo during gastrulation. Development 1990; 110:521–528.
28. Loffler KA, Koopman P. Charting the course of ovarian development in vertebrates. Int J Dev Biol 2002; 46:503–510.
29. Godin I, Wylie C, Heasman J. Genital ridges exert long-range effects on mouse primordial germ cell numbers and direction of migration in culture. Development 1990; 108:357–363.
30. Di Carlo A, De Felici M. A role for E-cadherin in mouse primordial germ cell development. Dev Biol 2000; 226:209–219.
31. Dudley BM, Runyan C, Takeuchi Y, et al. BMP signaling regulates PGC numbers and motility in organ culture. Mech Dev 2007; 124:68–77.
32. Adams IR, McLaren A. Sexually dimorphic development of mouse primordial germ cells: switching from oogenesis to spermatogenesis. Development 2002; 129:1155–1164.
33. Konishi I, Fujii S, Okamura H, et al. Development of interstitial cells and ovigerous cords in the human fetal ovary: an ultrastructural study. J Anat 1986; 148:121–135.
34. Ikeda Y, Tanaka H, Esaki M. Effects of gestational diethylstilbestrol treatment on male and female gonads during early embryonic development. Endocrinology 2008; 149:3970–3979.
35. Pepling ME. From primordial germ cell to primordial follicle: mammalian female germ cell development. Genesis 2006; 44:622–632.
36. Bowles J, Koopman P. Retinoic acid, meiosis and germ cell fate in mammals. Development 2007; 134:3401–3411.

37. Pepling ME, Spradling AC. Mouse ovarian germ cell cysts undergo programmed breakdown to form primordial follicles. Dev Biol 2001; 234:339–351.
38. Gondos B. Comparative studies of normal and neoplastic ovarian germ cells: 1. Ultrastructure of oogonia and intercellular bridges in the fetal ovary. Int J Gynecol Pathol 1987; 6:114–123.
39. Gondos B, Zamboni L. Ovarian development: the functional importance of germ cell interconnections. Fertil Steril 1969; 20:176–189.
40. McNatty KP, Fidler AE, Juengel JL, et al. Growth and paracrine factors regulating follicular formation and cellular function. Mol Cell Endocrinol 2000; 163:11–20.
41. Morrison LJ, Marcinkiewicz JL. Tumor necrosis factor α enhances oocyte/follicle apoptosis in the neonatal rat ovary. Biol Reprod 2002; 66:450–457.
42. Marcinkiewicz JL, Balchak SK, Morrison LJ. The involvement of tumor necrosis factor-α (TNF) as an intraovarian regulator of oocyte apoptosis in the neonatal rat. Front Biosci 2002; 7:d1997–d2005.
43. Nilsson EE, Stanfield J, Skinner MK. Interactions between progesterone and tumor necrosis factor-α in the regulation of primordial follicle assembly. Reproduction 2006; 132:877–886.
44. Kezele P, Skinner M. Regulation of ovarian primordial follicle assembly and development by estrogen an progesterone: endocrine model of follicle assembly. Endocrinology 2003; 144:3329–3337.
45. Chen Y, Jefferson WN, Newbold RR, et al. Estradiol, progesterone, and genistein inhibit oocyte nest breakdown and primordial follicle assembly in the neonatal mouse ovary in vitro and in vivo. Endocrinology 2007; 148:3580–3590.
46. Lamprecht SA, Kohen F, Ausher J, et al. Hormonal stimulation of oestradiol-17β release from the rat ovary during early postnatal development. J Endocrinol 1976; 68:343–344.
47. Terada N, Kuroda H, Namiki M, et al. Augmentation of aromatase activity by FSH in ovaries of fetal and neonatal mice in organ culture. J Steroid Biochem 1984; 20:741–745.
48. Weniger JP, Zeis A, Chouraqui J. Estrogen production by fetal and infantile rat ovaries. Reprod Nutr Dev 1993; 33:129–136.
49. Kurita T, Cooke PS, Cunha GR. Epithelial-stromal tissue interaction in paramesonephric (Müllerian) epithelial differentiation. Dev Biol 2001; 240:194–211.
50. Svechnikov K, Söder O. Ontogeny of gonadal sex steroids. Best Pract Res Clin Endocrinol Metab 2008; 22:95–106.
51. Greco TL, Payne AH. Ontogeny of expression of the genes for steroidogenic enzymes P450 side-chain cleavage, 3β-hydroxysteroid dehydrogenase, P450 17α-hydroxylase/C17-20 lyase, and P450 aromatase in fetal mouse gonads. Endocrinology 1994; 135:262–268.
52. Weniger JP, Chouraqui J, Zeis A. Conversion of testosterone and progesterone to oestrone by the ovary of the rat embryo in organ culture. J Steroid Biochem 1984; 21:347–349.
53. Weniger JP. Estrogen production by fetal rat gonads. J Steroid Biochem Mol Biol 1993; 44:459–462.
54. Greco TL, Furlow JD, Duello TM, et al. Immunodetection of estrogen receptors in fetal and neonatal female mouse reproductive tracts. Endocrinology 1991; 129:1326–1332.
55. Nielsen M, Björnsdóttir S, Høyer PE, et al. Ontogeny of oestrogen receptor α in gonads and sex ducts of fetal and newborn mice. J Reprod Fertil 2000; 118:195–204.
56. Couse JF, Korach KS. Estrogen receptor null mice: what have we learned and where will they lead us? Endocr Rev 1999; 20:358–417.
57. Dupont S, Krust A, Gansmuller A, et al. Effect of single and compound knockouts of estrogen receptors α (ERα) and β (ERβ) on mouse reproductive phenotypes. Development 2000; 127:4277–4291.
58. Couse JF, Curtis Hewitt S, Bunch DO, et al. Postnatal sex reversal of the ovaries in mice lacking estrogen receptors α and β. Science 1999; 286:2328–2331.
59. Lydon JP, DeMayo FJ, Funk CR, et al. Mice lacking progesterone receptor exhibit pleiotropic reproductive abnormalities. Gene Dev 1995; 9:2266–2278.
60. Du H, Taylor HS. Molecular regulation of Müllerian development by Hox genes. Ann N Y Acad Sci 2004; 1034:152–165.
61. Armstrong J, Pritchard-Jones K, Bickmore W, et al. The expression of the Wilms' tumour gene, Wt1, in the developing mammalian embryo. Mech Dev 1992; 40:85–97.

62. Kreidberg JA, Sariola H, Loring JM, et al. WT-1 is required for early kidney development. Cell 1993; 74:679–691.
63. Birk OS, Casiano DE, Wassif CA, et al. The LIM homeobox gene *Lhx9* is essential for mouse gonad formation. Nature 2000; 403:909–913.
64. Miyamoto N, Yoshida M, Kuratani S, et al. Defects of urogenital development in mice lacking Emx2. Development 1997; 124:1653–1664.
65. Pellegrini M, Pantano S, Lucchini F, et al. Emx2 developmental expression in the primordia of the reproductive and excretory systems. Anat Embryol (Berl) 1997; 196:427–433.
66. Katoh-Fukui Y, Tsuchiya R, Shiroishi T, et al. Male-to-female sex reversal in M33 mutant mice. Nature 1998; 393:688–692.
67. Vanio S, Heikkila M, Kispert A, et al. Female development in mammals is regulated by *Wnt-4* signaling. Nature 1999; 397:405–409.
68. Kispert A, Vainio S, McMahon AP. Wnt-4 is a mesenchymal signal for epithelial transformation of metanephric mesenchyme in the developing kidney. Development 1998; 125:4225–4234.
69. Brisken C, Heineman A, Chavarria T, et al. Essential function of Wnt-4 in mammary gland development downstream of progesterone signaling. Genes Dev 2000; 15:650–654.
70. Heikkilä M, Peltoketo H, Leppäluoto J, et al. Wnt-4 deficiency alters mouse adrenal cortex function, reducing aldosterone production. Endocrinology 2002; 143:4358–4365.
71. Uhlenhaut NH, Treier M. Foxl2 function in ovarian development. Mol Genet Metab 2006; 88:225–234.
72. Ottolenghi C, Pelosi E, Tran J, et al. Loss of Wnt4 and Foxl2 leads to female-to-male sex reversal extending to germ cells. Hum Mol Genet 2007; 16:2795–2804.
73. Uhlenhaut NH, Jakob S, Anlag K, et al. Somatic sex reprogramming of adult ovaries to testes by FOXL2 ablation. Cell 2009; 139:1130–1142.
74. Trombly DJ, Woodruff TK, Mayo KE. Suppression of Notch signaling in the neonatal mouse ovary decreases primordial follicle formation. Endocrinology 2009; 150:1014–1024.
75. Gorski RA. Sexual dimorphisms of the brain. J Anim Sci 1985; 61(suppl 3):38–61.
76. Simerly RB. Wired for reproduction: organization and development of sexually dimorphic circuits in the mammalian forebrain. Ann Rev Neurosci 2002; 25:507–536.
77. Vadász C, Baker H, Fink SJ, et al. Genetic effects and sexual dimorphism in tyrosine hydroxylase activity in two mouse strains and their reciprocal F1 hybrids. J Neurogenet 1985; 2:219–230.
78. Harlan RE, Shivers BD, Pfaff DW. Lordosis as a sexually dimorphic neural function. Prog Brain Res 1984; 61:239–255.
79. Nadler RD. Masculinization of female rats by intracranial implantation of androgen in infancy. J Comp Physiol Psychol 1968; 66:157–167.
80. Christensen LW, Gorski RA. Independent masculinization of neuroendocrine systems by intracerebral implants of testosterone or estradiol in the neonatal female rat. Brain Res 1978; 146:325–340.
81. MacLusky NJ, Naftolin F. Sexual differentiation of the central nervous system. Science 1981; 211:1294–1302.
82. McEwen BS, Lieberburg I, Chaptal C, et al. Aromatization: important for sexual differentiation of the neonatal rat brain. Horm Behav 1977; 9:249–263.
83. Andrews GK, Dziadek M, Tamaoki T. Expression and methylation of the mouse α-fetoprotein gene in embryonic, adult, and neoplastic tissues. J Biol Chem 1982; 257:5148–5153.
84. Bakker J, De Mees C, Douhard Q, et al. α-fetoprotein protects the developing female mouse brain from masculinization and defeminization by estrogens. Nat Neurosci 2006; 9:220–226.
85. González-Martínez D, De Mees C, Douhard Q, et al. Absence of gonadotropin-releasing hormone 1 and Kiss1 activation in α-fetoprotein knockout mice: prenatal estrogens defeminize the potential to show preovulatory luteinizing hormone surges. Endocrinology 2008; 149:2333–2340.
86. Bader D, Riskin A, Vafsi O, et al. Alpha-fetoprotein in the early neonatal period–a large study and review of the literature. Clin Chim Acta 2004; 349:15–23.
87. Lemmen JG, Broekhof JL, Kuiper GG, et al. Expression of estrogen receptor α and β during mouse embryogenesis. Mech Dev 1999; 81:163–167.

88. Kudwa AE, Michopoulos V, Gatewood JD, et al. Roles of estrogen receptors α and β in differentiation of mouse sexual behavior. Neuroscience 2006; 138:921–928.
89. Döhler KD, Hancke JL, Srivastava SS, et al. Participation of estrogens in female sexual differentiation of the brain; neuroanatomical, neuroendocrine and behavioral evidence. Prog Brain Res 1984; 61:99–117.
90. Ohtani-Kaneko R. Mechanisms underlying estrogen-induced sexual differentiation in the hypothalamus. Histol Histopathol 2006; 21:317–324.
91. Schwarz JM, McCarthy MM. Steroid-induced sexual differentiation of the developing brain: Multiple pathways, one goal. J Neurochem 2008; 105:1561–1572.
92. George FW, Ojeda SR. Changes in aromatase activity within rat brain during embryonic, neonatal, and infantile development. Endocrinology 1982; 111:522–529.
93. Bakker J, Honda S, Harada N, et al. The aromatase knock-out mouse provides new evidence that estradiol is required during development in the female for the expression of sociosexual behaviors in adulthood. J Neurosci 2002; 22:9104–9112.
94. Döhler KD, von zur Mühlen A, Döhler U. Pituitary luteinizing hormone (LH), follicle stimulating hormone (FSH) and prolactin from birth to puberty in female and male rats. Acta Endocrinol (Copenh) 1977; 85:718–728.
95. Grumbach MM. The neuroendocrinology of human puberty revisited. Horm Res 2002; 57(suppl 2):2–14.
96. Hage AJ, Groen-Klevant AC, Welschen R. Follicle growth in the immature rat ovary. Acta Endocrinol 1978; 88:375–382.
97. Andrews WW, Ojeda SR. A quantitative analysis of the maturation of steroid negative feedback controlling gonadotropin release in the female rat: the infantile-juvenile periods, transition from an androgenic to a predominantly estrogenic control. Endocrinology 1981; 108:1313–1320.
98. Kawagoe S, Hiroi M. Maturation of negative and positive estrogen feedback in the prepubertal female rat. Endocrinol Jpn 1983; 30:435–441.
99. Hirshfield AN. Development of follicles in the mammalian ovary. Int Rev Cytol 1991; 124:43–101.
100. Eppig JJ. Intercommunication between mammalian oocytes and companion somatic cells. Bioassays 1991; 13:569–583.
101. Goldenberg RL, Viatukatis JL, Ross GT. Estrogen and follicle stimulating hormone interactions on follicle growth in rats. Endocrinology 1972; 90:1492–1498.
102. Kessel B, Liu YX, Jia XC, et al. Autocrine role of estrogens in the augmentation of luteinizing hormone receptor formation in cultured rat granulosa cells. Biol Reprod 1985; 32:1038–1050.
103. Mitsushima D, Hei DL, Terasawa E. γ-Aminobutyric acid is an inhibitory neurotransmitter restricting the release of luteinizing hormone-releasing hormone before the onset of puberty. Proc Natl Acad Sci U S A 1994; 91:395–399.
104. Moguilevsky JA, Wuttke W. Changes in the control of gonadotrophin secretion by neurotransmitters during sexual development in rats. Exp Clin Endocrinol Diabetes 2001; 109:188–195.
105. Ma YJ, Junier MP, Costa ME, et al. Transforming growth factor-α gene expression in the hypothalamus is developmentally regulated and linked to sexual maturation. Neuron 1992; 9:657–670.
106. Gamba M, Pralong FP. Control of GnRH neuronal activity by metabolic factors: the role of leptin and insulin. Mol Cell Endocrinol 2006; 254-255:133–139.
107. Dhillo WS, Murphy KG, Bloom SR. The neuroendocrine physiology of kisspeptin in the human. Rev Endocr Metab Disord 2007; 8:41–46.
108. Seminara SB. Kisspeptin in reproduction. Semin Reprod Med 2007; 25:337–343.
109. Lee YR, Tsunekawa K, Moon MJ, et al. Molecular evolution of multiple forms of kisspeptins and GPR54 receptors in vertebrates. Endocrinology 2009; 150:2837–2846.
110. Shahab M, Mastronardi C, Seminara SB, et al. Increase hypothalamic GPR54 signaling: a potential mechanism for initiation of puberty in primates. Proc Natl Acad Sci U S A 2005; 102:2129–2134.
111. Kuohung W, Kaiser UB. GPR54 and Kiss-1: role in the regulation of puberty and reproduction. Rev Endocr Metab Disord 2006; 7:257–263.
112. Teles MG, Blanco SDC, Brito VN, et al. A GPR54-activating mutation in a patient with central precocious puberty. N Engl J Med 2008; 358:709–715.

4   Development of the mammalian nervous system

Kimberly Ehman

INTRODUCTION AND BACKGROUND
General Features of the Mammalian Nervous System

It is well established that the nervous system is not an especially vulnerable system relative to other organ systems; however, there are features of the developing nervous system that make it susceptible to developmental perturbations. With these vulnerabilities in mind, the objective of this chapter is to provide an overview of some of the critical aspects of mammalian nervous system development and function for the purpose of developing studies and interpreting data that are applicable to neurodevelopment. The subsequent sections will explore the structure and function of the mammalian nervous system and provide a comparison of developmental endpoints across species when those data are available. As there are entire books dedicated to this topic, the intent of this chapter is not to be all-inclusive; however, it serves to highlight some of the major developmental milestones that require consideration when testing compounds that could potentially disrupt the development of the nervous system. The primary division of the mammalian nervous system is between the central nervous system (CNS), which processes information, and the peripheral nervous system (PNS), which carries information to and from the CNS and the sensory, muscle, and gland cells. In the simplest terms, the CNS includes the brain and spinal cord, and the PNS consists of all the nervous tissue outside the CNS. There are two types of cells in the nervous system: neurons and neuroglial cells, or glia. The neuron is the structural and functional unit of the nervous system. Neurons are specialized for long-distance transmission of electrical stimuli throughout the body. Glial cells, which greatly outnumber neurons, have a variety of supportive functions within the nervous system. Peripheral nerves serve either somatic (skeletal, muscle, skin) or visceral (involuntary muscles and glands) tissues and carry sensory or motor information. Because of the extensive amount of literature on this topic, the focus of the subsequent sections will be development of the CNS, and more specifically, the developing brain.

There are a few generalities that can be applied across species with respect to general measures of neurodevelopment. For example, in humans and experimental animals, caudal (hindbrain) regions develop earlier than rostral (forebrain), and medial aspects of the structures mature earlier than lateral aspects [1,2]. The exception to the caudal to rostral pattern of development is the delayed development of the cerebellum relative to the development of other hindbrain structures [1,3,4]. Interestingly, in humans, the cerebellum continues to change for several months post parturition, which make it, and other brain regions with protracted development, prone to disruptions [3,4].

To more thoroughly understand potential toxicity to the developing nervous system, it is desirable to have experimental animal models that are representative of the human [5,6]. Although the caudal to rostral pattern of brain development is highly conserved across species, there are also many interspecies differences. For instance, humans have more in utero maturation of the nervous system compared to rodents [1,6]. There has been considerable work over the years focused on comparing critical neurodevelopmental periods in experimental animals to humans, and in the following sections, some of the major neurodevelopmental processes (i.e., proliferation, migration, differentiation, synaptogenesis, apoptosis, gliogenesis, and myelination) will be high-lighted along with a comparison of the developmental timelines in both the rodent and the human (when available). Additionally, the general structure and function of the nervous system will be outlined, including recent advances in our understanding of neurodevelopment at a cellular and molecular level, as well as a novel web-based tool that can be used to extrapolate brain development from experimental species to humans.

DEVELOPMENT OF THE NERVOUS SYSTEM

The mammalian brain is comprised of three main regions. The rhombencephalon (hindbrain) develops into the pons, cerebellum, and medulla; the mesencephalon (midbrain) forms the midbrain; and prosencephalon (forebrain) develops into the telencephalon and diencephalon. The telencephalon gives rise to the cerebral hemi-spheres and basal ganglia, and the latter produces the thalamus and hypothalamus. Development of the brain can be viewed as an increase in the number of internal compartments; initially, the process is controlled by neuronal cell bodies, followed by the addition of glial cells, and lastly cell processes (i.e., dendrites and synapses), to complete the creation of a complex, multicompartment system (7).

Morphogenesis of the nervous system begins when the mesoderm induces formation of the neural plate on the dorsal surface of the embryo. In the absence of mesodermal induction, proteins (growth factors) in the transforming growth factor-β (TGF-β) superfamily influence differentiation (8,9). Following induction, the neural plate folds and curls to form the neural tube that subsequently develops three evaginations that form the rhombencephalic, mesencephalic, and prosencephalic vesicles. As development continues, the neural fold becomes segmented; the segments occur in response to several morphogens. The above process is regulated by several homeobox genes such as Hox, Pax, and Barx (7), and the regulation of this segmentation process by transcription factors appears to be highly conserved in evolution. In rats, neural tube formation is complete at approximately gestational day (GD) 10.5 to 11, and in humans around GD 22 to 26 (1,4). Interruption of neural development during this early period can result in severe abnormalities of the brain and spinal cord. Spina bifida (divided spine) results from defective induction of the mesoderm around the notochord that forms osseous bone of the spine (1).

Beginning early in the second week of gestation in rodents (GD 7 in mice and GD 9.5 in rats), and the first month of gestation in humans, specific areas of the CNS begin to form with the neurogenesis and migration of the cells of the forebrain, midbrain, and hindbrain (1,10). With a few exceptions, most of the cells of the CNS are produced by one of two proliferative zones. The first is the ventricular zone and second is the subven-tricular zone (11). The majority of neurons in the brain arise in the primary germinal matrix formed by the neuroepithelium that lines the ventricles. In rats, most neurons are formed between GD 11 and 22. The bulk of neurogenesis in humans and most exper-imental species occurs in utero; however, in the postnatal period, there is a wave of secondary neurogenesis in the cerebellar cortex, the olfactory bulb, and the dentate gyrus of the hippocampus (6). In the rat, neurons continue to be generated in the secondary germinal matrices (i.e., those that give rise to the cerebellar cortex, olfactory bulb, and hippocampus) until the third postnatal week (12). Various regions of the brain develop at different times and, as such, have varying windows of vulnerability. As previously mentioned, the developmental processes of proliferation, migration, differentiation, synaptogenesis, apoptosis, gliogenesis, and myelination underlie nervous system devel-opment (1,4,13). All of the aforementioned processes occur in a structured sequence dependent on the brain region, cell type, and neurotrophic signals present.

Proliferation

Proliferation refers to the period of rapid cell growth in the brain that is dependent on the exponential proliferation of a multipotent population of pseudostratified cells localized in the ventricular zone (10,14). In rats, the most expansive phase of proliferation in the ventricular and subventricular regions occurs around GD 13 to 18; morphologically, this is approximately equivalent to human gestational weeks 5 to 9 (1,10,12). As mentioned earlier, most of the neurons of the adult CNS are produced during the perinatal period; however, temporal windows of development vary within different regions of the brain (1,4,11). Proliferation is a critical time period in development as the mammalian CNS has limited ability to replace missing neurons in the adult CNS.

Migration

In general, the zones where cell proliferation occurs are physically separate from the ultimate destination of the cells that they produce. As such, postmitotic young neurons must move from the site of their proliferation to their final position. At present, two mechanisms by which this can happen have been identified; passive cell displacement and neuronal migration. In some parts of the developing nervous system, the postmitotic neurons leave the proliferative population and move a short distance from the proliferative zone. Such cells are eventually displaced outward by newly produced cells (10,11,14). This process is referred to as passive cell displacement as there is no active locomotor activity. In contrast, neuronal migration requires that the neuron leaves the proliferative zone and travels a greater distance. This cell movement requires the active participation of the moving cell itself in its own displacement. If neuronal migration is disrupted, an abnormality in cell position results (11). When this happens, the neurons are said to be heterotopic or ectopic (15). Defects in neuronal migration have been associated with a variety of syndromes and diseases ranging from behavioral disorders to extremely severe mental retardation (11). In rats, by GD 16 and 17 (human gestational weeks 7–8, based on morphology), the first cells are arriving in the area that will form the laminae of the cortical plate (1,12). Generalizing about common mechanisms of migration is difficult due to the diverse strategies adopted by migrating neurons of different brain regions (16); yet, it is known that the conditions necessary for migration seem to be present only during development, and as such, if migration is not accomplished within an allotted developmental timeframe, it will likely remain incomplete (4).

Differentiation

Differentiation (phenotype expression) is one of the most highly specific developmental processes because it involves a specific loss of pluripotential status (10). The initial phase of differentiation likely begins as soon as neuronal precursors complete their last division and are prepared to migrate to the cortical plate (1). As neurons migrate, intra- and extracellular cues likely initiate expression that will ultimately influence the neuronal or glial phenotype. Once a neuron or glial cell has left the proliferative population and reached its final position, it enters the final phase of its life history (differentiation) (11); thus, differentiated neurons are postmitotic cells that are incapable of replication (17). The differentiation of neurons and glia is a complex process that is responsible for the generation of a large proportion of the diversity of the CNS. During differentiation, neurons produce their axon and dendrites. The dendrites then grow and form their characteristic pattern of arborization for the cell class. Moreover, each neuron expresses its own enzymes to produce neurotransmitters, and the postsynaptic sites express receptors necessary to receive input. Axons of some neurons become myelinated and concurrent with the maturation of neurons; glial cells also differentiate (11).

During differentiation and synaptogenesis (formation of functional synaptic connections), the circuitry of the nervous system is established.

Synaptogenesis

Synaptogenesis refers to the formation of synapses through an elaborate, precisely timed, process consisting of the establishment of biochemical and morphological elements followed by the competitive exclusion of inappropriate connections (6). Synapses are the neurobiological substrates of almost all intercellular communication (1,10). Synaptic junctions are asymmetric structures composed of three compartments: the presynaptic bouton, the synaptic cleft, and the postsynaptic reception apparatus (18). In the rat, synapses mature over a three-week period (i.e., the first 3 postnatal weeks), peaking during the first two postnatal weeks and continuing to mature through adolescence (6,10). During development, the biochemical and morphological changes occur in both pre- and postsynaptic elements. Within a region of the nervous system, the schedule by which synapses form is strictly followed: a rapid increase in synapse number followed by elimination of seemingly unnecessary synapses (10).

Apoptosis

Programmed cell death (apoptosis) has been so intensively investigated because this type of cell death is generally associated with physiological conditions as opposed to pathologic insults that lead to necrotic cell death (17). However, more recently, apoptosis has also been shown to play a role in physiological aging, many neurodegenerative disorders, and traumatic neuronal injuries (17). Proper development of the nervous system requires the systematic removal of large numbers of neurons through apoptosis, and the elimination of unnecessary dendritic processes and excess synapses (1,19,20). At birth, the human brain has an over production of neurons (~100 billion) that undergo selective apoptosis during postnatal life (3). This process, often referred to as "pruning," continues through adolescence. Selected cells are also removed during ontogeny via apoptosis, which ultimately results in the appropriate cell types in the correct brain region. Apoptotic cell death is distinguished from necrotic cell death by specific characteristics such as maintenance of membrane, chromatin condensation, cell shrinkage, oligonucleosomal fragmentation of DNA, and a general lack of inflammation (1,21).

Gliogenesis and Myelination

The glial, or supportive cells, develop last, and myelination is a protracted process. The period of rapid proliferation of glial cells is known as the brain growth spurt, during which time the developing brain is particularly vulnerable to insult (22). The glial cells of the CNS are astrocytes, which are the most abundant oligodendrocytes (myelin-producing cells in CNS), radial glia, ependymal cells, and microglia. Those of the PNS are Schwann cells (myelin-producing cells in PNS) and satellite cells.

In humans, myelination continues throughout adolescence, and white matter progressively increases in overall volume in a region-specific fashion (6). In contrast to the human and nonhuman primate, the rat acquires little myelin prenatally but undergoes rapid postnatal myelination (23). The rat brain acquires 50% of the adult myelin levels by PND 25 and completes myelination by approximately PND 90 to 100 (4). In comparison, the human brain has 50% of adult myelination at 18 months, and myelination is not complete until early adulthood (4).

Neurotransmitter Systems

Development of classical neurotransmitter systems (in humans) begins in utero and extends through approximately 10 years of age (24). The two apparent exceptions appear to be the nicotinic acetylcholine (ACh) and the γ-amino butyric acid (GABA)ergic systems (24). N-methyl-D-aspartate (NMDA), kainate, α-amino-3-hydroxy-5-methyl-4-isoxazolepropionic acid (AMPA), D1/D2 (dopamine receptors), muscarinic ACh, and serotonin systems begin to develop several months prior to birth and continue to develop for up to 10 years postnatally; however, the nicotinic ACh system is developed by approximately one year of age, and the GABAergic system does not begin to develop until a few months post parturition and continues to develop on the same timeline as the others (24). Although neuropeptides (e.g., opioids and galanin) and purines (e.g., adenosine) are also important modulators and/or components of the developing nervous system, the following sections will be restricted to a discussion of the development and roles of catecholamines, ACh, excitatory amino acids (EAAs), and GABA.

Catecholamines
Catecholamines (tyrosine hydroxylase, dopamine, noradrenaline, adrenaline, and serotonin) play a critical role in neurodevelopment. Catecholaminergic neurons are generated at the time of telencephalic vesicle formation in rodents and humans (25). Noradrenergic neurons (neurons that synthesize norepinephrine) appear at an early stage in the CNS; GD 12 to 14 in the rat, and 5 to 6 weeks in humans (25). Dopaminergic neurons appear early during development; GD 10 to 15 in the rat, and 6 to 8 weeks in the human (25,26). Serotonergic cells are generated around GD 10 to 12 in the mouse and appear between the gestational weeks 5 to 12 in the human (25).

Acetylcholine
The developing brain is vulnerable to neurotoxicants that perturb the cholinergic system because of the early appearance of ACh and cholinergic receptors, and their role in cell proliferation and differentiation (27). The cholinergic innervation of the cortex occurs later than the monoaminergic (i.e., catecholamines, above), around GD 19 in the mouse, and 20 gestational weeks in the human; however, the system is not mature until much later in postnatal development (approximately PND 55–60 in the mouse) (28).

Excitatory Amino Acids
Glutamate and aspartate are the dominating EAAs and major neurotransmitters of the mammalian forebrain (29). The NMDA receptor is sensitive to L-glutamate and responsible for excitatory synaptic transmission (6), and peak NMDA receptor density in the human at full term is comparable to the in rat at PND 6 to 9 (30).

γ-Amino Butyric Acid
GABA is reported to be the main inhibitory transmitter in the mature animal and acts as a trophic factor during development. During maturation, GABA switches from an excitatory to an inhibitory neurotransmitter. This change occurs at birth in the brain stem region, one-week postnatal in the hippocampus, and between one- and two-weeks postnatal in the cortex of the rat (31).

MORPHOLOGY OF THE NERVOUS SYSTEM
Neuronal Morphology
All neurons contain a nucleus, endoplasmic reticulum, a Golgi apparatus, mitochondria, microtubules, and other organelles typical of all cells. Neurons also contain a cluster of ribosomal complexes for the synthesis of proteins (i.e., Nissl substance), reflecting the unusual demand for protein synthesis. From a functional perspective,

neurons are specialized for integration and communication of activity. They integrate input from diverse sources, generate electrical signals, and transmit them to sites distant from the site of activation (32,33). Cell bodies can vary in size; however, one of the most notable features of a neuron is the arrangement of processes that arise from the cell bodies; dendrites (can be several) and the axon (only one). Dendritic morphology varies with respect to the numbers and length of the dendrites. Dendrites may branch many times (e.g., cerebellar tissue), or they may remain simple and unbranched. Axons consist of a thin, relatively unbranched process that may extend for very long distances (32). Branching patterns of axons, like dendrites, are characteristic of particular neuronal types and regions of the nervous system. Though simplified, chemical synapses can be classified as one of two morphological types: type I (asymmetrical) or type II (symmetrical). With respect to type I, the presynaptic axon terminals contain round synaptic vesicles and are separated from the postsynaptic surface by a relatively widened extracellular space or synaptic cleft (32). The postsynaptic surface contains an accumulation of a dense substance beneath the membrane, and the presence of this material in the postsynaptic, but not presynaptic element, contributes to its classification as asymmetrical. The second type, type II, contains flattened vesicles, and the postsynaptic element lacks a prominent postsynaptic density (32). With respect to electrical synapses, the synaptic cleft in these contacts is nonexistent and the two membranes meet with a small gap (gap junctions).

Glial Morphology
As mentioned earlier, neurons are greatly outnumbered by the second type of cell that populates the nervous tissue, the supporting cells (glia). Unlike most neurons, glial cells continue to divide throughout life. Astrocytes have small cell bodies from which extend a large number of sinuous processes (star-shaped) (32,34). Oligodendrocytes are small cells with long processes that form the myelin sheaths of axons in the CNS. The myelin-forming membrane consists of lipid and protein (32,34,35). Microglia are small cells that appear to be active only after injury or in disease, and ependymal cells are ciliated epithelial-like cells that line the ventricles of the brain (32,34). The tight junctions between the membranes of adjacent endothelial cells are thought to be the foundation of the blood-brain barrier (BBB) (6,32,34). In the PNS, the supporting cells are referred to as Schwann cells, which function to form myelin sheaths around peripheral nerve fibers (i.e., similar function to oligodendrocytes of the CNS) (32,34).

Blood-Brain Barrier
The hallmark characteristic of the BBB is the continuity of the cellular layer between the blood and the interstitial space of the brain (6,7,36). The barrier is formed by tight junctions between the endothelial cells (ependymal cells) that line the capillary of the brain, close apposition of astrocytes, and extensive association of microglia (7). Specializations characteristic of the BBB are induced in endothelial cells by contact with astrocytes. The establishment of the BBB requires specialized endothelial tight junctions, particular patterns of enzymatic activity, distinct electrochemical gradients, and specific BBB transporters (6,7,36).

PHYSIOLOGY OF THE NERVOUS SYSTEM
As highlighted in the previous sections, in the brain, there are two major classes of cells. The first is the class of excitable cells (neurons). Nerve cells contact each other at specialized sites known as synapses and as such form continuous functional networks that are highly organized. The second class of cells is the nonexcitable cells, or supporting cells (glia), which are several times more numerous than neurons.

Neuronal Communication

The main function of the mammalian brain is generation and transmission of impulses. The developmental processes reviewed earlier require various ion channels, receptors, growth factors, and signaling molecules to exhibit unique features in the immature brain that may not be present in adults (37,38). Information traveling through the nervous system is transmitted in the form of electrical and chemical signals. In the nervous system, impulses are conducted over great distances at rapid speed to provide information about the environment to the organism in a coordinated manner. Electrical signals are nerve impulses that travel within the plasma membrane of the neuron and are of two kinds: graded potential and action potential (37). The graded potential declines in magnitude as it travels along a nerve fiber. An action potential is an all-or-none phenomenon. Once initiated, it propagates without decrement along a nerve fiber. Action potentials are often used for long-distance signaling in the nervous system. In addition to electrical signals, chemical signals are also used for communication; chemical signals are generated at the synapses. Neurotransmitters diffuse across the synaptic junction and settle on the associated cellular process of the next neuron.

Neurotransmitter Systems

Neurotransmitters are defined as chemicals released from neurons that act on specific receptors (25), and the general role neurotransmitters play in neurodevelopment is not entirely clear. Neurotransmitters have qualitatively different functions during development of the nervous system than in the adult organism (38,39). In the adult nervous system, neurotransmitters mediate or modulate synaptic transmission across a cleft between two cells. During development, these factors interact with their associated receptors over a much greater distance and form a morphogenetic gradient that is important in pattern formation of different regions of the nervous system and differentiation of different neurotransmitter systems. The transmitters and modulators affect formation of synaptic contacts, maturation of synapses, and structural refinement of connectivity (40).

In general, neurotransmitters communicate by opening and closing ion gates. The ion channel may consist of transmembrane proteins, which are selective for either cations, that activate the channel, or anions, that inhibit the channel (25). For example, the ACh ion channel is a prototype; the binding of a ligand causes an allosteric change of the ion channel pore (25). ACh is synthesized in the cell body of the neuron and transported to the axon terminal. Following nerve stimulation, an action potential travels down the axon and when it reaches the terminal, there is an influx of Ca^{2+}, and ACh is released from the presynaptic membrane into the synapse and attaches itself to the ACh receptor. When this occurs at the neuromuscular junction, sodium ion channels are opened in the skeletal muscle, causing a contraction. ACh is immediately broken down by acetylcholinesterase (AChE) (41). Normal cholinergic development is seemingly important for cortical development, plasticity, and function (25). Moreover, development of the cholinergic system is important in a number of neurologic diseases including Huntington's and Alzheimer's disease (27).

Catecholamines such as dopamine, noradrenaline, adrenaline, and serotonin have numerous roles in the nervous system. Dopamine neurons are concentrated in the substantia nigra, the ventral tegmental area, basal ganglia, olfactory bulbs, limbic regions, hippocampus, and cortex (25). Dopamine has many roles including regulation of the endocrine system as it initiates the hypothalamus to produce hormones for storage in the pituitary gland and then trigger their release into circulation. Dopamine is also involved in the regulation of movement, cognition, and emotions (41). Disturbances of the development of the dopaminergic systems may lead to dyskinesia, dystonia, obsessive-compulsive disorders, and ADHD (25). The noradrenergic neurons are concentrated in the brain stem; however, five major noradrenergic tracts innervate the entire brain. Norepinephrine is present in many nerve fibers and is secreted by the adrenal gland in response to stress or events producing excitation. During development, disruption of noradrenergic function has been demonstrated to prevent glial cell proliferation, change dendritic outgrowth, and potentially alter cortical differentiation (28). The function of adrenaline, which is localized in the brain stem, is most likely related to neuroendocrine and blood pressure control. Serotonin (5-HT), which is localized in the midbrain, pineal gland, substantia nigra, hypothalamus, and brain stem, serves to coordinate complex sensory and motor patterns during various states (25). Serotonin is further involved in inducing strong contractions of smooth muscle and states of consciousness, mood, depression, and anxiety. During development, serotonin has been reported to impact neuronal proliferation, differentiation, migration, and synaptogenesis (42). It has been posited that autism may be related to hyposerotonism prenatally and hyperserotonism postnatally (43).

As mentioned earlier, glutamate and aspartate are part of the EAA and constitute the major transmitters of the pyramidal cells. Glutamate has been shown to act on a number of different receptors: NMDA receptors, which are prevalent in the immature brain when synaptic transmission is very plastic, and kainate and AMPA during maturation. In general, the NMDA receptors permit entry of sodium and calcium when opened, and play a role in long-term potentiation (LTP), and synaptic plasticity underlying learning and memory throughout life (25). During maturation, the AMPA and kainic ionotropic receptors predominate and carry much of the fast neuronal traffic in the brain (25). Excess activation of NMDA and non-NMDA receptors is implicated in the pathophysiology of brain injury including hypoxia-ischemia and seizures. In general, stimulation of NMDA is beneficial; however, overstimulation may cause neuronal damage or death (41). In the rat, NMDA receptor blockage during the perinatal period leads to widespread apoptotic neurodegeneration as well as arrested cell migration (44).

GABA is the dominating neurotransmitter in the nonpyramidal cells. As an inhibitory neurotransmitter, GABA inhibits the firing of neurons. Roughly 25% to 40% of all nerve terminals contain GABA. GABA is reported to be the main inhibitory transmitter in the mature animal but has a different role during development, at which point it acts as a trophic factor to influence events such as proliferation, migration, differentiation, synapse maturation, and apoptosis. GABA is a crucial transmitter for the human infant. When vitamin B_6 (which plays a crucial role in GABA synthesis) was mistakenly excluded from infant formula, it resulted in disastrous stories of deaths mainly due to a GABA deficiency leading to fatal seizures (3,25).

Glial Physiology

The primary glial cells include astrocytes, oligodendrocytes, microglia, ependymal cells, and Schwann cells, all of which are integral for the normal development and function of the nervous system. Astrocytes provide trophic and structural support to neurons of the CNS (i.e., maintain ionic and trophic balance of extracellular milieu), and oligodendrocytes produce myelin necessary to enwrap axons of the CNS and generate myelin sheaths for saltatory conduction of action potentials (34).

Oligodendrocytes and astrocytes develop later than the microglia, the resident immune cells of the CNS. Microglia function as brain macrophages and clean cellular debris following apoptosis (10). In the PNS, Schwann cells produce myelin that insulates axons; both myelinating and nonmyelinating Schwann cells serve to support peripheral nerves (34). Schwann cell production of myelin produces only one segment of myelin whereas oligodendrocytes can produce multiple myelin segments (35).

Both astrocytes and microglia become reactive following nervous system insult that has created a number of markers for neurotoxic insult based on gliosis (mainly in adults) (10,45). In the developing nervous system, if an insult to the nervous system precedes proliferation and differentiation of astrocytes, there may be no reactive astrocytes (10). Lastly, ependymal cells construct the walls of the ventricles and, as such, provide support to the BBB.

Blood-Brain Barrier

The BBB is specialized for the maintenance of the neuronal microenvironment and plays a critical role in tissue homeostasis, fibrinolysis and coagulation, vasotonous regulation, the vascularization of normal and neoplastic tissues, and blood cell activation and migration during physiologic and pathologic processes (36). The nervous system is protected from the adverse effects of many potential toxicants by the BBB. Molecules must pass through membranes of endothelial cells, rather than between them, as they do in other tissue, which creates an anatomic barrier (6,46). The absence of the BBB is a significant difference between the mature and developing CNS, and developmental maturation of the BBB includes tightening of the interendothelial junctions and changes in the expression of various transport proteins (4,37). It is thought that the structural and functional aspects of the BBB are similar in various species. The development of the BBB in humans is a gradual process, beginning in utero and acquiring capabilities similar to that of an adult rat at approximately six months of age. In the rat, the permeability of the BBB is very high at birth and decreases in the first few weeks following birth; the functional BBB is not fully developed until PND 24 (47). In general, the apparent protracted development and compromised integrity of the BBB should be carefully considered when administering compounds during critical developmental windows.

Overview of the Major Brain Regions and Functions

As previously mentioned, the development of the brain follows a caudal to rostral structural sequence, and functional development follows a hierarchical sequence from sensory to motor, and finally to associative functions. The previous sections have primarily addressed brain structure and function at the cellular level; however, to summarize, the Table 1 provides a general outline of the various regions of the brain and their primary functions from caudal to rostral development.

CELLULAR AND MOLECULAR DEVELOPMENT OF THE NERVOUS SYSTEM

The central dogma has taught us that genes are encoded in DNA, DNA is transcribed into mRNA, and that mRNA is translated into protein. With respect to cellular and molecular influences on neurodevelopment, there are still many unknowns and seemingly endless possibilities with respect to these dynamic processes. The mammalian CNS is an organized structure consisting of thousands of cell types, all of which require multiple cell-cell interactions to form a functional circuit (48). It is assumed that brain development requires changes in gene expression and protein production; however, our knowledge of the precise mechanisms is continuously evolving. The application of molecular techniques and transgenic technology has many advantages for elucidating the mechanistic role of specific molecules to affect the normal development of the nervous system. Molecular approaches (e.g., mRNA expression) may be useful for mechanistic studies and in the context of gene discovery, but unless they are used in conjunction with an examination of protein levels, they may not prove to be useful in predicting adverse effects of perturbation. Spontaneous mutations, knockouts, and

TABLE 1   Summary of the Major Brain Regions and Functions

Brain region	Function
Hindbrain (rhombencephalon)	In general, all bundles of axons carrying sensory information to, and motor instructions from, higher brain regions pass through the hindbrain; therefore, conduction of information is one of its most important functions; helps coordinate large-scale body movements (e.g., walking)
Medulla oblongata	Controls visceral (autonomic, homeostatic) functions, including breathing, heart and blood vessel activity, swallowing, vomiting, and digestion
Pons	Contains the pneumotaxic center that inhibits activity of respiratory centers; helps control respiration and heart rhythms
Cerebellum	Coordinates unconscious movement and balance, and fine motor movements; coordinates aspects of language and learning (e.g., classical conditioning such as eye blink); most sexually dimorphic part of the brain
Midbrain (mesencephalon)	Coordinates receipt and integration of several types of sensory (e.g., visual, auditory, and tactile) information; serves as a projection center, sending coded sensory information along neurons to specific regions of the forebrain
Forebrain (prosencephalon) (diencephalon/ telecephalon)	Neural processing; intricate networks of integrating centers and sensory and motor pathways that allow for image formation and associative functions (e.g., memory, learning, and emotions).
Diencephalon (lower region of prosencephalon)	Contains the 2 integrating centers: thalamus and hypothalamus
Thalamus	Integrates sensory signals to the cortex and motor output signal from the motor cortex and basal ganglia to cerebellum and spinal cord; involved in consciousness and attention
Hypothalamus	One of the most important regions for regulation of homeostasis; regulates appetite, satiation, thirst, maternal behavior, sexual response and mating behavior, stress via release of corticotrophin releasing factor (CRF) to the anterior pituitary; controls aspects of reward/motivation mechanisms; contains many signaling neuropeptides
Telencephalon (upper region of prosencephalon)	Most complex integrating system in the CNS
Basal ganglia	Important centers for motor coordination (degeneration of cells entering the basal ganglia occurs in Parkinson's disease)
Caudate Putamen Globus pallidus	Involved in voluntary motor control; goal-oriented learning
Nucleus accumbens	Involved in reward mechanisms
Hippocampus	Involved in spatial mapping, episodic, reference, and declarative memory
Amygdala	Important emotional center, emotional-based learning (e.g., conditioned fear and conditioned taste aversion); anxiety
Cerebrum Corpus striatum	Involved in the voluntary motor system; modulated by the extrapyramidal motor system
Cerebral cortex	Largest and most complex part of the brain which is divided into 5 lobes: **Frontal lobe**: thinking, reasoning, working memory, executive functions **Parietal lobe**: thinking, sensorimotor integration, perception **Temporal lobe**: auditory perception, spatial mapping, episodic and reference memory, path integration **Occipital lobe**: visual perception **The insula** (hidden within the fold separating the temporal and parietal lobes): speech production, touch, nociception

transgenic animals have been incredibly valuable for identifying some fundamental mechanisms underlying neurodevelopmental processes, and genetic mapping of mutations has led to identification of proteins essential to neuronal migration, differentiation, and survival (49,50).

The number of molecules involved in modulating neurogenesis is enormous. Furthermore, the signal transduction of trophic molecules can use both second and third messengers, which may lead to alterations in gene expression or epigenetic changes via phosphorylation and/or glycosylation of existing signals and structural proteins (1). The following is a brief discussion of a few molecules that have been reported to be important in the process of proliferation, migration, differentiation, and apoptosis.

Most of the following Hhs (hedgehog), Fgf (fibroblast growth factor), Sox (SRY-related HMG), and Hox (homeobox) families represent large gene families that encode transcription factors and serve assorted roles in signaling cascades and development. The functions of these genes, and the proteins that they encode, are diverse; as such, the following is an overview of the major role some of these families might play in development of the nervous system, and is by no means intended to be comprehensive. Similarly, the function of nerve growth factor (NGF) and glial cell line–derived neurotrophic factors (GDNF) will be reviewed as they pertain to development and function of the nervous system. Essentially, the subsequent sections will hopefully serve to highlight some of the major elements (e.g., genes, proteins, neurotransmitter-receptor systems) that have contributed to our understanding of the mechanisms involved in brain development.

Nerve Growth Factor

NGF is the prototypical neurotrophic ligand (protein) and has been shown to play a role in supporting neuronal survival, as it was the first neural survival factor to be identified (1,51). The discovery of NGF highlighted the power of understanding developmental events in terms of ligand-receptor interactions (52). NGF is a soluble peptide and a ligand for tyrosine kinase receptors and p75 receptors throughout the developing brain. NGF and other homologous neurotrophic factors [e.g., brain-derived neurotrophic factor (BDNF), neurotrophin-3 (Nt3), and neurotrophin-4 (Nt4)] are important to many processes in neural development. BDNF has been shown to play a role as a survival factor for peripheral neurons, and subsequently shown to be involved in axonal and dendritic growth as well as synaptic plasticity (53–55). Apoptosis during development has been shown to be controlled by NGF, BDNF, Nt3, and Nt4 (51). Additionally, manipulations to neurotransmitter systems lead to pronounced changes in neurotrophic factor expression (1,51).

Glial Cell Line–Derived Neurotrophic Factor Family Ligands

GDNF family ligands (GFLs) also play a critical role in the development and function of the nervous system (56). GDNF was originally characterized as a growth factor promoting the survival of ventral midbrain dopaminergic neurons, and later discovered to have a potent survival effect on motor neurons and other neuronal subpopulations in the central and peripheral nervous systems (56). In addition to its function as a survival factor, GDNF is essential for proliferation, migration, and differentiation of neurons. The GFLs consist of four members: GDNF, neurturin (NRTN), artemin (ARTN), and persphin (PSPN), all of which are involved in both the CNS and the PNS developmental processes. In the PNS, GFLs are crucial for developing enteric, sympathetic, and parasympathetic neurons (57). GFLs are the second family of neurotrophic factors (the other being NGF) where different members influence distinct events and support survival of various classes of peripheral neurons during development. GFLs also influence the development of sensory and sympathetic neurons and have a broader spectrum of developmental action than the NGF family (57).

Proliferation

Members of the hedgehog family (Hhs), which have a number of functions including signaling in neurogenesis, and the bone morphogenetic protein family (Bmps) act both as morphogens and mitogens to promote the proliferation of cells (48,51). The fibroblast growth factor (Fgf) family members possess broad mitogenic and cell survival activities, and are involved in a variety of biological processes, including embryonic development, cell growth, morphogenesis, and tissue repair. Ectopic expression of the midbrain organizing factor (Fgf8), has been shown to lead to increased size of proliferative zones, and complete loss of Fgf8 and sonic hedgehog (Shh) (i.e., a member of the hedgehog family) gene function causes early embryonic lethality, or severe and widespread morphological defects (48,51). Additionally, Fgf8 and Fgf17 have been shown to be critical to cerebellar development (51). Retinoic acid maintains Fgf8 and Shh expression in the forebrain and promotes proliferation. In the midbrain, Shh determines the size and shape of tissue patterns. Overall, Shh is a potent morphogen with patterning capabilities in the CNS; however, it also contributes to proliferation, differentiation, and axon growth (48). Around GD12.5 in the rat, the proliferative ventricular and subventricular zone expresses a number of Hox proteins including Mash1, Gsh2, Lhx2, and Vax1 (58). The Hox gene family has some of the most highly conserved sequences in our body and include a number of genes involved in brain development.

Migration

As previously mentioned, cell migration serves a crucial role during embryonic morphogenesis and in adults, functions in tissue repair and immune response (16). Recently, Hox genes, specifically Hoxa2, have been shown to influence neuronal migration, guiding neurons to their proper brain region (59). The Hoxa2 gene regulates the expression of the Robo and Slit receptors. Slit prevents neurons from being drawn toward other chemoattractants. As a result, the migrating neuron can ignore other influences and migrate directly to the precerebellar region (in this case). When the Hoxa2 gene is knocked out, the neurons are not able to resist other chemoattractants and often do not reach their proper destination (59). In addition to Hoxa2 genes, other molecules that appear to be involved in neuronal migration and the steps for which they are responsible are as follows: (i) P13K, Rac1, Cdc42, RhoA, and MT (microtubule)-actin interaction, and stathmin, γ-tubulin, and ninein play a role in leading-process extension; (ii) Cdc42, PARD6α, PKCζ, GSK3β, actin, and FAK contribute to the forward movement of the centrosome; (iii) Dynactin, LIS1, NDEL1, DISC1, DCX, and Ca^{2+}, KASH proteins, and neurofilaments are required for nucleokinesis; and (iv) PTEN, RhoA, and myosin II contribute to trainlin-process retraction (16). Sox genes, or SRY (sex-determining region Y)-related HMG (high-mobility group) box genes are also involved in neurodevelopment. In general, Sox genes have been demonstrated to play a role in neural induction, and Sox2 and Sox3 are involved in the migration processes in the cerebellum.

Differentiation

Changes in gene expression and signaling are reflected in cellular processes and events leading to differentiation. The intervening factors that can potentially modulate every step of the signaling cascade further complicate evaluation of neuronal differentiation. Gbx1 and Lhx7 have been identified in the adult rat ventral forebrain, specifically in the regions that harbor cell bodies of the basal forebrain cholinergic system. Therefore, these genes might be important for the synthesis of cholinesterase (58). In rats, the basal forebrain cholinergic neurons originate between GD12.5 and 17.5; therefore, Lhx7 and Gbx1 may prove to be useful markers for delineating the developmental timelines of cholinergic neurons in the ventral forebrain, including those in the striatum (58). A distinct feature of ventral forebrain development is the role of signaling cascades. All three major signaling families have centers of expression in the region: Shh, Fgf8, and Bmp-4 and Bmp-9. Fgf8 has been shown to induce Lhx6 and Lhx7 in the brachial arch and Gbx2 in the forebrain. Shh induces Nkx2.1 in the basal forebrain, and Lhx7 is expressed in cholinergic neurons and may be important for their development (58). Lhx2 and Foxg1 are necessary for specification and expansion of cortical precursor cells and terminal differentiation. In addition, Notch1 and Notch3 in multipotent progenitor cells induce astroglial differentiation (60).

Apoptosis

As mentioned before, apoptosis is an extremely important event in the normal development of the mammalian nervous system, and the basic mechanisms of apoptosis are highly conserved across mammalian cell types and throughout phylogenesis (17). The genes involved in neurodevelopment are often those responsible for apoptotic functions (61). Apoptosis is principally regulated by the Bcl2 (i.e., mammalian B-cell lymphoma-2) family of proteins, the adaptor protein Apaf1, and the cysteine protease caspase family (51,62–64). The different neuronal populations at various developmental stages express different combinations of Bcl2 and caspase family members. Additionally, the Jnk family of MAP kinases is a potential candidate mechanism for controlling regional specificity of cell death in nervous system tissue (51,65). Lim and Pou genes have also been shown to play a role in cell fate determination. It appears as if several different stimuli can initiate apoptotic cascades; however, there are morphological and biochemical alterations that converge on a restricted number of common effector pathways: (*i*) effector caspase is activated prior to mitochondrial alterations and (*ii*) cytochrome *c* is released from mitochondrial intermembrane space prior to caspase activation (17).

FUTURE DIRECTIONS

Neurodevelopment is one of the least well-understood areas of neurobiology. As outlined in this chapter, the developing nervous system is prone to insult from gestation through adolescence, and in some cases, adulthood. As technology advances, the number of assays available to assess neurodevelopment also evolves. However, in addition to advanced test methods, there needs to be an understanding that the brain does not develop and exist in isolation from the rest of the body (e.g., interactions with endocrine and immune systems).

It is inevitable that advances to our current understanding of nervous system development will involve cellular and molecular biology. Developmental endpoints are routinely described in terms of transcription factors and ligand-receptor interactions. The future will involve a better understanding of molecular mechanisms, the role of the environment, and gene-environment interactions. In terms of regulatory pathways, basic research has shifted from research on transcriptional regulation by studying sequence-specific transcription factors to investigation of mechanisms that regulate chromatin. This is due to new discoveries of histone modification, DNA methylation, and chromatin remodeling in transcriptional regulation.

Understanding how the circuitry of the brain develops is important for delineating the potential impact of early exposure on developmental outcomes, as well as potentially understanding how early exposure could alter neural mechanisms later in life (e.g., neurodegenerative diseases). Although cellular and molecular tools will greatly impact our continuing understanding of neurodevelopment, it is equally important to find experimental species to closely represent human development. Recent advances in the area of neuroinformatics will certainly contribute to our overall ability to design more appropriate studies.

Neuroinformatics has created a single statistical model, designed to compare, adjust for differences, and predict the dates of neural development (67). These data are presented on an interactive website (http://www.translatingtime.net/) that allows the user to access comparative and predictive brain development data and convert it across species in real time. Currently, the model includes data from 10 species (hamsters, mice, rats, rabbits, spiny mice, guinea pigs, ferrets, cats, macaques, and humans) for 102 neurodevelopmental events, though not all events have been studied in all 10 species (66–71). There are numerous publications describing both the concept and the data set (67–69). This website is being emphasized because a neuroinformatics approach can be used to relate neurodevelopment across species and, as such, is extremely useful for the developmental studies in a species that is ultimately supposed to model what will happen in the human. There are recognized limitations to the current model (e.g., data limitations regarding minimum and maximum age ranges); however, the neuroinformatics approach is certainly a positive step forward for cross-species comparisons in neurodevelopmental studies (66–71).

SUMMARY AND CONCLUSIONS

Implicit in the most recently advanced hypotheses that concern neurodevelopmental perturbations is the notion that spatially and temporally restricted alteration of CNS formation that occur prenatally will lead to permanent functional abnormalities throughout life. This further suggests that certain developmental events, if occurring during a sensitive period of development, might never spontaneously recover. As such, it is important that there is an understanding or, at the very least, recognition that these critical periods exist. It is further important to recognize that disturbances to the developing nervous system can manifest themselves in countless ways (e.g., neuro-transmitter dysfunction, morphologic changes, protein expression, cell death, or behavioral abnormalities).

Overall, the objective of this chapter was to highlight the major developmental processes that contribute to the unique complexity of the mammalian nervous system. To reiterate, the nervous system itself is not an unusually vulnerable organ system; however, there are numerous features of the developing nervous system that make it more prone to environmental insult during its development. For instance, cell sensitivity differs with the developmental stage that leads to critical windows of vulnerability. Although synaptogenesis can continue throughout life, proliferation occurs only in rare cases; therefore, the CNS is unique in that damaged neural cells are not readily replaced. Finally, there are kinetic differences in the developing organism that may profoundly influence its sensitivity, including slow formation of the BBB and lack of key metabolic enzymes necessary to protect the brain and eliminate toxicants.

Although the developing nervous system is seemingly more sensitive to insult than the adult nervous system, the high rate of proliferation and regeneration in the developing nervous system may also lead to greater recovery or plasticity, which could attenuate some injuries. It should be noted that even though some developmental changes may appear to be transient due to this plasticity and compensation, the possibility exists that the underlying changes in the nervous system may become evident with aging or exposure to environmental stressors.

REFERENCES

1. Rice D, Barone S Jr. Critical periods of vulnerability for the developing nervous system: evidence from humans and animal models. Environ Health Perpect 2000; 108(suppl 3): 511–533.
2. Rodier PM. Developing brain as a target of toxicity. Environ Health Perspect 1995;103 (suppl 6):73–76.
3. Kalia M. Brain development: anatomy, connectivity, adaptive plasticity, and toxicity. Metabolism 2008; 57(suppl 2):S2–S4.
4. Rodier PM. Vulnerable periods and process during central nervous system development. Environ Health Perpect 1994;102(suppl 2):121–124.
5. Morse AC, Barlow C. Unraveling the complexities of neurogenesis to guide development of CNS therapeutics. Drug Discov Today 2006; 3:495–501.
6. Watson RE, DeSesso JM, Hurtt ME, et al.. Postnatal growth and morphological development of the brain: a species comparison. Birth Defects Res B 2006; 77:471–484.
7. Jensen KF, Catalano SM. Brain morphogenesis of the nervous system. In: Slikker W Jr, Chang LW, eds. Handbook of Developmental Neurotoxicology. New York: Academic Press, 1998.
8. Hemmati-Brivanlou A, Melton D. Vertebrate neuronal induction. Annu Rev Neurosci 1997; 20:43–60.
9. Wilson PA, Lagna G, Suzuki A, et al. Concentration-dependent patterning of Xenopus ectoderm by BMP4 and it signal transducer Smad1. Development 1997; 124:3177–3184.
10. Barone S Jr., Das KP, Lassiter TL, et al. Vulnerable processes of nervous system development: a review of markers and methods. Neurotoxicology 2000; 21:15–36.
11. Nowakowski RS, Hayes NL. CNS development: an overview. Dev Psychopathol 1999; 11:395–417.
12. Bayer SA, Altman J, Russo RJ, et al. Timetables of neurogenesis in the human brain based on experimentally determined patterns in the rat. Neurotoxicology 1993;14: 83–144.
13. Knudsen EI. Sensitive periods in the development of the brain and behavior. J Cog Neurosci 2004; 16:1412–1425.
14. Brittis PA, Meiri K, Dent E, et al. The earliest patterns of neuronal differentiation and migration in the mammalian nervous system. Exp Biol 1995; 134:1–12.
15. Rakic P. Mode of cell migration to the superficial layers of fetal monkey cortex. J Comp Neurol 1972; 145:61–83.
16. Marín O, Valdeolmillos M, Moya F. Neurons in motion: same principles for different shapes? Trends Neurosci 2006; 29:655–661.
17. Lossi L, Cantile C, Tamagno I, et al. Apoptosis in the mammalian CNS: lessons from animal models. Vet J 2005; 170:52–66.
18. Garner CC, Zhai RG, Gundelfinger ED, et al. Molecular mechanisms of CNS synaptogenesis. Trends Neurosci 2002; 25:243–250.
19. Costa LG, Ascher M, Vitalone A, et al. Developmental neuropathology of environmental agents. Ann Rev Pharmacol Toxicol 2004; 44:87–110.
20. Toga AW, Thompson PM, Sowell ER. Mapping brain maturation. Trends Neurosci 2006; 29:148–159.
21. Bredesen DE. Neural apoptosis. Ann Neurol 1995; 38:839–851.
22. Dobbing J, Sands J. Comparative aspects of the brain growth spurt. Early Hum Dev 1979; 3:79–83.
23. Wiggins RC. Myelin development and nutritional insufficiency. Brain Res Dev 1982; 4:151–175.
24. Herschkowitz N, Kagan J, Ziles K. Neurobiological bases of behavioral development in the first year. Neuropediatrics 1997; 28:296–306.
25. Herlenius E, Lagercrantz H. Development of neurotransmitter systems during critical periods. Exp Neurol 2004; 190:S8–S21.
26. Sundstrom E, Kolare S, Souverbie F, et al. Neurochemical differentiation of human bulbospinal monoaminergic neurons during the first trimester. Dev Brain Res 1993; 75:1–12.
27. Jett DA. Central cholinergic neurobiology. In: Handbook of Developmental Neurotoxicology. New York: Academic Press, 1998.
28. Berger-Sweeney J, Hohmann CF. Behavioral consequences of abnormal cortical developments: insights into developmental disabilities. Behav Brain Res 1997;86:121–142.

29. Benitez-Diaz P, Miranda-Contreras L, Mendoza-Briceno RV, et al. Prenatal and post-natal contents of amino acid neurotransmitters in mouse parietal cortex. Dev Neurosci 2003; 25:366–374.
30. Haberny KA, Paule MG, Scallet AC, et al. Ontogeny of the N-methyl-D-aspartate (NMDA) receptor system and susceptibility to neurotoxicity. Toxicol Sci 2002; 68:9–17.
31. Miles R. Neurobiology. A homeostatic switch. Nature 1999; 397:215–216.
32. Cant NB. Cellular neuroanatomy. In: Neurotoxicology. Ann Arbor: CRC Press, 2000.
33. Khazipov R, Luhmann HJ. Early patterns of electrical activity in the developing cerebral cortex of humans and rodents. Trends Neurosci 2006; 29:415–418.
34. Freeman MR, Doherty J. Glial cell biology in Drosophilia and vertebrates. Trends Neurosci 2006; 29:82–90.
35. Harry GJ, Toews AD. Myelination, dysmyelination, and demyelination. In: Handbook of Developmental Neurotoxicology. New York: Academic Press, 1998.
36. Risau W. Differentiation of endothelium. FASEB J 1995; 9:926–933.
37. Erecinksa M, Cherian S, Silver IA. Energy metabolism in mammalian brain during development. Prog Neurobiol 2004; 73:397–445.
38. Ehman KD, Moser VC. Evaluation of cognitive function in weanling rats: a review of methods suitable for chemical screening. Neurotoxicol Teratol 2006; 28:144–161.
39. Buznikov GA, Shmukler YB, Lauder JM. From oocyte to neuron: do neurotransmitters function the same way throughout development? Cell Mol Neurobiol 1996; 16:537–559.
40. Zhang LI, Poo MM. Electrical activity and development of neural circuits. Nat Neurosci 2001; 4:1207–1214.
41. Kandel ER, Schwartz JH. Principles of Neural Science. 2nd ed. New York: Elsevier, 1985.
42. Gaspar P, Cases O, Maroteaux L. The developmental role of serotonin: news from mouse molecular genetics. Nat Rev Neurosci 2003; 4:1002–1012.
43. Chugani DC. Role of altered brain serotonin mechanisms in autism. Mol Psychiatry 2002; 7:S16–S17.
44. Ikonomidou C, Bosch F, Miksa M, et al. Blockade of NMDA receptors and apoptosis in the developing brain. Biochem Pharmacol 1999; 62:401–405.
45. O'Callaghan JP, Miller DB. Assessment of chemically induced alterations in brain development using assays of neuron- and glia-localized proteins. Neurotoxicology 1989; 10:393–406.
46. Kniesel U, Wolburg H. Tight junctions of the blood-brain barrier. Cell Mol Neurobiol 2000; 20:57–76.
47. Schulze C, Firth JA. Interendothelial junctions during blood-brain barrier development in the rat: morphological changes at the level of individual tight junctional contacts. Dev Brain Res 1992; 69:85–95.
48. Martí E, Bovolenta P. Sonic hedgehog in CNS development:one signal multiple outputs. Trends Neurosci 2002; 25:89–96.
49. Cogswell CA, Sarkisian MR, Leung V, et al. A gene essential to brain growth and development maps to the distal arm of rat chromosome 12. Neurosci Lett 1998; 251:5–8.
50. Pinhasov A, Mandel S, Torchinsky A, et al. Activity-dependent neuroprotective protein: a novel gene essential for brain formation. Dev Brain Res 2003; 144:83–90.
51. De Zio D, Giunta L, Corvaro M, et al. Expanding roles of programmed cell death in mammalian neurodevelopment. Sem Cell Devel Biol 2005;16:281–294.
52. Qiu Z, Ghosh A. A brief history of neuronal gene expression: regulatory mechanisms of cellular consequences. Neuron. 2008; 60:449–455.
53. Lohof AM, Ip NY, Poo MM. Potentiation of developing neuromuscular synapses by neurotrophins NT-3 and BDNF. Nature 1993; 363:350–353.
54. Kang H, Schuman EM. Long-lasting neurotrophin-induced enhancement of synaptic transmission in the adult hippocampus. Science 1995; 273:1402–1406.
55. Figurov A, Pozzo-Miller LD, Olafsson P, et al. Regulation of synaptic responses to high frequency stimulation and LTP by neurotrophins in the hippocampus. Nature 1996; 381:706–709.
56. Paratcha G, Ledda F. GDNF and GFRα: a versatile molecular complex for developing neurons. Trends Neurosci 2008; 31:384–391.
57. Baloh RH, Enomoto H, Johnson EM Jr., et al. The GDNF family ligands and receptors – implications for neural development. Curr Opin Neurobiol 2000; 10:103–110.

58. Asbreuk CHJ, Van Schaick HSA, Cox JJ, et al. The homeobox genes Lhx7 and Gbx1 are expressed in the basal forebrain cholinergic system. Neuroscience 2002; 109:287–298.
59. Geisen MJ, Meglio TD, Pasqualetti M, et al. Hox paralog group 2 genes control the migration of mouse pontine neurons through slit-robo signaling. PLoS Biology 2008; 6:e142.
60. Gaiano N, Fishell G. The role of notch in promoting glial and neural stem cell fates. Annu Rev Neurosci 2002; 25:471–490.
61. Kuan C-Y, Roth KA, Flavell RA, et al. Mechanisms of programmed cell death in the developing brain. Trends Neurosci 2000; 23:291–297.
62. Korsmeyer SJ. BCL2 gene family and the regulation of programmed cell death. Cancer Res 1999; 59:1693–1700.
63. Merry DE, Korsmeyer SJ. Bcl2 gene family in the nervous system. Annu Rev Neurosci 1997; 20:245–267.
64. Nicholson DW. Caspase structure, proteolytic substrates, and function during apoptotic cell death. Cell Death Differ 1999; 6:1028–1042.
65. Xia Z, Dickens M, Raingeaud J, et al. Opposing effects of ERK and JNK-p38 MAP kinases on apoptosis. Science 1995; 270:1326–1331.
66. Clancy B, Finlay BL, Darlington RB, et al. Extrapolating brain development from experimental species to humans. Neurotoxicology. 2007; 28:931–937.
67. Clancy B, Kersh B, Hyde J, et al. Web-based method for translating neurodevelopment from laboratory species to humans. Neuroinformatics 2007; 5:79–94.
68. Clancy B, Darlington RB, Finlay BL. Translating developmental time across mammalian species. Neuroscience 2001; 105:7–17.
69. Clancy B, Darlington RB, Finlay BL. The course of human events: predicting the timing of primate neural development. Devel Sci 2000; 3:57–66.
70. Darlington RB. Regression and Linear Models. New York: McGraw-Hill, 1990.
71. Finlay BL, Darlington RB. Linked regularities in the development and evolution of mammalian brains. Science 1995; 268:1578–1584.

FDA and ICH perspectives on reproductive and developmental toxicology

Rochelle W. Tyl

INTRODUCTION AND BACKGROUND

In the fourth century BC, Aristotle wrote the first treatise on embryology, including observations, arguments, and speculation. His belief that slime and decaying matter gave rise to living animals was not disproved until 1668 by Redi, although the final repudiation of "spontaneous generation" occurred in 1864 by Louis Pasteur. Until approximately 1800, it was believed that a fully formed organism in miniature was either in the egg (and only needed the sperm to activate it) or it was in the sperm (drawings of a human sperm cell with a miniature "homunculus" squatting in the sperm head, especially by Hartsoeker, were popular in the late 1600s) and only needed the egg for nutrients. This doctrine of "preformation" was seriously questioned initially by Harvey in the 17th century, followed by Wolff in the 18th century, and finally disproved by Driesch in 1900. Physical identification of the human sperm (by Leeuwenhoek in 1677) and the human egg (by von Baer in 1827) and the recognition that the cell was the structural and functional unit of the organism (by Schwann in 1839) occurred in the 19th century.

The observation and appreciation of the events involved in fertilization of the egg by the sperm (by Hertwig in 1875), the recognition that the egg and sperm contribute the same number of chromosomes to the fertilized egg (by Van Benedin in 1883), the perpetuation of chromosomes by mitosis, and their persistent individuality (by Boveri in 1888 and 1909), all moved the science of embryology and reproduction rapidly forward (1). Developmental and reproductive biology began as descriptive, then comparative, and finally experimental and analytical. This most recent experimental stage was pioneered by Roux and Spemann working with amphibians in Europe and by Morgan and Harrison working with the *Drosophila melanogaster* (fruit fly) in the United States in the early 1900s (2).

In the late 19th century, Gregor Mendel (an Austrian monk), while working in the monastery garden with the common garden pea, established the first genetic laws with his "hereditary characters." The identification of these hereditary characters with genes, situated in specific locations in particular chromosomes, was achieved by Morgan in 1912 using the huge reduplicated pachytene salivary gland chromosomes of the fruit fly after the rediscovery of Mendel's work in peas. The elucidation of the interrelated roles of heredity (internal to the nucleus of the cells) and the environment (external to the nucleus, including the extragenetic components such as histones on the chromosomes, the nuclear and cytoplasmic environment, other cells and organs, blood, lymph, organisms, nutrition, etc.) began almost immediately in the early 1900s.

The study of genes, especially in fruit fly mutants early on, provided the initial explanation of aberrant development. The view then was that genes were everything, and the conceptus developed safe and secure in the egg or womb based solely on his/her genes. However, almost immediately, it became obvious that there were environmental causes for malformations, including dietary deficiencies [vitamin A deficiency in the feed resulting in anophthalmic piglets (3,4)] and excesses, radiation (5,6), chemicals per se [e.g., dinitrophenol, trypsin, ficin (7)], heat (8), and even oxygen (9), etc. The

German measles (rubella) epidemics in Australia (10) and the United States (11), resulting in congenital cataracts and deafness, exposure to herpes simplex virus or cytomegalovirus resulting in blindness, deafness, cerebral palsy, mental retardation, and death in children exposed in utero (11), implicated viruses and diseases as causative. The worldwide (almost) epidemic of phocomelia syndrome in newborns from maternal usage of thalidomide in the late 1950s and early 1960s (see later in text) also implicated drugs as causative.

The research in normal and abnormal developmental biology has progressed rapidly, with current focus on the problems of differentiation, morphogenesis, reproduction, and evolution from organismic, organ, and tissue levels to molecular, genetic, and epigenetic aspects (11; chap. 1, p. 34). The role(s) of genes throughout development, conserved across phyla or brand new (or the same genes with new functions), and the interaction of genes are being rapidly elucidated. The inappropriate methylation or demethylation of DNA bases and/or certain amino acids in histones, altering gene expression at the wrong time(s) and/or in the wrong place(s), resulting in reproductive effects and/or birth defects (with no change to the DNA base sequence; i.e., not genotoxic but epigenetic), is now also recognized as causative. Newbold at National Institute of Environmental Health Sciences (NIEHS) reported such epigenetic transgenerational effects from diethylstilbestrol in mice (12–15), and Skinner and coworkers at the University of Washington (Pullman, Washington, U.S.) reported transgeneration effects on four generations of offspring male rats (F1, F2, F3, and F4) from initial (F0) maternal exposure to vinclozolin during embryonic gonadal differentiation (16–18) and methoxychlor (19). Other laboratories (20) have been unable to replicate the vinclozolin effects. Postnatal carcinogenesis can also occur from prenatal exposure and presumed in utero initiation (12,13,21).

The epidemic of a rare limb defect syndrome (phocomelia) in 1959–1963, from maternal intake of the drug thalidomide during the first trimester of pregnancy, implicated drugs as teratogens. This worldwide epidemic of thalidomide embryopathy was also responsible for the convergence of developmental biology and developmental toxicology, and for the initiation of regulatory developmental and reproductive toxicology [DART (22)] by the promulgation of the FDA's regulations for preclinical testing of drugs to specifically evaluate possible reproductive and/or developmental effects in 1966. The regulatory aspect of DART, therefore, began when this new "wonder drug" went on the market (ultimately almost all over the world), beginning in 1958. It was called thalidomide, Contergan, Distaval, Grippex, etc., and nicknamed the "German babysitter," since it was used to get small children to sleep and to stay asleep. It was also highly touted and prescribed for women in the first trimester of pregnancy to treat typical nausea and vomiting.

In 1961, reports began to appear in medical journals of children born with a previously very rare (or unheard of) constellation of malformations, characterized by (depending on the timing of maternal exposure) external ear malformations, upper and/or lower limb long bone shortening, abnormal hands/fingers, feet/toes, fused legs, and anorectal stenosis, with the syndrome designated as phocomelia ("seal limbs"), first observed as early as 1956 in Germany (23). It was initially reported in Germany (24–28) and Australia (29,30), and in other countries as the drug was released there: Italy (31), Norway (32,33), Denmark (34), Japan (35), Mexico (36), Canada (37), France (38), the United Kingdom, Brazil, Sweden, Finland, Austria, Belgium, Ireland, the Netherlands, Portugal, Spain, Switzerland, Argentina, Taiwan, etc. All told, there were over 8000 affected children (with varying national mortality rates averaging ∼50% worldwide) in at least 25 countries in Europe, Asia, North America (Canada), Mesoamerica, South America, Africa, etc.

The two earliest and enlightening brief notes were from Dr Lenz in Germany (24–27,39) and Dr McBride in Australia (30,31) in *Lancet,* who reported on the children they had recently delivered in their respective practices with this new malformation syndrome, something they had only previously seen in textbooks. The only commonality they noted among the affected pregnancies in their practices was that the mothers of the affected children had taken thalidomide (or Contergan) during their first trimester of pregnancy. The link between maternal exposure to thalidomide and phocomelia in their newborn children was proposed very quickly (28), mainly because the malformation syndrome was so rare, although consensus was rather delayed (40). The delay in consensus was due in large part because there was no such epidemic in the United States, and the European countries were "sure" that thalidomide was marketed in the United States. It was also because mice and rats did not exhibit malformations from in utero exposure to thalidomide (40) (see later in text). In fact, thalidomide was not marketed in the United States due to the actions of Dr Frances Kelsey (new to the FDA in 1960), who denied the producers' requests to market it in the United States. She demanded basic toxicity tests be done before FDA approval, tests not requested by any other country (41,42). Also, two American drug companies (SmithKline and French and Lederle Laboratories) refused to be the American licensees for the German manufacturer because they found no evidence that the drug was "a particularly good tranquilizer" (43; p. 75). Dr Kelsey was initially vilified by disappointed American women (41,42).

The drug houses producing and marketing thalidomide (or Contergan) had followed the existing FDA rules and regulations at that time, which were minimal and did not include evaluations of drugs under development for possible reproductive or developmental toxicology. The producers did not violate any laws, although they did not report to regulatory agencies that they were receiving letters from obstetricians/gynecologists/pediatricians noting the "new" malformation syndrome and the apparent connection between maternal use of thalidomide and the phocomelia syndrome (44). Perhaps 1% of the pregnant women in Europe and Australia were prescribed thalidomide, resulting in a 175-fold increased risk of phocomelia in their newborn children. However, it took six years and thousands of affected children before the drug-defect connection was established. There was no routine surveillance system in place anywhere for birth defects.

In reaction to the thalidomide disaster and the absence of any mechanism to have identified it prior to use in humans, conferences were held (45,46), and Dr Edwin I. Goldenthal, Chief of the Drug Review Branch (Division of Toxicological Evaluation, Bureau of Scientific Standards and Evaluation of the FDA), convened a group of members of the then new Teratology Society (established in 1960) to discuss and develop new testing requirements for registration of drugs to prevent anything like this from happening again. On March 1, 1966, Dr Goldenthal sent a letter to the chief medical directors of all the American pharmaceutical companies, notifying them of the new FDA preclinical testing requirements for drugs, to specifically evaluate possible reproductive and/or developmental toxicity effects of their products under development. This letter essentially detailed the FDA Segment I, II, and III studies (47) and began the governmental requirement for evaluating reproductive and developmental risk from drugs (under FDA), followed by pesticides [EPA Federal Insecticide, Fungicide, Rodenticide Act (FIFRA) (48,49)] and other chemicals [EPA Toxic Substances Control Act (TSCA) (50–53)] and combined EPA FIFRA and TSCA [EPA Office of Prevention, Pesticides and Toxic Substances (OPPTS) (54–57)]. Other countries followed, with Canada (58), United Kingdom (59), the OECD (60), and Japan (61), etc., establishing their own testing guidelines. Wilson (62) presented the early evolution of "teratological" testing. Organized birth defects monitoring quickly followed suit (63).

FDA TESTING GUIDELINES

After the thalidomide disaster, and with the knowledge that there were no specific testing paradigms for the appraisal of safety of new drugs for use during pregnancy and in women of childbearing potential, Dr Goldenthal's March 1, 1966, letter established the "FDA Guidelines for Reproductive Studies for Safety Evaluation of Drugs for Human Use." Three sequential segments (or phases) were proposed as follows.

Segment (Phase) I

This study in rodents involves prebreed exposure to F0 males only (for one full spermatogenic cycle, ten weeks in rats and eight weeks in mice), to F0 females only (two weeks for either species to encompass two to three estrous cycles), or to both sexes, continuing exposure during mating (originally two to three weeks), necropsy of the F0 males after the mating period, continuation of exposure to the F0 females and their F1 litters to gestational day (gd) 5 for mice or gd 6 for rats (the time of embryonic implantation into the uterine lining), then continuation of the pregnant females on study with no exposure to the test material to mid-pregnancy (gd 13–15). The F0 dams and their F1 litters are then necropsied at mid-pregnancy. This study provides information on production of gametes (spermatogenesis, ovulation of the preformed oocytes), mating, fertility, nidation (implantation), and embryonic pre- and postimplantation development (or demise) to mid-pregnancy (Fig. 1A).

A suggested modification is presented in Figure 2. The exposure to the F0 parents and F1 offspring is the same as in the original study design, but the F0 dams and their F1 litters are retained, not exposed, to the end of gestation, 1 to 1.5 days prior to parturition (gd 20–21 in rats, gd 17–18 in mice). At this necropsy at term, the same maternal and embryonic information can be obtained as in the original study design, but a more complete assessment of the prenatal offspring is also possible, such as numbers of total, live, resorbed, dead implants, fetal body weight, crown-rump length, and fetal external, visceral, and skeletal examinations (see Segment II study design) to detect offspring effects from parental and/or preimplantation exposure.

Segment (Phase) II

This study for rodents and nonrodents initially involved exposure to the pregnant F0 females only during major organogenesis, from conceptual implantation into the uterine lining to closure of the secondary palate (gd 6–15 in rodents, gd 6–18 in rabbits). There is a maternal and fetal recovery period (termed the fetal period) from the end of major organogenesis (termed the embryonic period) to term necropsy. At necropsy, information is collected on conceptual loss: preimplantation loss from the time of ovarian corpora lutea (number of ovulated eggs) to implantation (number of uterine implantation sites), postimplantation loss from number of uterine implantation sites to total, live, dead, resorbed conceptuses at term, fetotoxicity (litter size, fetal body weights, sex ratio, external, visceral, and skeletal evaluations), and teratogenicity (specific and general malformations and variations). In-life and necropsy observations will characterize maternal toxicity as well: mortality (up to 10% is acceptable in the top dose), reduced body weights, weight gains, feed and/or water consumption (in grams/day and in grams/kg body weight/day), and clinical observations (daily, prior to and after the dosing period and at least twice daily during the dosing period, at and after dosing). At necropsy, weights are recorded for maternal body, gravid uterus, liver, paired kidneys (absolute and relative to terminal body weight), and any known target tissues or abnormal organs/tissues. Optional endpoints would be histopathologic evaluation of the maternal organs of interest, hematology, endocrinology, urinalysis, etc (Fig. 1B).

A. FDA SEGMENT I: FERTILITY STUDY (Rodent)

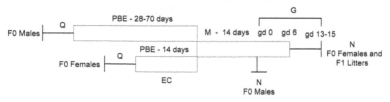

Information on: breeding, fertility, nidation (implantation), and pre- and postimplantation development

B. FDA SEGMENT II: DEVELOPMENTAL TOXICITY STUDY

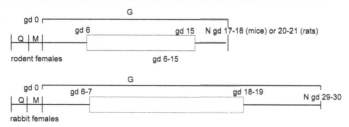

Information on: maternal toxicity, embryotoxicity, fetotoxicity, teratogenicity

C. FDA SEGMENT III: PERINATAL AND POSTNATAL STUDY (Rodent)

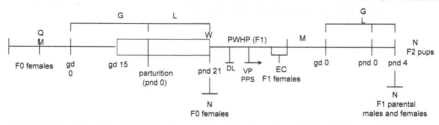

Information on: maternal toxicity, late *in utero* growth and development, parturition, lactation, postnatal growth and development, puberty, reproductive functions of offspring

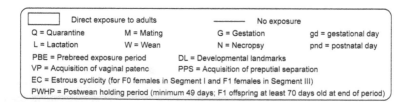

□ Direct exposure to adults	—— No exposure		
Q = Quarantine	M = Mating	G = Gestation	gd = gestational day
L = Lactation	W = Wean	N = Necropsy	pnd = postnatal day
PBE = Prebreed exposure period		DL = Developmental landmarks	
VP = Acquisition of vaginal patenc		PPS = Acquisition of preputial separation	
EC = Estrous cyclicity (for F0 females in Segment I and F1 females in Segment III)			
PWHP = Postwean holding period (minimum 49 days; F1 offspring at least 70 days old at end of period)			

FIGURE 1   (**A**) FDA Segment I: Fertility study (rodent). (**B**) FDA Segment II: Developmental toxicity study. (**C**) FDA Segment III: Perinatal and postnatal study (rodent).

The required use of a second species (rabbits) only in the FDA Segment II testing guideline was due to the fact that rat and mouse fetuses did not exhibit malformations from maternal exposure to thalidomide [one of the main reasons that Dr Warkany initially doubted whether thalidomide was the cause of the phocomelic epidemic in

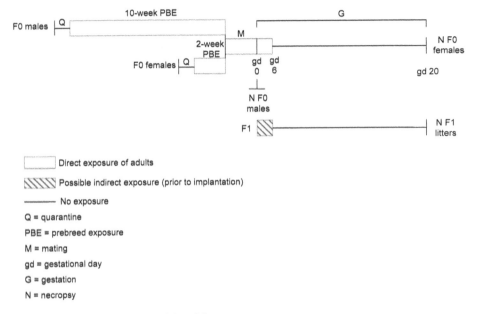

FIGURE 2 Modified Segment (phase) I.

1959–1961 (40)]. However, rabbit fetuses did present with the characteristic phocomelia syndrome from exposure to thalidomide, so rabbits were added as the second mandatory species to be used in Segment II testing right from the start (47).

For years, this study design was considered definitive to assess in utero effects of agents on development. Relatively recently, effects of test agents on the reproductive system, especially on the developing male reproductive system (e.g., antiandrogens), were observed in multigeneration reproductive toxicity studies (with continuous exposure over two generations) but not in Segment II studies. It turns out that the initially specified exposure for rodents (gd 6–15) does not cover the period of male sexual differentiation, and thus the dams were not exposed during the critical developmental window for male reproductive differentiation (e.g., gd 15–19 for antiandrogenic phthalates). Therefore, the EPA (52,54,56), the FDA (64), and the OECD (65) extended the dosing period in their Segment II testing guidelines to gd 20 in the rat, gd 16 to 17 in the mouse, or gd 29 in the rabbit to encompass the critical window of prenatal male reproductive development (Fig. 3). However, there was no change in the time of fetal termination and examination just prior to term in any of these updated guidelines, which makes determination of internal malformations of the prostate or other organs in the reproductive tract, or even hypospadias (external failure of the urogenital folds to completely fuse into the penile tube) difficult, if not impossible, by gross examination, even under a microscope, in the term fetus. Detailed histopathologic examination would reveal such effects, but this is not required in the guidelines as currently written, and therefore not usually done. In addition, Segment II studies examine only structural parameters, such as organ weights, shape, color, lobes, etc., but clearly not functional parameters.

Segment (Phase) III

This study was designed to evaluate postnatal consequences of late gestational exposure. It follows in sequence from prebreed, mating, and preimplantation gestational

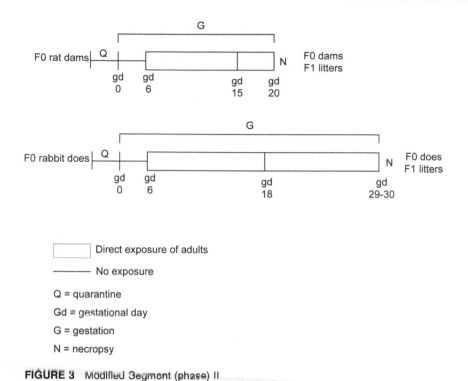

FIGURE 3 Modified Segment (phase) II

exposure (phase I), and embryonic (from implantation to closure of the secondary palate) gestational exposure (phase II) to fetal (from closure of secondary palate to parturition) gestational exposure through parturition to weaning on postnatal day (pnd) 21 in rodents for phase III. This third segment was provided to respond to the need for evaluation of postnatal sequelae to in utero structural or functional insult, as well as recognition that exposure to a developing system may result in qualitatively and/or quantitatively different effects than exposure to an adult organism. The study design consists of exposure of pregnant rats or mice to the test chemical from the end of major organogenesis (gd 15) through histogenesis (the fetal stage), through parturition, through lactation, until the offspring are weaned on pnd 21. The offspring are never directly dosed but are possibly indirectly exposed via transplacental and/or translactational routes. At weaning, the F0 dams are necropsied, and selected F1 offspring are retained with no exposures for the postwean/prebreed period, then mated (the F1 parental males are necropsied after completion of mating). The F1 females are retained through their pregnancies and deliveries, and the F1 females and their F2 litters are necropsied on pnd 4 of the F2 litters. Developmental landmarks can be assessed in the F1 offspring during lactation, for example, surface righting reflex, pinna detachment, eye opening, auditory startle, incisor eruption, as well as testis descent (typically during pnd 14–21), and during the postwean prebreed period, for example, acquisition of puberty (vaginal patency in females, preputial separation in males), motor activity, and learning and memory. Mating of the F1 animals evaluates any effects on offspring reproductive and development function from in utero and lactational exposure. Maternal functions such as F0/F1 delivery, F0/F1 maternal and F1/F2 pup interactions such as pup retrieval, nursing, nest building, F1 (F2 briefly) pup postnatal growth and development, etc., are also evaluated (66). This is the only segment with a postnatal component, which makes it very powerful. This design also encompasses male in utero reproductive system development and any postnatal effects. However, the F1 pups are not exposed during major organogenesis, so only effects from possible fetal exposure are tracked postnatally (66). This concern has led to many pharmaceutical companies and/or their contract research laboratories to perform Segment III studies as generally described, but with F0 maternal dosing beginning on gd 6 (the time of conceptual implantation) rather than on gd 15 (at the end of major organogenesis). Therefore, this modified Segment III has the more inclusive period of possible embryo-fetal exposure and the long-term postnatal evaluation period (Fig. 1C).

These FDA study designs have changed very little since their introduction in 1966 (67). The EPA, OECD, and the Japanese regulators incorporated these study designs into their testing guidelines. However, the other national regulatory agencies recognized that even Segment III (as modified) did not evaluate the consequences of continuous exposure of a test material (in the environment, in the work force, from medications) to reproduction and development over more than one generation. Therefore, they included a two-generation reproductive toxicity study as part of the regulatory requirements (see chap. 2 on EPA and OECD regulations).

The FDA responded in 2000 (64) with a two-generation reproduction and teratology study (Fig. 4). The basic design is comparable to the revised EPA [870.3800 (56,57)] and OECD [OECD 416 (65,68,69)] two-generation testing guidelines with a ten-week (for rats) prebreed exposure for the initial F0 animals, with exposure continuing through mating (F0 males necropsied after the two-week mating period), gestation, parturition, and lactation. At weaning of the F1(a) litters on pnd 21, the F0 dams are necropsied. The F1 offspring also go through a ten-week prebreed, two-week mating,

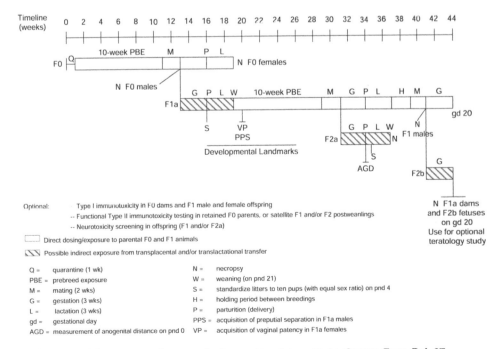

FIGURE 4 FDA two-generation reproduction and teratology study. *Source*: From Ref. 67.

three-week gestation, and three-week lactation. The F2a offspring are necropsied at weaning on pnd 21 as in the EPA OECD guideline. In the FDA version, F1 parents are held for approximately 10 to 14 days (with exposure continuing) and rebred to produce F2b litters. The F1(a) parental males are necropsied after the second mating period. The F1(a) dams and their F2b litters are terminated on gd 20 for a teratological evaluation of the F2b fetuses. During the 10-week prebreed period for the F1(a) offspring, acquisition of puberty is determined and developmental landmarks can be assessed. Optional immunotoxicity evaluations can be done for F0 parents, satellite F1 offspring, or F2a postweanlings (i.e., if not terminated at weaning), and neurotoxicity in F1 and/or F2a offspring.

There are two concerns (at least by this author) about the FDA combined multigeneration reproductive toxicity and developmental toxicity ("teratology") test on the F2b progeny. One concern is that not every paired female becomes positive for vaginal sperm or a copulation plug (i.e., becomes mated). Not every mated female is, in fact, pregnant (typical pregnancy rate is $\sim 95\%$ in control rats). Not every female who is pregnant delivers a live litter that survives lactation. This may be especially relevant for certain test materials and at higher doses. Therefore, the number of F1 females (and/or males) per group available for the F2b mating may be considerably less than the preferred number of at least 25, to obtain at least 20 pregnant females per group at term, for rodents (and now rabbits), with which the study began. These smaller numbers would reduce the sensitivity and statistical power of the system to detect an effect. The other concern is that each generation in a multigeneration reproductive toxicity study is self-selecting. Treatment-related effects on one or more of the stages from oogenesis/ spermatogenesis, mating, fertilization, implantation, organogenesis, in utero survival and development, parturition, nursing/lactation, growth, etc., will impact the number of animals that reach adulthood and are available for (and competent in) reproduction. Therefore, each succeeding generation may, in fact, be populated by the least affected or unaffected animals, and the last generation used for the teratology assessment may be very different from (and less affected than) the initial parental generation(s). For example, if the test material is an antiandrogen (e.g., vinclozolin that binds to the androgen receptor or certain phthalates that inhibit fetal testicular testosterone biosynthesis), then the F0 parents appear unaffected, the F1 generation offspring are affected and many cannot mate (depending on the extent of effects), and the F2 generation offspring are therefore from the least affected F1 animals. Note, however, that these effects would have been detected in the F1 and subsequent generations during the reproductive phase of the study.

"INTERNATIONAL CONFERENCE ON HARMONISATION (ICH)" TESTING GUIDELINES FOR DETECTING REPRODUCTIVE AND DEVELOPMENTAL TOXICITY FOR MEDICINAL PRODUCTS (1994) (70)

The recognition has been growing that the market for pharmaceuticals is worldwide, that drug companies are also international, and that the regulatory requirements for preclinical testing of drugs under development differ across nations. Many important initiatives have been undertaken by regulatory authorities and pharmaceutical industry associations to identify and then reduce such differences in the preclinical testing requirements. The goal in these efforts is to promote international "harmonization" of regulatory requirements.

The ICH was organized to encourage tripartite harmonization initiatives, developed with input from both regulatory and industry representatives. The six ICH sponsors are the European Commission, the European Federation of Pharmaceutical

Industry Associations, the Japanese Ministry of Health and Welfare, the Japanese Pharmaceutical Manufacturers Association, the U.S. FDA, and the U.S. Pharmaceutical Research and Manufacturers of America. The focus of the ICH is to harmonize the technical requirements for registration of pharmaceutical products among the European Union (EU), Japan, and the United States. The ICH Secretariat, which coordinates the preparation of documentation, is provided by the International Federation of Pharmaceutical Manufacturers Association (IFPMA). The ICH Steering Committee includes representatives from each of the ICH sponsors and the IFPMA, as well as observers from the World Health Organization (WHO), the Canadian Health Protection Branch, and the European Free Trade Area.

Harmonization of reproductive toxicology testing was selected as a "priority topic" during the early stages of the ICH initiative. In 1993, the FDA published a draft tripartite guideline titled *Guideline on Detection of Toxicity in Reproduction for Medicinal Products* in the *Federal Register* (58FR 21074, April 16, 1993), with a request for comments (71). After consideration of the comments received and revisions to the draft guideline, a final guideline was endorsed by the three participating regulatory agencies. It is important to note that this guideline is only applicable to sponsors submitting applications to either the FDA Center for Drug Evaluation and Research (CDER) and to the FDA Center for Biologics Evaluation and Research (CBER). The final guideline notice was published in 1994 (72). An addendum on the toxicity to male fertility was recommended for adoption by the ICH Steering Committee in 1995 (73) and published in the *U.S. Federal Register* in 1996 (74).

The three study designs (and their combination) are very similar to the original FDA (47) Segment I study (evaluation of fertility and early embryonic development), the Segment II study (evaluation of embryo-fetal development), and the Segment III study (evaluation of late pre- and postnatal development, including maternal function). The 1994 FDA document describes these proposed ICH studies in detail, with notes and comments. These study designs are described and discussed in the following sections.

ICH 4.1.1 Study of Fertility and Early Embryonic Development to Implantation

This ICH study design (Fig. 5) is very similar to the FDA Segment I study, except that the duration of the F0 male prebreed exposure period is four weeks, "unless data from other studies suggest that this should be modified" (72; p. 48746). The rationale is also that with continuing the two- to three-week mating period and until F1 implantation in the F0 females on gd 6, the total exposure period for the males should be "at least seven to nine weeks dosing" (72; p. 48747).

The FDA further explains (72; p. 48747) that:

> The design of the fertility study, especially the reduction in the premating period for males, is based on evidence accumulated and reappraisal of the basic research on the process of spermatogenesis that originally prompted the demand for a prolonged premating treatment period. Compounds inducing selective effects on male reproduction are rare; mating with females is an insensitive means of detecting effects on spermatogenesis; good pathological and histopathological examination ... of the male reproductive organs provides a more sensitive and quicker means of detecting effects on spermatogenesis; compounds affecting spermatogenesis almost invariably affect postmeiotic stages; there is no conclusive example of a male reproductive toxicant the effects of which could be detected only by dosing males for 9 to 10 weeks and mating them with females.

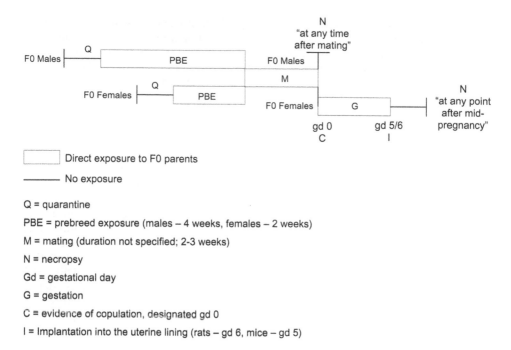

Direct exposure to F0 parents

No exposure

Q = quarantine

PBE = prebreed exposure (males – 4 weeks, females – 2 weeks)

M = mating (duration not specified; 2-3 weeks)

N = necropsy

Gd = gestational day

G = gestation

C = evidence of copulation, designated gd 0

I = Implantation into the uterine lining (rats – gd 6, mice – gd 5)

FIGURE 5   ICH 4.1.1 Study of Fertility and Early Embryonic Development to Implantation.

Information on potential effects on spermatogenesis can be derived from repeated dose toxicity studies. This allows the investigations in the fertility study to be concentrated on other, more immediate, causes of effect. It is noted that the full sequence of spermatogenesis (including sperm maturation) in rats lasts 63 days. When the available evidence, or lack of it, suggests that the scope of investigations in the fertility study should be increased, or extended from detection to characterization, appropriate studies should be designed to further characterize the effects.

Nonetheless, this shortened prebreed exposure period for the F0 males in the ICH protocol (versus the 10 weeks for rats and 8 weeks for mice in the FDA Segment I study design) is of concern to pharmaceutical companies and to the contract research laboratories who serve them. It is not uncommon to use the ICH protocol but retain the 10-week prebreed exposure period for the F0 males, to encompass one full spermatogenic cycle prior to mating.

ICH 4.1.2 Study for Effects on Prenatal and Postnatal Development, Including Maternal Function

This study (Fig. 6) is very similar to the FDA Segment III study, except that the dosing of the F0 females begins at the time of implantation of the F1 conceptuses and the start of major organogenesis (i.e., the embryonic period) rather than at the closure of the secondary palate at the end of the period of organogenesis and the start of the fetal period of histogenesis. This difference in the ICH study design (for the FDA Segment III study design) is laudable since the ICH prenatal exposure encompasses both the embryonic and fetal period, as well as the lactational period, and the F1 offspring are continued on study through adulthood and their reproductive activity to pnd 4 of their

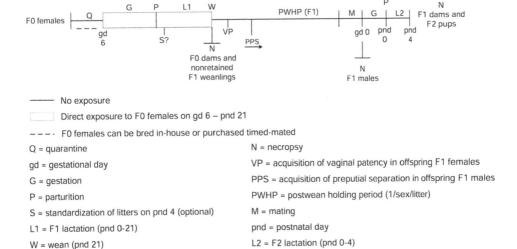

FIGURE 6   ICH 4.1.2 Study of Effects on Prenatal and Postnatal Development, Including Maternal Function.

F2 offspring. This earlier onset of gestational dosing has also been suggested for the FDA Segment III study design (see earlier text).

ICH 4.1.3 Study for Effects on Embryonic and Fetal Development

This study (Fig. 7) is identical to the original FDA 1966 (47) Segment II study. Interestingly, although the FDA Redbook 2000 description of the revised Segment II study specifies maternal dosing beginning on gd 6 and continuing to the end of gestation, which matches the revised EPA [OPPTS 870.3700 (56)] and OECD (65) study designs, the ICH study design does not continue maternal dosing to the end of gestation but stops at the end of major organogenesis (at the closure of the secondary palate, the end of the embryonic period), with no maternal exposures (and potential recovery and rebound for dams and/or fetuses) during the fetal period of histogenesis, after the closure of the secondary palate, thereby missing the period of prenatal male reproductive system development (see earlier text).

ICH 4.3 Two-Study Design in Rodents

In the ICH 4.3 study design (a combination of ICH 4.1.1 and 4.1.2 in rodents; Fig. 8), there is an F0 male and female prebreed exposure as in a Segment I study, and one-half of the F0 dams and F1 litters are necropsied on gd 20 as in a Segment II study. The other half of the F0 dams and their F1 litters are necropsied on pnd 21, or selected F1 pups are retained (1/sex/litter, with no exposure) to adulthood and are mated as in a Segment III study. F1 dams and their F2 litters are necropsied at the end of gestation (gd 20 in rats).

This two-study design is similar to a combined FDA Segment I, II, and III study design except for the following:

1. The F0 prebreed exposure period is four weeks for the F0 males (not ten weeks for rats or eight weeks for mice).
2. Mating of F0 animals is for three weeks (while the multigenerational reproductive toxicity protocols specify two weeks).

    No exposure

    The maternal animals can be bred in-house or purchased timed mated

    Direct exposure to F0 dams from implantation to closure of secondary palate

I = implantation into the uterine lining (gd 5 in mice, gd 6 in rats and rabbits)
gd = gestational day
CL = closure of secondary plate (gd 15 in rats and mice, gd 18 in rabbits)
N = necropsy
E = external evaluation of term fetuses
V = visceral evaluation of term fetuses
S = skeletal evaluation of term fetuses

FIGURE 7   ICH 4.1.3 Study of Effects on Embryo–Fetal Development.

3. The F1 dams and their F2 litters are terminated on gd 20 (not on pnd 4 as in the FDA Segment III and ICH 4.1.2).

    The strengths of this combined study design are as follows:

1. There is a prebreed exposure period for both F0 parental males (although only four weeks of duration) and females.
2. Maternal exposures continue from prebreed, through mating, gestation, and lactation. The F1 offspring are therefore at least potentially exposed throughout gestation (preimplantation, embryonic, and fetal periods) and lactation.
3. The termination of one-half pregnant F0 dams and their F1 litters on gd 20 allows for teratologic evaluation (external, visceral, and skeletal) on the fetuses.
4. The continuation of the remaining F0 dams and their F1 litters with maternal exposures through parturition, through lactation to weaning and the maternal termination, and F1 litters (all or all but 1/sex/litter) on pnd 21 allows for prewean evaluations, such as anogenital distance, nipple retention in males (typically on pnd 11–13), testes descent, as well as the standard parameters of survival, growth, and acquisition of prewean developmental landmarks.

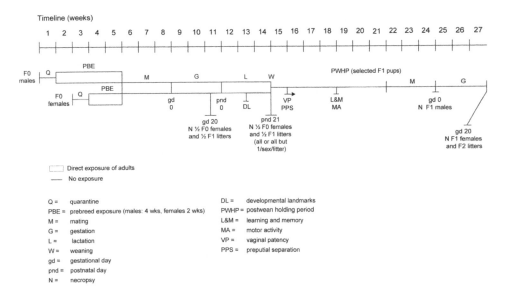

FIGURE 8 ICH 4.3 Two-Study Design in Rodents.

5. The retention of F1 pups (1/sex/litter) after weaning (with no exposures) allows for evaluation of F1 postwean development, including acquisition of puberty (vaginal patency in females and preputial separation in males) and apical evaluations such as learning and memory and motor activity.
6. The retained F1 offspring are mated, with the F1 males necropsied after the mating period, so reproductive behaviors, structures, and functions can be assessed.
7. The pregnant F1 females are retained throughout gestation and terminated on gd 20 so that a teratologic evaluation is possible on the F2 fetuses.

The weaknesses of this combined study design are as follows:

1. The short F0 male prebreed exposure period.
2. The lack of direct exposures to the F1 offspring.
3. The lack of possible indirect exposures to the F2 offspring.
4. The lack of evaluation of parturition/delivery of the F1 dams and their F2 litters.
5. The lack of evaluation of postnatal survival and growth of the F2 offspring (at least to pnd 4 as in the FDA Segment III study).

DISCUSSION AND CONCLUSIONS

The testing guidelines for FDA (47,64,67,71,75,76) and for ICH (72,74) have been presented with the strengths and weaknesses of each study design as perceived by the chapter author and other practitioners in the field. However, it must be stated, as ICH did (72; p. 48750), that "these guidelines are not mandatory rules; they are a starting point rather than an endpoint. They provide a basis from which an investigator can devise a strategy for testing according to available knowledge of the test material and the state-of-the-art." The authors of the testing guidelines understood and provided for scientific flexibility. Fine details of study design and technical procedures are intentionally omitted from the FDA testing guidelines. The expectation is that the investigator is best qualified and will select techniques, procedures, endpoints, etc., that are appropriate for the staff, resources, and laboratory; "human attributes of attitude, ability and consistency are more important than material facilities" (72; p. 48750). The testing guidelines are considered minimum standards; use of additional endpoints, use of satellite animals for PK/TK, more animals per group, more test groups, use of a positive control group, etc., are acceptable (and welcomed) if they strengthen the study design to detect any indication of toxicity from the test material to reproduction and/or development. In addition, new endpoints not validated at the time of the promulgated test guidelines, but subsequently validated, may be a useful addition to the study design. Use of a different rat strain (e.g., Wistar Han rather than caesarean derived (Sprague Dawley) [CD (SD)] rat) or different species (e.g., mouse rather than rat) may also be employed if justified. To make sure that the study data are recorded, manipulated, and retained completely and correctly, all such studies will comply with the appropriate good laboratory practice (GLP) regulations, principles, or requirements.

A final note is that all testing guideline studies are essentially apical in that an effect observed may have one or more different causes, and that it is assumed that the fundamental processes and biochemical, structural, and functional attributes are equivalent in the initial adult subject animals across all groups prior to study start; for example, it is assumed that the adult test animals have limbs, so motor activity can be evaluated; they have (had) normal livers so absorption, distribution, metabolism, and excretion (ADME) can be evaluated. They can (could) ovulate (females) or produce sperm (males), and they have had the appropriate reproductive organs and functions to gestate, lactate, and raise offspring prior to study start. In addition, for effects with a normal low incidence in control animals, determination of a treatment-related effect versus a coincidental occurrence is especially dependent on the association of the "initial" effect with other effects (and, of course, with dose). Additional analyses or even additional studies may be necessary to determine the associations, the nature, scope, and origin(s) of the critical effect and to better define the dose-response curve to facilitate risk assessment. This last point differentiates these secondary tests from the initial apical test where the presence or absence of a dose-response pattern of effects is used to initially determine whether the effect is or is not treatment related.

In conclusion, testing guidelines for reproductive and developmental toxicity testing under FDA and ICH have evolved somewhat over the years (as have EPA and OECD testing guidelines; see chap. 2) since their inception. However, their role remains the same—to provide minimal standardized testing study designs to detect any indication of toxicity from the test material to reproduction and/or development, and to provide GLP-compliant data for governmental formal risk assessment.

ACKNOWLEDGMENTS
The author wishes to thank all the people in her life and career who have been (and are) family, friends, colleagues, collaborators, mentors, mentees, teachers, students; all of the dedicated, experienced, and competent technicians who actually do the work, especially Ms Missy Marr and Ms Chris Myers who have been by her side for decades; and to Ms Cathee Winkie for her patient and professional typing (and retyping) of this manuscript. The author also wishes to thank the RTI International Fellows Program that provided the time for her to read voraciously and to write this chapter.

REFERENCES
1. Arey LB. Chapter I. The historical background. In: Developmental Anatomy: A Textbook and Laboratory Manual of Embryology. 7th ed. (revised). Philadelphia, PA: WB Saunders Co., 1974.
2. Balinsky BI. Part One. The science of embryology. Chapter 1: The scope of embryology and its development as a science. In: An Introduction to Embryology. 4th ed. Philadelphia, PA: WB Saunders Co., 1975:1–15.

3. Hale F. Pigs born without eyeballs. J Heredity 1933; 24:105–106.
4. Hale F. The relation of vitamin A to anophthalmos in pigs. Am J Ophthalmol 1935; 18:1087–1093.
5. Warkany J, Schraffenberger E. Congenital malformations induced in rats by roentgen rays. Am J Roentgenol Radium Ther Nucl Med 1947; 57:455–463.
6. Macht SH, Lawrence PS. Congenital malformations from exposure to roentgen radiation. AJR Am J Roentgenol 1955; 73:442–466.
7. Horstadius S. Vegetalization of the sea-urchin egg by dinitrophenol and animalization by trypsin and ficin. J Embryol Exp Morphol 1953; 1:327–348.
8. Walsh D, Klein NW, Hightower LE, Edwards MJ. Heat shock and thermotolerance during early rat embryo development. Teratology 1987; 36:181–191.
9. Miller JA. Some effect of oxygen on polarity in *Tubularia crocea*. Biol Bull Mar Biol Lab Woods Hole 1937; 73:369.
10. Gregg NM. Congenital cataract following German measles in the mother. Trans Ophthalmol Soc Aust 1941; 3:35–46.
11. Gilbert SF. Developmental Biology. 5th ed. Sunderland, MA: Sinauer Associates, Inc., Publishers, 1997.
12. Newbold RR, Hanson RB, Jefferson WN, et al. Increased tumors but uncompromised fertility in the female descendants of mice exposed developmentally to diethylstilbestrol. Carcinogenesis 1998; 19(9):1655–1663.
13. Newbold RR, Hanson RB, Jefferson WN, et al. Proliferative lesions and reproductive tract lesions in male descendants of mice exposed developmentally to diethylstilbestrol. Carcinogenesis 2000; 21(7):1355–1363.
14. Newbold RR, Padilla-Banks E, Jefferson WN. Adverse effects of the model environmental estrogen diethylstilbestrol are transmitted to subsequent generations. Endocrinology 2006; 147(6):511–517.
15. Newbold RR. Review: lessons learned from perinatal exposure to diethylstilbestrol. Toxicol Appl Pharmacol 2004; 199:142–150.
16. Anway MD, Cupp AS, Uzumcu M, et al. Epigenetic transgenerational activities of endocrine disruptors and male fertility. Science 2005; 308:1466–1469; plus Science supplemental on-line material, June 3, 2005.
17. Skinner MK, Anway MD. Seminiferous cord formation and germ cell programming: epigenetic transgenerational actions of endocrine disruptors. Ann N Y Acad Sci 2005; 1061:18–32.
18. Uzumcu M, Suzuki H, Skinner MK. Effect of the anti-androgenic endocrine disruptor viclozolin on embryonic testis cord formation and postnatal testis development and function. Reprod Toxicol 2004; 18:765–774.
19. Cupp AS, Uzumcu M, Suzuki H, et al. Effect of transient embryonic in vivo exposure to the endocrine disruptor methoxychlor on embryonic and postnatal testis development. J Androl 2003; 24(5):736–745.
20. Gray LE, Furr J. Vinclozolin induces reproductive malformations and infertility when administered during sexual but not gonadal differentiation. Toxicologist 2008; 102(1):11 (abstr 59).
21. Rice JM. Carcinogenesis: a late effect of irreversible toxic damage during development. Environ Health Perspect 1976; 18:133–139.
22. Tyl RW. Chapter 53. Developmental toxicology. In: Ballantyne B, Marrs T, Syversen T, eds. General and Applied Toxicology. Vol. 2, 2nd ed. London: Macmillan Reference Ltd., 2000:1167–1201.
23. Fraser FC. Thalidomide retrospective: what did we learn? Teratology 1988; 38(3):201–202.
24. Lenz W. Diskussionsbemerkung zu dem vortag von R.A. Pfeiffer and K. Kosenow; zur frage der exogenen entstehung, schwerer extremitatenmissbindungen. Tagung der Rheinish-Westfalischen Kinderrarztevereinigung in Dussldorf 1961; 19:11.
25. Lenz W. Kindliche Missbildungen nach Medikament-Einnahme der gravität? Dtsch Med Wochenschr 1961; 86:2555–2556.
26. Lenz W. Thalidomide and congenital abnormalities. Lancet 1962; i:45.
27. Lenz W. Epidemiology of congenital malformations. Ann N Y Acad Sci 1965; 123:228–236.
28. Taussig HB. A study of the German outbreak of phocomelia—the thalidomide syndrome. J Am Med Assoc 1962; 180:1106–1114.
29. McBride WG. Thalidomide and congenital malformations. Lancet 1961; i:912.

30. McBride WG. Thalidomide and congenital abnormalities. Lancet 1961; ii:1358.
31. Gomirato-Sandrucci M, Ceppelini R. Considerazioni cliniche e pathogenetiche su alcuni chsi di focomelia. Min Ped 1962; 14:1181–1202.
32. Kvalle K. Iatrogen misdannelse. Tidsskr Nor Laegeforen 1962; 907–908.
33. Sundal A. Iatrogen misdannelse. Tidsskr Nor Laegenforen 1962; 379–380.
34. Lütken P. Neosedyn-misdannet barn iagttaget. Denmark Ugeskr Laeg 1962; 124:367.
35. Kajii T. Thalidomide and congenital deformities. Lancet 1962; 2:151.
36. Mateos Candano M. Un caso de syndrome talidomidico. Ginecol Obstet Mex 1963; 18:73–85.
37. Webb JT. Canadian thalidomide experiences. Can Med Assoc J 1963; 89:987–992.
38. Ancel P. La Chimioteratogenese Realisation des Monstruosities par des Substances Chemiques chez les Vertebes. Paris: Doin, 1950.
39. Lenz W. Thalidomide embryopathy in Germany, 1959–1961. In: Marois M, ed. Prevention of Physical and Mental Congenital Defects. Part C. Basic and Medical Science, Education, and Future Strategies. New York, NY: Alan R. Liss, 1985:77–83.
40. Warkany J. Why I doubted that thalidomide was the cause of the epidemic of limb defects of 1959 to 1961. Teratology 1988; 38(3):217–219.
41. Kelsey FO. Regulatory aspects of teratology: role of the food and drug administration. Teratology 1982; 25:193–199.
42. Kelsey FO. Thalidomide update: regulatory aspects. Teratology 1988; 38(3):221–226.
43. Norwood C. At Highest Risk. Environmental Hazards to Young and Unborn Children. New York, NY: McGraw-Hill Book Company, 1980.
44. Sunday Times Investigative Team. Suffer the Children: The Study of Thalidomide. London, U.K.: Viking Press, 1979.
45. Commission on Drug Safety. Report, Conference on Prenatal Effects of Drugs. Chicago, March 29–30, 1963. Federation of American Societies for Experimental Biology and Medicine, Washington, D.C., 1963.
46. Commission on Drug Safety. Report. Federation of American Societies for Experimental Biology and Medicine, Washington, D.C., 1964.
47. Goldenthal EI. (Chief, Drug Review Branch, Division of Toxicological Evaluation, Bureau of Scientific Standards and Evaluation). Guidelines for Reproduction Studies for Safety Evaluation of Drugs for Human Use, March 1, 1966.
48. EPA. Environmental Protection Agency. Teratogenicity Study. Pesticide Assessment Guidelines. Subdivision F. Hazard Evaluation: Human and Domestic Animals. EPA-540/9-82-025, 1982:126–130.
49. EPA. Environmental Protection Agency. Pesticides assessment guidelines (FIFRA), subdivision F. Hazard Evaluation: Human and Domestic Animals, Section 83-3 (Final Rule). Available from NTIS (PB86-108958), Springfield, VA, 1984.
50. EPA. Environmental Protection Agency. Toxic Substances Control Act (TSCA), Section 798.4700, Reproduction and Fertility Effects (September 23, 1985). Fed Regist 1985; 50(188):39432–39433.
51. EPA. Environmental Protection Agency. Toxic Substances Control Act (TSCA) test guidelines: final rule. Fed Regist 1987; 50:39412.
52. EPA. Environmental Protection Agency. Toxic Substances Control Act Test Guidelines; Final Rule 40 CFR Part 799.9370 TSCA prenatal developmental toxicity. (August 15, 1997) Fed Regist 1997; 62(158):43832–43834.
53. EPA. Environmental Protection Agency. Toxic Substances Control Act Test Guidelines; Final Rule 40 CFR Part 799.9380 TSCA reproduction and fertility effects. (August 15, 1997) Fed Regist 1997; 62(158):43834–43838.
54. EPA. Environmental Protection Agency. OPPTS (Office of Prevention, Pesticides and Toxic Substances) Health Effects Test Guidelines, OPPTS 870.3700, Prenatal Developmental Toxicity Study (Public Draft, February 1996), 1996.
55. EPA. Environmental Protection Agency. OPPTS (Office of Prevention, Pesticides and Toxic Substances) Health Effects Test Guidelines, OPPTS 870.3800, Reproduction and fertility effects (Public Draft, February 1996). 1996.
56. EPA. Environmental Protection Agency. Office of Prevention, Pesticides and Toxic Substances (OPPTS). Health Effects Test Guidelines OPPTS 870.3700, Prenatal Developmental Toxicity Study; Final Guideline, August 1998.

57. EPA. Environmental Protection Agency. Office of Prevention, Pesticides and Toxic Substances (OPPTS). Health Effects Test Guidelines OPPTS 870.3800, Reproduction and Fertility Effects; Final Guideline, August 1998.
58. Canada Ministry of Health and Welfare, Health Protection Branch. The Testing of Chemicals for Carcinogenicity, Mutagenicity and Teratogenicity. The Ministry of Ottawa, 1973.
59. United Kingdom. Committee on Safety of Medicines: Notes for Guidance on Reproduction Studies. London: Department of Health and Social Security, 1974.
60. OECD. (Organisation for Economic Cooperation and Development). Guideline for Testing of Chemicals: Teratogenicity, OECD No. 414, Director of Information, Paris, France, 1981.
61. Japanese Guidelines of Toxicity Studies. Notification No. 118 of the Pharmaceutical Affairs Bureau, Ministry of Health and Welfare. 2. Studies of the Effects of Drugs on Reproduction. Tokyo, Japan: Yakagyo Jiho Co., Ltd., 1984.
62. Wilson JG. The evolution of teratological testing. Teratology 1979; 20(2):205–211.
63. Klingsberg MA, Papier CM, Hart J. Birth defects monitoring. Am J Ind Med 1983; 4:309–328.
64. FDA. (Food and Drug Administration). Center for Food Safety and Applied Nutrition Redbook 2000. IV.C.9.b. Guidelines for Developmental Toxicity Studies. July 20, 2000:1–6.
65. OECD. OECD Guideline for the Testing of Chemicals; Proposal for Updating Guideline 414: Prenatal Developmental Toxicity Study, adopted January 22, 1001, Paris: OECD, 2001:1–11.
66. Tyl RW, Marr MC. Chapter 7. Developmental toxicity testing—methodology. In: Hood RD, ed. Developmental and Reproductive Toxicology—A Practical Approach. 2nd ed. Boca Raton, FL: CRC Press, 2006:201–262.
67. FDA. (Food and Drug Administration). Center for Food Safety and Applied Nutrition Redbook 2000. IV.C.9.a. Guidelines for Reproduction Studies. July 20, 2000:1–11.
68. OECD. OECD Guideline for the Testing of Chemicals; Proposal for Updating Guideline 415: One-generation Reproduction Toxicity Study, adopted January 22, 2001, OECD, Paris, 2001:1–8.
69. OECD. OECD Guideline for the Testing of Chemicals; Proposal for Updating Guideline 416: Two-generation Reproduction Toxicity Study, adopted January 22, 2001, OECD, Paris, 2001:1–13.
70. ICH. International Conference on Harmonisation. Guideline on Detection of Toxicity to Reproduction for Medicinal Products. Fed Regist 1994; 59:48746–48752.
71. FDA. (Food and Drug Administration). International Conference on Harmonization (ICH): Draft guideline on detection of toxicity to reproduction for medical products; availability. Fed Regist 1993; 58(72):21074–21080.
72. FDA. (Food and Drug Administration). International Conference on Harmonization; guideline on detection of toxicity to reproduction for medicinal products; availability. Notice. Fed Regist 1994; 59(183):48746–48752.
73. ICH. International Conference on Harmonisation of Technical Requirements for Registration of Pharmaceuticals for Human Use. ICH Harmonised Tripartite Guideline. Toxicity to Male Fertility. An Addendum to the ICH Tripartite Guideline on Detection of Toxicity to Reproduction for Medicinal Products. Recommended for Adoption at Step 4 of the ICH Process on 29 November, 1995 by the ICH Steering Committee, 2 pages, 1995.
74. FDA. (Food and Drug Administration). International Conference on Harmonization: Guideline on detection of toxicity to reproduction for medical products: addendum on toxicity to male fertility; availability. Fed Regist 1996; 61(67):15359–15361.
75. FDA. Toxicological Principles for the Safety Assessment of Direct Food Additives and Color Additives Used in Food. Washington, D.C.: U.S. Food and Drug Administration, Bureau of Foods, 1982.
76. FDA. Good Laboratory Practice Regulations for Nonclinical Laboratory Studies. Code of Federal Regulations (CFR), April 1, 1988, Washington, D.C.: U.S. Food and Drug Administration, 1988:229–243.

6 EPA and OECD perspectives on reproductive and developmental toxicity testing

Susan L. Makris

INTRODUCTION—DEVELOPMENTAL AND REPRODUCTIVE RISK ASSESSMENT FOR ENVIRONMENTAL CHEMICALS

In 1962, *Silent Spring* was published. In this carefully researched book, Rachel Carson discussed the indiscriminate use of pesticides and their resulting impacts on the environment, wildlife, and human health (1). This book initiated a revolution of public opinion on environmental conservation, which played a significant role in the establishment of the Environmental Protection Agency (EPA) in 1970 under the Nixon administration (2). At its inception, the EPA subsumed duties previously held under the purview of other diverse Federal entities and was initially organized into offices for the regulation of water quality, air pollution, pesticides, radiation, and solid wastes. Although primarily involved in pollution control in its initial activities, the agency and its functions have evolved over the years. In part, this included a burgeoning focus on human health risk assessment, which continues to the present.

Consideration of the effects of exposure to environmental pollutants on reproductive health and subsequent generations was strongly represented in the initial concerns described in *Silent Spring*. In the formative years of the EPA, extensive consideration was given to the methods for evaluation of risk to reproduction and development from environmental exposures and to the determination of levels deemed "safe" for humans. Important efforts of the National Academies of Science described a paradigm for risk assessment that included five primary components: hazard identification, dose response evaluation, exposure evaluation, risk characterization, and risk management (3,4). The broad-risk assessment concepts embodied in these publications were refined by the EPA in risk assessment guidelines for developmental and reproductive toxicity (5,6). In these guidelines, the concepts of developmental and reproductive toxicity are clearly defined. Developmental toxicity is defined as adverse effects on the developing organism that may result from exposure prior to conception (to either parent), during prenatal development, or postnatally to the time of sexual maturation. An important aspect to the definition is that developmental effects may be detected at any point during the life span of the organism. Major manifestations of developmental toxicity are recognized to include death, altered growth, structural alterations, and functional deficits. Reproductive toxicity is defined as the occurrence of biologically adverse effects on the reproductive system, which may be expressed as alterations to the female or male reproductive organs, the related endocrine system, or pregnancy outcomes. Manifestation of reproductive toxicity may include adverse effects on onset of puberty, gamete production and transport, reproductive cycle normality, sexual behavior, fertility, gestation, parturition, lactation, developmental toxicity, premature reproductive senescence, or modifications of other functions that are dependent on the integrity of the reproductive systems.

Over the past decade and a half, specific consideration of the risks of environmental exposures to children, a potentially susceptible life stage characterized by ongoing postnatal development, has been strongly advocated. In 1993, a report on *Pesticides in the Diets of Infants and Children* was published by the National Research Council of the National Academies of Science (7). This publication highlighted the inadequacy of risk assessment policies and procedures in place at that time for the protection of children from environmental exposures when the basis for decision-making primarily relied on information derived from studies in adult organisms. Influenced by this report, the U.S. Congress passed the Food Quality Protection Act (FQPA), an amendment to the Federal Insecticide, Fungicide and Rodenticide Act (FIFRA), and the Safe Drinking Water Act (SDWA) Amendments (8,9). Both of these laws mandated the evaluation of exposures and toxicities to children for risk assessment and specifically indicated the need to screen all pesticides for endocrine-disrupting potential, which can have extensive impacts on development. In addition, the FQPA required that the EPA use an additional 10-fold margin of safety in setting pesticide tolerances, to take into account the potential for pre- and postnatal toxicity and the completeness of the toxicology and exposure database. According to the language of the FQPA, this 10-fold factor could be replaced with a different FQPA factor only if reliable data demonstrated that the resulting level of exposure would be safe for infants and children.

Subsequent to the raised awareness that the NRC report and the child-protective legislation engendered, the focus on children's risk assessment issues increased exponentially. Presidential Executive Order 13045, *Protection of Children from Environmental Health Risks and Safety Risks*, was issued by President Clinton, requiring all governmental agencies to consider potential risks from environment exposures during childhood (10). The EPA released a *Policy on Evaluating Health Risks in Children* issued by the Administrator (11), established an Office of Children's Health Protection (12), issued a *Strategy for Research on Environmental Risks to Children* (13), developed guidance for the application of FQPA-mandated child-specific safety factors in risk assessment of food use pesticides (14), and implemented a number of other important initiatives (reviewed in Ref. 15). There is a continued focus on the consideration of children in risk assessment at the U.S. EPA, which is illustrated by documents such as the *Framework for Assessing Health Risks of Environmental Exposures to Children* (16), the *Supplemental Guidance for Assessing Susceptibility for Early-Life Exposure to Carcinogens* (17), and the *Child-Specific Exposure Factors Handbook* (18). Additionally, the U.S. EPA has sponsored a number of important workshops on issues relevant to the consideration of susceptible early life populations in hazard evaluation and risk assessment. These include workshops on critical windows of exposure for children's health risk assessment (19), the timing and progression of puberty (20), and children's inhalation dosimetry and health effects for risk assessment (21). Guidance on risk assessment procedures that directly address childhood susceptibilities and that encourage the use of life stage-specific data and approaches has been issued internationally (15,16,22–25).

The Organisation for Economic Co-operation and Development (OECD) was established by convention at a 1961 meeting in Paris. The OECD addresses global issues in the areas of economy (e.g., economics and growth, agriculture, and trade), society (e.g., education, employment, migration, and health), development issues, finance, governance (e.g., corporate and public), innovation (e.g., biotechnology and information/communication technologies), and sustainability (e.g., fisheries, energy, and environment). The sustainability program on the environment includes a number of diverse topic areas, among which is chemical testing guidelines.

SCREENING AND TESTING FOR DEVELOPMENTAL AND REPRODUCTIVE TOXICITY FOR ENVIRONMENTAL CHEMICALS

Reproductive and developmental toxicity testing for environmental chemicals has its historical origins and theoretical basis in the testing paradigm designed for the preclinical toxicological evaluation of pharmaceuticals (described in the previous chapter). Standardized protocols (guidelines) for toxicology testing in animal models that had been developed by the Food and Drug Administration (FDA) were adopted by EPA and revised for the evaluation of pesticides and industrial chemicals. Such testing was required under Part 158 of the Federal Insecticide, Fungicide and Rodenticide Act (FIFRA) for pesticides [first published in 1985 and recently updated (26)] and by establishment of chemical-specific test rules under the Toxic Substances Control Act (TSCA) for industrial chemicals. These studies are also recognized as critical to developing high-confidence risk assessments for chemicals regulated under other U.S. environmental laws such as the Clean Water Act (amended in 2006) and the Clean Air Act.

International regulatory programs also developed toxicity testing guidelines for use in developmental and reproductive toxicity assessment. For example, in Japan, guidelines for the assessment of pesticide toxicity were issued by the Ministry of Agriculture, Forestry and Fisheries (MAFF). However, OECD efforts in this area are of even greater international influence due to the deliberate multinational collaboration that results in the development of testing guidelines and risk assessment guidance, for example, OECD GD 43, "Guidance Document on Mammalian Reproductive Toxicity Testing and Assessment" (27). The OECD guidelines for the testing of chemicals are a compilation of internationally agreed on test methods used by government, industry, and independent laboratories to determine the safety of chemicals and chemical preparations. Among the guidelines, which were initiated in 1981 and are undergoing continuous revision and update (28), are those that address health effects (29). The OECD Test Guideline Programme is managed by the Working Group of the National Coordinators of the Test Guidelines Programme (WNT), composed of representatives that are nominated by governments of member countries, and by some nonmember economies and the European Commission. This body oversees guideline and guidance development and update, as well as validation status (30).

OECD guidelines are harmonized with the EPA Office of Pollution Prevention and Toxic Substances (OPPTS) guidelines for the testing of health effects for pesticides and industrial chemicals. A concerted effort to harmonize a number of toxicity testing guidelines, including the prenatal developmental toxicity study and the reproduction and fertility effects study, was finalized in 1998 (31). The use of harmonized test methods is addressed in the OECD Council Decision on Mutual Acceptance of Data. Mutual data acceptance means that "data generated in the testing of chemicals in an OECD member country (or some nonmember economies) in accordance with OECD Test Guidelines and OECD principles of Good Laboratory Practice shall be accepted in other member countries (or nonmember economies) for purposes of assessment and other uses relating to the protection of man and the environment" (28). This principle of harmonization between OECD and EPA testing guidelines has important consequences in that it facilitates consistency in the test methods across multiple laboratories, and it conserves animal resources (since chemical registrants no longer find themselves repeating toxicology studies simply to meet the testing requirements of diverse regulatory entities). The harmonized guidelines have been used for many years in screening industrial chemicals for developmental and reproductive toxicity under the OECD Screening Information DataSets (SIDS) and EPA High Production Volume (HPV) programs, and will be used for more extensive screening under the Registration, Evaluation, Authorisation, and restriction of CHemicals (REACH) chemical registration process that has been implemented by the European Union (EU) (32).

EPA AND OECD TESTING GUIDELINES

EPA and OECD guideline studies that assess developmental and reproductive toxicity endpoints in mammalian species are listed in Table 1. This list includes studies (e.g., subacute, subchronic, chronic/carcinogenicity, and mutagenicity) that are not often

TABLE 1 EPA and OECD Published Guidelines for the Testing of Developmental and Reproductive Toxicity in Mammalian Species

Guideline	EPA	OECD
Prenatal developmental toxicity study	OPPTS 870.3700	GL 414
One-generation reproduction study	c,d	GL 415
Reproduction and fertility effects (two-generation reproduction study)	OPPTS 870.3800	GL 416
Reproduction/developmental toxicity screening test	OPPTS 870.3550	GL 421
Combined repeated dose toxicity study with the reproduction/developmental toxicity screening test	OPPTS 870.3650	GL 422
Developmental neurotoxicity study	OPPTS 870.6300	GL 426
Uterotrophic assay[a]	OPPTS 890.1600	GL 440
Hershberger assay[a]	OPPTS 890.1400	GL 441
Juvenile/peripubertal male assay[a]	OPPTS 890.1500	c,d
Juvenile/peripubertal female assay[a]	OPPTS 890.1450	c,d
28-day oral toxicity in rodents[b]	OPPTS 870.3050	GL 407
21/28-day dermal toxicity[b]	c	GL 410
28-day inhalation toxicity[b]	c	GL 412
90-day oral toxicity in rodents[b]	OPPTS 870.3100	GL 408
90-day oral toxicity in nonrodents[b]	OPPTS 870.3150	GL 409
90-day dermal toxicity[b]	OPPTS 870.3250	GL 411
90-day inhalation toxicity[b]	OPPTS 870.3465	GL 413
Chronic toxicity[b]	OPPTS 870.4100	GL 452
Carcinogenicity[b]	OPPTS 870.4200	GL 451
Combined chronic toxicity/carcinogenicity[b]	OPPTS 870.4300	GL 453
Combined chronic toxicity/carcinogenicity testing of respirable fibrous particles[b]	OPPTS 870.8355	c
Rodent dominant lethal assay[b]	OPPTS 870.5450	GL 478
Mammalian spermatogonial chromosomal aberration test[b]	OPPTS 870.5380	GL 483
Spot test[b]	c	GL 484
Heritable translocation assay[b]	OPPTS 870.5460	GL 485

[a]Endocrine screening assay.
[b]While not traditionally considered to be a reproductive toxicity assay, the protocol contains reproductive system endpoints that should be considered in a weight-of-evidence evaluation of developmental and reproductive toxicity.
[c]No comparable assay listed on website.
[d]EPA and OECD collaborative effort on guideline development.
Source: http://www.epa.gov/opptsfrs/home/guidelin.htm; http://titania.sourceoecd.org/vl=926302/cl=20/nw=1/rpsv/cw/vhosts/oecdjournals/1607310x/v1n4/contp1-1.htm.

thought of as tests that specifically address reproduction or development. Yet these studies do include the assessment of endpoints that are important to the weight-of-evidence evaluation and interpretation of developmental or reproductive toxicity for a chemical database. It is also important to note that there are a number of study protocols (e.g., endocrine disruption assays) that are currently being developed collaboratively by EPA and OECD.

Descriptions of the listed guideline studies follow.

Prenatal Developmental Toxicity Study

The prenatal developmental toxicity study evaluates embryo-fetal toxicity resulting from in utero exposures. The historical basis of this study design lies in the exceptional work of pediatricians and researchers who studied congenital malformations, growth retardation, and fetal death [described by Wilson (33) and discussed at length in chap. 5].

Originally called a "teratology" study, the name was eventually revised to better encompass the broader spectrum of possible developmental outcomes. Both EPA and OECD guidelines for prenatal developmental toxicity assessment (OPPTS 870.3700 and TG 414, respectively) have been available since the earliest days of guideline development, and they were updated and enhanced during the process of guideline harmonization (34,35). One of the most important updates at that time entailed the extension of the dosing period to the entire duration of gestation; earlier versions of the guideline focused on dosing only during the period of major organogenesis (e.g., GD 6-15 in the rat), which would not facilitate detection of effects on development that occur in late gestation (e.g., male reproductive system development) (31).

This study can be conducted in rodents (e.g., rat and mouse) or nonrodents (e.g., rabbits). The choice of species can be dictated to some extent by regulatory requirements, but should be based as much as possible on information relative to the interpretation and use of the data in a risk assessment context. For example, toxicokinetic (TK) data may indicate similarities or differences between test species and humans that could result in a determination regarding the most appropriate species (and/or strain) of animal model to predict potential human response to in utero developmental exposure.

The prenatal developmental toxicity study design is illustrated in Figure 1. In this study, young mature virgin female animals are artificially inseminated or mated with males of the same strain and then placed on study once there is positive evidence of mating. The dams are assigned to control and treatment groups (preferably 20/group, although group sizes of 16 may be considered adequate for detection of many adverse developmental outcomes). Maternal animals are administered the test substance throughout gestation, via a route that most closely approximates potential human exposure. If there is evidence (e.g., from preliminary or dose-range finding studies) of excessive treatment-related preimplantation loss, the dosing period can be initiated at the time of implantation [e.g., gestation day (GD) 6 in the rodent or GD 7 in the rabbit]. Daily dosing is continued to the day prior to expected delivery (e.g., GD 18 in the mouse, GD 19–20 in the rat, GD 28–30 in the rabbit, dependent on strain differences). At that time, the dams are killed and necropsied, and the fetuses are removed from the uteri for subsequent evaluation.

Assessments on maternal animals standardly include weekly clinical observations, body weight, and food consumption. Postmortem observations include necropsy findings, ovarian corpora lutea counts, and a detailed evaluation of uterine content. The

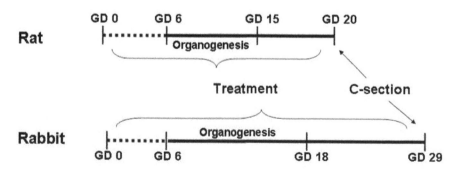

•••••••• Dosing optional if treatment-related preimplantation loss is demonstrated

FIGURE 1   Prenatal developmental toxicity study (OPPTS 870.3700; OECD 414).

number and placement of implantation sites, live and dead fetuses, and early and late resorptions are assessed. Placental tissues are also examined for any abnormalities. The sex of each fetus is determined, and each fetus is individually weighed and examined for external abnormalities. Fetuses from each litter are then humanely killed, assigned for visceral and/or skeletal evaluation, and processed accordingly. Visceral evaluation includes an assessment of the state of development and conformation of cranial, thoracic, and abdominal organs. This is accomplished with methods of fresh specimen dissection and/or with serial sectioning of specimens preserved in Bouins solution. Skeletal evaluation entails an evaluation of the conformation and rate of development of ossified bone and/or cartilage precursor elements of the skull, hyoid, sternebrae, pectoral girdle, ribs, vertebrae, pelvic girdle, forelimbs, and hindlimbs. Preparation of fetal specimens for skeletal evaluation entails chemical maceration of tissues, staining with alizarin red S and/or alcian blue (which selectively stain ossified bone or cartilage, respectively), and clearing of excess stain from soft tissues. Standardized nomenclature for recording external, visceral, and skeletal fetal anomalies (36) and the availability of historical control databases facilitate the interpretation of findings and their use in a regulatory context.

Two-Generation Reproduction and Fertility Effects Study

The reproduction and fertility effects study, that is, the two-generation reproduction study (OPPTS 870.3800; OECD GL 416), is designed to assess the effects of a test substance on the integrity and performance of the male and female reproductive systems, and to provide an evaluation of the growth and development of the offspring. The guideline study design originated in early FDA Segment III studies and remained similar in design to these studies until 1998, when EPA and OECD updated and harmonized Health Effects testing guidelines (37,38). At that time, an effort was made to incorporate a number of sensitive endpoints into the study, including some that are indicators of endocrine disruption (31).

The test (shown in Figure 2) is conducted in rodents (typically the rat). Test substance is administered to first generation (parental) animals during growth, mating, pregnancy, and lactation to the weaning of the first generation (F1) litters; F1 offspring

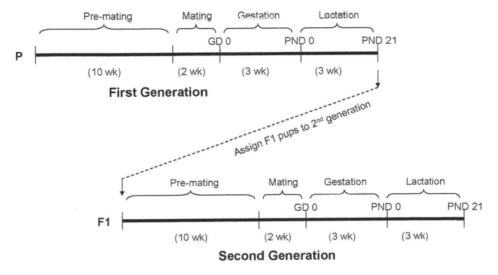

FIGURE 2 Multigeneration reproduction and fertility effects study (OPPTS 870.3800; OECD TG 416).

are selected to continue on study through growth, mating, gestation, and lactation phases to the weaning of the second generation (F2) litters, at which time the study is terminated. A sufficient number of young animals (5–9 weeks of age) are assigned to a control and three dose groups to yield approximately 20 pregnant females at or near the time of parturition. The test substance is administered by a route that is similar to expected human exposure, oral routes (e.g., diet, drinking water, or gavage) being preferred.

In parental animals, clinical observation, body weight, and food consumption data are collected during a 10-week growth (pre-mating) period, and during gestation and lactation in females. Estrous cyclicity is evaluated in parental females prior to mating. At the end of the premating period, male and female animals are paired (1:1) and evaluated for positive evidence of copulation. Observations are recorded on the course and duration of gestation, on parturition, and on the care and nurturing of litters. The litters are examined and monitored throughout the lactation period. For individual pups, the number and sex, clinical findings, body weights, and survival are carefully recorded. In F1 pups, anogenital distance can be measured, triggered on evidence such as skewing of sex ratio or other endocrine-related effects. Litter sizes are adjusted to four to five pups per sex on postnatal day (PND) 4 (this task is optional in the OECD guideline), and pups are weaned on approximately PND 21, at which time weanlings are randomly selected to become parents for the next generation or to be terminated with necropsy. F1 offspring selected for the second generation are evaluated for the age of sexual maturation (vaginal patency or balanopreputial separation). Parental animals of either generation are killed after it is evident that they will not be required for production of additional litter(s).

Postmortem observations in parental animals include organ weights, gross pathology findings, and microscopic pathology of reproductive and other target organs. Organ weights are measured for the uterus, ovaries, testes, epididymides, seminal vesicles, prostate, brain, pituitary, liver, kidneys, adrenal glands, spleen, and known target organs. Histopathological evaluation is conducted for the vagina, uterus/oviducts, cervix, ovaries, testis, epididymis (caput, corpus, and cauda), seminal vesicles, prostate, coagulating gland, pituitary, adrenal glands, target organs, and grossly abnormal tissue. Control and high-dose tissues are initially examined, and the low- and intermediate-dose groups are subsequently evaluated if a treatment-related effect is identified. In the females, a vaginal smear is taken at the time of necropsy and ovarian pathological assessment includes methodologies for quantifying primordial and growing follicles. In the males, an assessment of epididymal or vas deferens sperm count, progressive motility, and morphology is conducted; additionally, testicular spermatid counts are evaluated.

Postmortem evaluation in weanling offspring includes necropsy of three pups per sex per litter (when available), and organ weights of brain, spleen, and thymus [as potential indicators of the need for subsequent developmental neurotoxicity (DNT) or developmental immunotoxicity testing]. Histopathological assessment is conducted selectively on tissues for which there is macroscopic evidence of treatment-related abnormality.

One-Generation Reproduction Study

OECD 415 is a one-generation reproductive toxicity study (39) that has limited use for dose range–finding purposes and for initial screening and prioritization of test substances. This study is, in essence, composed of the first generation of a two-generation reproduction study, with a shortened (2-week duration) premating period for parental females (Figure 3). Since this guideline was not updated at the time of EPA and OECD harmonization efforts in 1998, many of the endocrine-sensitive endpoints (e.g., estrous cyclicity, sperm measures, ovarian follicle counts, anogenital distance, sexual maturation) and postmortem assessments that were added to the two-generation study guideline at

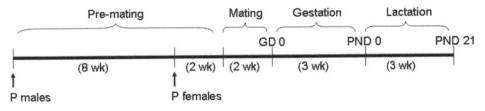

FIGURE 3 One-generation reproduction study (OECD TG 415).

that time are not included in this one-generation study design. Currently, there is no correlate EPA guideline, although collaborative efforts are ongoing in EPA and OECD to develop an extended one-generation reproduction study (40).

The concept of an extended one-generation reproduction study was raised in an International Life Sciences Institute (ILSI) Agricultural Chemical Safety Association (ACSA) effort (41) that focused on creating an alternative reproductive/developmental toxicity screening study that maximized the amount of relevant information generated across specific critical and potentially susceptible life stages. Additional goals of the effort were to reduce the number of animals and other resources required and limit the redundancy that inevitably arose when utilizing multiple study protocols for toxicological screening. As a result of extensive discussion and international efforts to further develop this study design for regulatory purposes, a collaborative OECD/EPA draft extended one-generation study design has been released for review and comment (40). This study, which is illustrated in Figure 4, is conducted using three cohorts of F1

Necropsy
P animals

Dosing			
Pre-mating	Mating	Post-mating	

P ♂	2 weeks	2 weeks	6 weeks	
P ♀	2 weeks	2 weeks	Pregnancy	Lactation

	F₁	In-utero development	Pre-weaning	Post-weaning

Parental generation	Cohort	Designation	Animals/Cohort	Sexual Maturation	Approximate age at necropsy (weeks)
Target is 20 litters per group	1A	Reproductive	20 M +20 F	Yes	13
	1B	Reproductive	20 M +20 F	Yes	14 or 20 if triggered
	2A	Neurotoxicity	10 M +10 F@	Yes	9
	2B	Neurotoxicity	10 M +10 F@	No	3
	3	Immunotoxicity	10 M +10 F@	Yes	8
	Surplus	Spares		No	3

@ one per litter and representative of 20 litters in total where possible

FIGURE 4 Extended one-generation reproduction study (proposed draft, October 28, 2009). *Source*: From Ref. 29.

animals. In cohort 1, reproductive and developmental endpoints are assessed. Cohort 1 animals are exposed to the test substance for approximately 13 weeks, and this cohort may be extended to include an F2 generation, dependent on available background data and/or observations recorded during the in-life phase of the study. Cohort 2 assesses the potential impact of the test substance on the developing nervous system, while Cohort 3 assesses the potential impact on the developing immune system. In concept, this extended one-generation study might, in some situations, replace the two-generation reproduction and fertility effects study and the stand-alone guideline DNT study in a future regulatory context. The inclusion of a developmental immuno-toxicity (DIT) cohort is unique and important because (i) there is currently no stand-alone EPA or OECD DIT study guideline, although a framework for this type of study has been developed (42–45) and (ii) the assessment of DIT is recognized as important to the adequate evaluation of children's health risk (25,46,47). Another important aspect of the extended one-generation reproduction study is the explicit directive to consider TK data in designing the study, selecting dose levels, and interpreting results. While the collection and use of TK data in studies that include early life stage exposures have been encouraged for many years (48,49), inclusion in the extended one-generation study protocol provides additional prominence to the concept. The extended one-generation reproduction study protocol also includes enhanced assessment of endocrine endpoints, including thyroid hormone levels (which are not evaluated in the typical two-generation reproduction study).

Reproductive/Developmental Toxicity Screening Tests

The ability to screen a large number of chemicals quickly for potential effects on developmental and/or reproductive endpoints has been an issue of concern for many years. To address this issue, two screening studies were designed, which were shorter in duration, used fewer resources, and were more limited in scope than standard prenatal developmental toxicity and reproductive toxicity tests. These are the "Reproductive/Developmental Toxicity Screening Test" (OPPTS 870.3550 and OECD TG 421 (50,51) and the "Combined Repeated Dose Toxicity Study with the Reproductive/Developmental Toxicity Screening Test" (OPPTS 870.3650 and OECD TG 422) (52,53). The first protocol is a reduced one-generation reproduction study, while the second is a combination of a 28-day toxicity study and a reduced one-generation reproduction study. The purpose of these tests is to generate limited information on the effects of a test substance on male and female reproductive performance, including mating behavior, gonadal function, conception, parturition, and offspring development. However, the tests are not considered to be alternatives or replacements for prenatal developmental toxicity or reproductive toxicity studies (27). The studies have been traditionally utilized for screening and prioritization of HPV chemicals by EPA and OECD, in part through the implementation of the SIDS program. The EPA Office of Pesticide Programs (OPP) has also used these studies in the screening of pesticide "inert" ingredients in an attempt to meet specific requirements of FQPA in an expeditious manner.

The studies, which are diagrammed in Figure 5, are generally conducted in rats and include dosing of the parental animals for two weeks prior to mating and then continuing during mating. In the females, dosing continues to PND 4, and in males it continues for a total dosing period of at least 28 days. At least 10 animals per sex are assigned to control and treated groups, in an attempt to ensure at least 8 pregnant dams/group. Effects on fertility, pregnancy, parturition, maternal and suckling behavior, and growth and development of the offspring are noted. Parental clinical observations, body weight, and food consumption data are recorded. Live pups are counted, sexed, and weighed on PND 1 and 4. Pups are killed at PND 4, examined for gross

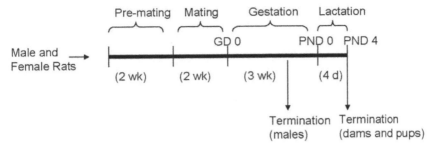

FIGURE 5 Developmental and reproductive toxicity screening studies (OPPTS 870.3550 and 870.3650; OECD TG 421 and 422).

abnormalities, and discarded without necropsy. Postmortem assessments in parental adults include gross necropsy observations, organ weights (testes and epididymides), and detailed histopathological evaluation of the ovaries, testes, and epididymides of control and high-dose animals, with the potential to evaluate the low- and mid-dose groups if a treatment-related effect is noted.

Developmental Neurotoxicity Testing

While the concept of DNT testing originated in scientific publications appearing in the early 1960s, the science and the study design have continued to develop over the past 40 to 50 years. The background scientific literature provides a strong basis for the development, implementation, and validation of the DNT guideline. There were a number of key scientific contributions to the development of the DNT guideline, including studies that evaluated physical, pharmaceutical, and environmental agents for their potential to affect the development and function of the nervous system after prenatal and early postnatal exposure (54).

The first regulatory protocol specifically designed to evaluate effects of chemical exposures on the developing nervous system was developed and implemented by the U.S. EPA in support of hazard evaluation for a group of solvents (55). The U.S. EPA DNT testing guideline was developed by the Office of Toxic Substances [now named the Office of Pollution Prevention and Toxics (OPPT)] and the OPP, and first proposed and published for public review and comment in the United States in 1986. The DNT guideline was finalized in 1991 (56). Then, in 1998, it was revised (57) and titled OPPTS 870.6300 as part of the comprehensive effort to harmonize EPA testing guidelines across OPPT and OPP, and also with OECD.

Following the recommendations of an OECD Working Group on Reproduction and Developmental Toxicity (58), the development of an OECD DNT test guideline was initiated in 1995. The 1991 U.S. EPA DNT guideline was used as the template for the study design, but the OECD guideline (TG 426) also addressed a number of important issues and incorporated improvements recommended through Expert Consultation Meetings held in 1996, 2000, and 2005. TG 426 was accepted and finalized by the OECD Council in 2007 (59). Because the EPA guideline was not revised since 1998, there are some differences between the EPA and OECD DNT study protocols. These are summarized in Table 2. In actual practice, EPA scientists generally recommend that DNT studies be conducted according to the general principles of the more progressive OECD guideline.

The DNT study design is illustrated in Figure 6. In a study conducted according to EPA OPPTS 870.6300 or OECD GL 426, pregnant rats (approximately 20/group) are

TABLE 2   Differences Between the U.S. EPA and OECD DNT Test Guidelines

	Guideline recommendations	
Test parameter	U.S. EPA	OECD
Dosing period	Gestation day 6 through PND 11 Note: PND 21 recommended[a]	Gestation day 6 through lactation (PND 21)
Functional observations	Specific days recommended (PND 4, 11, 21, 35, 45, and 60)	Specifies assessment weekly preweaning and biweekly postweaning
Minimum group size for pup behavioral assessments	10/sex/dose for most tests	20/sex/dose for most tests
Neuropathology assessment at early timepoint	PND 11, with immersion fixation Note: PND 21 accepted, with perfusion fixation[a]	Between PND 11 and 22, either perfusion or immersion fixation
Behavioral ontogeny[b]	Not discussed	At least two measures recommended
Motor activity	Specific days recommended (PND 13, 17, 21, and around 60)	Specifies assessment 1–3 times preweaning, once during adolescence (around PND 35), and once for young adults (PND 60–70)
Motor and sensory function	Auditory startle habituation specified	Quantitative sampling of sensory modalities and motor functions specified, auditory startle habituation listed as example
Neuropathology—number of animals	6/sex/dose specified Note: 10/sex/dose recommended[a]	10/sex/dose
Direct dosing to preweaning pups	Not discussed Note: Recommended in some situations[a]	Should be considered for some situations

[a]EPA practical testing recommendations reflect procedural enhancements that have been incorporated into the OECD guideline.
[b]Behavioral ontogeny is not specifically discussed in the EPA guideline; however, two measures are required in the OECD guideline, one of which can be an assessment of preweaning motor activity (an assessment that is included in the EPA guideline). Other examples cited in the OECD guideline are righting reflex and negative geotaxis.

administered the test substance, generally via an oral route, from gestation day 6 through to PND 10 (or 21) (57,59). Thus, the offspring, which are the primary subjects of this study, are potentially exposed to the chemical, through maternal circulation and/or milk, during in utero and early postnatal development for approximately 25 to 36 days. Maternal observations are limited. They consist of detailed clinical observations conducted outside of the home cage on approximately half of the dams in each group twice during gestation and twice during lactation, and weekly (at least) measures of maternal body weight. The offspring are assessed for evidence of alterations in functional development. On PND 4, litters are randomly standardized to yield four to five pups per sex per litter, and the pups are assigned for testing; alternative methods of assignment are described in OECD GL 426. Between birth and PND 60, endpoints assessed include measures of physical development, reflex ontogeny, motor activity, motor function, sensory function, and cognitive function (learning and memory). Cage-side observations are conducted daily, and additionally, 10 pups/sex/group are

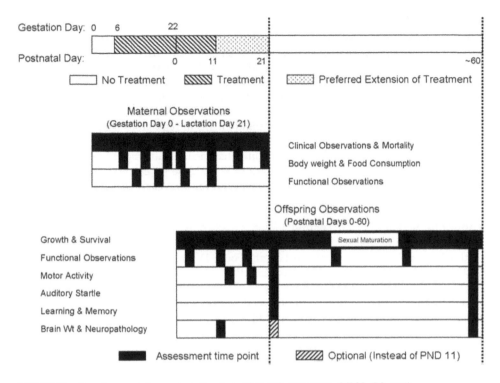

Gestation Day: 0   6   22

Postnatal Day: 0   11   21   ~60

☐ No Treatment   ▨ Treatment   ▦ Preferred Extension of Treatment

Maternal Observations
(Gestation Day 0 - Lactation Day 21)

Clinical Observations & Mortality

Body weight & Food Consumption

Functional Observations

Offspring Observations
(Postnatal Days 0-60)

Growth & Survival   Sexual Maturation

Functional Observations

Motor Activity

Auditory Startle

Learning & Memory

Brain Wt & Neuropathology

■ Assessment time point   ▨ Optional (Instead of PND 11)

FIGURE 6   Developmental neurotoxicity study (OPPTS 870.6300; OECD TG 426).

examined outside the home cage on days 4, 11, 21, 35, 45, and 60. Pups are counted and weighed individually at birth; on PNDs 4, 11, 17, 21; and at regular internals after weaning. Developmental landmarks are recorded; at a minimum they consist of the age of sexual maturation (vaginal opening in females or balanopreputial separation in males). Motor activity is evaluated at several early time points and at termination; the EPA guideline specifies the use of an automated activity-recording apparatus on PNDs 13, 17, 21, and 60 ($\pm$2), thereby facilitating an assessment of the ontogeny of motor activity in the offspring. Tests of auditory startle habituation (preferably using prepulse inhibition) and associative learning and memory are conducted in the offspring around the time of weaning (day 21) and around day 60. At an early time point (PND 11 or 21) and at study termination, cohorts of offspring are subjected to extensive neuropathological examination, which includes not only a qualitative assessment of neurological tissues but also simple morphometric analysis of specified areas of the brain. On PND 11 or 21, 1 pup per sex per litter is killed, and of these, 6 to 10 pups per sex per group are assigned to neuropathological evaluation. The brains of PND 11 pups are removed and immersed in an aldehyde fixative, while PND 12 pups are prepared with in situ perfusion of appropriate fixatives (paraformaldehyde and gluteraldehyde). At study termination, all remaining offspring are killed; 6 to 10 rats per sex per group are perfused in preparation for neuropathological evaluation. At both the early time point (PND 11 or 21) and at study termination (around PND 60), brain weight is recorded. Qualitative neuropathological examination of specified central and/or peripheral nervous system tissues is conducted for the control and high-dose groups. If a treatment-related finding is identified, the low- and mid-dose groups are also evaluated. The guideline provides guidance regarding the regions of the brain to be examined and the types of alterations on which to focus, particularly emphasizing structural changes indicative of developmental insult. Simple morphometric analysis, performed on off-spring killed at the early time point and at termination, is defined as consisting, at a minimum, of a reliable estimate of the thickness of major layers at representative locations within the neocortex, hippocampus, and cerebellum.

In the context of toxicological screening and testing conducted and evaluated in support of human health risk assessment and chemical regulatory activities, the DNT study provides information that is not addressed by other EPA or OECD test guidelines. Most importantly, it is the only test guideline that includes functional, behavioral, and anatomical evaluations of the nervous system at multiple time points, in subjects that were exposed to test substance during critical pre- and early postnatal periods of nervous system development.

Endocrine Screening and Testing
The EPA Endocrine Disruption Screening Program (EDSP) was established in response to the 1996 FQPA revisions of the Federal Food, Drug, and Cosmetic Act (FFDCA), Section 408(p). This regulation directed the EPA to "develop a screening program ... to determine whether certain substances may have an effect in humans that is similar to an effect produced by a naturally occurring estrogen, or such other endocrine effect as the Administrator may designate." The development of this screening and testing program has utilized substantive resources within the EPA Program Offices and research laboratories; has involved participation by stakeholders, contract research laboratories, nongovernment organizations, public interest groups, and academia; and has included interagency collaborations (e.g., with OECD). An Endocrine Disruptor Screening and Testing Advisory Committee (EDSTAC) was initially formed (i.e., in 1996) to provide advice to the EPA. Through this process, three fundamental recommendations were established for the development of the requisite testing assays. First was that the assays should expand the evaluation to include test systems designed to detect not only estrogen disrupting modes of action, but also those that target androgen and thyroid disruption, both directly and through the hypothalamic-pituitary-gonadal (HPG) axis and the hypothalamic-pituitary-thyroid (HPT) axis. Secondly, it was recommended to expand the target population to include not only humans but also animal wildlife. Finally, a two-tiered approach to testing was recommended. Conceptually, the Tier 1 tests would consist of a battery of assays designed to efficiently and effectively screen chemicals for interactions with the estrogen, androgen, or thyroid hormonal systems. If a weight-of-evidence evaluation of the results from the Tier 1 assays were to indicate a positive potential for interaction with these hormonal systems, then additional more comprehensive screening would be implemented in Tier 2 testing. On the basis of these recommendations, the EPA, in a joint collaborative effort with OECD, developed a list of Tier 1 and 2 endocrine assays (Table 3). Extensive validation efforts were applied to ensure that the study protocols will result in data that are sensitive and reliable in the detection of endocrine-related effects, predictive for the outcomes of concern, relevant to potential endocrine disruptive effects in humans and wildlife, and reproducible across multiple laboratories.

The Tier 1 assay approach, including study guidelines, was reviewed by a FIFRA Scientific Advisory Panel, and implementation was initiated in 2009 for a subset of environmental chemicals. The battery of in vitro and in vivo tests included in Tier 1 is designed to function as a complementary unit. Assays include those that will detect estrogen- and androgen-mediated effects by various modes of action, including receptor binding (agonist and antagonist) and transcriptional activation, steroidogenesis, and

TABLE 3   Endocrine Disruptor Screening Assays

Assay	Species	Description
Tier 1 testing assays		
Estrogen receptor Binding	In vitro	Binding assay using rat uterine cytosol
Estrogen receptor	In vitro	Transcriptional activation using human cell line (HeLa-9903)
Androgen receptor binding	In vitro	Binding assay using rat prostate cytosol
Steroidogenesis	In vitro	Human cell line (H295R) assay to detect interference with production of steroid sex hormones
Aromatase	In vitro	Human recombinant microsome assay to assess disruption of biosynthetic pathway that converts androgens into estrogens, estradiol, and estrone
Amphibian metamorphosis	Frog	Use of tadpoles to determine effects on thyroid and subsequent development during metamorphosis
Fish short-term reproduction	Fish	Screen for abnormalities associated with survival, reproductive behavior, secondary sex characteristics, histopathology, and fecundity
Uterotrophic	Rat	Estrogenic effects in ovariectomized or immature female rats
Hershberger	Rat	Androgenic and antiandrogenic effects in castrated or immature males
Pubertal male	Rat	Androgenic, antiandrogenic, and thyroid activity in males during sexual maturation
Pubertal female	Rat	Estrogenic and thyroid activity in females during sexual maturation
Tier 2 testing assays		
Amphibian development, reproduction	Frog	Adverse reproductive and developmental effects and dose-response characteristics
Avian 2-generation	Japanese quail	Adverse reproductive and developmental effects and dose-response characteristics
Fish life cycle	Fish	Adverse reproductive and developmental effects and dose-response characteristics
Invertebrate life cycle	Mysid shrimp	Adverse reproductive and developmental effects and dose-response characteristics
Mammalian 2-generation	Rat	Adverse reproductive and developmental effects and dose-response characteristics
Tier to be determined		
In utero through lactation	Rat	Use of pregnant rats to assess postnatal development of neonates after in utero and lactational exposure

Sources: Tier 1: http://www.epa.gov/scipoly/oscpendo/pubs/assayvalidation/tier1battery.htm; Tier 2: http://www.epa.gov/scipoly/oscpendo/pubs/assayvalidation/consider.htm.

HPG feedback. Additionally some rodent and amphibian in vivo assays included in the Tier 1 testing scheme are designed to detect direct and indirect effects on thyroid function through HPT feedback. The assessment of complementary modes of action in the Tier 1 assays is shown in Table 4. The Tier 1 rodent in vivo assays are described below.

Uterotrophic Assay

The uterotrophic assay (OPPTS 890.1600; OECD GL 440) is an in vivo short-term screening test used for the detection of estrogenic agonists (or antagonists) (60,61). It

TABLE 4   Complementary Modes of Action Evaluated in Tier 1 Endocrine Screening Assays

Screening assays	Receptor binding E	Anti-E	A	Anti-A	Steroidogenesis E	A	HPG axis	HPT axis
In vitro								
ER binding	x	x[a]						
ERα transcriptional activation	x							
AR binding			x	x				
Steroidogenesis H295R					x	x		
Aromatase recombinant					x			
In vivo								
Uterotrophic	x							
Hershberger			x	x		x		
Juvenile/peripubertal male			x	x		x	x	x
Juvenile/peripubertal female	x	x[a]			x		x	x
Amphibian metamorphosis								x
Fish short-term Reproduction	x	x[a]	x	x	x	x	x	

[a]Assays are expected to detect anti-estrogens although this outcome was not evaluated during the validation process.
Abbreviations: E, Estrogen; A, Androgen; HPG, Hypothalamic-Pituitary-Gonadal; HPT, Hypothalamic-Pituitary-Thyroidal.
Source: http://www.epa.gov/scipoly/oscpendo/pubs/assayvalidation/tier1battery.htm.

relies on an increase in uterine weight (i.e., the uterotrophic response), which is the result of water imbibition and cell growth that are a response to estrogenic substances, in an animal model without a functional hypothalamic-pituitary-ovarian axis (i.e., with a low level of endogenous circulating estrogen). There are two estrogen-sensitive states in the female rodent that meet this criterion: (*i*) immature females between the ages of weaning and puberty and (*ii*) young adult ovariectomized (ovx) females that have had sufficient time (a minimum of 14 days) for uterine tissues to regress. In the study, which is shown in Figure 7, either immature (e.g., immediately following early weaning at PND 18 in rats) or ovx female rats or mice of 8 to 10 weeks of age (at least 6/group) are

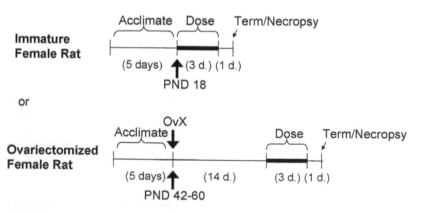

FIGURE 7   Uterotrophic assay (OPPTS 890.1600; OECD TG 440).

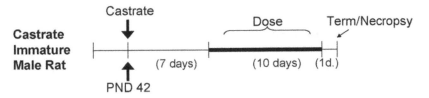

FIGURE 8 Hershberger assay (OPPTS 890.1400; OECD TG 441).

administered the test substance daily for three consecutive days by subcutaneous injection or by another route that is deemed more relevant to potential human exposure. Clinical observations, body weight, and food consumption data are recorded daily. The animals are killed 24 hours after the last treatment, and the uteri are carefully removed and weighed (wet and blotted weights). A significant increase in uterine weight, observed in treated animals as compared to control, is considered indicative of a positive estrogenic response to treatment. For this study, a reference response group, treated with 17α-ethinyl estradiol, is conducted by the performing laboratory to serve in the validation of laboratory procedures and to provide baseline positive control responses.

Hershberger Assay
The Hershberger assay (OPPTS 890.1400; OECD GL 441) is a short-term in vivo screening test that evaluates the ability of a chemical to result in biological responses that are consistent with androgen agonists, androgen antagonists, or 5α-reductase inhibitors (62,63). The bioassay (illustrated in Fig. 8) is reliant on treatment-related changes in weight of five androgen-dependent tissues in castrate peripubertal male rats, with minimal endogenous androgen production. Castration is conducted at approximately PND 42 (dependent on strain characteristics), and the animals (at least 6/group) are assigned to study after at least seven days of recovery. The rats are assigned to two or three treated groups, a negative control, and two reference positive control groups (i.e., testosterone propionate, an androgen agonist and flutamide, an androgen antagonist). The test substance is administered daily for 10 consecutive days by oral gavage or subcutaneous injection; testosterone propionate is administered by subcutaneous injection and flutamide by oral gavage according to the same schedule. Clinical observations, body weight, and food consumption data are recorded daily, and the animals are killed 24 hours after the last treatment. At necropsy, the following tissues are excised, trimmed, and weighed: (i) ventral prostate, (ii) seminal vesicle (plus fluids and coagulating glands), (iii) levator ani-bulbocavernosus muscle, (iv) paired Cowper's glands, and (v) glans penis. A statistically significant increase or decrease in the weights of two out of five tissues in treated groups as compared to control is considered indicative of a positive androgenic or antiandrogenic response, respectively.

Juvenile/Peripubertal Male Assay
The pubertal development and thyroid function assay in intact juvenile/peripubertal male rats (OPPTS 890.1500) is designed to detect chemicals with antithyroid, androgenic, or antiandrogenic (androgen receptor or steroid-enzyme-mediated) activity or agents that alter pubertal development via changes in gonadotropins, prolactin, or hypothalamic function (64). In this study, juvenile male rats (15/group; at least two treatment levels plus controls) are administered test substance daily by gavage from PND 23 to 53, as shown in Figure 9. Daily clinical observations and body weights are

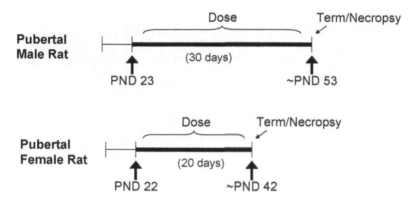

FIGURE 9 Juvenile/peripubertal male and female assays (OPPTS 890.1500 and OPPTS 890.1450, respectively).

recorded, and the age of balanopreputial separation is carefully determined for each subject. The animals are terminated at PND 53. At the time of sacrifice, blood samples are collected for standard clinical chemistry analyses (including creatinine and blood urea nitrogen), as well as for measurements of serum hormone levels, specifically total testosterone, total thyroxine (T4), and thyroid-stimulating hormone. Organs weights are recorded, including seminal vesicle plus coagulating gland (with and without fluid), ventral prostate, dorsolateral prostate, levator ani plus bulbocavernosus muscle complex, epididymides (left and right separately), testes (left and right separately), thyroid (postfixation), liver, kidneys (paired), adrenals (paired), and pituitary. Histopathology is conducted on one epididymis, one testis, the thyroid (colloid area and follicular cell height), and the kidney.

Juvenile/Peripubertal Female Assay
The pubertal development and thyroid function assay in intact juvenile/peripubertal female rats (OPPTS 890.1450) is capable of detecting chemicals with antithyroid, estrogenic, or antiestrogenic activity (including agents that act via alterations in receptor binding or steroidogenesis), or agents that alter pubertal development via changes in luteinizing hormone, follicle-stimulating hormone, prolactin, or growth hormone levels, or via alterations in hypothalamic function (65). This study is similar in design to the pubertal male assay described earlier (Fig. 9). The pubertal female assay can be conducted concurrently with the male assay, thus facilitating the efficient use of littermates of both sexes from the same litters that are used as a source for weanlings for assignment to testing groups. Juvenile female rats (15/group; at least two treatment levels plus controls) are administered the test substance daily by gavage from PND 22 to 42. Daily clinical observations and body weight data are recorded, and the rats are examined for vaginal opening, starting at PND 22. After the day of vaginal opening is determined, vaginal smears are taken daily. These data are used in evaluating the age at first estrus after vaginal opening, length of estrous cycle, percent of animal cycling, and percent of animals cycling regularly. At study termination on PND 42, blood samples are collected for standard clinical chemistry analyses (including creatinine and blood urea nitrogen), as well as for measurements of serum hormone levels, that is, total thyroxine (T4) and thyroid-stimulating hormone. Organ weights include uterus (blotted), ovaries (paired), thyroid (postfixation), liver, kidneys (paired), pituitary, and

adrenal (paired). Histopathological evaluation of the uterus, one ovary, the thyroid (colloid area and follicular cell height), and the kidney is conducted.

28-Day Repeated Dose, 90-Day Subchronic, Chronic, and Carcinogenicity Studies

Guidelines exist for a number of general toxicity studies of various durations, in rodent and nonrodent species, and by various routes of administration (Table 1). They include studies of 3 or 4 weeks duration, 90-day subchronic studies, chronic studies that can last anywhere from 4 months to 1-year duration, and carcinogenicity studies that approximate the lifetime of the rodent test models used. In these studies, clinical observations conducted at regular scheduled intervals throughout the duration of the in-life phase may identify abnormalities (e.g., lesions or masses) of reproductive organs or tissues. Likewise, at the time of necropsy, gross pathological findings in reproductive organs and organ weight data (when required) may direct further investigation of the reproductive system. Gross lesions identified at necropsy are generally fixed for subsequent histopathological examination, which may provide important information regarding reproductive system outcomes. Unique to such postmortem assessments, OECD GL 407 (the repeated dose 28-day oral toxicity study in rodents) includes a specific directive to consider reproductive tissues, including the male mammary gland, as potential indicators for endocrine-related effects (66).

Mutagenicity Studies

Among the many guideline protocols available to assess mutagenicity, several can be particularly relevant to the evaluation of developmental and reproductive toxicity or the assessment of life stage–related susceptibility, due to specific aspects of study conduct or endpoint assessment. These are the dominant lethal test, the mammalian spermatogonial chromosome aberration test, the mouse spot test, and the mouse heritable translocation assay. A brief description of each test follows.

Dominant lethal effects consist of alterations to the germinal tissue of a test species, which do not cause dysfunction of the gamete, but which are lethal to the fertilized egg or developing embryo. They are generally accepted to be the result of chromosomal aberrations, but may be due to gene mutations or other toxic effects. The primary demonstration of a dominant lethal effect is embryonic or fetal death. In the dominant lethal guideline study (OPPTS 870.5450; OECD GL 478), either rats or mice can be used as the test species (67,68). Male animals are assigned to control and treatment groups. A single oral or intraperitoneal dose of the test substance is administered to the males, and then they are mated to cohorts of untreated virgin females at sequential mating intervals that encompass the duration of germ cell maturation. The females are killed in the second half of pregnancy (e.g., at approximately GD 15); the ovaries are examined for number of corpora lutea, and the contents of the uteri are examined to determine the numbers of implantation sites, and live and dead embryos. A significant increase in pre- and/or postimplantation embryo loss might indicate a dominant lethal effect. This might be observed as an increased number of resorptions or decreased litter size in reproduction studies conducted with the same test substance.

The spermatogonial chromosome aberration test (OPPTS 870.5380; OECD GL 483) is expected to be predictive of the induction of heritable mutations in germ cells (69,70). In this study, male Chinese hamsters or mice are assigned to control and treatment groups (at least 5/group) and are administered the test substance, generally as a single or two treatments, by gavage or intraperitoneal injection. Samples are collected at approximately 1.5 cell cycle lengths after treatment. Approximately three to five hours (for mice) or four to five hours (for hamsters) prior to scheduled sacrifice, the animals are injected with a metaphase-arresting substance. Immediately after sacrifice, cell suspensions are obtained from the testes, fixed, spread on slides, and stained. At least 100 well-spread metaphases are analyzed per animal. Increased numbers of cells with chromosomal aberrations are indicative of a positive response.

The mouse spot test (OECD 484) is designed to detect presumed somatic mutations in fetal cells following transplacental absorption and exposure of developing embryos to the test substance (or its metabolites) (71). Melanoblasts in the developing embryos are the target cells of interest; a mutation of melanoblast target genes that control hair coat pigmentation will result in changed color spots in the coat of the offspring. In this study, T-strain mice are mated with HT, C57BL, or another appropriate strain. The pregnant mice are treated on GD 8, 9, and/or 10 by gavage or intraperitoneal injection. The resulting F1 offspring are coded and scored for spots at three to four weeks postpartum. The presence of randomly distributed pigmented and white spots on the coat is considered to be indicative of somatic mutation. Because of the need for the use of a specific, sometimes fragile, strain of mouse in this study, it is not conducted extensively.

The mouse heritable translocation study (OPPTS 870.5460; OECD 485) detects structural and numerical chromosome changes (reciprocal translocations) in germ cells of first generation progeny (72,73). The test substance is administered as a single dose or as repeated doses for 35 days. Translocation heterogeneity is identified either (i) through fertility testing of F1 offspring and subsequent cytogenetic verification of possible translocation carriers or (ii) by cytogenetic analysis of all male F1 offspring without prior selection by fertility testing. This study requires a large number of animals (e.g., 500 F1 males per dose level) to detect a treatment-related effect; therefore, its use is curtailed in practice. An increase in the number of translocations observed is considered a positive response to treatment. Male carriers of translocations and females with a single X chromosome loss demonstrate reduced fertility, evidence of which may also be observed in the two-generation reproduction study.

SUMMARY

Through collaborative programs and activities, issues of toxicology screening and testing, study performance requirements (guidelines), and scientific data interpretation are harmonized between EPA and OECD. Mutual acceptance of guideline toxicity data by EPA and OECD is mandated.

There are a number of testing guidelines that specifically address developmental and reproductive toxicity, including guidelines that have been specifically designed to screen for endocrine disrupting effects. These guidelines are readily available through the EPA and OECD websites and are utilized extensively in the assessment of environmental chemicals.

A weight-of-evidence evaluation of developmental and reproductive toxicity data is encouraged by both organizations. Thus, the consideration of information on reproductive organ toxicity from subacute, subchronic, and chronic/carcinogenicity studies as well as information from mutagenicity studies that may affect gametes or early life stages are also considered in the interpretation of developmental or reproductive consequences for an environmental chemical under assessment.

REFERENCES

1. Carson RL. Silent Spring. New York: Houghton Mifflin Co., 1962.
2. U.S. EPA. EPA Order 1110.2, December 4, 1970. Available at: http://www.epa.gov/history/org/origins/1110_2.htm
3. NRC (National Research Council). Risk Assessment in the Federal Government: Managing the Process. Committee on the Institutional Means for Assessment of Risks to Public Health, Commission on Life Sciences. Washington, DC: National Academy Press, 1983.

4. NRC (National Research Council). Science and Judgment in Risk Assessment. Committee on Risk Assessment of Hazardous Air Pollutants, Board on Environmental Studies and Toxicology, Commission on Life Sciences. Washington, DC: National Academies Press, 1994.

5. U.S. EPA. Guidelines for Developmental Toxicity Risk Assessment (EPA/600/FR-91/001). Fed Regist 1991; 56(234):63798–63826. Available at: http://cfpub.epa.gov/ncea/raf/recordisplay.cfm?deid=23162

6. U.S. EPA. Guidelines for Reproductive Toxicity Risk Assessment (EPA/630/R-96/009). Fed Regist 1996; 61(212):56274–56322. Available at: http://cfpub.epa.gov/ncea/raf/recordisplay.cfm?deid=2838

7. NRC (National Research Council). Pesticides in the Diets of Infants and Children. Washington, DC: National Academy Press, 1993.

8. U.S. 104th Congress. Food Quality Protection Act (FQPA) PL 104-170, 1996.

9. U.S. 104th Congress. Safe Drinking Water Act (SDWA) Amendments. PL 104-182, 1996.

10. Executive Order. Executive order 13045—protection of children from environmental health risks and safety risks. Fed Reg 1997; 62(78):19883–19888.

11. U.S. EPA. Policy on Evaluating Health Risks to Children. Office of the Administrator, Washington, DC, 1995.

12. U.S. EPA. The EPA Children's Environmental Health Yearbook. Office of Children's Health Protection. EPA 100-R-98-100, Washington, DC, 1998.

13. U.S. EPA. Strategy for Research on Environmental Risks to Children. Office of Research and Development, EPA/600/R-00/068, Washington, DC, 2000.

14. U.S. EPA. Determination of the Appropriate FQPA Safety Factor(s) for Use in the Tolerance-Setting Process. February 28, 2002. Office of Pesticide Programs; Office of Prevention, Pesticides, and Toxic Substances; U.S. Environmental Protection Agency, Washington, DC, 2002.

15. Brown RC, Barone S Jr., Kimmel CA. Children's health risk assessment: incorporating a lifestage approach into the risk assessment process. Birth Defects Res B Dev Reprod Toxicol 2008; 83:511–521.

16. U.S. EPA. A Framework for Assessing Health Risk of Environmental Exposures to Children. National Center for Environmental Assessment, EPA/600/R-05/093F, Washington, DC, 2006.

17. U.S. EPA. Supplemental Guidance for Assessing Susceptibility from Early-Life Exposure to Carcinogens. Risk Assessment Forum. EPA/630/R-03/003F, Washington, DC, 2005.

18. U.S. EPA. Child-Specific Exposure Factors Handbook (Final Report). EPA/600/R-06/096F, Washington, DC, 2008.

19. Selevan SG, Kimmel CA, Mendola, P. Identifying critical windows of exposure for children's health. Environ Health Perspect 2000; 108(S3):451–555.

20. Euling SY, Herman-Giddens ME, Lee PA, et al. Examination of US puberty-timing data from 1940 to 1994 for secular trends: panel findings. Pediatrics 2008; 121(S3):S172–S191.

21. Foos B, Sonawane B. Overview: workshop on children's inhalation dosimetry and health effects for risk assessment. J Toxicol Environ Health A 2008; 71(3):137–148.

22. WHO (World Health Organization). Principles for evaluating health risks in children associated with exposure to chemicals. International Programme on Chemical Safety (IPCS), Environmental Health Criteria document (EHC-237). Geneva, Switzerland, 2006.

23. Wolterink G, van Engelen JGM, van Raaij MTM. Guidance for risk assessment of chemicals for children. RIVM report 320012001/2007. RIVM National Institute for Public Health and the Environment, Bilthoven, Netherlands, 2007. Available at: http://www.rivm.nl/bibliotheek/rapporten/320012001.html

24. Cohen Hubal EA, Moya J, Selevan SG. A lifestage approach to assessing children's exposure Birth Defects Res B Dev Reprod Toxicol 2008; 83:522–529.

25. Makris S, Thompson CM, Euling SY, et al. A lifestage-specific approach to hazard and dose-response characterization for children's health risk assessment. Birth Defects Res B Dev Reprod Toxicol 2008; 83:530–546.

26. U.S. EPA. 40 CFR Parts 9 and 158. Pesticides; Data Requirements for Conventional Chemicals; Final Rule. October 26, 2007. Fed Regist 2007; 72(207):60934–60988.

27. OECD. OECD Series on Testing and Assessment, Number 43: Guidance Document on Mammalian Reproductive Toxicity Testing and Assessment. Organisation for Economic

Co-operation and Development. ENV/JM/MONO(2008)16, July-2008, Paris, 2008. Available at: http://www.oecd.org/document/30/0,3343,en_2649_34377_1916638_1_1_1_1,00.html

28. OECD. Questions & Answers regarding the OECD Test Guidelines Programme (TGP). May 2008. Available at: http://www.oecd.org/dataoecd/52/33/40728679.doc

29. OECD. OECD Guidelines for Testing of Chemicals. Full List of Test Guidelines, September 2009. Available at: http://www.oecd.org/dataoecd/8/11/42451771.pdf

30. OECD. OECD Series on Testing and Assessment, Number 34: Guidance Document on the Validation and International Acceptance of New or Updated Test Methods for Hazard Assessment. ENV/JM/MONO(2005)14, 18-Aug-2005, Paris, 2005. Available at: http://www.oecd.org/document/30/0,3343,en_2649_34377_1916638_1_1_1_1,00.html

31. Kimmel CA, Makris SL. Recent developments in regulatory requirements for developmental toxicology. Toxicol Lett 2000; 120:73–82.

32. EU (European Union). Regulation (EC) No 1907/2006 of the European Parliament and of the Council of 18 December 2006 concerning the Registration, Evaluation, Authorisation and Restriction of Chemicals (REACH), 2006. Available at: http://eur-lex.europa.eu/LexUriServ/LexUriServ.do?uri=CONSLEG:2006R1907:20071123:EN:pdf or: http://ec.europa.eu/environment/chemicals/reach/reach_intro.htm

33. Wilson, JG. The evolution of teratological testing. Teratology 1979; 20(2):205–211.

34. U.S. EPA. OPPTS 870.3700, Prenatal Developmental Toxicity Study, Health Effects Test Guidelines, EPA 712-C-00-367, Washington, DC, 1998. Available at: http://www.epa.gov/opptsfrs/home/guidelin.htm

35. OECD. Test Guideline 414. OECD Guideline for Testing of Chemicals. Prenatal developmental toxicity study. Organisation of Economic Co-operation and Development, Paris, France, 2001. Available at: http://miranda.sourceoecd.org/vl=5375569/cl=26/nw=1/rpsv/cw/vhosts/oecdjournals/1607310x/v1n4/contp1-1.htm

36. Makris SL, Solomon H, Clark R, et al. Terminology of developmental abnormalities in common laboratory mammals (Version 2). Birth Defects Res B Dev Reprod Toxicol 2009; 86:227–327.

37. U.S. EPA. OPPTS 870.3000, Reproduction and Fertility Effects, Health Effects Test Guidelines, EPA 712-C-98-208, Washington, DC, 1998. Available at: http://www.epa.gov/opptsfrs/home/guidelin.htm

38. OECD. Test Guideline 416. OECD Guideline for Testing of Chemicals. Two-generation reproduction toxicity study. Organisation of Economic Co-operation and Development, Paris, France, 2001. Available at: http://miranda.sourceoecd.org/vl=5375569/cl=26/nw=1/rpsv/cw/vhosts/oecdjournals/1607310x/v1n4/contp1-1.htm

39. OECD. Test Guideline 415. OECD Guideline for Testing of Chemicals. One-generation reproduction toxicity study. Organisation of Economic Co-operation and Development, Paris, France, 1983. Available at: http://miranda.sourceoecd.org/vl=5375569/cl=26/nw=1/rpsv/cw/vhosts/oecdjournals/1607310x/v1n4/contp1-1.htm

40. OECD. Draft extended one-generation reproductive toxicity test guideline. OECD guideline for the testing of chemicals. Draft version 28, October 2009. Paris, France, 2009a. Available at: http://www.oecd.org/document/55/0,3343,en_2649_34377_2349687_1_1_1_1,00.html

41. Cooper RL, Lamb JC, Barlow SM, et al. A tiered approach to life stages testing for agricultural chemical safety assessment. Crit Rev Toxicol 2006; 36:69–98.

42. Holsapple, MP. Developmental immunotoxicology and risk assessment: a workshop summary. Hum Exp Toxicol 2002; 21:473–478.

43. Luster MI, Dean JH, Germolec DR. Consensus workshop on methods to evaluate developmental immunotoxicity. Environ Health Perspect 2003; 111:579–583.

44. Holsapple MP, Burns-Naas LA, Hastings KL, et al. A proposed testing framework for developmental immunotoxicology (DIT). Toxicol Sci 2005; 83(1):18–24.

45. Burns-Naas LA, Hastings KL, Ladics GS, et al. What's so special about the developing immune system? Int J Toxicol 2008; 27(2):223–254.

46. Dietert RR, Piepenbrink MS. Perinatal immunotoxicity: why adult exposure assessment fails to predict risk. Environ Health Perspect 2006; 14(4):477–483.

47. Dietert RR, Dietert JM. Early-life immune insult and developmental immunotoxicity (DIT)-associated diseases: potential of herbal- and fungal-derived medicinals. Curr Med Chem 2007; 14:1075–1085.

48. Dorman DC, Allen SL, Byczkowski JZ, et al. Methods to identify and characterize developmental neurotoxicity for human health risk assessment: pharmacokinetics. Environ Health Perspect 2001; 109(suppl 1):93–100.
49. Barton HA, Pastoor TP, Baetcke K, et al. The acquisition and application of absorption, distribution, metabolism, and excretion (ADME) data in agricultural chemical safety assessments. Crit Rev Toxicol 2006; 36:9–35.
50. U.S. EPA. OPPTS 870.3550, Reproduction/Developmental Toxicity Screening Test, Health Effects Test Guidelines, EPA 712-C-00-368, Washington, DC, 2000. Available at: http://www.epa.gov/opptsfrs/home/guidelin.htm
51. OECD. Test Guideline 421. OECD Guideline for Testing of Chemicals. Reproduction/developmental toxicity screening test. Organisation of Economic Co-operation and Development, Paris, France, 1995. Available at: http://miranda.sourceoecd.org/vl=5375569/cl=26/nw=1/rpsv/cw/vhosts/oecdjournals/1607310x/v1n4/contp1-1.htm
52. US EPA. OPPTS 870.3650, Combined Repeated Dose Toxicity Study with the Reproduction/Developmental Toxicity Screening Test, Health Effects Test Guidelines, EPA 712-C-98-207, Washington, DC, 2000. Available at: http://www.epa.gov/opptsfrs/home/guidelin.htm
53. OECD. Test Guideline 422. OECD Guideline for Testing of Chemicals. Combined repeated dose toxicity study with the reproduction/developmental toxicity screening test. Organisation of Economic Co-operation and Development, Paris, France, 1996. Available at: http://miranda.sourceoecd.org/vl=5375569/cl=26/nw=1/rpsv/cw/vhosts/oecdjournals/1607310x/v1n4/contp1-1.htm
54. Makris SL, Raffaele K, Allen S, et al. A retrospective performance assessment of the developmental neurotoxicity study in support of OECD test guideline 426. Environ Health Perspect 2009; 117(1):17–25.
55. U.S. EPA. Triethylene glycol monomethyl, monoethyl and monobutyl ethers: proposed test rule, 1986:17883–17894.
56. U.S. EPA. Developmental Neurotoxicity Study, Series 83-6, Addendum 10 (Neurotoxicity), Subdivision F: Hazard Evaluation: Human and Domestic Animals. EPA Publication No. 540/09-91-123; NTIS Publication No. PB 91-154617. Washington, DC, 1991. U.S. EPA, Washington, DC.
57. U.S. EPA. OPPTS 870.6300, Developmental Neurotoxicity Study, Health Effects Test Guidelines, EPA 712-C-98-239, Washington, DC, 1998. Available at: http://www.epa.gov/opptsfrs/home/guidelin.htm
58. OECD. Draft Report of the OECD Ad Hoc Working Group on Reproduction and Developmental Toxicity. Copenhagen, Denmark, 13th-14th June 1995, 1995.
59. OECD. Test Guideline 426. OECD Guideline for Testing of Chemicals. Developmental Neurotoxicity Study. Organisation of Economic Co-operation and Development, Paris, France, 2007. Available at: http://miranda.sourceoecd.org/vl=5375569/cl=26/nw=1/rpsv/cw/vhosts/oecdjournals/1607310x/v1n4/contp1-1.htm
60. U.S. EPA. OPPTS 890.1600, Uterotrophic Assay, Endocrine Disruptor Screening Program Test Guidelines, EPA 740-C-09-010, Washington, DC, 2009. Available at: http://titania.sourceoecd.org/vl=926302/cl=20/nw=1/rpsv/cw/vhosts/oecdjournals/1607310x/v1n4/contp1-1.htm
61. OECD. Test Guideline 440. OECD Guideline for Testing of Chemicals. Uterotrophic Bioasay in rodents: A short-tem screening test for oestrogenic properties. Organisation of Economic Co-operation and Development, Paris, France, 2007. Available at: http://titania.sourceoecd.org/vl=26648990/cl=31/nw=1/rpsv/ij/oecdjournals/1607310x/v1n4/s34/p1
62. U.S. EPA. OPPTS 890.1400, Hershberger Bioassay, Endocrine Disruptor Screening Program Test Guidelines, EPA 740-C-09-008, Washington, DC, 2009. Available at: http://titania.sourceoecd.org/vl=926302/cl=20/nw=1/rpsv/cw/vhosts/oecdjournals/1607310x/v1n4/contp1-1.htm
63. OECD. Test Guideline 441. OECD Guideline for Testing of Chemicals. Hershberger Bioassay in Rats: A Short-term Screening Assay for (Anti)Androgenic Properties. OECD Guideline for the Testing of Chemicals. Organisation of Economic Co-operation and Development, Paris, France, 2009. Available at: http://titania.sourceoecd.org/vl=26648990/cl=31/nw=1/rpsv/ij/oecdjournals/1607310x/v1n4/s56/p1

64. U.S. EPA. OPPTS 890.1500, Pubertal Development and Thyroid Function in Intact Juvenile/Peripubertal Male Rats, Endocrine Disruptor Screening Program Test Guidelines, EPA 740-C-09-012, Washington, DC, 2009. Available at: http://titania.sourceoecd.org/vl=926302/cl=20/nw=1/rpsv/cw/vhosts/oecdjournals/1607310x/v1n4/contp1-1.htm

65. U.S. EPA. OPPTS 890.1450, Pubertal Development and Thyroid Function in Intact Juvenile/Peripubertal Female Rats, Endocrine Disruptor Screening Program Test Guidelines, EPA 740-C-09-009, Washington, DC, 2009. Available at: http://titania.sourceoecd.org/vl=926302/cl=20/nw=1/rpsv/cw/vhosts/oecdjournals/1607310x/v1n4/contp1-1.htm

66. OECD. Test Guideline 407. OECD Guideline for Testing of Chemicals. Repeated Dose 28-Day Oral Toxicity Study in Rodents. Organisation of Economic Co-operation and Development, Paris, France, 2005. Available at: http://puck.sourceoecd.org/vl=3183829/cl=38/nw=1/rpsv/ij/oecdjournals/1607310x/v1n4/s7/p1

67. U.S. EPA. OPPTS 870.5450, Rodent Dominant Lethal Assay, Health Effects Test Guidelines, EPA 712-C-98-227, Washington, DC, 1998. Available at: http://www.epa.gov/opptsfrs/home/guidelin.htm

68. OECD. Test Guideline 478. OECD Guideline for Testing of Chemicals. Genetic Toxicology: Rodent Dominant Lethal Test. OECD Guideline for the Testing of Chemicals. Organisation of Economic Co-operation and Development, Paris, France, 1984. Available at: http://lysander.sourceoecd.org/vl=1760478/cl=11/nw=1/rpsv/ij/oecdjournals/1607310x/v1n4/s43/p1

69. U.S. EPA. OPPTS 870.5380, Mammalian Spermatogonial Chromosome Aberation Test, Health Effects Test Guidelines, EPA 712-C-98-224, Washington, DC, 1998. Available at: http://www.epa.gov/opptsfrs/home/guidelin.htm

70. OECD. Test Guideline 483. OECD Guideline for Testing of Chemicals. Mammalian Spermatogonial Chromosome Aberation Test. OECD Guideline for the Testing of Chemicals. Organisation of Economic Co-operation and Development, Paris, France, 1997. Available at: http://oberon.sourceoecd.org/vl=2003112/cl=23/nw=1/rpsv/ij/oecdjournals/1607310x/v1n4/s48/p1

71. OECD. Test Guideline 484. OECD Guideline for Testing of Chemicals. Genetic Toxicology: Mouse Spot Test. OECD Guideline for the Testing of Chemicals. Organisation of Economic Co-operation and Development, Paris, France, 1986. Available at: http://oberon.sourceoecd.org/vl=2003112/cl=23/nw=1/rpsv/ij/oecdjournals/1607310x/v1n4/s49/p1

72. U.S. EPA. OPPTS 870.5460, Rodent Heritable Translocation Assays, Health Effects Test Guidelines, EPA 712-C-98-228, Washington, DC, 1998. Available at: http://www.epa.gov/opptsfrs/home/guidelin.htm

73. OECD. Test Guideline 485. OECD Guideline for Testing of Chemicals. Genetic Toxicology: Mouse Heritable Translocation Assay. Organisation of Economic Co-operation and Development, Paris, France, 1986. Available at: http://oberon.sourceoecd.org/vl=2003112/cl=23/nw=1/rpsv/ij/oecdjournals/1607310x/v1n4/s50/p1

Disclaimer: The views expressed in this chapter are those of the author and do not necessarily reflect the views or policies of the U.S. Environmental Protection Agency.

7 Reproductive study evolution and IND submissions for the Food and Drug Administration

Robert W. Kapp, Jr.

OVERVIEW OF FDA
Introduction

The Food and Drug Administration (FDA) is a federal agency that is part of the U.S. Department of Health and Human Services (DHHS). All food, drug, medical devices, and cosmetics legally available for animals and man in the United States are regulated under the authority provided by Congress in the Food, Drug, and Cosmetic Act of 1938 and its subsequent amendments. The act itself also established the regulatory agency that today is known as the FDA. As with many government regulations and regulatory agencies, the FDA and its regulatory authority were created in response to significant human health events. The regulatory powers granted to the FDA by Congressional Acts since 1902 vary considerably depending on the product, the potential for risk, and the evolution of the laws of the product being regulated. The FDA's authority has evolved over many years in response to developments in scientific research, human tragedy as a result of misuse of drugs and devices, and laws enacted by Congress generally in response to public outcry of suffering from either some new malady or a misguided attempt to cure some disease entity that sometimes can appear counterintuitive because of the political nature of creating and passing such legislation.

Over the past 100 years, the FDA has evolved into a critical arm of the federal government charged with protecting and promoting the nation's public health and is specifically responsible for safety regulation of foods, drugs, dietary supplements, vaccines, biological medical products, medical devices, blood products, radiation-emitting devices, cosmetics, veterinary products, sanitation regulations for interstate travel, and monitoring and control of disease on a variety of products. The degree of control over each category varies significantly. While dietary supplements are regulated as foods rather than drugs, there is generally no safety or efficacy testing or FDA approval requirements. Other items such as microwave ovens and X-ray machines have to meet performance standards as opposed to standards for effectiveness or safety. Cosmetics are regulated primarily on labeling and safety. However, the FDA regulates virtually every aspect of the process of new drug, biologics, and medical device development including testing, manufacturing, labeling, safety, advertising, efficacy, follow-up, and monitoring. These products must be proven safe and effective before they can be placed on the market.

To properly oversee the health and safety of its citizens and this vast array of responsibilities, the FDA has been provided a broad regulatory authority enacted over the years by the Congress. As various crises would develop, the Congress would eventually try to appease the public's opinion by passing laws to try to prevent or limit the reoccurrence of such calamities.

The FDA's regulatory oversight of drugs and biologics is based on the concept that human health risk can be estimated by examining animal data and extrapolating that data to humans and subsequently managing risk versus benefit (1). The FDA invokes a science-based risk management process with a primary mission of allowing for scientific progress yet attempting to minimizing risk to the public. Regulating such products is not risk-free; therefore, the FDA closely examines any risk-benefit analyses in an effort to make as prudent decisions for human health as possible. Toward that end, the FDA requires sponsors to provide properly designed studies from multiple species and monitors those aspects of the manufacture and importing of a substance that present the greatest public health risks by providing the consumer science-based information about the product and by closely monitoring safety post approval. Even though these and many other precautions are taken, the variables that can impact a developmental study are many indeed, including physiological and biochemical differences affecting all aspects of ADME (absorption, distribution, metabolism and excretion), differences in placental barriers among species, and differences in timing and sequencing of gestation. To further complicate matters, developmental effects are a spectrum of many different endpoints coalescing as an apparent single event that can be extremely difficult to decipher (1–5).

Although the FDA does not develop or test products itself, the agency reviews data from chemistry laboratories, animal laboratories, and from human clinical testing, which are presented to the FDA to assess effectiveness and safety by the sponsoring company. Besides assuring that the food supply is free from contaminants and disease-causing substances, the FDA also approves new food additives prior to their entry into the foodstuff. In addition, the FDA staff also inspects domestic and foreign manufacturers by checking selected shipments of imported products, and collecting and testing product samples for signs of contamination.

This chapter will concentrate primarily on new drugs that mandate the most rigorous requirements since the products have no previous human experience.

Operations and Budget

The FDA's 10,500 employees are assigned to one of eight FDA subdivisions to oversee the safety of much of the nation's consumer products (6,7). These divisions are as follows:

1. Office of the Commissioner (OC)
2. Center for Food Safety and Applied Nutrition (CFSAN)
3. Center for Drug Evaluation and Research (CDER)
4. Center for Biologics Evaluation and Research (CBER)
5. Center for Devices and Radiological Health (CDRH)
6. Center for Veterinary Medicine (CVM)
7. National Center for Toxicological Research (NCTR)
8. Office of Regulatory Affairs (ORA)

The CDER is charged with drug evaluations and has different requirements for the three main types of drug products: new drugs, generic drugs, and over-the-counter (OTC) drugs. CDER's approximately 1300 employees review applications for new and generic pharmaceuticals, manages U.S. cGMP (Current Good Manufacturing Practices) regulations for pharmaceutical manufacturing, determines which medications require prescriptions, monitors advertising of approved medications, and collects, monitors, and analyzes safety data about pharmaceuticals that are already on the market. A drug is considered a "new drug" if it is made by a different manufacturer, undergoes any substantial change from its original use, uses different excipients or is used for a different purpose.

The CBER is responsible for ensuring the safety and efficacy of biological therapeutic agents. CBER's approximately 1000-person staff reviews a variety of new entities including a number of items such as blood and blood products, allergens, cell and tissue-based products, gene therapy products, and vaccines. As with drugs that are managed by CDER, new biologics require a rigorous premarket approval process.

The FDA has regulatory authority over more than $1.5 trillion of consumer goods that includes about $400 billion in drugs, $200 billion in cosmetics, $100 billion in dietary supplements, and over $800 billion in food sales. The fiscal year (FY) 2008 budget was $2.1 billion while the FY 2009 budget request was $2.398 billion. Of the $2.398 billion, $1.77 billion was proposed from direct federal funding while $628 million was proposed from increased user fees. User fees were established under the Prescription Drug User Fee Act (PDUFA) of 1992 to help shorten the review time of drug and biologic applications by enhancing the funding for resources to speed the reviewers of applications (8). PDUFA authorized the FDA to collect fees from pharmaceutical companies that manufacture certain human drug and/or biological products. Although no fees are collected by the FDA with the submission of an Investigational New Drug (IND), a fees is collected when the marketing or New Drug Application (NDA) of a new drug or biologic is submitted to support the review process. In addition, the company also pays annual fees for each manufacturing establishment and for each prescription drug sold. The first fees were collected in FY 1995. For prescription drug applications containing clinical data, the fee was $208,000, the annual establishment fee for that same time period was $126,000, and the annual per product fee was set at $12,500. The prescription drug, annual establishment, and product fees for FY 2009 are now $1,247,200, $425,600, and $71,520, respectively (7,9,10). These fees are only a small fraction of the cost of performing studies and preparing documents for submission. Some argue that the process is simply too cumbersome and keeps some stellar drugs from ever being sold, while others argue that the process should methodically examine every facet of the drug candidate's potential for good as well as harm before it is allowed to be marketed. While both of these viewpoints are important, the FDA is charged with balancing the risk of not enough data versus stifling important drug development.

Legal Regulatory Authority

Change comes slowly to government as evidenced by the fact that a host of human tragedies occurred before measures were passed by Congress in an effort to protect the public from tainted foods and drugs that had no value or were toxic. In 1900, there were only a few federal laws passed to regulate drug and food sales in the United States. Beginning in 1883, chemist Harvey Wiley began a series of project evaluating adulteration and misbranding of food and drug products on the market. The U.S. FDA was formed in 1906 due, in part, to the public outcry to Upton Sinclair's novel titled "The Jungle" that sensationalized the problems of the meatpacking industry at the turn of the century. Although Sinclair originally intended to expose industrial labor and working conditions, food safety became the driving force of the publication. Because of the loss of public trust, the meat packers lobbied the federal government to pass legislation for enhancement of the inspection and certification of meat packaged in the United States (11). As a result of the lobbying, public pressure, and support by Harvey Wiley, the Meat Inspection Act and the Pure Food and Drug Act were passed. The Pure Food and Drug Act—also known as the "Wiley Act"—was signed into law by President Teddy Roosevelt, which gave authority to the Bureau of Chemistry of the U.S. Department of Agriculture (12). The Bureau of Chemistry eventually became the FDA, an agency of the DHHS.

In 1911, the Supreme Court ruled that the 1906 Act did not apply to false therapeutic claims—contrary to the Bureau's interpretation. An amendment to increase the Bureau's capacity to regulate false claims was added in 1912; however, it remained difficult for the Bureau to prove false claims in a court of law.

During the 1920s and 1930s, numerous questionable products appeared in the U.S. marketplace. The products created health tragedies that lead the way toward further regulations of drugs. Among the most notorious of these products was Banbar, a "cure" for diabetes. This elixir was concocted from horse-tail weeds and was completely nonfunctional, but many patients used Banbar in lieu of insulin and died as a direct result of not taking insulin. Lash-Lure was an eyelash dye containing synthetic aniline dye belonging to the paraphenylenediamine group that blinded some women. There are numerous examples of foods deceptively packaged or labeled; Radithor, a radium-containing tonic that claimed to enhance sexual process and cured mental illness and retardation and a host of other ailments, but, in reality ultimately exposed the users to a slow and painful death. Another such item was a "mechanical aid to deep breathing"—the Wilhide Exhaler, which falsely promised to cure tuberculosis and other pulmonary diseases (13).

In 1933, a bill was introduced into the Senate recommending a complete revision of the 1906 Food and Drugs Act for just the reasons noted in the previous paragraph. This bill launched a five-year legislative battle in the Congress. During the political debate, sometime in 1937, S.E. Massengill Company, a Tennessee drug company, marketed a form of a new sulfa "wonder" drug that was aimed primarily at children. It was a liquid form of sulfonamide called Elixir Sulfanilamide to which the chemists had added a solvent that had a sweet, raspberry flavor. Unfortunately, however, the solvent used in this elixir was diethylene glycol, a chemical analogue of antifreeze, which ultimately poisoned over 100 people, most of whom were children. This health disaster was the lynch pin that pushed the revision legislation through Congress. In the wake of this health disaster, President Franklin Delano Roosevelt signed into law the Food, Drug, and Cosmetic Act on June 25, 1938 (14). The critical drug provisions in this legislation included the following:

1. Requiring new drugs to be shown safe before marketing
2. Eliminating the requirement to prove intent to defraud in drug misbranding cases
3. Providing that safe tolerances be set for unavoidable poisonous substances
4. Authorizing factory inspections
5. Adding the remedy of court injunctions to the previous penalties of seizures and prosecutions
6. Extending control to therapeutic devices

Over the next few years, several administrative changes occurred including transferring of the FDA from the Department of Agriculture to the Federal Security Agency, with a Commissioner of Food and Drugs in 1940. The vast majority of the FDA's regulatory activities from the 1940s through the 1960s were directed toward abuse of amphetamines and barbiturates. The new legislation resulted in the submission of 13,000 new drug applications between 1938 and 1962 (15,16).

In 1944, the Public Health Service Act (17) consolidated and revised much of the preexisting legislation relating to the Public Health Service including regulation of biological products and control of communicable diseases. In 1950, a court of appeals ruling in Alberty Food Products Co. v. United States mandated that the directions for use on a drug label must include the purpose for which the drug is offered. This case assured that a worthless remedy could not escape the law by not stating the condition it is supposed to treat. With the passage of the 1938 Act, what drugs would be OTC and what drugs needed to be prescribed resulted in a debate that lasted up until 1951 when the Durham-Humphrey Amendment was passed. This amendment differentiated OTC drugs from prescription medications by restricting acquisition of prescription medications only through a licensed practitioner (18). The vast majority of the laws administered by FDA are found in the Food, Drug, and Cosmetic Act [Title 21, Chapter 9 of the U.S. Code (14)]. Table 1 displays the critical events that ultimately provide the FDA with authority to regulate food and drugs in the United States and other critical events in the evolution of the FDA.

TABLE 1   FDA Has Specific Regulatory Authority Through the Following Legislation, Litigation, and Rulings

1. 1902—Biologics Control (Virus)Act
2. 1906—Pure Food and Drug Act
3. 1911—United States v. Johnson—Supreme Court rules that FDA does not prohibit false therapeutic claims
4. 1912—False and Fraudulent Amendment to 1906 Act—Sherley Amendment
5. 1927—Bureau of Chemistry reorganized under the Food, Drug, and Insecticide Administration
6. 1930—Food, Drug, and Insecticide Bureau renamed the Food and Drug Administration
7. 1938—Federal Food, Drug, and Cosmetic Act (14)
8. 1944—Public Health Services Act (17)
9. 1949—FDA publishes guidance to industry for toxicity of Chemicals in Foods—document became known as the "Black book"
10. 1950—Alberty Food Products Co. v. U.S._ label must state the purpose for which the drug is offered
11. 1951—Food, Drug, and Cosmetics Act Amendments PL 82–215—Durham-Humphrey Amendment
12. 1958—FDA publishes first list of Generally Recognized as Safe (GRAS)
13. 1959—Published three generation protocol guidance for foods and drugs (19)
14. 1960—Federal Hazardous Substances Labeling Act (20)
15. 1962—Food, Drug, and Cosmetics Act Amendments PL 87–781—Kefauver-Harris Amendment
16. 1966—Fair Packaging and Labeling Act PL 890755
17. 1966—FDA Guidelines for 3 Segment Reproduction Studies (2,3)
18. 1967—Principles for the Testing of Drugs for Teratogenicity (21)
19. 1968—Drug Efficacy Study Implementation (DESI) to review drugs approved from 1938 through 1962 for effectiveness
20. 1970—Controlled Substance Act (PL 91–513, 84 Stat. 1236, enacted October 27, 1970, codified at 21 U.S.C. § 801 et. seq.)
21. 1970—FDA requires first package insert to accompany drugs
22. 1971—National Center for Toxicological research established in Pine Bluffs, AK
23. 1972—Regulation of Biologics including serums, vaccines, and blood products transferred to FDA from NIH
24. 1973—Supreme Court endorses FDA action to control classes of products by regulations rather than through litigation
25. 1976—Medical Device Regulation Act PL 94–295
26. 1979—FDA requires Pharmaceutical Pregnancy Risk Categories in selected drugs
27. 1981—Recommended Guidelines for Teratogenicity Studies in the Rat, Mouse, Hamster, or Rabbit (22)
28. 1982—FDA Redbook Guidelines
29. 1983—Federal Anti-Tampering Act—in response to Tylenol tampering deaths
30. 1984—Drug Price Competition and Patent Term Restoration Act of 1984—(Waxman-Hatch Act)
31. 1986—Interagency Regulatory Liaison Group Workshop on Reproductive Toxicity Risk Assessment (23)
32. 1987—Investigational drug regulations revised to expand access to experimental drugs to patients with life-threatening diseases with no alternatives
33. 1988—Prescription Drug Marketing Act
34. 1990—Nutrition Labeling and Education Act PL 101–535
35. 1990—Safe Medical Devices Act—required reporting of medical device malfunctions to FDA
36. 1991—Subcommittee on Risk Assessment (SRA)—Federal Coordinating Council on Science Engineering and Technology (FCCSET) formed Working Party on Reproductive Toxicology (WRPT) to set assumptions to be made when evaluating reproductive data (24)
37. 1991—Accelerated Drug Approval Regulations
38. 1992—Prescription Drug User Fee Act PL 102–571
39. 1993—Draft Guideline on the Detection of Toxicity in Reproduction for Medical Products—ICH-S5A (25)

(Continued)

TABLE 1   FDA Has Specific Regulatory Authority Through the Following Legislation, Litigation, and Rulings (Continued)

40. 1994—Final Guideline on the Detection of Toxicity in Reproduction for Medical Products—ICH-S5A (26)
41. 1994—Dietary Supplement Health and Education Act (DSHEA)
42. 1996—Addendum on Toxicity to Male Fertility—Guideline on the Detection of Toxicity in Reproduction for Medical Products—ICH-S5B (27)
43. 1996—Guidance for Industry—E6 Good Clinical Practice: Consolidated Guidance—ICH (28)
44. 1997—Guidance for Industry—S6 Preclinical Safety Evaluation of Biotechnology-Derived Pharmaceuticals—ICH-S6 (29)
45. 1997—Food and Drug Administration Modernization Act PL 105–115
46. 1998—Pediatric Rule requires manufacturers to test for safety and efficacy in children
47. 2000—Redbook—Guidelines for Reproduction Studies (30) and Guidelines for Developmental Studies (31)
48. 2001—Reviewer Guidance—Integration of Study Results to Assess Concerns About Human Reproductive and Developmental Toxicities (32)
49. 2001—CTD Format Guidances (33–36)
50. 2002—Bioterrorism Act PL 107–188
51. 2002—Office of Combination Products is formed at FDA
52. 2002—Medical Device User Fee and Modernization Act PL 107–250
53. 2003—Animal Drug User Fees Act PL 108–130
54. 2005—Drug Safety Board is formed to advise on drug safety issues
55. 2008—Animal Generic Drug User Fee Act PL 110–316 (37)
56. 2008—FDA issued proposed rule to Change Pharmaceutical Pregnancy Risk Categories (38)

Source: From Ref. 18, except where noted.

EVOLUTION OF DEVELOPMENTAL AND REPRODUCTIVE TOXICITY TESTING

In general, reproduction studies are designed to examine effects of a chemical on both male and female reproductive systems, the postnatal maturation, reproductive capacity of the offspring, and any possible cumulative effects over generations. The purpose of reproduction toxicity studies—in terms of the FDA—is to provide information for clinicians and patients regarding the potential of unintentional exposure to an embryo or fetus and to provide information and to provide information for the package insert (label) per 21 CFR 201.57 (f)(6) (39). Duration of exposure and disease population are critical factors in deciding whether or not a reproductive toxicology study is indicated. This would include exposure scenarios of women of child-bearing potential and where repeat drug administration is necessary. Generally, these studies are performed in rats or mice and rabbits are used as the second species in the developmental studies. What is important in the guidelines is that any potential for reproductive toxicity be revealed during these preclinical studies. The combination of studies selected for the thorough examination of reproductive toxicity should provide for exposure of mature male and female adults as well as all stages of development from conception to sexual maturity. Therefore, there must be continuous observations through at least one complete life cycle.

The historical perspective of DART testing development as well as the specifics of Segment I, II, and III study designs are described in considerable detail in chapter 5 and will not be further discussed here. It should be noted however, that much of the initial impetus for reproductive testing was determined by the World Health Organization Scientific Group on Principles for the Testing of Drugs for Teratogenicity, which met in

1966 in Geneva to evaluate safety testing in animals. The proceedings were published in 1967 (21) with the critical findings of the report summarized in the following:

1. It should be possible to greatly reduce the risk of teratological reaction in human pregnancy by striving to improve preclinical screening methods with attention given to the choice of appropriate species, time of testing, and effective dose levels
   a. The predictive reliability of teratological tests can be improved by utilizing basic mechanistic studies on teratogenesis.
   b. There is a need for internationally accepted criteria for testing for teratogenicity through multidisciplinary research.
   c. Strive for teratological studies in primates.
   d. Continued drug surveillance after approval should be performed since prediction of human teratogenicity is not well understood with animal tests alone.
   e. Epidemiological studies of malformation should complement drug-monitoring activities.
   f. Additional efforts should be made to inform the medical profession of teratogenic risks that may not be apparent in current testing methods.
   g. Careful assessment of the balance between teratogenic risk and therapeutic benefit in women during the reproductive years should be done.

Since their publication, the general consensus is that these conclusions remain valid with the possible exception of utilizing primates. Although reproduction toxicology studies should be conducted in pharmacologically relevant species, if the only relevant species is nonhuman primates (NHPs), there are many challenges and limitations. Among them include the fact that availability of NHPs are quite limited, the gestation period is five to seven times that of rats and rabbits, spontaneous abortion rates are high and they appear to be more heterogeneous compared with other lab animals. Although CDER has not provided any specific guidance on reproduction testing in NHPs, they do suggest an "embryo-fetal development study" that includes implantation toward the end of pregnancy; however, prenatal and postnatal development studies cannot be reasonably performed in NHPs (40). In addition, effects on fertility may be addressed by incorporating appropriate endpoints in an NHP repeated dose toxicity study. In general though, the costs of performing studies in NHPs, the lengthy gestational period, and the limited number of offspring per pregnancy minimize the amount of useful data that can be derived from these studies (5,21).

While the science of toxicology was only beginning to come into its own at the outset of this era, rapid advances in experimental assays for food additive and drug safety testing were made during this period by the FDA and other regulatory agencies. It was soon discovered, however, that the reproductive guidelines created by the U.S. agencies had inadvertently developed slightly different testing protocols. In an effort to correct these discrepancies, the FDA, the Consumer Product Safety Commission (CPSP), the Environmental Protection Agency (EPA), and the Occupational Safety and Health Administration (OSHA) formed the Interagency Regulatory Liaison Group (IRLG) in 1977 that resulted in the publication of the IRLG Testing Standards and Guidelines Work Group Recommended Guideline for Teratogenicity Studies in the Rat, Mouse, Hamster, or Rabbit (22).

The IRLG expanded this effort and sponsored a workshop organized by the IRLG Reproductive Toxicity Risk Assessment Task Group with an assignment to develop criteria to support consistent interagency interpretation of reproductive and teratogenic data for determining human reproductive risk. There were six workgroups that examined endpoints such as teratogenic endpoints, mechanisms of teratogenicity, endpoints of reproductive toxicity, mechanisms of reproductive toxicity, pharmacokinetics, and the risk of environmental agents (23).

Beginning in 1990, a group of regulatory authorities in the United States, Europe, and Japan met to form the International Conference on Harmonization of Technical requirements for Registration of Pharmaceuticals for Human Use (ICH) whose task was to "make recommendations on ways to achieve greater harmonization on the interpretation and application of technical guidelines and requirements for product registration in order to reduce or obviate the need to duplicate the testing carried out during the research and development of new medicines" (41). This was a concerted effort to obviate the need to conduct duplicate testing on new drugs, to understand each region's regulatory requirements, and decrease the need for unnecessary animal experimentation with the idea that harmonized guidelines would be issued (see chap. 5) (41).

In 1991, further examination of developmental risk assessment was published in the Federal Register (24) by the Subcommittee on Risk Assessment. The purpose of the Subcommittee on Risk Assessment (SRA) was to more closely coordinate the methods among the U.S. agencies in assessing health risk. The SRA formed the Working Party on Reproductive Toxicology (WPRT) to strengthen the coordination of reproductive risks. The WPRT developed assumptions that were to be utilized only when scientific data related to uncertainties were missing.

The ICH provided guidelines for reproductive studies that was published for comments in 1993 (5,42). The finalized version was published in the Federal Register in 1994 (26) and was endorsed by the United States, European Union (EU), and Japan. The ICH guidelines are covered in considerable detail elsewhere in this book, therefore only an overview is provided herein (see chap. 5). The ICH study designs specified dose selection methodology, the species (rat and rabbit being the two prominent species), and number of animals (litters = 16 to 20). For most drugs these three studies designs [(i) premating through postnatal development included fertility and early development; (ii) embryo-fetal development, and (iii) pre- and postnatal development)] are sufficient to assess reproduction and developmental effects. However, the guidance also indicates that there is no need to perform studies in a second species if the data show that a single species is sufficient or that a second species would provide limited value because of dissimilarities to humans. An additional guideline was published to provide clarification of the testing concepts with respect to flexibility, premating treatment duration, and observations.

The regulation of biologics was transferred from the National Institutes of Health to the FDA in 1972. Since then there have been several developments that involve DART studies with biologics. Among them was the publishing of the ICH S6 guideline (29) that described reproduction study designs that could be modified on the basis of the biologic's species specificity, biological activity, and immunogenicity. Another critical change in S6 was permitting the use of relevant species as well as the use of any extensive literature on a related material. In such cases where the ADME data indicate that similar effects were likely to be caused by related chemicals, the possibility of limited or no reproduction studies was now a legitimate pathway (43). Flexibility in study design has always been a critical factor in the ICH guidance documents.

Table 2 lists the general guidance documents and other critical publications that have been notable events in the development of the approach FDA has utilized since the outset of regulatory reproductive testing in the United States.

In addition to the above specific technical guidelines all studies should follow the following basic study design recommendations (30,31):

1. Be Good Laboratory Practice (GLP)-compliant (57)
2. Animals must receive standard animal care (58)

TABLE 2 Critical Events in the Evolution of FDA Reproduction and Developmental Testing Guidances

Item	Evaluation	Year	Reference
Three-generation protocol	Cumulative effects of foods	1949	44
Three-generation protocol found in "Appraisal of the safety of chemicals in foods, drugs and cosmetics"	Three-generation study guidance for foods and drug	1959	19
President's Science Advisory Committee Report on Use of Pesticides	Need for multigenerational studies in two species	1963	45
Goldenthal Letter	Guidance for 3-segment studies	1966	2 (See chap. 5)
WHO—International Developmental Guidance	Testing drugs for teratogenic effects—November 14–19, 1966	1967	21
Review of three-generation procedures	Three-generation study requirement in food petition submissions	1968	46
IRLG Recommended Guidelines for Teratogenicity Studies	Teratogenic guidance for rat, mouse, hamster, and rabbit	1981	22
FDA Redbook	General study guidelines for food additives and colorants	1982	18
Safety factors applied to NOELs	Applications to type I and type II characterization of study results	1983, 1988	5,47,48
One- vs. two-generation study effects	Determination that alone one-generation study is inadequate	1988	49
Subcommittee on Risk Assessment (SRA)—Federal Coordinating Council on Science Engineering and Technology (FCCSET)	Working Party on Reproductive Toxicology (WPRT) coordinated the methods and data in estimating reproductive health risk	1992	5,50
Subcommittee on Risk Assessment (SRA)—Federal Coordinating Council on Science Engineering and Technology (FCCSET)	Working Party on Reproductive Toxicology (WPRT) coordinated the methods and data in estimating reproductive health risk	1992	5,50
International Committee on Harmonization	Final guideline on the detection of toxicity in reproduction for medical products. ICH-S5A	1994	26
International Committee on Harmonization	Guideline on the detection of toxicity in reproduction for medical products—addendum on toxicity to male fertility—ICH-S5B	1996	27
International Committee on Harmonization	Guidance for Industry—E5 Good clinical practice: consolidated guidance	1996	28
International Committee on Harmonization	Preclinical safety evaluation of biotechnology-derived pharmaceuticals, ICH-S6	1997	29
International Committee on Harmonization	Addendum to guidelines on toxicity to male fertility	2000	51
FDA Redbook Revised	Guidelines for reproduction studies (IV.C.9.a)	2000	30
FDA Redbook Revised	Guidelines for developmental toxicity studies (IV.C.9.b)	2000	31
Excipient testing	Nonclinical studies for the safety evaluation of pharmaceutical excipients	2005	52
Vaccines	Considerations for developmental toxicity studies for preventive and therapeutic vaccines for infectious disease indications	2006	53
Pediatric drug products	Nonclinical safety evaluation of pediatric drug products	2006	54
Combination drugs	Nonclinical safety evaluation of drug and biological combinations	2006	55
Metabolites	Guidance for Industry—Safety testing of drug metabolites	2008	56

See chapter 5.

3. Utilize only untreated, healthy animals with nulliparous females
4. Animal group assignment by stratified randomization
5. Animals individually identified
6. Dose range–finding studies should be performed to set study dose levels

Over the years, the FDA's regulatory authority has continued to evolve in an effort to assure that products are safe as well as effective with respect to what the drug maker claims. The agency also must ensure adherence to the comprehensive Good Clinical Practice Standards as described in FDA regulations and in the International Testing Standards (ICH E6) (28).

REPRODUCTIVE STUDY DATA ANALYSIS
Preclinical Reports
The reports for each of the reproduction studies should contain basic information such as a copy of the protocol and amendments, absolute values for all parameters, individual pup data, summarized tables of litter data. Data should be calculated as incidence by litter—that is, number and percentage of litters by endpoint. The dose levels should be reported as mg/kg/day. The average number of pups surviving at various intervals should also be presented and analyzed. Any historical data that are used to further clarify study results must be properly identified for the reviewer (4).

A basic concept in evaluating preclinical studies is that one can estimate human risk by extrapolating from animal studies. One cannot be assured that the animal model will, in fact, be reflective of the human risk, therefore a well-designed animal study should provide an indication of the effects and risk to humans. Understandably, the factors that are examined in the interpretation of animal reproduction and developmental studies to the human experience are complex. Among the factors considered are as follows (1):

1. Effects on ADME
2. Differences in placental barrier physiology
3. Variable toxic mechanisms at cell, tissue, and organ levels
4. Effects of various disease states
5. Variability in gestational development

Drug Labeling
Because of these and other factors, the FDA requires several protocols on multiple species in an effort to cover as many possibilities as is practical to provide a sound basis on which to make safety decisions (59). As there is considerable uncertainty in the determination of human risk as it applies to various drugs, the FDA introduced a risk-ranking system (A, B, C, D, and X) in 1979 to classify drug products according to the potential risks posed to the fetus and the potential benefits posed to the pregnant woman (and fetus). The descriptions of categories A and B (and to some degree C) are based on increasing risk. The descriptions of categories D and X (and to some degree C) are based on a comparison of potential risk versus benefit (60,61). According to the statute, certain pregnancy information must be provided in the package insert precautionary section. To provide information on drug administration to providers, manufacturers have been required by law to designate the pregnancy categories in all reference materials and package inserts. The level of risk is determined through the use of "a matrix of human and animal reproduction toxicity data" (5). Through this matrix, five categories of risk have been identified as shown in Table 3.

TABLE 3   U.S. FDA Pharmaceutical Pregnancy Risk Categories

Pregnancy Category A	Adequate and well-controlled studies have failed to demonstrate a risk to the fetus in the first trimester of pregnancy (and there is no evidence of risk in later trimesters).
Pregnancy Category B	Animal reproduction studies have failed to demonstrate a risk to the fetus and there are no adequate and well-controlled studies in pregnant women. OR Animal studies have shown an adverse effect, but adequate and well-controlled studies in pregnant women have failed to demonstrate a risk to the fetus in any trimester.
Pregnancy Category C	Animal reproduction studies have shown an adverse effect on the fetus and there are no adequate and well-controlled studies in humans, but potential benefits may warrant use of the drug in pregnant women despite potential risks.
Pregnancy Category D	There is positive evidence of human fetal risk based on adverse reaction data from investigational or marketing experience or studies in humans, but potential benefits may warrant use of the drug in pregnant women despite potential risks.
Pregnancy Category X	Studies in animals or humans have demonstrated fetal abnormalities and/or there is positive evidence of human fetal risk based on adverse reaction data from investigational or marketing experience, and the risks involved in use of the drug in pregnant women clearly outweigh potential benefits.

Source: From Ref. 60.

The categories have been criticized as being too simplistic and misleading in some respects and have created some confusion. Concerns about the current categorical ranking system include the following:

1. The categories provide insufficient information for providers and patients seeking to make informed decisions about drug therapy because they do not distinguish fetal developmental toxicities according to severity, incidence, and type.
2. The current categories imply that the degree of risk increases with each letter, which is not an accurate description.
3. The current categories imply that the drugs within a given category pose an equal risk, which is not the case.
4. The criteria also permit drugs with known risks and drugs with no known risks to be placed in the same category, which is especially true in pregnancy category C.
5. The categories do not indicate whether risks are posed in degrees, according to the dosage and the duration of use of a drug, or gestational age at the time of exposure.
6. Information on the safety of drug use during lactation is not included in the current labeling format.

On May 29, 2008, the FDA issued a proposed rule that would address the issues with respect to pregnancy and lactation (62). This proposal replaces these five categories with pregnancy and lactation subsections containing a summary of the risks posed to a fetus and newborn infant. The intent is to provide more complete information for physicians and consumers on the possible effects of drugs used during pregnancy. Each new subsection would contain three principal components:

1. A risk summary
2. Clinical observations
3. A data section describing available animal and human data, including data acquired in postmarketing experience

The proposed rule also contains a discussion as to how risk is to be conveyed including five risk designations:

1. No predicted increase in risk
2. Low likelihood of increased risk

3. Moderate likelihood of increased risk
4. High likelihood of increased risk
5. Insufficient data

If implemented, the format and content of labeling of all drugs that were required to have a pregnancy category as described in the Physician Labeling Rule (63) will be affected. The timing of the implementation of the revised labeling will be phased in depending on the drug's approval date. The FDA has assumed that the proposed labeling change will become effective from June 30, 2010 (62).

Risk Assessment

The complexity of reproduction and developmental toxicity data can be overwhelming since there are so many variables. As previously noted, the IRLG began an effort to develop criteria to support consistent interagency interpretation of reproductive and teratogenic data for determining human reproductive risk (22). One of the workshops sponsored by the IRLG Reproductive Toxicity Risk Assessment Task Group was designed to develop consistency in interpretation of reproductive and teratogenic data. The findings from this effort reinforced the notion that extraneous factors, randomization of experimental units to treated groups, and adequate sample size to detect measurable and meaningful changes are all critical to the statistical power of the study. Teratogenic effects are complex events that manifest themselves as patterns of abnormalities rather than single easily measured defects (5). This produces considerable variation among affected fetuses that can be partially explained by the following (5,23):

1. Dose
2. Timing of exposure
3. Differences in maternal susceptibility
4. Differences in fetal susceptibility
5. Possible interactions with environmental factors

It has been shown that indicators of teratogenic potential include agents that are associated with the following:

1. Prenatal onset growth deficiency
2. Fetal wastage and/or infertility
3. Abnormalities in behavioral performance
4. Carcinogenicity
5. Mutagenicity

The next critical development in reproduction risk assessment occurred with the Subcommittee on Risk Assessment's WPRT report titled Guidelines for Developmental Toxicity Risk Assessment (24). The WPRT developed a set of assumptions that could be utilized when there was insufficient human data, and assumptions had to be made about human safety from animal data. As noted previously, the assumptions were to be applied as default in the absence of data, but should not exclude additional research. The basic assumptions were proposed in a draft FDA document in 1991 and then finalized in 1996 (24) and include the following:

1. For metabolism or toxicokinetics use data from the most sensitive species.
2. If there is an effect in animals, assume it will have a similar effect in humans.
3. Adverse effects on reproductive systems will be the same across mammalian species.

4. Developmental effects in animals will not necessarily be the same in humans.
5. The agencies assume that a threshold exists for the dose-response curve with respect to reproductive and developmental effects.

A chemical that produces an adverse reproductive effect in animals is assumed to be a potential reproductive toxin to humans based on comparative animal and human experimental data. Further, since there are similar mechanisms in many male and female mammals, reproductive processes are generally assumed to be similar across species. However, it is assumed that specific outcomes noted in experimental animal studies do not necessarily predict those seen in humans. This assumption is based on species-specific differences seen in timing of exposure relative to critical developmental periods, metabolism, placentations, and/or modes of action. However, the apparent dichotomy is, therefore, that not only do we assume that a favorable outcome in animal data might not be reflected in humans, we also assume that any adverse developmental effects noted in animal studies are predictive of adverse effects in humans.

In an effort to be more consistent and to provide a common framework in the interpretation of reproductive and developmental toxicity data, the FDA formed a Pregnancy Data Integration Working Group in the late 1990s (64). This group was formed to develop a method to judge the adequacy of preclinical reproductive toxicity data and to organize study findings for clarity. This objective also encompassed the integration of reproductive and developmental toxicity study data with all other available pharmacological and toxicological data and was to provide consistency in the approach and evaluation of the reproductive and developmental endpoints. The idea was to memorialize the evaluative process using all data—reproductive/developmental and all other data—in all possible combinations and still arrive at the same reproductive and developmental endpoint using the best scientific judgment.

The process begins by examining the following endpoints:
Reproductive toxicity endpoints

- Fertility and fecundity
- Parturition
- Lactation

Developmental toxicity endpoints

- Developmental mortality
- Dysmorphogenesis
- Alterations to growth
- Functional toxicity

These results of above endpoints were then placed into a data process that has been divided into three components:

1. Overall decision tree for evaluation of all endpoints noted above to decide which of the two decision processes below to follow.
2. Decision tree for reproductive developmental studies that were negative.
3. Integration flowchart for reproductive developmental toxicities that were positive.

The integration flowchart proceeds to review the level of concern for each positive signal finding by examining the following factors:

1. Signal strength with respect to cross-species concordance (if data are available), the presence of multiple effects, and/or the presence of adverse effects at different time points.

2. Signal strength with respect to the coexistence of maternal toxicity, the presence of a dose-response relationship, and the observation of rare events.
3. Pharmacodynamics with respect to the therapeutic index, biomarkers as a possible benchmark, and similarity between pharmacological and toxicological mechanisms.
4. Concordance between the test species and humans evaluated with respect to metabolic and drug distribution profiles, general toxicity profiles, and biomarker data.
5. Relative exposures—this factor should be emphasized when there appears to be a link between the reproductive or developmental effect and the biomarker.
6. Class alerts—this factor is based on prior human experience for a drug with related chemical structure or related pharmacological effects or with known reproductive or developmental outcomes in humans.

In evaluating the level of concern for each of the above noted factors, the analysis should reflect weight of evidence assessing the quality and type of available data. The arithmetic sum of each of the six factors is made to determine if

1. the drug is predicted to increase reproductive risk—net value of +3 or more,
2. the drug may increase risk—net value of (+)2 to (−)2,
3. the drug does not predict an increased reproductive risk—net value of $\leq (-)3$.

The complete data integration process is beyond the scope of this chapter and can be found in its entirety on the FDA website (32). Neither the comments on this document nor the document itself were finalized as of July 2009. Therefore, these are not definitive assessment processes; however, it appears that the basic concept will be adopted in the future.

INVESTIGATIONAL NEW DRUG SUBMISSION
Overview
The development of a drug is a stepwise process involving evaluation of safety information from both humans and animals and includes a variety of pharmacological and toxicological studies including reproductive and developmental studies as just one part of a very large safety characterization picture. As previously described, prior to the introduction of any such new chemical into humans, a series of testing protocols must be performed that are directed at characterizing the potential for risk. This is accomplished in various animal and laboratory studies that examine the toxic effects on target organs, dose dependence, relationships to exposure, and any potential for reversibility. There are substantial requirements and conditions for sponsors who wish to develop and market new drugs, including detailed laboratory and clinical testing procedures, extensive analytical and sampling procedures, and other intricate examinations that are quite involved.

At the outset of early drug development, the sponsor company's goal is to determine whether or not the compound in question has pharmacological activity that warrants commercialization of the product. Once a target medical condition or disease is identified, researchers examine how the product might act to prevent, cure, or treat that malady. This entails a variety of tests that attempt to establish that the product does, in fact, prevent, cure, or treat the ailment (efficacy), as well as collecting other data that would establish that the material poses no unreasonable risks to humans in early stage clinical studies (safety).

Therefore, the IND application is the result of a successful preclinical development program. The IND becomes the mechanism by which a sponsor formally and with government approval advances to the next stage of drug development known as clinical trials (human trials). The remainder of this section explores the overall IND process of which reproductive and developmental studies constitute a single but significant part. The activities do not lend themselves to discuss only the reproductive and developmental sections; therefore, the entire IND process will be outlined and described.

IND Process

To gain sufficient and scientifically based data on which the FDA can rely to make decisions about human trials, the agency mandates significant requirements and conditions in the development of drugs, including detailed animal testing procedures and chemistry thereby laying the foundation for ultimately securing FDA's permission to initiate human trials. The sponsor must develop evidence that the product may be used safely by humans. This phase of drug development is appropriately termed the "preclinical investigation" as it occurs before human testing is permitted. Preclinical investigations are an all-encompassing term that basically includes collecting laboratory and animal testing data that would provide insight as to its safety and effectiveness in humans. The FDA becomes formally involved in the drug development process at the point where the sponsor company has developed a drug that has shown the appropriate pharmaceutical activity and, in their view, has not shown significant toxicity at the intended dose level in preclinical animal testing. At that point, the sponsor submits an IND application to seek FDA approval to test the entity in small-scale studies in humans. Because of the evolution of the legal authority of the FDA, an IND is required before any clinical trials may be conducted in any of the following situations (65):

1. The drug is a new chemical
2. The drug is not approved for the sponsor's proposed use
3. The drug is in a new dose form
4. The drug is at a different dose level
5. The drug is in combination with another drug and the combination per se has not been approved

The IND is a formal submission to the FDA requesting approval to conduct a limited clinical study on a new drug. However, the Federal Food, Drug, and Cosmetic Act and its amendments also require that all drugs have an approved marketing application before the drug can be shipped in interstate commerce. Therefore, the IND is also a request to the FDA for an exemption from the act's requirement. The IND permits the sponsor to legally ship or import the new drug without the approved NDA (66).

Therefore, the primary goal of the IND submission, however, is to provide the FDA with the data necessary to either permit or deny the sponsoring company to proceed with a proposed clinical trial on an unapproved drug without subjecting the human subjects to unreasonable risk.

There are five types of INDs (66,67):

1. *Commercial IND*, which are applications that are submitted generally by pharmaceutical companies seeking marketing approval of a drug for sale to the public.
2. *An Investigator or Research IND*, which is reserved for a physician who is personally conducting an investigation of an unapproved drug or an approved drug in a new use or for a new patient population.
3. *Emergency Use IND* permits FDA to quickly authorize the use of an experimental drug in potential lifesaving situations where there has not been sufficient time to submit an IND in accordance with the regulations or where the patient does not meet the criteria of an existing protocol or where no approved study protocol exists.

4. *Treatment IND* is reserved for drugs show promise in clinical testing for serious or immediately life-threatening situations while the final phases of clinical testing are completed.
5. *Screening or Exploratory IND* permits exploratory studies to be conducted on closely related molecules to determine compound selection.

The Common Technical Document Format

The Common Technical Document or "CTD" is a set of specifications for an application dossier for the marketing registration of drugs, which was designed to be used across Europe, Japan, and the United States. It was developed by the European Medicines Agency (Europe), the Ministry of Health, Labour and Welfare (Japan), and the FDA (United States). It is maintained by the ICH. In the United States, use of the CTD format was initiated on a volunteer basis in 2001 for marketing applications, and at present is mandated for marketing applications for submissions to the United States, EU, Japan, Australia, Switzerland, and Canada.

The aim of the CTD is to create a general format for filing of global applications which created a single set of technical requirements for the registration of new drug products. The documents not only streamline the drug development process, but increase the consistency of the reviews (67). In addition, the CTD format can save resources, reduce costs, can prevent the omission of critical data or analyses that are necessary for the FDA to examine the data and readily permit the simplified exchange of technical data among regulatory authorities. Further, it is the basis for supporting electronic submissions. Using the CTD format for the IND permits the applicant to build a marketing application (NDA) from the outset so that when the NDA is prepared for submission, the general format is already in place. The only change in converting the IND requirements to the CTD format is the numbering within the sections. In the CTD document, pagination is on a section-by-section basis and not the entire submission.

The CTD is divided into five major sections or modules.

1. Administrative and prescribing information
2. Overview and summary of Modules 3 to 5
3. Quality (pharmaceutical documentation)
4. Safety (toxicology studies)
5. Efficacy (clinical studies)

The CTD format is outlined in guidance documents that are available on the FDA website. Module 1 is not part of the CTD because it contains authority-specific information for the receiving agency. In the United States, Module 1 contains all of the administrative documents necessary for a submission and how they should be identified. This module also should contain a comprehensive table of contents as well as an index for the submission package (33).

Module 2 contains a complete table of contents for Modules 2, 3, 4, and 5. In addition, this module should contain a one page general overview of the product such as its pharmacological classification, mode of action, and proposed clinical use. In addition, the module should contain a brief overview of the chemistry, manufacturing, and controls (CMC) process (which is provided in detail in Module 3). Module 2 also should contain a nonclinical overview that would include an interpretation of the data, the clinical relevance of the findings, and the implications for the safe use of the drug submission. This module also contains written and tabular summaries of the nonclinical data. Lastly, Module 2 should contain a brief clinical overview and discussion of the

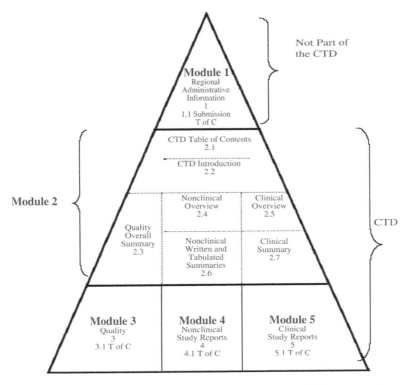

FIGURE 1 Diagrammatic representation of the Common Technical Document. *Source*: From Ref. 33.

clinical findings as well as pertinent animal data and factual summaries of the clinical information in the application.

Module 3 contains the detailed CMC information including its own table of contents and a detailed data on the drug substance and pertinent literature references.

Module 4 contains its own table of contents, the nonclinical study reports, and the pertinent literature references.

Module 5 also contains its own table of contents and a tabular listing of all clinical studies, clinical study reports, as well as pertinent literature references.

Modules 2 through 5 are harmonized and contain the technical information mandated by the authorities. M4 is the organization of the CTD.

Throughout the CTD, the display of information should be unambiguous and transparent. A schematic of the CTD modules is presented in Figure 1.

Content and Format of an IND Application

There are several iterations of IND submissions. The content and format of the initial submission, which requests testing of the drug product in humans for the first time, are critical. Once phase I is submitted, it is continually updated as new data become available and as the drug moves through to phases II and III. Because of the importance of the data on which decisions are made to permit testing in humans, the FDA requires not only sound science, but a particular format. Although one of the hall marks of FDA management of drug products is flexibility, maintaining a consistent format for FDA reviewers is also critical in an effort to expedite the product into clinical testing (68). In

TABLE 4 Phase I Study Content Summary

Type of information	Type of data	Purpose
Animal pharmacology and toxicology studies	Preclinical data + previous human experience, i.e., foreign use	Assessment of safe use for initial testing in humans
Manufacturing information	Composition, manufacturer, stability, monitoring control data	Evaluation that the product is consistent in content and reproducible
Clinical protocols	Detailed clinical protocols	Assessment of potential risk for initial phase human subjects
Investigator information	Qualifications of clinical investigators	Assessment of the individual medical doctors overseeing the clinical trials and their experience
Informed consent	Institutional Review Board (IRB) reviews, Consent forms	Assessment of the sponsor's adherence to the investigational new drug guidelines

Source: From Ref. 66.

general, phase I applications must contain information in the following basic areas as presented in Table 4.

All INDs submitted to the FDA are expected to have the same format with the following major sections (68). The relevant CTD modules for converting to the CTD format follow the CFR notation:

1. *Cover Documents* [21 CFR 312.23 (a)(1)] [CTD Module 1, Sections 1.2 1.1.1]
   This consists of a cover letter and FDA Form 1571 that identifies the drug by name, sponsor, the project manager, personnel responsible for monitoring the upcoming clinical testing and the review of the safety evaluation, phase of the submission, and appropriate signatures of responsible regulatory affairs persons.
2. *Table of Contents* [(21 CFR 312.23 (a)(2)]
   This is a comprehensive listing that provides the agency with an index of the necessary submission sections as well as the supporting documents.
3. *Introductory Statement* [21 CFR 312.23 (a)(3)] [CTD Module 2, Section 2.2]
   This is a two to three page summary to provide the FDA with a view toward the sponsor's developmental plan for the drug in order that the FDA may anticipate what it should expect from this drug submission. This statement includes the name of the drug, active ingredients, pharmacological class, structural formula, dose form, route of administration, and general objectives during the clinical investigations.
   A brief summary of previous clinical testing, any other INDs submitted on this chemical, and any relevant marketing, investigational, and/or safety data from foreign countries.
   Any potentially negative information from foreign submissions, especially with respect to safety concerns.
   A brief description of the drug investigation over the next year, including study rationale, studies to be run, general approach to drug evaluation, estimate of number of subjects to be examined, and any anticipated risks based on the animal toxicity data or previous human studies in this or related chemical studies.
4. *Reserved for future requirements* [21 CFR 312.23 (a)(4)]
5. *Investigator's Brochure* [21 CFR 312.23 (a)(5)] [CTD Module 4, Section 1.14.4.1]

This brochure includes the following:

- A brief description of the drug
- A summary of the pharmacological and toxicological effects in animals and humans, if known
- A summary of the pharmacokinetics and biological disposition of the drug in animals and humans, if known
- A summary of safety and effectiveness information in humans from previous clinical studies in the literature
- A description of potential risks and side effects on the basis of previous studies or studies in related chemicals

6. *Protocols* [21 CFR 312.23 (a)(6)] [CTD Module 5, Section 5.3]

Outlines of each planned phase I clinical study including numbers of individuals, controls, safety exclusions, dosing plan, observations and measurements to be made to fulfill the study objectives, and vital signs and blood chemistry measurements to be made relating to the safety of the participants. Phase I study protocols may be "less detailed and more flexible than the phase 2 and 3 studies" with a particular concern for safety and changes that might ensue for safety concerns. Detailed phase 2 and 3 protocols are submitted, which include

- Objectives
- Identification of responsible personnel and the research facility involved
- Criteria for subject selection and exclusion
- Type of study and type of control group employed and methods used to control bias
- Methodology for determination of dose, planned maximum dosage, and duration of subject exposures
- Observations and measurements to be made to fulfill the objectives of the study
- Description of the clinical procedures, laboratory tests, and any other measurement taken to monitor drug effects and minimize subject risk

7. *Chemistry, Manufacturing, and Control (CMC) Information* [21 CFR 312.23(a)(7)] [CTD Module 2, Section 2.3]

The CMC section describes the composition, manufacture, and control of the drug product. Emphasis in phase I submission is placed on identification and control of the raw materials with final specifications at the end of the investigational process.

The amount of CMC necessary depends on the scope and duration of the clinical investigation. The longer the study, the more CMC data are necessary.

As the scope of the investigation expands, the sponsor is expected to amend the initial CMC to reflect the expanded scope of the studies.

The following additional data are required for the *drug substance*:

- Description of the substance including identification of the manufacturer
- Physical, chemical, and biological characteristics
- Method of preparation and analytical methods to assure consistency during the toxicological and planned clinical studies

The following additional information is required for the *drug product*:

- List of all components that are in the product as well as those that are used in the manufacture process but do not appear in the final product
- A quantitative composition of the drug product with anticipated variations
- Identification of the manufacturer
- Description of the manufacturing and analytical methods to assure consistency during the planned clinical studies

[21 CFR 312.23(a)(7)(iv)(c)]—A description of the composition, manufacture and control of any placebo to be used in the clinical trial must be included.

[21 CFR 312.23(a)(7)(iv)(d)]—A copy of all labels and labeling must be provided to each investigator.

[21 CFR 312.23(a)(7)(iv)(e)]—An environmental assessment as outlined under 25.40 must be performed and submitted or the sponsor must claim a categorical exclusion under *21 CFR 25.30 or 21 CFR 25.31.*

8. *Pharmacology and Toxicology Disposition* [21 CFR 312.23(a)(8)] [CTD Module 2, Sections 2.4 and 2.6; CTD Module 4, Section 4.2]

This section presents the pharmacological and toxicological study data on which the sponsor has concluded that it is reasonably safe to conduct the proposed clinical studies. The nature, duration, and scope of the preclinical studies necessary is dependent on the type and duration of proposed clinical studies.

All data related to the ADME generated on this drug is presented in this section.

A summary of the toxicological data including acute, subacute, chronic, reproduction, developmental and special tests such as inhalation, or dermal toxicity that may have been preformed for this drug is presented in this section. All studies performed for submission to the FDA must be conducted under standard GLP compliance (or must be submitted with an explanation as to why they were not. The data from each of the toxicology studies is presented in a tabular form suitable for detailed review by the FDA, and can be found in guidance documents on the FDA website.

9. *Previous Human Experience with the Investigational New Drug* [21 CFR 312.23(a)(9)] [CTD Module 2, Sections 2.5 and 2.7; Module 5, Section 5.3]

Any previous animal data and human experience including any information about the drug if it had been part of a previously marketed drug either in the United States or in a foreign country, any clinical trial data, and/or studies published in the literature should be included in this section. Any previous animal and human experience with this drug in a combination product where the data relate to the safety of this drug should also be included in this section. Any information regarding the marketing of this drug outside the United States, including a list of countries, if any, where the drug has been withdrawn from the market for safety reasons, must be included in this section.

10. *Additional Information* [21 CFR 312.23(a)(10)] [CTD Module 2, Section 2.7.4]

Depending on the nature of the drug and the submission, additional information may also be required in this section as follows:
- Any potential drug dependence and/or abuse potential
- Sufficient data on radioactive drugs to permit safety dosimetry calculations
- Plans for pediatric assessment studies
- Adequate references to any previously submitted data needed for review

11. *Relevant Information* [21 CFR 312.23(a)(11)] [Modules 1–5, as applicable]

If requested by the FDA, any or all of the following items will need to be submitted with the IND:
- Information previously submitted need not be resubmitted; however, it must be identified by file name, reference number, accompanied by a written statement authorizing the references from the previous submitter.
- Reference material in a foreign language must be translated to English and copies of both the original and the translation become part of the submission package.
- The sponsor must submit an original and two copies of all submissions to the IND file, including not only the original submission but also all amendments and reports.

- The IND submission is required to be serially numbered. The first submission is required to be numbered 000 and each subsequent submission is to be numbered chronologically in sequence.
- If the investigation involves an exception from informed consent, the sponsor shall prominently identify on the cover sheet that the investigation is subject to the exception requirements as described in 21 CFR 50.24.

The IND application also contains a detailed section about the pharmacology and toxicology and any other studies performed by the sponsor to determine that it is safe to conduct small-scale studies in humans. In their IND submission, the sponsors can support their safety claims depending on whether or not the drug has had previous studies in different ways. If there has been widespread use of the drug outside the United States, the sponsor can

1. assemble existing nonclinical data from well-documented in vitro or in vivo laboratory studies,
2. assemble well-documented data from a submission to another country where there are similarities to the U.S. market, or
3. perform their own preclinical studies.

In any case these data must be sufficient for the sponsor to be able to evaluate the drug's toxic and pharmacological characteristics.

The above-noted CTD format can be used when submitting an IND. The changes that occur in that formatting are the section numbers. In the CTD document, pagination is on a section-by-section basis and not the entire submission.

Meeting with the FDA

Direct contact with the FDA scientists is an important part of the drug approval process. It is critical that there is an active forum for the sharing of information that can clarify the safety, efficacy, and product content of a new drug to the FDA. A successful meeting depends on sound science, understanding of the regulatory process, and following a meeting protocol (69).

There are three categories of meetings between FDA and IND submitters; each type of meeting is described in Table 5.

TABLE 5 Types of Meetings with the FDA

Type	Purpose	Notes	Schedule
Type A	Critical path meeting	Dispute resolutions such as clinical holds, special protocol assessment meeting after an FDA evaluation of protocols in assessment letter	Within 30 days of request. Briefing Package 14 days in advance of meeting
Type B	(*i*) Pre-IND meeting, (*ii*) end of phase 1, (*iii*) end of phase 2/pre-phase 3 meeting, (*iv*) pre-NDA/ BLA meeting	Requestor should try to anticipate future needs since FDA expects to grant only one type B meeting for each potential application. The FDA may grant exceptions to this depending on the circumstance	Within 60 days of request. Briefing Package 30 days in advance of meeting
Type C	Any other meeting than type A or B	For example, meetings regarding advertising and promotional labeling	Within 75 days of request. Briefing Package 2 wk in advance of meeting

Source: From Ref. 69.

Any meeting with the FDA is formal since their time is limited and usually does not exceed one hour in length. The meeting objective is the discussion of the drug and its application to human health—limited primarily to scientific and technical discussions. The meeting has no sales or financial discussion and should be limited to the science at hand with a very specific agenda with a closely limited timeframe. The sponsor should provide the FDA with thoughtful positions for discussions and not ask for guidance as to what should be done next (70).

Briefing (Information) Package

All meetings with the FDA must have a Briefing Packages submitted to the FDA either two or four weeks in advance depending on what type of meeting has been requested. The Briefing Package is critical to the success of the meeting. It should set the entire agenda for the meeting by providing the FDA with issues to be discussed and the sponsor's critical thinking on these topics. There must be adequate background information so that all the FDA staffers in attendance are aware of the issues at hand. To assist the sponsor and the FDA with their consideration of the product, the FDA suggests that the Briefing Package be fully paginated with a table of contents, indices, appendices, cross-references, and section tabs for ease of locating various topics. The FDA suggests that enough hard copies be provided for each FDA attendee plus an additional five hard copies for the FDA consultants. Although the contents of the Briefing Package will vary, the FDA has provided the following guidance for the contents and organization (69):

1. Product name and application number
2. Chemical name and structure
3. Proposed use
4. Dosage form, route of administration, and dosing regimen
5. A brief statement of the purpose of the meeting with some background and the data the sponsor intends to discuss at the meeting and any critical questions to be asked
6. A list of specific objectives and expected outcomes expected from the meeting
7. Questions grouped by discipline
8. Clinical data summary, if appropriate
9. Preclinical data summary, if appropriate
10. Chemistry, manufacturing, and controls information, if appropriate

REFERENCES

1. Frankos VH. FDA perspectives on the use of teratology data for human risk assessment. Fund Appl Toxicol 1985; 5:615–625.
2. Goldenthal EI. (Chief, Drug Review Branch, Division of Toxicological Evaluation, Bureau of Scientific Standards and Evaluation). Guidelines for Reproduction Studies for Safety Evaluation of Drugs for Human Use, March 1, 1966.
3. Goldenthal EI. Current views on safety evaluation of drugs. FDA papers 1968; (May):13–19.
4. Collins TFX, Sprando RL, Shackleford ME, et al. Food and drug administration proposed testing guidelines for reproduction studies—revision committee. Regul Toxicol Pharmacol 1999; 30:29–38.
5. Collins TFX. History and evolution of reproductive and developmental toxicology guidelines. Curr Pharm Des 2006; 12:1449–1465.
6. Schmit J. FDA receives modest boost in budget plan. Updated 2/17/2008. USA Today, AP, 2008. Available at: http://www.usatoday.com/money/industries/food/2008-02-04-fda-budget-food-safety_N.htm.
7. FDA. FDA Field Activities—Office of Regulatory Affairs. 2009. (organization/budget). Available at: http://www.fda.gov/downloads/AboutFDA/ReportsManualsForms/Reports/BudgetReports/UCM153559.pdf.

8. FDA. User fee rates archive—PDUFA. 2008. Available at: http://www.fda.gov/ForIndustry/UserFees/PrescriptionDrugUserFee/ucm152775.htm.
9. Food and Drug Administration Amendments Act (FDAAA) of 2007. Law strengthens FDA. 2007. Available at: http://www.fda.gov/oc/initiatives/advance/fdaaa.html.
10. FDA. President's FY 2009 budget advances food and medical product safety, and the safety of FDA-regulated imports. 2008. Available at: http://www.fda.gov/NewsEvents/Newsroom/PressAnnouncements/2008/ucm116850.htm.
11. Reed LW. Where's the Beef? Mackinaw Center for Public Policy Web site, August 11, 2005. Available at http://www.mackinac.org/article.aspx?ID=7229.
12. FDA. Federal Food and Drugs Act of 1906 (The "Wiley Act") Public Law No. 59-384 34 STAT. 768 (1906). Available at: http://www.fda.gov/RegulatoryInformation/Legislation/ucm148690.htm.
13. FDA. FDA history part II, USFDA. 1981. Available at: http://www.fda.gov/AboutFDA/WhatWeDo/History/Origin/ucm054826.htm.
14. Justia US Laws. US Chapter 9 Federal Food, Drug, and Cosmetic Act, June 5, 1938. 2008. Available at: http://law.justia.com/us/codes/title21/chapter9_.html.
15. FDA. Drugs and foods under the 1938 act and its amendments. 2008. Available at: http://www.fda.gov/AboutFDA/WhatWeDo/History/Origin/ucm055118.htm.
16. FDA. Summary of NDA approvals & receipts, 1938 to the present. 2009. Available at http://www.fda.gov/AboutFDA/WhatWeDo/History/ProductRegulation/SummaryofNDAApprovalsReceipts1938tothepresent/default.htm.
17. FDA. Public Health Services Act, US FDA. 1944. Available at: http://www.fda.gov/RegulatoryInformation/Legislation/ucm148717.htm.
18. FDA. Milestones in U.S food and drug law history. 2005. Available at: http://www.fda.gov/AboutFDA/WhatWeDo/History/Milestones/default.htm.
19. AFDO. Association of Food and Drug Officials of the United States. (Food and Drug Administration). Appraisal of the safety of chemicals in foods, drugs and cosmetics. U.S. FDA, Division of Pharmacology, 1959.
20. OSHA. Occupational Safety and Health Standards, U.S. Department of Labor. 2008. Available at: http://www.osha.gov/pls/oshaweb/owastand.display_standard_group?p_toc_level=1&p_part_number=1910.
21. WHO. WHO Scientific Group on principles for the testing of drugs for teratogenicity. Technical Report Series, No. 364, Geneva, November 14–19, 1966. 1967. Available at: http://whqlibdoc.who.int/trs/WHO_TRS_364.pdf.
22. IRLG (Interagency Regulatory Liaison Group). IRLG Recommended Guidelines for Teratogenicity Studies in the Rat, Mouse, Hamster or Rabbit. Washington, D.C.: IRLG Testing Standards and Guidelines Workshop, 1981.
23. IRLG (Interagency Regulatory Liaison Group). IRLG Workshop on Reproductive Toxicity Risk Assessment. Environ Health Perspect 1986; 66:193–221.
24. FDA. Draft guidelines for developmental toxicity risk assessment. Finalized in 1996. 1996. Available at: http://www.epa.gov/nceawww1/raf/pdfs/repro51.pdf.
25. FDA. Guidelines for reproductive toxicity assessment. Fed Regist 1996; 61:56274–56322.
26. FDA. International Conference on Harmonisation. Guideline on detection of toxicity to reproduction for medicinal products—ICH-S5A. 1994. Available at: http://www.fda.gov/downloads/RegulatoryInformation/Guidances/UCM129282.pdf.
27. FDA. International Conference on Harmonization. Guideline on detection of toxicity to reproduction for medical products: addendum on toxicity to male fertility—ICH S5B. 1996. Available at: http://www.fda.gov/downloads/Drugs/GuidanceComplianceRegulatoryInformation/Guidances/ucm074954.pdf.
28. FDA. Guidance for Industry—E6 good clinical practice: consolidated guidance. 1996. Available at: http://www.fda.gov/downloads/RegulatoryInformation/Guidances/UCM129515.pdf.
29. FDA. Guidance for Industry—S6 preclinical safety evaluation of biotechnology-derived pharmaceuticals. 1997. Available at: http://www.fda.gov/downloads/RegulatoryInformation/Guidances/ucm129171.pdf.
30. FDA. Guidelines for Reproduction Studies (IV.C.9.a). Toxicological Principles for the Safety Assessment of Food Ingredients Redbook 2000, USFDA. 2000. Available at: http://www.fda.gov/Food/GuidanceComplianceRegulatoryInformation/GuidanceDocuments/FoodIngredientsandPackaging/Redbook/ucm078311.htm

31. FDA. Redbook 2000: IV.C.9.b. Guidelines for Developmental Toxicity Studies. Toxicological Principles for the Safety Assessment of Food Ingredients Redbook 2000, USFDA. 2000. Available at: http://www.fda.gov/Food/GuidanceComplianceRegulatoryInformation/GuidanceDocuments/FoodIngredientsandPackaging/Redbook/ucm078399.htm.

32. FDA. Reviewer Guidance. Integration of study results to assess concerns about human reproductive and developmental toxicities. Rockville, MD., Center for Drug Evaluation and Research, October, 2001. 2001. Available at: http://www.fda.gov/downloads/Drugs/GuidanceComplianceRegulatoryInformation/Guidances/ucm079240.pdf.

33. FDA. Guidance for Industry—M4: Organization of the CTD. 2001. Available at: http://www.fda.gov/downloads/Drugs/GuidanceComplianceRegulatoryInformation/Guidances/ucm073257.pdf.

34. FDA. Guidance for Industry—M4E: The CTD—efficacy. 2001. Available at: http://www.fda.gov/downloads/Drugs/GuidanceComplianceRegulatoryInformation/Guidances/ucm073290.pdf.

35. FDA. Guidance for Industry—M4Q: The CTD—quality. 2001. Available at:http://www.fda.gov/downloads/Drugs/GuidanceComplianceRegulatoryInformation/Guidances/ucm073280.pdf.

36. FDA. Guidance for Industry—M4S: The CTD—safety. 2001. Available at: http://www.fda.gov/downloads/Drugs/GuidanceComplianceRegulatoryInformation/Guidances/ucm073299.pdf.

37. FDA. Important information from the FDA concerning the passage of the animal generic drug user fee act of 2008. 2008. Available at: http://www.fda.gov/ForIndustry/UserFees/AnimalGenericDrugUserFeeActAGDUFA/ucm045206.htm.

38. FDA. Summary of proposed rule on pregnancy and lactation labeling. 2008. Available at: http://www.fda.gov/Drugs/DevelopmentApprovalProcess/DevelopmentResources/Labeling/ucm093310.htm.

39. Justia US Laws. 21 CFR § 201.57 Specific requirements on content and format of labeling for human prescription drug and biological products described in § 201.56(b)(1). 2006. Available at: http://law.justia.com/us/cfr/title21/21-1.0.1.1.2.2.1.5.html.

40. Ghantous H. FDA regulatory perspective on reproductive and developmental testing for biopharmaceuticals. Presentation made at the Annual Meeting of the Society of Toxicology, Charlotte, NC, March 26–29, 2008.

41. Ohno Y. ICH guidelines—implementation of the 3Rs (refinement, reduction, and replacement): incorporating best scientific practices into the regulatory process. ILAR J 2002; 43(suppl):S95–S98.

42. ICH. Harmonized Tripartite Guideline. Detection of toxicity to reproduction for medicinal products, endorsed by the ICH Steering Committee at Step 4 of the ICH process on June 24, 1993.

43. Martin P. Reproductive toxicity testing for biopharmaceuticals. In:Cavagnaro JA, ed. Preclinical Safety Evaluation of Biopharmaceuticals—A Science-Based Approach to Facilitating Clinical Trials. Hoboken, NJ: John Wiley & Sons, 2008:357–377.

44. Lehman AJ, Laug EP, Woodard G, et al. Procedures for the appraisal of the toxicity of chemicals in foods. Food Drug Cosmet Law Quart 1949; 4(3):412–434.

45. President's Science Advisory Committee. Use of Pesticides. Washington, D.C.: Government Printing Office, 1963.

46. Fitzhugh OG. Reproductive tests. In: Boyland E, Goulding R. eds. Modern Trends in Toxicology. London: Butterworth, 1968:75–85.

47. Jackson BA. Bureau of Food Development Toxicity Guideline. A progress report presented at the Tripartite Meeting on January 12, 1983.

48. Jackson BA. Report presented at the 1988 Annual Winter Meeting, February 22–24, 1988, Bethesda, MD. The Toxicology Forum, Inc., 1988:207–216.

49. Frances EZ, Kimmel GL. Proceedings of the workshop on one-vs-two-generation reproductive effects studies. J Am Coll Toxicol 1988; 7:911–925.

50. Bromley DA. The President's Scientists: Reminiscences of a White House Science Advisor (The Silliman Memorial Lecture Series). New Haven, CT: Yale University Press, 2004.

51. ICH. ICH S5B—International Conference on Harmonization of Technical Requirements for Registration of Pharmaceuticals for Human Use]. ICH Harmonized Tripartite Guideline (S5B): Toxicity to Male Fertility. An addendum to the ICH tripartite guideline

on detection of toxicity to reproduction for medicinal products. Recommended for Adoption at Step 4 of the ICH Process on 29 November 1995 and amended on 8 November 2000 by the ICH Steering Committee, 2000.

52. FDA. Guidance for Industry—Nonclinical studies for the safety evaluation of pharmaceutical excipients. 2005. Available at: http://www.fda.gov/ohrms/dockets/98fr/2002d-0389-gdl0002.pdf.

53. FDA. Guidance for Industry—Considerations for developmental toxicity studies for preventive and therapeutic vaccines for infectious disease indications. 2006. Available at: http://www.fda.gov/downloads/BiologicsBloodVaccines/GuidanceCompliance RegulatoryInformation/Guidances/Vaccines/ucm092170.pdf.

54. FDA. Guidance for Industry—Nonclinical safety evaluation of pediatric drug products. 2006. Available at: http://www.fda.gov/downloads/Drugs/GuidanceCompliance RegulatoryInformation/Guidances/ucm079247.pdf.

55. FDA. Guidance for Industry—Nonclinical safety evaluation of drug and biological combinations. 2006. Available at: http://www.fda.gov/downloads/Drugs/Guidance ComplianceRegulatoryInformation/Guidances/ucm079243.pdf.

56. FDA. Guidance for Industry—Safety testing of drug metabolites. 2008. Available at: http://www.acciumbio.com/FDA-MIST2008.pdf.

57. FDA. Good Laboratory Practice Regulations. 21 CFR Part 58. 1978. Available at: http://www.accessdata.fda.gov/scripts/cdrh/cfdocs/cfCFR/CFRSearch.cfm?CFRPart=58.

58. ILAR. Institute of Laboratory Animal Resources—Guidelines for the Care and Use of Laboratory Animals. Washington, D.C.: National Academy Press, 1996.

59. FDA. Advisory committee on protocols for safety evaluation. Panel on reproduction studies in the safety evaluation of food additives and pesticide residues. Toxicol Appl Pharmacol 1970; 16:264–296.

60. FDA. Food and drug administration. United States FDA Pharmaceutical Pregnancy Categories. Fed Regist 1980; 44:37434–37467.

61. Manson JM. Testing of pharmaceutical agents for reproductive toxicity. In:Kimmel CA, Buelke-Sam J, eds. Developmental Toxicology. 2nd ed. New York: Raven Press, 1994:379–402.

62. Federal Register. Content and format of labeling for human prescription drug and biological products; requirements for pregnancy and lactation labeling. Fed Regist 2008; 73:30831–30868. May 29, 2008. Available at: http://edocket.access.gpo.gov/2008/pdf/E8-11806.pdf.

63. Lal R, Kremzner M. Introduction to the new prescription drug labeling by the food and drug administration. Am J Health Syst Pharm 2007; 64(23):2488–2494.

64. FDA. An Integrated Approach to the Evaluation of Non-Clinical Reproductive Toxicity Data. Presentation made by david Morse, July 28, 1999. Available at: http.//www.fda. gov/OHRMS/DOCKETS/dockets/99n2079/ts00002/index.htm.

65. Hamrell MR. What is an IND? In: Pisano DJ, Mantus DS, eds. FDA Regulatory Affairs— A Guide for Prescription Drugs, Medical Devices, and Biologics. 2nd ed. New York: Informa Healthcare, 2008:33–68.

66. FDA. Investigational New Drug (IND) Application. 2007. Available at: http://www. fda.gov/cder/Regulatory/applications/ind_page_1.htm.

67. Brown-Tuttle M. Chapter 22. In: IND Submissions: A Primer. Needham, MA: Cambridge Healthtech Institute, 2009.

68. FDA. Guidance for Industry—Content and format of investigational new drug applications (INDs) for phase I studies of drugs, including well-characterized, therapeutic, biotechnology-derived products. CDER, CBER, FDA. 1995. Available at: http://www. fda.gov/downloads/Drugs/GuidanceComplianceRegulatoryInformation/Guidances/ucm071597.pdf.

69. FDA. Guidance for Industry—Formal meetings with sponsors and applicants for PDUFA products. CDER, CBER, FDA. 2000. Available at: http://www.fda.gov/downloads/Drugs/GuidanceComplianceRegulatoryInformation/Guidances/UCM079744.pdf.

70. Grignolo A. Meeting with the FDA. In: Pisano DJ, Mantus DS, eds. FDA Regulatory Affairs—A Guide for Prescription Drugs, Medical Devices, and Biologics. 2nd ed. New York: Informa Healthcare, 2008:109–124.

8 Developmental immunotoxicology

Mark Collinge, Leigh Ann Burns-Naas, and Thomas T. Kawabata

ABSTRACT

The immune system is composed of a number of organs and cell types, and its development requires the orchestration of a complex series of events. There is concern that exposure to immunotoxic drugs and nondrug chemicals during pre- or postnatal periods of development may result in greater, unique and more persistent effects on the immune system in comparison to exposure as adults. This chapter will address the current landscape of developmental immunotoxicology (DIT), with respect to current regulatory guidance and triggers for developmental immunotoxicity testing, primarily from a pharmaceutical industry perspective. An overview of immune development is presented, focusing on processes common to all mammalian species, and key differences between the immune system of human and that of two major preclinical test species, the rat and monkey, are highlighted. The description of the morphological and physiologic development of the immune system provides a framework to identify processes that may be altered by immunotoxicants. Current methods available for DIT testing are discussed along with specific areas requiring further assay development. Compounds that have been determined to be developmental immunotoxicants are addressed, with a particular emphasis on those where the developing immune system appears more sensitive than that of the adult. Major gaps in our knowledge in the area of DIT are highlighted along with perspectives for the future of DIT.

INTRODUCTION AND BACKGROUND

The great concern and driver to assess DIT is that there may be both quantitative and/or qualitative differences in the response of the adult and developing immune systems to immunomodulatory drugs. The immature immune system may be more sensitive at lower doses, or may have delayed or irreversible recovery from immune modulation, relative to the adult. Given the ongoing processes of cellular differentiation and selection during immune ontogeny it is possible that immunomodulatory effects could be unique during development compared to the mature immune system.

The field of DIT has received increasing attention over the last 10 to 15 years. Significant progress has been made in identifying gaps in our current knowledge and state of the science, and attempts have been made to come to consensus on a number of key issues that should be addressed now and in the future. Numerous reviews have been published and highlight the scientific basis for potential concerns. Participation in recent workshops and round table discussions in the area of DIT have involved contributions from both the pharmaceutical and environmental/chemical sectors. However, this chapter will be written from a decidedly pharmaceutical perspective. In this regard, background will be provided, which emphasizes the regulatory guidelines that govern the assessment of therapeutic compounds that are immunomodulators, and situations or causes for concern that would trigger DIT assessments in the pharmaceutical industry. The current status of the field will then be considered, and immune system development in humans and animal species used in DIT studies will be discussed. Methods currently used for DIT assessment will be highlighted, along with an overview of compounds known to induce developmental immunotoxicity. Finally, a perspective on future directions based on gaps in the current state of the field will be provided.

Regulatory Guidance on DIT Testing

Before considering the current status of DIT as a science, it is important to put into perspective the regulatory environment that governs DIT in the pharmaceutical industry. Just as routine assessment of immunotoxicity in adult animals is not required during the development of therapeutic drugs, there are also no required routine assessments of DIT. There are no specific guidelines focused in the area of DIT testing. However, more general guidance can be obtained from guidelines on immunotoxicity testing and developmental toxicology. For pharmaceuticals, there are two general study designs to assess DIT: (*i*) administering test drug to the mother for prenatal (in utero) and postnatal (lactation) exposure (P&P study) and (*ii*) administering test drug directly to the developing animal (juvenile toxicity study). Currently, the practice is to add immunotoxicity endpoints to P&P or juvenile toxicity studies to assess DIT. The primary goal for the inclusion of immunotoxicity endpoints is to determine if the developing immune system is more sensitive, which components of the immune system are altered (e.g., macrophages, B-lymphocytes), and if the effects are reversible. The decision to include immunotoxicity endpoints and the type of endpoint to add is driven by a weight of evidence review of the data. For most situations, the primary driver is if the test compound is immunotoxic in adults. All compounds found to produce DIT in the published literature to date are known to be immunotoxic in adults. Guidance on this case-by-case decision process may be obtained from the principles described in the ICH S8 Note for Guidance on Immunotoxicity Studies for Human Pharmaceuticals (1). This guidance is focused on unintended immunosuppression and immunoenhancement of drugs and does not include allergenicity, drug-specific autoimmunity, or DIT testing. A weight of evidence review of potential causes for concern such as data from standard nonclinical toxicity studies (hematology, lymphoid organ weights, and histopathology), presence of the drug target on the cells of the immune system, and the target patient population are evaluated to determine if additional immunotoxicity testing is needed. The additional immunotoxicity tests may include immunophenotyping of lymphocyte subsets, the in vivo T-dependent antibody response (TDAR), and/or natural killer (NK) cell activity. It is recommended that the additional immunotoxicity testing be completed before large human populations are exposed in clinical trials.

For drugs intended to modulate the immune system, such as anti-inflammatory drugs or immunosuppressive drugs, the decision to conduct nonclinical immunotoxicity testing is more complex and is driven by numerous factors such as the indication, length of treatment and the immune targets. It has been argued that since the drug is designed to alter the immune response, why would nonclinical immunotoxicity testing be needed? This issue was recently addressed in an ILSI/HESI-sponsored workshop of nonclinical and clinical scientists from regulatory agencies, academia, and industry (2). There was consensus that immunomodulatory drugs may be evaluated for potential off-target or unexpected effects on the immune system. Given the wide variety of situations, the nonclinical immunotoxicity testing strategy should be determined on a case-by-case basis. Strategies around DIT were not discussed; however, some of the principles obtained from this workshop, along with the ICH S8 guidance, could be applied to DIT testing.

The ICH S8 document supersedes overlapping guidance from the 2002 FDA (3) and 2000 EMEA (4) immunotoxicology guidance documents. DIT testing was briefly discussed in the FDA's 2002 "Guidance for industry on immunotoxicology of investigational new drugs." It was noted that for new drugs there is a need to assess DIT if immunosuppression was observed in adult animals, and if the intended patient population included pregnant women. It is also recommended that methods to determine developmental immunotoxicity should be incorporated into a reproductive toxicity study. However, no recommendation was made concerning the appropriate studies to determine the effect of fetal and/or perinatal exposure on immune function.

General guidance for DIT is also found in pediatric nonclinical testing guidelines from the FDA (2006) (5) and EMEA (2008) (6). The FDA "Guidance on nonclinical safety evaluation of pediatric drug products" highlighted that developmental differences exist between the immune system of human neonates/infants and adults until the age of 12. Specifically, adult levels of IgG and IgA are not achieved until 5 and 12 years of age, respectively. In addition, the guidelines noted that data from adult studies may not always appropriately assess any possible drug effects on developmental processes specific to the pediatric age group, and that some therapeutics may have different safety profiles in pediatric and adult patients. It was also highlighted that the immune system is among those organ systems most highly at risk for drug toxicity in view of its significant postnatal development.

In the EMEA "Guideline on the need for nonclinical testing in juvenile animals of pharmaceuticals for pediatric indications" (2008), it is noted that standard nonclinical studies with adult animals or safety information from adult animals cannot always predict safety for all pediatric age groups, especially for various immature systems. The effects on the immune system were noted as one of these systems. DIT studies are only required if there is significant cause for concern for an effect on the developing immune system based on the chemical/pharmacological class of compounds or previous animal or human data. The ICH S8 guidance on immunotoxicity is referenced for the type of immune assays to use.

The aforementioned documents provide guidance on preclinical assessments that should be performed on a case-by-case basis to support pediatric and juvenile clinical trials for new drug entities. In addition, there is legislation in both Europe and the United States that requires pediatric studies for new drugs, indications, formulations, and routes of administration under certain circumstances. While these do not directly address issues of DIT, compliance with these directives and future interactions with the regulatory agencies are likely to lead to an increased need for nonclinical and clinical pediatric immunotoxicity assessments in the future.

In Europe, Regulation (EC) 1901/2006 of the European Parliament and of the council on medicinal products for pediatric use (7) came into effect in early 2007. This law requires pharmaceutical companies to conduct pediatric studies and to submit a Pediatric Investigation Plan (PIP) usually at the end of phase 1 trials. In the United States, the Pediatric Research Equity Act (PREA) of 2003 requires sponsors to conduct pediatric studies for certain drugs and biological products following submission of a proposed pediatric study request (PPSR) to the FDA Pediatric Review Committee. In either case, it is expected that any pediatric plan would adequately assess the safety and effectiveness of the drug for the claimed indications in all relevant pediatric subpopulations, and support the dosing and administration of each pediatric subpopulation for which the drug or biological product is safe and effective. Any nonclinical juvenile animal studies should be designed to support the clinical development plan and would most likely include a dose-range finding study prior to a definitive juvenile animal study. It is expected that these studies would be completed prior to long-term exposures in children. These studies could include DIT endpoints as deemed appropriate, and it is anticipated that these requirements will likely lead to an increased need for nonclinical pediatric/juvenile DIT testing.

In summary, guidance for immunotoxicity testing for immunosuppression and immunoenhancement using a weight of evidence review is provided by the ICH S8

TABLE 1 Guidances and Laws with Relevance to DIT of Pharmaceuticals

Year	Regulatory body	Guidance title	Comments
Guidelines			
2000	EMEA	Note for guidance on repeated dose toxicity	General toxicity testing guidance with specific recommendations in immunotoxicity sections. The immunotoxicity guidance was superseded by the ICH S8 guidance in 2006
2002	U.S. FDA	Guidance for industry. Immunotoxicology evaluation of investigational new drugs	Covers a wide range of areas in immunotoxicity testing including a few statements around DIT. Certain sections were superseded by the ICH S8 guidance
2006	ICH	Guidance for industry. S8 immunotoxicity studies for human pharmaceuticals	Focuses on immunosuppression and immunoenhancement of compounds not intended to alter the immune system. Did not discuss DIT but the general principles in identifying causes for concern and a weight of evidence review could be applied to DIT
2006	U.S. FDA	Guidance for industry. Nonclinical safety evaluation of pediatric drug products	General guidance for juvenile toxicity testing. Did not provide specific DIT guidance
2008	EMEA	Guideline on the need for nonclinical testing in juvenile animals of pharmaceuticals for pediatric indications	General guidance for juvenile toxicity testing. Did not provide specific DIT guidance. Noted that immunotoxicity testing in adults may not predict toxicity in the developing immune system
Laws			
2003	U.S. Federal Government	USA Pediatric Research Equity Act (PREA)	Requires sponsors to conduct pediatric studies following submission of a PPSR (proposed pediatric study request) to the FDA Pediatric Review Committee. Nonclinical juvenile toxicity studies are required to support clinical pediatric studies
2007	European Parliament and Council of the European Union	Regulation (EC) 1901/2006 of the European Parliament and of the council on medicinal products for pediatric use	Requires pharmaceutical companies to conduct pediatric studies and to submit a Pediatric Investigation Plan (PIP) usually at the end of phase 1 trials

guidance. General guidance is also provided by the FDA and EMEA pediatric guidelines. With immunomodulatory drugs, the workshop on nonclinical and clinical immunotoxicity strategies may also provide some guidance. A summary of the different guidelines that apply to DIT for pharmaceuticals are listed in Table 1.

CURRENT STATUS
Morphological and Physiological Development of the Immune System
The development of the mammalian immune system consists of a complex series of carefully timed and coordinated events. While the events and processes by which the immune system develops are essentially similar when compared across mammalian

species, the timing of these events clearly differs. Immune system development begins early in fetal life and continues throughout early postnatal development, and has been extensively described in a number of recent reviews (8–17).

The rodent has traditionally been the primary species for toxicology assessments in the pharmaceutical industry. However, relevant to immunotoxicology, the development of the immune system and attainment of immunocompetence is delayed in rodents relative to human. Neonatal rodents are not fully immunocompetent at birth, and immune system ontogeny continues into early postnatal life. However, like humans where the immune system is quite well developed by 13 weeks gestation, dogs and nonhuman primates (NHP) are born with functioning immune systems (with the exception of TDAR, discussed later in more detail). Nevertheless, the immune systems of humans, NHP, and dogs demonstrate developmental differences in utero. The ontogeny of the immune system is quite well defined in rodents, and less so in human and NHP. However, each involves a series of common processes, including cell production and migration into primary and secondary lymphoid organs, in addition to cellular differentiation and maturation within these organs (16,18–20). An extensive description of immune system development is beyond the scope of this chapter. However, a brief introduction is necessary to highlight the complexity of immune system development, which is important to understand where risks for DIT may lie, highlight key species differences that impact the choice of animal models for assessment of DIT, and identify gaps in our knowledge and challenges for future research.

Prenatal Immune System Development

Prenatal development of the immune system of human and mouse is outlined in Figure 1A and B, respectively. While the short gestation period of rodents is a factor in delayed immune system development at birth, relative to human and NHP, there are clear similarities in the overall sequence of developmental events between mammalian species. In humans, immune system development begins with hematopoietic stem cell (HSC) formation in the yolk sac around gestation weeks (GW) 2 to 6. In mice, uncommitted mesenchymal stem cells in the intraembryonic tissues surrounding the heart give rise to pluripotent HSC around gestation day (GD) 8. In either case, HSC migrate to the fetal liver where differentiation into various myeloid and lymphoid progenitor lineages occurs. These progenitor cells seed various tissues as well as the developing primary (thymus, bone marrow) and secondary [spleen, lymph nodes (LN)] immune organs where their development and differentiation are supported. The bone marrow and thymus are the primary sites of lymphopoiesis, and it is here that development of immunocompetent cells primarily occurs. Generally, all mammalian species follow this program of immune development. However, there are key species-specific developmental differences. Since the rodent has traditionally been the species of choice to assess immunotoxicity, the development of its immune system has been the subject of a number of recent reviews (9,16,21). The liver is the primary hematopoietic tissue throughout gestation in rodents and supports the formation of progressively more differentiated and lineage restricted stem cells. While hematopoeitic cells expand in the liver, few morphologically or functionally identifiable mature leukocytes are found in the embryo until birth. By comparison, during development of the human immune system (17,22), lymphocyte progenitors appear in the fetal liver seven to eight weeks post conception, and stem cells and T-cell progenitors move from the liver to the thymus nine weeks post conception. At GD 18 in rodents, shortly before parturition, the number of hematopoietic cells in the fetal liver declines and the bone marrow becomes the primary site for hematopoiesis, and it is here that expansion of lineage-restricted progenitor cells for leukocytes occurs. In contrast, in humans, bone

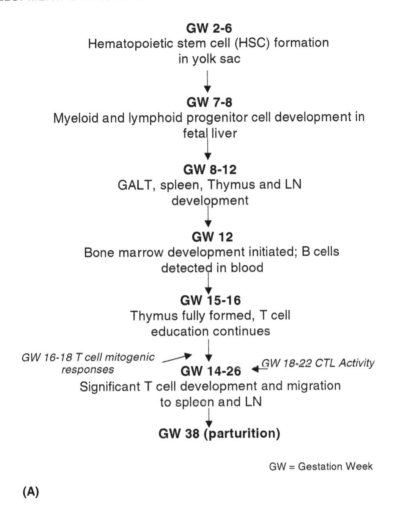

FIGURE 1 (A) Human fetal immune system development; (B) mice prenatal immune system development. (*Continued*)

marrow lymphopoiesis begins much earlier in development, at around GW 12 (17). In the rodent, fetal spleen myeloid and erythrocyte development continues into postnatal life. The thymus is detectable as early as GD 11 and is immediately colonized by HSC, and hematopoiesis in the developing thymus is largely limited to leukocyte production. Thymic selection leads to the deletion of both autoreactive and nonreactive cells, and perturbations here may manifest later in life in autoimmune conditions. T-lymphocytes are found in peripheral lymphoid organs soon after birth in rodents but in the human are detectable in the spleen as early as GW 14, and B-lymphocytes are apparent in the blood by GW 12.

Much remains to be understood about the development of both the rodent and human immune systems. However, it is clear even from the brief outline in the preceding text that significant differences exist, with immune ontogeny in the rodent being significantly delayed compared to that of human. While the dog represents

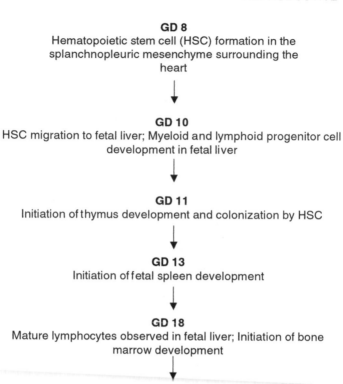

GD 8
Hematopoietic stem cell (HSC) formation in the
splanchnopleuric mesenchyme surrounding the
heart

GD 10
HSC migration to fetal liver; Myeloid and lymphoid progenitor cell
development in fetal liver

GD 11
Initiation of thymus development and colonization by HSC

GD 13
Initiation of fetal spleen development

GD 18
Mature lymphocytes observed in fetal liver; Initiation of bone
marrow development

GD 21 (parturition)
Low levels of T and B cells in spleen; red and white pulp
development not observed in the spleen; no germinal
centers

GD = Gestation

(B)

FIGURE 1   (*Continued*)

another species widely used in toxicology studies, relatively little is known about immune system development in this species (12). Given the recent increased focus on biologic therapeutics in the pharmaceutical industry in general, a species phylogenetically closer to human is likely to be a more appropriate species for some DIT studies, given the need for cross-reactivity of antibody and protein therapeutics. Immune system development in NHP has been the subject of a number of recent publications (14,23–27). As might be expected, there are many parallels in the morphological and functional development of the NHP and human immune systems. Both species are immune competent at birth, although both require immune maturation during postnatal development. While immune system development in these two species closely parallels each other, there may be subtle differences during both the pre- and postnatal periods that will only be revealed following more detailed comparative studies of the structure and function of immune system development in each species. This information can then be used in the translation of preclinical studies with NHP to aid in the design and interpretation of clinical pediatric/juvenile studies.

Postnatal Development and Acquisition of Functional Competence

The relatively short, 21-day gestation period in rats, compared to human, largely accounts for their comparatively immature immune system at birth. Rodent and human immune organ development primarily occurs in utero. However, organogenesis continues into postnatal life in the rodent (e.g., demarcation of splenic architecture), unlike human where it is essentially complete at birth. However, in both species immune system development does not cease at birth. The bone marrow continues to be a reserve of HSC, and a location where blood cell development and production of immune cells continue for the remainder of postnatal life. In addition, during early postnatal development, the thymus continues to be a site for production of immuno-competent cells. From both of these sites, cells migrate to the secondary immune organs, particularly the spleen, LN, and mucosal lymphoid tissues. In rodents thymic cell production wanes following sexual maturity with the virtual absence of thymocyte production at one year of life (16). In humans the adult thymus retains some capacity for cellular production, although the level of T-lymphocyte production is minor, with this role being transferred to the periphery (28,29).

Some of the key postnatal immune developmental landmarks in humans and rodents are compared in Figure 2. At birth, in rodents, relatively low B- and T-cell numbers are observed. Phenotypic analysis of 10-day-old rat pups indicates the lack of T- and B-lymphocyte subsets in peripheral blood (30) and the absence of splenic germinal centers. At weaning, the 21-day-old rat has comparable numbers of B cells to the adult but there are fewer relative numbers of T cells. In contrast, humans and NHP display a relatively well-developed immune system at birth. Splenic hematopoi-esis has already ceased at birth, and there are only relatively small differences in the numbers of T and B cells between newborns and adults. There are some changes in lymphocyte populations following birth in humans (17) where the proportion of lymphocytes represented by T cells is lowest at birth and increases over time. T cells present at birth have the phenotypic and functional properties of naive cells with fewer memory T cells present, a consequence of little intrauterine contact with environmental antigens, and B cells are generally immature.

Splenic architecture changes with development and highlights another difference between rodents and human. In rodents, the spleen demarcates into red and white pulp around postnatal day 6, but in human it has demarcated prior to birth, at around GW 26. Germinal centers are observed at postnatal weeks 6 to 8 in human spleens and postnatal days 21 to 28 in rodent spleens, respectively.

Although the adaptive immune response in humans and NHP is functional at birth, it has a limited repertoire. Maturation of functional immune responses continues

Process	Human	Rodent
Birth	GW 38	GD 22
Splenic Demarcation	GW 26 (splenic hematopoiesis ends @ birth)	PND 6
TIAR	TI-1 @ birth; TI-2 @ 2 years	TI-1 @ birth;TI-2 3-4 weeks
Splenic Germinal Centers	PNW 6-8	PND 21-28
Detectable TDAR	<1 year	PND 21
Maximal TDAR	1-2 years	6-8 weeks
Adult levels of IgG	4-6 years	2-4 months
Adult levels of IgM	1-2 years	3-5 months
Adult levels of IgA	12 years	4-6 months
Maturation of MZ and MZB cells	2 years	2-3 weeks

FIGURE 2 Postnatal immune system development.

to develop to the adult level, and the ability to generate immunologic memory is being established. Functional assessments show that innate immune function is demonstrable in rodents during the immediate postnatal stage. However, the adaptive immune system of the 10-day-old rat is too immature to elicit a primary immune response, and the magnitude of the TDAR is lower in 21-day-old rats than in young adults (6–8 weeks old). Both humans and rodents are capable of mounting responses to some T-independent (TI) antigens soon after birth, but there is a delay in developing immune responses to other TI antigens. TI antigens are classified as either type I (TI-1) or type 2 (TI-2), and responses to these antigens differ in their timing during development. Immune responses to TI-1 antigens can be mounted from birth, but humans have been shown to display a delayed immune response to TI-2 antigens. Significant neonatal infant mortality can be attributed to infections with encapsulated bacteria, including *Streptococcus pyogenes*, *Staphylococcus aureus*, *Neisseria meningitidis*, and *Haemophilus influenzae*. A major component of the capsules of these bacteria are polysaccharides, most of which are TI-2 antigens that poorly, if at all, induce immunologic memory (31,32), and neonates display an inadequate response to these types of antigens. This delayed response to TI-2 antigens is thought to be due in large part to delayed development of marginal zone (MZ) B cells. Polysaccharide antigens localize preferentially to the MZ of the spleen. The MZ becomes well demarcated from four months of age in human, but MZ B cells whose function is to respond to such antigens appear to be largely absent until at least two years of age. The B cell specific marker CD21, involved in binding of polysaccharide antigen/C3b complement complexes, is absent from the MZ for the first two years of life (33). CD27, a marker of memory B cells, also first becomes apparent in the MZ at two years of age. While CD27+ B cells may be present before age 2 they may be unable to migrate to the MZ where they can encounter the polysaccharide antigens and mount a functional response. Concomitant with the lack of MZ B cells, neonates are also deficient in the production of IgG_2, which is thought to be the most effective isotype against polysaccharide antigens. Similar delayed responses to TI antigens have been described in rodent systems, where responses to TI-2 antigens have been reported not to develop fully until three to four weeks of age (34). Equivalent studies in NHP have not been performed.

Potential Mechanisms for Immunotoxicity During Immune System Development

There are no specific examples to date of compounds that can be considered uniquely immunotoxic to the neonate since most also demonstrate immunotoxicity following adult exposure. However, there is some evidence (10,35) that the developing immune system can be disrupted by immunotoxicants at lower doses than those needed to produce a similar effect in the adult. This differential sensitivity could be manifested in a number of ways (8). There may be a qualitative difference, such that there are specific effects on the developing immune system that are not observed in the adult. Differences may be quantitative so that drug perturbations affect the developing immune system at lower doses than in the adult. Finally, there may be temporal differences. Toxicant exposures that might produce only transient immune effects in adults have the potential to result in persistent immunotoxicity following early life exposures.

It has been suggested that immune system development is best viewed as a continuum of alterations (36) and is a sequence of exquisitely timed and coordinated events that begin early in prenatal life and continue through the early postnatal period (9). It has been proposed that there may be critical windows of development in the immune system of greater vulnerability (37), which may differ substantially from those observed in the mature immune system. However, there are very little data with developmental immunotoxic chemicals in which alteration to the immune system can be attributed to unique effects during a specific critical window. It should not be assumed that the earliest events in development are those most susceptible to immunotoxicant exposure. Each developmental window is likely to have unique risks and outcomes associated with it.

There are a number of developmental processes that have helped to define the concept of "critical windows of development," where the relative risk associated with immunotoxicant exposure is suspected to be different. Five discrete windows of immune maturation have been proposed on the basis of landmark events in immune system development, three prebirth and two postbirth to sexual maturity, and these have been recently defined and described in recent reviews and symposia (8,9,17,19,37,38). In addition to defining such windows, these have been compared across species including rodents, canines, NHP, and humans (8,10,15,16,39,40). These windows highlight key processes of cell development and differentiation that may contribute to this differential immunotoxic sensitivity. There are cell types undergoing development and proliferation in the developing animal that are not present in the adult and offer targets for unique sensitivity. (1) The first of these windows encompasses the timing of HSC formation from undifferentiated mesenchymal cells. During this process disruption of the immune system could lead to aberrant HSC formation, abnormalities in the production of hematopoietic lineages, abnormal hematopoiesis, and partial or complete immune failure (2). HSC migrate to fetal tissues such as the liver and thymus to undergo early hematopoiesis, and macrophages migrate to tissues. Immune disruption here could lead to thymic atrophy, impaired postnatal T-cell function, impaired innate immune function, and inflammation in organs where macrophages play a key role in development (3). The bone marrow is established as the primary organ of hematopoiesis, and the bone marrow and thymus as primary sites for lymphopoiesis of B and T cells, respectively. The proposed impact of perturbations during this period would include increased risk of cancer later in life, autoimmunity, and allergy (4). During the perinatal period the functional development and maturation of immunocompetence is achieved and is a period where the Th1/Th2 bias of the immune system is shifted. Disruption at this point could lead to increased likelihood of premature termination of the pregnancy, increased childhood viral infections, and reduced responsiveness to vaccination (5). Perturbation of mature immune responses and establishment of immunologic memory may lead to increased risk of infections and cancer, allergy, and atopy.

Exposure and Dosing Paradigms for DIT

A number of different approaches to include immunotoxicology endpoints within DIT studies have been proposed (19,37,41,42). However, before discussing specific methodologies and endpoints for assessing DIT, it is important to consider key issues relating to dosing regimens for such studies. A number of workshops and recent reviews (8,9,19,36,38,43) have discussed this topic and only a brief overview is provided here. For pharmaceuticals, the exposure period of the DIT study design should reflect the specific question that one is asking. For example, if the test drug will be used in pediatric populations, drug exposure and testing of animals should only be conducted during that period of development with a similar immunological development as the intended human population. In contrast, if one is interested in evaluating immunotoxicity with in utero and postnatal lactation exposure, the study should focus on these periods. One of the key issues to consider when designing a dosing scheme, particularly in rodents, is that immune system development in rodents is delayed compared to that of human. Given this knowledge, the majority of rodent DIT publications to date have exposed dams to compounds during both gestation and lactation, and the offspring evaluated later as young adults, typically beyond PND 42, where functional immune assays can be performed. While there are some studies involving neonatal exposures to immunotoxicants, very few juvenile toxicity studies with DIT endpoints have been published.

While there are clear differences in the timing of the development of rodent and human immune systems, development in NHP and humans are thought to largely parallel one another. Both are born with a competent immune system, although as described earlier, immune system maturation is still required in the neonate. While the relative postnatal development of the immune system is believed to be comparable between human and NHP (23), there are little data comparing these two species following parturition. This is important since a determination may need to be made to identify the appropriate age of NHP to use when performing preclinical studies to support pediatric/juvenile drug studies. For example, if one is testing a drug that is intended for use in children of two years of age, what is the appropriate age NHP to study? There are some immune function data for the juvenile NHP that allow for limited comparisons to human. This includes data from lymphocyte subset analysis and serum immunoglobulin levels from both human and NHP. Analysis of human lymphocyte populations (44–46) demonstrate that total T, B, and NK cells numbers are essentially at adult levels at birth. However, the ratio of helper (CD4+):cytotoxic (CD8+) T cells is greater in newborns than in adults. During development there is a reduction in CD4+ T cells and a corresponding increased percentage of CD8+ T cells, and the ratio decreases from 2.9 at birth to approximately 1.6 in the adult. Similar alterations in the CD4+:CD8+ ratio are also observed in NHP (14,47). Similar comparisons can be made for serum immunoglobulins in human and NHP (14,23,48). Immunoglobulin levels increase for the first 5 (IgG) to 10 (IgA and IgM) years of development in human and corresponding values are available for NHP. While much work remains to be done to compare the postnatal development of the NHP and human immune systems, both lymphocyte subset analysis and serum immunoglobulin levels can be used to begin to approximate the equivalent "immunologic ages" of humans and NHP.

One of the key issues for DIT is the potential for persistent changes, and it is critical that recovery groups are included. In addition, as discussed later in this chapter, there is an emerging body of data describing differential effects of compounds in males and females, and so it is important that both sexes are included in any DIT studies.

Another key issue is consideration of biologics versus small molecular weight compounds. The chemical and drug compounds established to date to produce DIT have been small molecules. However, with the increased investment in biologic therapeutics in recent years a range of other issues are involved. For example, small molecules more readily cross the placental barrier, and dosing of pregnant females can be expected in most cases to lead to embryonic/fetal exposure. However, although protein therapeutics, such as antibodies, can cross the placental barrier, particularly during the third trimester, it may not be necessary to dose females during the entire pregnancy and thus across all immune developmental windows. Species selection is particularly important in this regard. Placentation in monkeys is similar to that of human, and this could be a key consideration for some therapeutics, as may assessment of transfer of drug to the breast milk. For biologics, not only are the issues of species selection critical to assess for placental or lactational exposure, but they are also key with regard to tissue cross-reactivity. It is possible that the only relevant species may be NHP, and in some cases there may be no relevant species.

Few DIT studies with biologic therapeutics have been published. While most DIT assessments performed with protein therapeutics have been performed with monoclonal antibodies, therapeutic peptides have also been studied. For example, Barusiban, an oxytocin receptor antagonist peptide was administered to pregnant monkeys from GD85 through to parturition (49). Postnatal assessments of the offspring indicated no alterations following a number of structural and functional immune assessments. Similarly, Carlock et al. (50) demonstrated that when pregnant Cynomolgus monkeys were administered human-soluble IL-4 receptor, to inhibit the activity of IL-4, no abnormalities were noted in any immunologic parameters evaluated in the offspring.

Martin and coworkers (51,52) have examined the effect of antitumor necrosis factor α (TNFα) monoclonal antibodies on the developing mouse and Cynomolgus macaque immune systems. TNFα knockout mice show increased susceptibility to infection and reduced delayed-type hypersensitivity (DTH) and humoral responses (53). Administration of an anti-TNFα monoclonal (golimumab) to cynomolgus macaques during pregnancy and lactation resulted in neonates being exposed to high concentrations of maternally derived antibody during lactation, and detectable antibody was detected in the neonate serum. However, no effects on the structural and functional development of the immune system were observed in the offspring. Similarly, in studies where a rat anti-mouse TNFα surrogate monoclonal antibody was administered to pregnant and lactating females, no structural or functional effects on the immune system of neonates were observed.

Embryo/fetal development (54) and postnatal developmental studies (55) demonstrated immune system effects in embryos following treatment of pregnant Cynomolgus monkeys with natalizumab, an antibody to α4 integrin. Thymic atrophy, extramedullary hematopoiesis in the spleen, as well as reductions in T- and B-cell lymphocyte subsets were all observed in fetuses. However, no such immunologic structural and functional effects were observed in the offspring of natalizumab treated dams, demonstrating the reversibility of the effects seen prenatally.

Methods for DIT Assessment

As discussed earlier, immunotoxicity endpoints are usually added on P&P and juvenile toxicity studies to assess DIT. The majority of DIT studies reported in the literature have focused on P&P studies, while more studies using the latter design will be conducted in the future to support pediatric or juvenile clinical studies. For both of these approaches, endpoints used to assess immunotoxicity in standard toxicity studies (as described in ICH S8) may be included in developmental toxicology studies to assess DIT. The additional immunotoxicity studies (also described in ICH S8) such as immunophenotyping and the TDAR may also be included in juvenile toxicity studies. The selection of endpoints should be based on the mechanism by which immunotoxicity is expected and changes observed in the adult studies.

In addition to identifying immune suppression, which has been the primary focus of research to date, there may also be a need to assess the overall balance and potential for Th1/Th2 skewing of immune function. Therefore, there is a need to validate a spectrum of assays. While there are a number of assays that have been developed to assess immunotoxicity in adult systems, many of these may need to be modified and optimized for the developing immune system, and new methods may need to be developed.

Assessments of Immune System Pathology

Immune organ pathology and its relevance to DIT have recently been extensively reviewed (8). Assessments of hematology, lymphocyte subset analysis (immunophenotyping), lymphoid organ weights, and histology of lymphoid tissues may be incorporated into standard developmental toxicity studies. For situations in which the mechanism of modulation is closely related to changes in cell numbers in the circulation and in lymphoid organs, immunophenotyping may be an important endpoint since it is easily translatable in clinical studies to monitor potential concerns. While the histological features of adult immune organs are well known, much less is known about the structure of developing immune organs, particularly in NHP. There is a need for further evaluation of comparative immune organ pathology during development in different species in response to immunotoxicants. However, additional histopathology evaluations may also reveal immune-mediated effects, including autoimmune-related inflammation.

T-Dependent Antibody Response

This assay involves administering a foreign antigen to the test animals and evaluating the antibody response in serum or lymphoid organs. The TDAR is particularly informative because of its holistic nature, since a response requires a number of processes and cell types, with a requirement for antigen presentation, T cell help, and B cell function. In adult studies, animals have been challenged with a variety of immunogens, including sheep red blood cells (SRBC), keyhole limpet hemocyanin (KLH), and tetanus toxoid (TTX), and the production of antigen-specific antibodies (IgM or IgG) assessed either by ELISA or antibody-forming cell assays. However, the usefulness of this assay in juvenile rodents is compromised by the delayed development of the rodent immune system. Studies have demonstrated that no TDAR response can be measured at PND 10 by ELISA and is suboptimal at weaning (PND 21) (30,56). It is not until young adulthood (PND 42–49) that a maximal response can be detected. In view of this limitation, studies have typically involved exposure of pregnant females and/or newborn pups to immunotoxicants with continued exposure to adulthood, with the TDAR performed in these young adults. It has also been recommended that a multi-isotype TDAR be performed (i.e., IgM and IgG) due to potential differential sensitivities to immunotoxicants and to assess any effects on functional immune skewing that may occur during development (36,57,58). Despite the holistic nature of the TDAR, it may be necessary to combine the TDAR with other assessments of immune parameters. Although the TDAR can be detected at PND 21, a reduced response is observed relative to that in the adult, potentially making determinations of immune suppression difficult. While it may be problematic to detect a TDAR response by ELISA with samples prior to weaning, it may be possible using a plaque-forming cell assay (30). The TDAR has been evaluated only to a limited extent in infant Cynomolgus macaques using either TTX or KLH as the antigen, and on PND 140 both antigen-specific IgM and IgG production was observed (27). However, while a measurable TDAR can be demonstrated in NHP preweaning, to date there have been no assessments of the TDAR in NHP following exposure to known developmental immunotoxicants.

Assays of Cell-Mediated Immunity

Various assays of cellular immunity including DTH, cytotoxic T-lymphocyte (CTL) activity, and NK cell activity assays will be important in the overall assessment of DIT. This view is consistent with recent reviews (8,36,38,43) that stress the need to broaden the scope of DIT testing beyond the TDAR. It is recommended that a cellular immunity assay be included in DIT protocols (19,36,38,41). The importance of this is highlighted by the demonstration that some developmental immunotoxicants fail to affect the TDAR response but can alter cellular immunity (59–63).

For assessments of cellular immunity, DTH responses to KLH or BSA, CTL activity, and NK cell assays have been most frequently used, and typically in combination with other immune endpoints, including a TDAR and/or immunohisto-pathology. It is possible that certain cellular immune responses (e.g., CTL responses or NK cell activity) may be performed earlier in postnatal development than the TDAR, at least in rodents. For example, in vitro induction of CTL responses from five-day-old mouse LN cells to EL4 cell targets have been shown to be comparable in magnitude to those of adults (18,64). However, while NK cell activity in mice is absent at birth and can be measured at three weeks of age with adult levels attained by four weeks (65), this is not significantly earlier than when the TDAR can be detected. Despite these advantages, it is important with DTH responses that the endpoint being evaluated is not confounded by production of antibody against the antigen used. Assays have typically used protein antigens such as KLH (T-dependent), where animals are sensitized and then rechallenged, followed by determinations of footpad thickness. The problem lies in the potential to produce antigen-specific antibodies (humoral response), which may result in type III hypersensitivity reactions and confound measures of DTH. While antigens typically used to date do have the potential to produce antibody, no consensus has been reached as to the best methodology and immunogens to use (8). However, there is a clear need for a validated assay specific to cellular immunity.

Other Approaches to the Assessment of DIT
Assessment of alterations in immune cell subsets by flow cytometry is used widely to assess immunotoxicity. However, in the absence of any changes in specific cellular subsets it is possible that cell function may be impaired (depending on the mechanism of action of the test drug), which would not be detected unless additional functional assays were conducted. Studies comparing immunophenotypic profiles in rats at PND 10 and 21 have been performed (30). B and T cells generally represent less than 20% of those seen in the adult at PND 10. B cells generally approximate those observed in the adult at PND 21, with T cells higher than adult at this point. More immunophenotyping studies are required using known immunosuppressant compounds to determine the sensitivity of assessing immune populations for DIT testing. However, adding immunophenotyping as an endpoint will likely aid in the interpretation of data obtained from functional DIT assays (36,66).

Classical host resistance assays have only been used in assessments of DIT to a limited extent, primarily because these studies require significant numbers of animals per dose and challenge group and may not be the best approach as an initial screen for immunosuppression (41). Host resistance assays could be used if other approaches indicate functional immune alteration and if there is sufficient understanding of the mechanism to drive the use of an appropriate host-resistance model.

Cytokine expression displays distinctive Th1 (e.g., interferon-gamma (IFNγ), interleukin (IL)-12) and Th2 (IL-4, IL-5, IL-13) profiles, and given the general shift from Th2 to Th1 profiles in early development, determination of cytokine levels may be useful in DIT assessment (36,66). However, few DIT studies have analyzed cytokine expression in detail, or across species, and there is a need for additional cytokine data to determine the general predictive value of this approach.

Compounds Indicated to Induce Developmental Immunotoxicity
One of the basic tenets underlying the field of DIT is that the developing immune system is uniquely sensitive to immunotoxicants, compared to that of the adult, either quantitatively or qualitatively. There is some evidence that supports this and will be discussed later. However, there is an overall paucity of data in general that addresses immunotoxicity in the developing immune system, and much of the work has been

TABLE 2   Compounds Demonstrated to be Developmental Immunotoxicants

Developmental immunotoxicant	References
Environmental chemicals	
Atrazine	67–69
Cadmium	70
Heptachlor	71
Lead	11, 35, 59, 60, 63, 72, 73
Mercury	74
Methoxychlor	75–77
Nonylphenol	62, 75
Propylthiouracil	78
TCDD	35, 61, 79
Tributyltin oxide (TBTO)	35
Pharmaceutical agents	
Aciclovir	80
Cyclosporine	58, 81
Dexamethasone	57
Diazepam	35, 82, 83
Diethylstilbestrol	35, 84
Genistein[a]	75, 85

[a]Genistein is a natural product and has been classified as a pharmaceutical agent.

TABLE 3   Immune Assays Used in Immunotoxicity Assessments

Immune endpoints	References
Structural analyses	11, 58, 60, 71, 76, 77, 78, 80, 81, 84
Assessment of immune cell populations	11, 57, 58, 61–63, 69, 71, 76, 78, 81, 84, 85
Cytokine analysis	57, 58, 60, 63, 69, 74
NK, CTL assays	62, 63, 67, 70, 75, 76, 78, 85
Lymphoproliferative responses	70, 83, 74
Delayed-type hypersensitivity(DTH)	57, 58, 59–61, 63, 67, 71, 73
T-dependent antibody response(TDAR)	57–60, 62, 63, 67, 69, 71, 76, 81, 85
Host resistance assay	80, 82
Phagocytosis	67

driven toward understanding the effects of environmental chemicals, with relatively little data with pharmaceutical agents.

A number of environmental chemicals and therapeutic compounds have been described as developmental immunotoxicants in various animal models. Some of these studies are listed in Table 2, and since these have been the subject of several recent reviews these will not be addressed in detail here. These studies have been performed exclusively in rodents and a variety of immunologic endpoints have been analyzed. These include "structural" as well as functional assessments and representative examples are shown in Table 3. While this does not provide a comprehensive list of all immune endpoints used in the assessment of DIT, it does provide an indication of the array of methods that can be used; more detailed descriptions can be found elsewhere (36). Some of these P&P studies with lead have demonstrated significant changes in functional immunity with only small, or no, change in immune cell populations or lymphoid organs (11,72). Therefore, additional DIT P&P or juvenile toxicity studies are needed with known immunosuppressive drugs to assess the need for structural and functional immune assessments.

The developing immune system is proposed to be more sensitive to lower doses of immunotoxicant exposure than that of the adult. There is some evidence to support this, although the pool of evidence is small. There are reports of immunotoxicants that produce perturbations in the developing immune system, but at the same dose have little or no effect on the adult immune system. For example, pregnant rats administered lead in their drinking water showed no differences in a number of immune endpoints compared to control animals, although 13-week-old female offspring demonstrated increased cytokine production and depression of cell-mediated immune function (DTH) (63). Similarly, dexamethasone treatment led to a decreased DTH response in offspring, where at the same dose the response in adult animals was unaffected (57). Cadmium administration to dams leads to decreased proliferation of thymocytes from offspring in response to Con-A stimulation, but at doses much lower than those which produce similar effects in adults (70). While there have been limited data describing the increased sensitivity of the developing immune system to immunotoxicant insult, there is no evidence for any compound having effects only on the developing immune system, but not on the adult at any dose. However, given the processes unique to the developing immune system it is possible that the reason for this is that insufficient compounds have been investigated in this regard.

In addition to increased sensitivity of the developing immune system to immunotoxicant exposure, embryonic or neonatal exposures to a number of compounds have been demonstrated to produce more persistent effects on the immune system, compared to adult exposures. This has been demonstrated for some chemicals (70) but a number of examples of pharmaceutical therapeutics have also been described, and include cyclosporine, diethylstilbestrol (DES), dexamethasone, diazepam, and aciclovir.

In utero exposure (GD 6-21) to cyclosporine has been compared to identically dosed adults (58). While exposure to cyclosporine produced acute immunotoxicity in adults with minimum long-term effects, offspring exposed in utero demonstrated persistent alterations, which included reductions in functional immune responses. While these persistent reductions in specific immune responses occurred at doses that also led to reduced anogenital distance in rat pups earlier in development, this morphologic change did not persist. Following prenatal exposure, production of anti-KLH-specific IgG in the TDAR assay was reduced at both 5 and 13 weeks postpartum, whereas adult responses recovered to control levels following 13 weeks of recovery. In addition, a slightly reduced DTH response and decreases in splenic B cells were observed in 5-week-old offspring, although not in 13-week-old offspring, suggesting that persistent immune suppression varies depending on the immune process studied. One caveat is that in these assays the reductions in DTH response in the 5-week-old offspring were minimal, and trends toward decreases were observed in adult animals following 13 weeks of recovery. In contrast to these studies, others have demonstrated that prenatal exposure of pregnant rats to cyclosporine led to no immunotoxic effects in offspring (81). However, postnatal exposure (PND 4–28) led to immune suppression and persistent impairment of the immune system, including lymphoid hyperplasia in the spleen, reduced CD4+ and CD8+ T cells, and reduced primary response to SRBC at 10 weeks of age. While some of the differences in these studies may be related to the dose of cyclosporine and the dosing regimen, it is clear that some effects on the developing immune system may take longer to reverse than in the adult and may depend on exposure during specific critical windows of development.

The timing of exposure to immunotoxicants is important, and such a critical window in development appears to occur during PND 1 to 5 in rats during which exposure to DES produces persistent immune defects (35). Female mice dosed with DES from PND 1 to 5 demonstrated decreased mitogen-induced splenocyte proliferation several months after treatment (86). In addition, DTH responses were diminished when measured 6 months after treatment (87), and at 16 weeks a reduced primary in vivo antibody response to SRBC was observed (88). While similar immune defects are noted in adult animals, the adult immune responses examined recovered to normal levels relatively quickly. Fetal exposure to dexamethasone has also been shown to produce a persistent functional loss (DTH) in 5- and 13-week-old rats following administration to pregnant rats and juvenile offspring (5 weeks) (57). A significantly reduced DTH response was observed in the offspring that continued into adulthood, whereas the DTH response in adults was unaltered. Treatment with the antiviral compound aciclovir leads to abnormal thymic development in rats (80), with reduced thymi apparent in 21-day-old fetuses, and persisting into adulthood. Functional consequences of prenatal aciclovir exposure included decreased host-resistance to Trichinella in 12-week-old rats. Finally, both male and female offspring of pregnant rats treated with the benzodiazepines diazepam or clonazepam exhibited severe and long-lasting depression of cellular immune responses (82,83), and those offspring displayed impaired host defense (82).

The timing of drug exposure and age-dependent differences in the immunotoxic effects of a number of compounds have been reviewed (35). For example, 2,3,7,8-tetrachlorodibenzo-p-dioxin (TCDD) exposure has been shown to suppress thymus-dependent immunity (79) and inhibit thymocyte maturation. These effects are more severe when administered during the pre- and postnatal period rather than solely in the postnatal period and tend to be longer lasting (61). Addition of lead in the drinking water of pregnant rats at GD3-9 or GD15-21 has also been demonstrated to produce differential effects on certain immune responses (60).

Finally, with regard to immune sensitivity during development, an increasing number of compounds have been shown to have differential developmental immuno-toxic effects on the immune system of males and females, and these have been summarized (9). These chemicals include lead (59,60), TCDD (61), atrazine (67–69), nonylphenol (62), DES (84,89), and methoxychlor (71,76). Bunn et al. (59) demonstrated that in utero exposure to lead reduced the DTH response in females but not in males at 5 and 13 weeks of age. The DTH response was also suppressed by TCDD, but males were more sensitive than females to lower doses (61). However, given that some of these compounds, for example, DES, are estrogenic or endocrine disruptors, such sex-specific effects are not unexpected. DES is a potent estrogen and has been demonstrated to be a developmental immunotoxicant in a number of studies (84,89,90). In utero exposure to DES during the third trimester in rodents leads to reductions in both T and B cell–mediated responses only in the female offspring (35). However, TI responses in males are enhanced versus suppressed in females.

FUTURE DIRECTIONS

DIT is an emerging field. While progress is being made in understanding the challenges of nonclinical DIT evaluation of pharmaceuticals, it is clear that there are significant gaps in the science. This impacts our ability to select the most relevant species, develop dosing and exposure regimens, and conduct appropriate assays to evaluate immune endpoints. Clearly more comparative species analysis of the developing immune system is needed. While there is a considerable amount of data regarding rodents, much of this has been generated in the mouse, and it has generally been assumed that the rat and mouse are largely similar. Relatively little is known regarding the developing human immune system, and less still for NHP. While for each species we have a broad understanding of the timing of some major immune developmental landmarks, there remain large gaps in our knowledge of immune system development in both structural and functional terms. For example, in efforts to design appropriate non-clinical assays of relevance to juvenile human studies, we can only estimate the "immunologic age" of a rat or monkey that will be comparable to the human.

There is a need for additional assessment of current methods and the development of new methods to assess immune function in juvenile animals. It cannot be assumed that assays developed for assessment of adult immunotoxicity are optimal for studies in young animals. For example, the TDAR, recommended for use in adult immunotoxicity assessments, is clearly limited in its application to neonatal and juvenile rodent studies. While cell-mediated immunity (CMI) assays have the potential to be useful functional assays in developmental immunotoxicity, and such responses may be detectable before those seen by TDAR, a DTH assay is needed that does not involve type III hypersensitivity reactions. Another concern with the KLH DTH assay is the need to immunize and challenge the animal to obtain a response. This is not needed for the primary TDAR since a single immunization will result in a measurable response (e.g., serum antibodies or antibody-secreting cells).

As new methods to evaluate the adult immune system are developed, these could be further optimized as DIT endpoints. In addition, there is a need for endpoints that can be translated to clinical testing to monitor DIT.

There has been a clear focus to date on immune suppression, but there will be a need to determine if similar assays can be used to assess immune enhancement. There is also a need for studies to determine if drugs are able to induce autoimmune disease with peri- or postnatal exposure. Methods to assess hypersensitivity or drug allergy with systemic exposure have not been adequately validated for adults. Once these methods are available, they may be evaluated for potential use in DIT studies.

CONCLUSIONS

Progress has been made in the field of immunotoxicology in recent years in the area of identifying gaps and needs. However, many more DIT studies with known immunosuppressive drugs are needed to better understand the concerns regarding increased sensitivity and reversibility of the immune system and the best approaches to assess DIT. Specific DIT guidelines are not available but the current immunotoxicology and pediatric guidelines provide some general guidance. Currently, the general practice is to include DIT endpoints to P&P or juvenile toxicity studies when immunotoxicity (i.e., immunosuppression) is demonstrated in the adult. Studies may be designed according to the specific questions being asked (pre- and postnatal exposure or support of pediatric trials). Many of the nonclinical immunotoxicity assays used for adult testing may be used to evaluate the developing immune system but may need further optimization. The rat has been the primary species used for most adult immunotoxicology studies with environmental chemicals and low molecular weight drugs. The delayed immune development in the rat compared to human needs to be taken into consideration when designing studies with DIT endpoints. While NHP immune system development appears largely parallel to that of human, further comparative structural and functional studies are needed.

For nonclinical DIT testing for pharmaceuticals, a case-by-case approach is used to determine if DIT testing is needed and what study design is most appropriate. The study design should be based on the potential exposure period to the test drug (P&P or pediatric). Immunotoxicity or immunopharmacology data in the adults (nonclinical and clinical data) along with mechanism of action data may be used to determine potential targets of the developing immune system and used to determine the best methods to assess DIT.

There are many gaps in the area of DIT that have been the subject of numerous reviews and workshops. These gaps include the need for a greater understanding of the development of the immune system in the different toxicity species and the development or optimization of DIT endpoints. Clearly, from what we already know about immune system development, it cannot be assumed that assays validated in adult animals will be directly applicable to DIT studies. Finally, since most of the emphasis to date has been placed on immune suppression, more attention must be paid to conditions where immune enhancement or dysregulation is a concern, such as auto-immunity, allergy, and hypersensitivity.

REFERENCES

1. ICH. Guidance for Industry. S8 Immunotoxicity Studies for Human Pharmaceuticals, 2006. U.S. Department of Health and Human Services, Food and Drug Administration, Center for Drug Evaluation and Research, Center for Biologics Evaluation and Research.
2. Piccotti JR, Lebrec HN, Evans E, et al. Summary of a workshop on nonclinical and clinical immunotoxicity assessment of immunomodulatory drugs. J Immunotoxicol 2009; 6(1):1–10.
3. FDA (United States Food and Drug Administration). Guidance for Industry-Immunotoxicology evaluation of investigational new drugs. Office of Training and Communication. Division of Drug Information, Center for Drug Evaluation and Research, U.S. Department of Health and Human Services, 2002.
4. EMEA (European Medicines Agency, Committee for Proprietary Medicinal Products). Note for Guidance on Repeated Dose Toxicity. EMEA/CPMP/SWP/1042/99corr, 2000.
5. FDA (United States Food and Drug Administration). Guidance for Industry-Nonclinical Safety Evaluation of Pediatric Drug Products. Office of Training and Communication. Division of Drug Information, Center for Drug Evaluation and Research, U.S. Department of Health and Human Services, 2006.
6. EMEA (European Medicines Agency, Committee for Human Medicinal Products). Guideline on the need for non-clinical testing in juvenile animals of pharmaceuticals for paediatric indications. EMEA/CHMP/SWP/169215/2005, 2008.
7. European Parliament,. Regulation (EC) No. 1901/2006 of the European Parliament and of the Council of 12 December 2006 on medicinal products for paediatric use and amending Regulation (EEC) No. 1768/92, Directive 2001/20/EC, Directive 2001/83/EC and Regulation (EC) No. 726/2004, 2006.
8. Burns-Naas LA, Hastings KL, Ladics GS, et al. What's so special about the developing immune system? Int J Toxicol 2008; 27(2):223–254.
9. Dietert RR, Burns-Naas LA. Developmental Immunotoxicity in Rodents. In: Herzyk DJ, Bussiere JL, eds. Immunotoxicology Strategies for Pharmaceutical Safety Assessment. Hoboken, NJ: John Wiley & Sons, Inc., 2008:273–297.
10. Dietert RR, Piepenbrink MS. Perinatal immunotoxicity: why adult exposure assessment fails to predict risk. Environ Health Perspect 2006; 114(4):477–483.
11. Dietert RR, Piepenbrink MS. Lead and immune function. Crit Rev Toxicol 2006; 36(4): 359–385.
12. Felsburg PJ. Overview of immune system development in the dog: comparison with humans. Hum Exp Toxicol 2002; 21(9–10):487–492.
13. Haley PJ. Species differences in the structure and function of the immune system. Toxicology 2003; 188(1):49–71.
14. Hendrickx AG, Peterson PE, Makori NM. The nonhuman primate as a model of developmental immunotoxicity. In: Holladay SD, ed. Developmental Immunotoxicology. Boca Raton, FL: CRC Press, 2005:117–136.
15. Holsapple MP, West LJ, Landreth KS. Species comparison of anatomical and functional immune system development. Birth Defects Res B Dev Reprod Toxicol 2003; 68(4):321–334.
16. Landreth KS. Critical windows in development of the rodent immune system. Hum Exp Toxicol 2002; 21(9–10):493–498.
17. West LJ. Defining critical windows in the development of the human immune system. Hum Exp Toxicol 2002; 21(9–10):499–505.
18. Fadel S, Sarzotti M. Cellular immune responses in neonates. Int Rev Immunol 2000; 19(2–3):173–193.

19. Luster MI, Dean JH, Germolec DR. Consensus workshop on methods to evaluate developmental immunotoxicity. Environ Health Perspect 2003; 111(4):579–583.
20. Marshall-Clarke S, Reen D, Tasker L, et al. Neonatal immunity: how well has it grown up? Immunol Today 2000; 21(1):35–41.
21. Landreth KS, Dodson VM. Development of the rodent immune system. In: Holladay SD, ed. Developmental Immunotoxicology. Boca Raton, FL: CRC Press, 2005:3–19.
22. Leibnitz R. Development of the human immune system. In: Holladay SD, ed. Developmental Immunotoxicology. Boca Raton, FL: CRC Press, 2005:21–42.
23. Buse E. Development of the immune system in the cynomolgus monkey: the appropriate model in human targeted toxicology? J Immunotoxicol 2005; 2(4):211–216.
24. Buse E, Habermann G, Osterburg I, et al. Reproductive/developmental toxicity and immunotoxicity assessment in the nonhuman primate model. Toxicology 2003; 185(3): 221–227.
25. Buse E, Habermann G, Vogel F. Thymus development in Macaca fascicularis (Cynomolgus monkey): an approach for toxicology and embryology. J Mol Histol 2006; 37(3-4):161–170.
26. Hendrickx AG, Makori N, Peterson P. The nonhuman primate as a model of developmental immunotoxicity. Hum Exp Toxicol 2002; 21(9–10):537–542.
27. Martin PL, Buse E. Developmental immunotoxicity in nonhuman primates. In: Hoboken, NJ: Herzyk DJ, Bussiere JL, eds. Immunotoxicology Strategies for Pharmaceutical Safety Assessment. Hoboken, NJ: John Wiley & Sons, Inc. 2008:299–317.
28. Hakim FT, Memon SA, Cepeda R, et al. Age-dependent incidence, time course, and consequences of thymic renewal in adults. J Clin Invest 2005; 115(4):930–939.
29. Petrie HT. Role of thymic organ structure and stromal composition in steady-state postnatal T-cell production. Immunol Rev 2002; 189:8–19.
30. Ladics GS, Smith C, Bunn TL, et al. Characterization of an approach to developmental immunotoxicology assessment in the rat using SRBC as the antigen. Toxicol Methods 2000; 10:283–311.
31. Klein Klouwenberg P, Bont L. Neonatal and infantile immune responses to encapsulated bacteria and conjugate vaccines. Clin Dev Immunol 2008; 2008:628963.
32. Landers CD, Chelvarajan RL, Bondada S. The role of B cells and accessory cells in the neonatal response to TI-2 antigens. Immunol Res 2005; 31(1):25–36.
33. Timens W, Boes A, Rozeboom-Uiterwijk T, et al. Immaturity of the human splenic marginal zone in infancy. Possible contribution to the deficient infant immune response. J Immunol 1989; 143(10):3200–3206.
34. Mosier DE, Mond JJ, Goldings EA. The ontogeny of thymic independent antibody responses in vitro in normal mice and mice with an X-linked B cell defect. J Immunol 1977; 119(6):1874–1878.
35. Luebke RW, Chen DH, Dietert R, et al. The comparative immunotoxicity of five selected compounds following developmental or adult exposure. J Toxicol Environ Health B Crit Rev 2006; 9(1):1–26.
36. Dietert RR, Holsapple MP. Methodologies for developmental immunotoxicity (DIT) testing. Methods 2007; 41(1):123–131.
37. Holsapple MP, Paustenbach DJ, Charnley G, et al. Symposium summary: children's health risk—what's so special about the developing immune system? Toxicol Appl Pharmacol 2004; 199(1):61–70.
38. Holsapple MP, Burns-Naas LA, Hastings KL, et al. A proposed testing framework for developmental immunotoxicity (DIT). Toxicol Sci 2005; 83(1):18–24.
39. Dietert RR, Etzel RA, Chen D, et al. Workshop to identify critical windows of exposure for children's health: immune and respiratory systems work group summary. Environ Health Perspect 2000; 108(suppl 3):483–490.
40. Holladay SD, Smialowicz RJ. Development of the murine and human immune system: differential effects of immunotoxicants depend on time of exposure. Environ Health Perspect 2000; 108(suppl 3):463–473.
41. Holsapple MP. Developmental immunotoxicology and risk assessment: a workshop summary. Hum Exp Toxicol 2002; 21(9–10):473–478.
42. Ladics GS, Chapin RE, Hastings KL, et al. Developmental toxicology evaluations—issues with including neurotoxicology and immunotoxicology assessments in reproductive toxicology studies. Toxicol Sci 2005; 88(1):24–29.

43. Holsapple MP, Van der Laan JW, Van Loveren H. Development of a framework for developmental immunotoxicity (DIT) testing. In: Luebke R, House R, Kimber I, eds. Immunotoxicology and Immunopharmacology. Boca Raton, FL: CRC Press, 2007:347–361.

44. Shearer WT, Rosenblatt HM, Gelman RS, et al. Lymphocyte subsets in healthy children from birth through 18 years of age: the Pediatric AIDS Clinical Trials Group P1009 study. J Allergy Clin Immunol 2003; 112(5):973–980.

45. Hannet I, Erkeller-Yuksel F, Lydyard P, et al. Developmental and maturational changes in human blood lymphocyte subpopulations. Immunol Today 1992; 13(6):215, 218.

46. Hulstaert F, Hannet I, Deneys V, et al. Age-related changes in human blood lymphocyte subpopulations. II. Varying kinetics of percentage and absolute count measurements. Clin Immunol Immunopathol 1994; 70(2):152–158.

47. Weinbauer GF, Frings W, Fuchs A, et al. Reproductive/Developmental toxicity assessment of biopharmaceuticals in nonhuman primates. In: Cavagnaro JA, ed. Preclinical Evaluation of Biopharmaceuticals: A Science-Based Approach to Facilitating Clinical Trials. Hoboken, NJ: John Wiley and Sons, Inc., 2008:379–397.

48. Isaacs D, Altman DG, Tidmarsh CE, et al. Serum immunoglobulin concentrations in preschool children measured by laser nephelometry: reference ranges for IgG, IgA, IgM. J Clin Pathol 1983; 36(10):1193–1196.

49. Rasmussen AD, Nelson JK, Chellman GJ, et al. Use of barusiban in a novel study design for evaluation of tocolytic agents in pregnant and neonatal monkeys, including behavioural and immunological endpoints. Reprod Toxicol, 2007; 23(4):471–479.

50. Carlock LL, Cowan LA, Oneda S, et al. A comparison of effects on reproduction and neonatal development in cynomolgus monkeys given human soluble IL-4R and mice given murine soluble IL-4R. Regul Toxicol Pharmacol 2009; 53(3):226–234.

51. Martin PL, Cornacoff JB, Treacy G, et al. Effects of administration of a monoclonal antibody against mouse tumor necrosis factor alpha during pregnancy and lactation on the pre- and postnatal development of the mouse immune system. Int J Toxicol 2008; 27(4):341–347.

52. Martin PL, Oneda S, Treacy G. Effects of an anti-TNF-alpha monoclonal antibody, administered throughout pregnancy and lactation, on the development of the macaque immune system. Am J Reprod Immunol 2007; 58(2):138–149.

53. Pasparakis M, Alexopoulou L, Episkopou V, et al. Immune and inflammatory responses in TNF alpha-deficient mice: a critical requirement for TNF alpha in the formation of primary B cell follicles, follicular dendritic cell networks and germinal centers, and in the maturation of the humoral immune response. J Exp Med 1996; 184(4):1397–1411.

54. Wehner NG, Shopp G, Oneda S, et al. Embryo/fetal development in cynomolgus monkeys exposed to natalizumab, an alpha4 integrin inhibitor. Birth Defects Res B Dev Reprod Toxicol 2009; 86(2):117–130.

55. Wehner NG, Shopp G, Osterburg I, et al. Postnatal development in cynomolgus monkeys following prenatal exposure to natalizumab, an alpha4 integrin inhibitor. Birth Defects Res B Dev Reprod Toxicol 2009; 86(2):144–156.

56. Kimura S, Eldridge JH, Michalek SM, et al. Immunoregulation in the rat: ontogeny of B cell responses to types 1, 2, and T-dependent antigens. J Immunol 1985; 134(5):2839–2846.

57. Dietert RR, Lee JE, Olsen J, et al. Developmental immunotoxicity of dexamethasone: comparison of fetal versus adult exposures. Toxicology 2003; 194(1–2):163–176.

58. Hussain I, Piepenbrink MS, Fitch KJ, et al. Developmental immunotoxicity of cyclosporin-A in rats: age-associated differential effects. Toxicology 2005; 206(2):273–284.

59. Bunn TL, Dietert RR, Ladics GS, et al. Developmental immunotoxicology assessments in the rat: age, gender, and strain comparisons after exposure to lead. Toxicol Methods 2001; 11:41–58.

60. Bunn TL, Parsons PJ, Kao E, et al. Exposure to lead during critical windows of embryonic development: differential immunotoxic outcome based on stage of exposure and gender. Toxicol Sci 2001; 64(1):57–66.

61. Gehrs BC, Smialowicz RJ. Persistent suppression of delayed-type hypersensitivity in adult F344 rats after perinatal exposure to 2,3,7,8-tetrachlorodibenzo-p-dioxin. Toxicology 1999; 134(1):79–88.

62. Karrow NA, Guo TL, Delclos KB, et al. Nonylphenol alters the activity of splenic NK cells and the numbers of leukocyte subpopulations in Sprague-Dawley rats: a two-generation feeding study. Toxicology 2004; 196(3):237–245.

63. Miller TE, Golemboski KA, Ha RS, et al. Developmental exposure to lead causes persistent immunotoxicity in Fischer 344 rats. Toxicol Sci 1998; 42(2):129–135.

64. Piguet PF, Irle C, Kollatte E, et al. Post-thymic T lymphocyte maturation during ontogenesis. J Exp Med 1981; 154(3):581–593.

65. Cook JL, Ikle DN, Routes BA. Natural killer cell ontogeny in the athymic rat. Relationship between functional maturation and acquired resistance to E1A oncogene-expressing sarcoma cells. J Immunol 1995; 155(12):5512–5518.

66. Dietert RR, Dietert JM. Early-life immune insult and developmental immunotoxicity (DIT)-associated diseases: potential of herbal- and fungal-derived medicinals. Curr Med Chem 2007; 14(10):1075–1085.

67. Rooney AA, Matulka RA, Luebke RW. Developmental atrazine exposure suppresses immune function in male, but not female Sprague-Dawley rats. Toxicol Sci 2003; 76(2): 366–375.

68. Rowe AM, Brundage KM, Barnett JB. Developmental immunotoxicity of atrazine in rodents. Basic Clin Pharmacol Toxicol 2008; 102(2):139–145.

69. Rowe AM, Brundage KM, Schafer R, et al. Immunomodulatory effects of maternal atrazine exposure on male Balb/c mice. Toxicol Appl Pharmacol 2006; 214(1):69–77.

70. Pillet S, Rooney AA, Bouquegneau JM, et al. Sex-specific effects of neonatal exposures to low levels of cadmium through maternal milk on development and immune functions of juvenile and adult rats. Toxicology 2005; 209(3):289–301.

71. Smialowicz RJ, Williams WC, Copeland CB, et al. The effects of perinatal/juvenile heptachlor exposure on adult immune and reproductive system function in rats. Toxicol Sci 2001; 61(1):164–175.

72. Dietert RR, Lee JE, Hussain I, et al. Developmental immunotoxicology of lead. Toxicol Appl Pharmacol 2004; 198(2):86–94.

73. Faith RE, Luster MI, Kimmel CA. Effect of chronic developmental lead exposure on cell-mediated immune functions. Clin Exp Immunol 1979; 35(3):413–420.

74. Silva IA, El Nabawi M, Hoover D, et al. Prenatal HgCl2 exposure in BALB/c mice: gender-specific effects on the ontogeny of the immune system. Dev Comp Immunol 2005. 29(2):171–183.

75. Guo TL, Germolec DR, Musgrove DL, et al. Myelotoxicity in genistein-, nonylphenol-, methoxychlor-, vinclozolin- or ethinyl estradiol-exposed F1 generations of Sprague-Dawley rats following developmental and adult exposures. Toxicology 2005; 211(3):207–219.

76. White KL Jr., Germolec DR, Booker CD, et al. Dietary methoxychlor exposure modulates splenic natural killer cell activity, antibody-forming cell response and phenotypic marker expression in F0 and F1 generations of Sprague Dawley rats. Toxicology 2005; 207(2):271–281.

77. Chapin RE, Harris MW, Davis BJ, et al. The effects of perinatal/juvenile methoxychlor exposure on adult rat nervous, immune, and reproductive system function. Fundam Appl Toxicol 1997; 40(1):138–157.

78. Rooney AA, Fournier M, Bernier J, et al. Neonatal exposure to propylthiouracil induces a shift in lymphoid cell sub-populations in the developing postnatal male rat spleen and thymus. Cell Immunol 2003; 223(2):91–102.

79. Vos JG, Moore JA. Suppression of cellular immunity in rats and mice by maternal treatment with 2,3,7,8-tetrachlorodibenzo-p-dioxin. Int Arch Allergy Appl Immunol 1974; 47(5):777–794.

80. Stahlmann R, Korte M, Van Loveren H, et al. Abnormal thymus development and impaired function of the immune system in rats after prenatal exposure to aciclovir. Arch Toxicol 1992; 66(8):551–559.

81. Barrow PC, Horand F, Ravel G. Developmental immunotoxicity investigations in the SD rat following pre- and post-natal exposure to cyclosporin. Birth Defects Res B Dev Reprod Toxicol 2006; 77(5):430–437.

82. Schlumpf M, Lichtensteiger W, van Loveren H. Impaired host resistance to Trichinella spiralis as a consequence of prenatal treatment of rats with diazepam. Toxicology 1994; 94(1–3):223–230.

83. Schlumpf M, Ramseier H, Lichtensteiger W. Prenatal diazepam induced persisting depression of cellular immune responses. Life Sci 1989; 44(7):493–501.

84. Fenaux JB, Gogal RM Jr., Ahmed SA. Diethylstilbestrol exposure during fetal development affects thymus: studies in fourteen-month-old mice. J Reprod Immunol 2004; 64(1–2):75–90.

85. Guo TL, White KL Jr., Brown RD, et al. Genistein modulates splenic natural killer cell activity, antibody-forming cell response, and phenotypic marker expression in F(0) and F(1) generations of Sprague-Dawley rats. Toxicol Appl Pharmacol 2002; 181(3):219–227.
86. Kalland T, Strand O, Forsberg JG. Long-term effects of neonatal estrogen treatment on mitogen responsiveness of mouse spleen lymphocytes. J Natl Cancer Inst 1979; 63(2): 413–421.
87. Kalland T, Forsberg JG. Delayed hypersensitivity response to oxazolone in neonatally estrogenized mice. Cancer Lett 1978; 4(3):141–146.
88. Kalland T. Alterations of antibody response in female mice after neonatal exposure to diethylstilbestrol. J Immunol 1980; 124(1):194–198.
89. Fenaux JB, Gogal RM Jr., Lindsay D, et al. Altered splenocyte function in aged C57BL/6 mice prenatally exposed to diethylstilbestrol. J Immunotoxicol 2005; 2(4):221–229.
90. Holladay SD, Blaylock BL, Comment CE, et al. Selective prothymocyte targeting by prenatal diethylstilbesterol exposure. Cell Immunol 1993; 152(1):131–142.

9 Developmental neurotoxicology

Merle G. Paule and William Slikker, Jr.

INTRODUCTION
In recent years, the topic of developmental neurotoxicology has garnered increased attention in parallel with the growing realization that the developing nervous system can be very sensitive to a variety of toxic insults. This heightened awareness has been fostered by published reports of relevant toxicity and exposure (1–4). Certainly, the incidences of behavioral abnormalities in children seem to have risen at a rapid rate: the prevalence of autism spectrum disorders and attention deficit/hyperactivity disorder, to name just two that are widely recognized, serve as primary examples. While it is likely that some of the risk factors for these and other behavioral/functional disorders are genetic, it is almost certain that environmental factors and gene-environment interactions contribute to the expression of these clinical entities (phenotypes). As part of the environmental component, exposure to agents (e.g., drugs, environmental chemicals, allergens, bacteria, viruses) that affect nervous system development are, by definition, expected to impact the trajectory of nervous system maturation. The environmental component, specifically chemical exposure, will be the focus of this chapter.

As chemicals are introduced into the market place, various regulatory bodies can require data concerning the developmental neurotoxicity of those agents and various components of the U.S. Environmental Protection Agency (EPA), the U.S. Food and Drug Administration (FDA), and European Union have published guidelines to be followed in such assessments (5–7). FDA has also offered guidelines for establishing the preclinical safety of pediatric drugs (8) and a recent workshop was held to discuss current and future needs for developmental toxicity testing [including developmental neurotoxicity testing (9)]. The rapid developments being made in the fields of genomics, proteomics, metabolomics, and other areas of molecular biology hopefully will provide opportunity to inform important toxicological processes going forward. Thus, a critical look at the state of the art is helping to identify needs and focus effort for future assessments.

The telling observation that "the objectives of academic science are not conducive to the establishment of guidelines for developmental neurotoxicity" (10) serves to highlight the differences between academic and regulatory science. Whereas innovation and novelty are often specific goals in academia, the development of specific set guidelines involves establishing validated, reliable, and reproducible findings. Whereas academic science has been well funded [e.g., National Institutes of Health (NIH)], the science needed for producing sound regulatory guidelines and decisions has not benefited to the same degree. Regardless of funding source or ultimate goal, the basic issues concerning developmental toxicity/neurotoxicity apply and include issues of cross-species extrapolation (translation), mechanisms/modes of action; cumulative versus acute exposures, critical periods of sensitivity, latency to onset of response, and functional versus structural outcomes (9).

CROSS-SPECIES EXTRAPOLATION—TRANSLATION
In the ideal world, the data obtained to determine the developmental neurotoxicity of a compound are obtained from animal and/or other nonhuman models. It is, thus, critical that the endpoints monitored in these model systems are relevant to the human condition. These might include metrics of reflex ontogeny, sexual differentiation,

motor and cognitive function, and responses to environmental challenges that can take a variety of forms including chemical, social, behavioral, and thermal. The need to relate findings from preclinical models to human outcome necessitates that endpoints used in our models should be validated, that is, shown to demonstrate some direct correlation with human findings. That this has not been the case historically, likely accounts for the past failures of animal and other preclinical models to always predict adverse human outcome. Future efforts should involve concerted validation efforts, particularly in the case of in vitro and in silico systems, to maximize their utility and predictivity. In the case of whole animal models, it will be critical to utilize the same (homologous) endpoints of importance to humans, where possible, or to at least use endpoints (analogues) that correlate with human endpoints or that are otherwise predictive.

MECHANISM/MODE OF ACTION DATA

It is critical to use models that metabolize test compounds in a fashion relevant to humans and that exhibit the appropriate pathways and sites of action (e.g., receptors, biochemical pathways). If, for example, a compound acts to render a specific pathway(s) (e.g., oxidative phosphorylation, protein synthesis, apoptosis) or receptor(s) (e.g., glutamate, dopamine, aromatic hydrocarbon) dysfunctional, then it will be important that those pathways and/or receptors also occur in humans.

CUMULATIVE EXPOSURE ISSUES

Humans are often exposed to a given chemical repeatedly and throughout various life stages. Single acute exposures to high doses of chemicals can lead to frank poisonings and are of obvious concern. However, repeated and/or continuous exposures to relatively low doses of compounds or to mixtures of chemicals are often the rule rather than the exception and pose more difficult problems for risk assessment. Substances with short half-lives will not accumulate, whereas those with long half-lives will. Regardless of whether a compound accumulates, experimental designs must take into consideration the cumulative effects of exposure, many of which can vary depending on the life cycle during which such exposures occur. Some substances may have no adverse effect if exposure occurs during one stage of the life cycle but be devastating if exposure occurs during a different stage.

CRITICAL WINDOWS OF EXPOSURE AND EFFECT

There are developmental periods during which organisms may be extremely sensitive to chemical exposure. Different parts of the nervous system/brain develop at different times and rates and often vary considerably in their sensitivity to chemicals depending on the developmental stage at exposure. Thus, exposure to compounds such as anesthetic agents, alcohol, lead, and other compounds during gestation and/or early postnatal life when the brain is rapidly developing can have enduring and devastating consequences, whereas similar exposures in adults may, by comparison, be relatively benign.

LATENCY OF RESPONSE

Unfortunately, the effects of a toxic insult(s) may not be evident until long after an exposure occurs. Efforts to determine the developmental effects of many compounds to which children are exposed (i.e., anesthetics, methyl mercury) have demonstrated clear, primarily functional, adverse effects that become evident only as the child develops. Adults exposed to chemicals that destroy neurons responsible for the synthesis of the neurotransmitter, dopamine, can eventually develop Parkinson's disease, an outcome that may not be evident for years or decades.

STRUCTURAL VS. FUNCTIONAL OUTCOMES

Alterations in the structure of the nervous system (e.g., vacuoles, demyelination, dead cells) are relatively easy to detect and measure. Changes in function, however, can be much more difficult to detect. This is particularly true for the nervous system since function can be altered sometimes in the absence of detectable structural and/or chemical changes. It is very clear that animals (and humans) that are physically normal may suffer significant functional aberrations. Thorough functional assessments require the use of a broad range of instruments or test batteries to provide the monitoring of a wide array of functional domains.

MOVING FORWARD

One major change in the way developmental neurotoxicity is being studied involves the scientific approaches being used: in the post-genomic world, a systems biology/toxicology approach (11) is being more widely employed. This chapter will therefore focus on recent advances in describing the discipline of developmental neurotoxicology and offer a case study on pediatric anesthetic agents using a systems toxicology approach.

Systems toxicology involves the iterative and integrative study of biological systems as they respond to perturbations or challenges (11). In a toxicological context, the approach involves the study of chemical effects by monitoring alterations in gene and protein expressions that are firmly linked to an identifiable toxicological outcome (12). This approach requires the understanding of key interactions between cellular components, organs, and systems and how these are affected by toxicant exposure or disease (13). The goal of systems toxicology then is to predict functional outcomes of component-component relationships using computational models when possible that allow for a quantitative description of the reaction of the complete organism in response to environmental perturbations (12).

The success of the systems toxicology approach lies in the establishment of cross-disciplinary teams of scientists including toxicologists, molecular biologists, mathematicians, modelers, and risk assessors. Integration of rapidly growing biological databases, including models of cells, tissues, and organs, with the use of powerful computing systems and algorithms will be necessary (13). Because of the complexity and temporal aspects associated with developmental neurotoxicity, no area of toxicology will benefit more from the systematic application of a systems toxicology approach. Neurotoxicity is broadly defined here as any adverse effect on the structure or function of the central and/or peripheral nervous system caused by a biological, chemical, or physical agent that diminishes the ability of an organism to survive, reproduce, or adapt to its environment (14). Neurotoxic effects may be permanent or reversible, produced by the neuropharmacological or neurodegenerative properties of a toxicant, and result from either direct or indirect action on the nervous system (15). These effects can often be measured using neurobiological, neurophysiological, neuropathological, or behavioral techniques. Extrapolation across species is clearly an aim, but such efforts must take into account species differences in the ontogeny of the nervous system. Neurotoxicity may take various forms and be quite subtle (16), and its manifestations may change with age. The developing nervous system may be more or less susceptible to neurotoxic insult depending on the stage of development, from embryonic implantation through senescence.

A recent description of developmental neurotoxicity has established the importance of deranged apoptotic neuronal death in several regions of the developing rat pup brain following early postnatal exposure to the dissociative general anesthetic, ketamine (17,18). When ketamine—a noncompetitive N-methyl-D-aspartate (NMDA) receptor antagonist— is administered to rat pups during the brain growth spurt [e.g., postnatal day (PND) 7], an increase in apoptosis occurs in several brain regions. Excessive suppression of neuronal activity by ketamine during this window of vulnerability is thought to trigger neurons to commit "suicide" by apoptosis. It has been suggested that the enhanced susceptibility of the brain during this period of rapid growth derives from the increased expression of specific NMDA receptor (NR) subunits (19). Because the NR system plays a critical role in nervous system/brain development, perturbation of this system can have profound, long-lasting, and detrimental effects (20) that result from fewer and/or nonfunctional neuronal connections caused by such perturbation.

The original observations that apoptosis in the nervous system is associated with exposure to the noncompetitive NMDA antagonists MK-801 and ketamine (17) are consistent with a mechanism involving blockade of NR-mediated neurotransmission. Ikonomidou et al. (21) have replicated and expanded their earlier neonatal rat work to include nitrous oxide, isoflurane, propofol, midazolam, halothane, barbiturates, benzodiazepines, and ethanol as suspect apoptotic agents, when used either alone or in combination (21,22). These data suggest that most anesthetic agents will also be able to cause similar apoptotic effects because they either block the NR or stimulate the inhibitory GABA receptor system (21,22). Several other groups have also reported that MK-801 and ketamine increase apoptosis both in vivo (18,23,24) and in embryonic chick motor neurons (25) or in cultured neurons (26).

This chapter (27) will discuss the application of a systems toxicology approach (11,28) to the developmental neurotoxicity produced by ketamine, phencyclidine (PCP), and related NR antagonists. Initially, the available information (both in vitro and in vivo) on the developing rodent brain was gathered and a preliminary model of the data was proffered in descriptive and graphical terms. Where possible, the genes and proteins expressed in suspected pathways were defined. Second, the system was challenged with chemicals (PCP and ketamine), and temporal experiments providing information for various developmental periods were conducted. Third, the model was then improved by incorporating new information when additional challenges (exposures) were effected. And finally, the model was further refined by repeating key steps.

PROGRAMMED CELL DEATH: APOPTOSIS

Apoptosis is a physiological process in which molecular pathways are activated to cause cell death. It is a process that is vital for all multicellular organisms; is crucial for normal development, organ morphogenesis, and tissue homeostasis (29); and serves as a defense mechanism against infection or cellular damage (30). However, excess apoptosis may lead to conditions such as immunodeficiency and neurodegenerative diseases (31). The machinery of cellular suicide is maintained in a critical balance between being active and inactive by a variety of regulatory factors. Many such factors play double duty, having opposite effects in different contexts or in different cell types (e.g., NF-κB or nuclear factor kappa-light-chain-enhancer of activated B cells). The dynamic interplay among these factors determines whether cells commit suicide in response to stimuli. Key apoptotic mediators include the proteolytic caspases. Caspase activation is responsible for apoptotic events including mitochondrial damage, nuclear membrane breakdown, DNA fragmentation, chromatin condensation, and eventually, cell death. Additionally, several caspase-independent apoptotic pathways are known (32). The Bcl-2 (B-cell lymphoma 2) family of proteins, for example, is also involved in the regulation of apoptosis. Key anti-apoptotic members include Bcl-2 and Bcl-X_L, whereas pro-apoptotic members include Bax and Bak. Two distinct apoptotic pathways have been described: an *extrinsic* pathway that is mediated by cell membrane receptors and an *intrinsic* pathway that is controlled by mitochondria (33). A simplified schema of the different apoptotic pathways can be found in Ref. 27.

GLUTAMATERGIC TRANSMISSION AND NR INVOLVEMENT
IN ANESTHETIC-INDUCED NEURODEGENERATION

As an excitatory neurotransmitter, glutamate promotes specific aspects of nervous system development including nerve cell growth, migration, and differentiation. Glutamate is also thought to function in the maintenance of neuronal plasticity throughout life (34). Abnormal neuronal development and synaptic plasticity and neurodegeneration are thought to be associated with anesthetic-induced neurotoxicity. Glutamate receptors are classified into three families based on their pharmacological responses to the agonists α-amino-3-hydroxy-5-methylisoxazolepropionic acid (AMPA), kainic acid (KA), and NMDA. It is now known that critical aspects of nervous system development and function (e.g., synaptic formation and plasticity) are critically dependent on NR function and that neurological damage caused by a variety of pathological states can result from exaggerated activation of NR activated ion channels (35,36).

NRs are widely distributed in the mammalian central nervous system and function as ligand-activated ion channels. NRs are composed primarily of two subunit families: NR1 with eight splice variants and NR2 (A–D) (37–43). The NR1 subunit is essential for receptor function, whereas the specific subunit composition affects specific ligand-binding affinities (44). During the first two weeks of life in the rat, NRs are hypersensitive and neurons bearing NRs are very sensitive to excitotoxic degeneration (45). Paradoxically, it has been shown that NR antagonists are very effective in triggering apoptosis in the rat forebrain during this sensitive period (PND 7) (17,18). Possible explanations for this observation were thought to include either a pathological upregulation of NRs or a deprivation of critical NR-mediated signals.

Previous works showed that, in adult rats, repeated administration of the non-competitive NR antagonist phencyclidine (PCP) results in a substantial loss of cells in the olfactory, cingulate, and dorsolateral cortices (46,47). This effect was shown to involve apoptotic mechanisms associated with increased NR synthesis and altered function (46,48). Behaviorally, this treatment was accompanied by robust locomotor sensitization to subsequent PCP challenge (47,49). It was postulated that upregulation of NRs was responsible for the increased sensitivity to later exposure to PCP. Perinatal treatment of rat pups with PCP altered prepulse inhibition responses, enhanced locomotor sensitization to PCP challenge, and delayed acquisition of tasks with both learning and memory components. It was hypothesized that these effects of PCP would also be preceded by increases in NR synthesis. Since other anesthetic agents also have NR blocking properties, it was, therefore, hypothesized that such agents (e.g., ketamine) would also have effects related to NR upregulation. A likely scenario is that exposure of developing brains to the NR blockers causes a compensatory upregulation of NRs, making the cells bearing these receptors more vulnerable to the excitotoxic effects of exogenous glutamate, once the anesthetic has been washed out.

In situ hybridization studies using rat brain organotypic cultures and a labeled oligonucleotide probe that targets NR1 subunits have shown that PCP treatment does, in fact, upregulate NR1 subunits in frontal cortex and nucleus accumbens (27). PCP was also shown via electron microscopic evaluation to induce apoptosis as evidenced by the presence of nuclear condensation typical of that process. The observation of an increase in the number of NR receptors leads to the hypothesis that the excitotoxic effects of glutamate would be largely mediated by an increase in Ca^{2+} influx through activation of the now more numerous NRs (50,51). Increases in reactive oxygen species (ROS) of suspected mitochondrial origin are known to be associated with Ca^{2+} influx (52). Glutamate-induced Ca^{2+} overloading of mitochondria beyond their buffering capacity reduces their membrane potentials and disrupts electron transport, resulting in the increased production of reactive free radicals (superoxide anions) (53). Several experiments have shown that inhibition of the electron transport chain by rotenone (complex I), antimycin A (complex III), or oligomycin (complex V) prevents ROS formation and in some cases can be neuroprotective (53).

Although apoptosis may result from an excitotoxic insult, the pathways leading from mitochondrial dysfunction and ROS generation to cell death are not completely understood. The use of metalloporphyrins such as manganese tetrakis (4-benzoyl acid) porphyrin has implicated the oxygen radical in glutamate-induced exicitotoxicity (53,54). However, these reagents lack specificity for the oxygen radical compared to M40403, a selective nonpeptidyl superoxide dismutase mimetic (54,55), and are much less potent. Recently (56,57), it was demonstrated that M40403 significantly attenuates the toxicity associated with both NR agonists and antagonists at concentrations about 100-fold lower than those needed for the metalloporphyrins (54). Thus, M40403 appears to be a valuable tool for exploring the specific role of oxygen radicals in NR antagonist–induced apoptosis and suggests its potential therapeutic utility in central nervous system diseases that involve the production of these radicals.

Consequences of increased oxidative stress include the activation and inactivation of redox-sensitive proteins (58). The transcription factor NF-κB is known to respond to changes in the cytoplasmic redox state and has been shown to translocate in response to NR antagonist–induced cellular stress (59). NF-κB is normally sequestered in the cytoplasm, bound to the regulatory protein IκB. In response to a wide range of stimuli including oxidative stress, infection, hypoxia, extracellular signals, and inflammation, IκB is phosphorylated making it a target for ubiquitination and subsequent proteasome degradation (60). This results in the release of the NF-κB dimer, which is then free to translocate into the nucleus. The ability of M40403 to prevent NR antagonist–induced nuclear translocation of NF-κB and neurotoxicity strongly suggests that the oxygen radical is a key player in this pathway.

Previous research has shown a complex and often contradictory role for NF-κB in neuronal apoptosis. Studies examining the effects of hypoxia and reoxygenation, serum withdrawal, and extracellular signaling proteins have generally found a protective role for NF-κB (61). On the other hand, NF-κB translocation appears to be a necessary step in apoptosis induced by cyanide and excitotoxic stimuli (62). The mechanism by which NF-κB translocation induces apoptosis is not completely clear, but it is assumed that it involves the regulation of one or more genes that play a role in apoptosis. However, since NF-κB is known to regulate both anti-apoptotic and pro-apoptotic proteins depending on cell type and nature of stimulus (60–62), it is possible that the ultimate effect will be the sum of several downstream regulators. In neurons, astrocytes, and glia, these regulators include the anti-apoptotic Bcl family members, Bcl-X and Bcl-2 (61,63), the important antioxidant manganese superoxide dismutase, and the potentially detrimental proteins p53, inducible nitric oxide synthase, and cyclooxygenase-2 (64).

The prevention of NR antagonist–induced cell death by SN50, a peptide inhibitor of NF-κB translocation, clearly suggests that the overall effect of NF-κB translocation is pro-apoptotic (56). The ability of SN50 to prevent NMDA-induced apoptosis and increased expression of Bax (56) demonstrated that NF-κB is crucial to this process. Furthermore, the antagonistic effect of M40403 on NR antagonist–induced increases in NF-κB translocation and Bax strengthens the argument that increased NF-κB is critical in Bax upregulation.

It is hypothesized that superoxide generation via mitochondrial dysfunction and perhaps other Ca^{2+}-sensitive enzymatic pathways occurs upstream of NF-κB activation and acts as an intracellular signal by changing the redox state of the cell. The resulting translocation of NF-κB is postulated to be pro-apoptotic, either directly or indirectly, leading to a relative change in the expression of the Bcl family proteins Bax (pro-apoptotic) and Bcl-X_L (anti-apoptotic). On the basis of these data, pharmacological interruption of these events at one or several places would seem to have therapeutic merit.

NR DYSFUNCTION, NEURODEGENERATION, AND SYNAPTOGENESIS

Previous studies in rats demonstrated that perinatal PCP administration results in subsequent behavioral abnormalities (observed in adolescence) and that these effects were likely related to enhanced apoptotic cell death observed in frontal cortical neurons (64,65). These cortical deficits could have a significant impact on the function of subcortical structures, such as the nucleus accumbens, which serves as an important regulatory center by integrating the functions of the basal ganglia and the limbic system.

In a later study, organotypic brain slice cultures were used to examine the relationship between the neurotoxic effects of PCP in the frontal cortex and the striatum and nucleus accumbens (57). Organotypic slices allow a more direct investigation of mechanisms underlying toxicity by providing a simplified system that retains its original cytoarchitecture including connections between brain areas. In these experiments, PCP treatment of corticostriatal slices caused a significant increase in apoptosis as indicated by TUNEL-positive neurons in the cortex (57). There was also a concomitant decrease in the neuronal surface marker, polysialic acid neural cell adhesion molecule (PSA-NCAM) in the striatum. To focus in on the synapses from striatal-nucleus accumbens slices, synaptoneurosomes were isolated and immunostained using an antibody to PSA-NCAM. Intense PSA-NCAM immunostaining was observed on both pre- and post-synaptic membrane surfaces in control cultures but greatly diminished (~50%) in PCP-treated preparations (57). This decrease in PSA-NCAM could be the direct result of local NR blockade or the indirect result of cortical neurotoxicity accompanied by decreases in NMDAergic input to the striatum. The observation that M40403, the superoxide dismutase mimetic, blocked cortical apoptosis as well as the loss of PSA-NCAM immunoreactivity in the ventral striatum supports the latter mechanism.

The results from several experiments show that PCP causes significant cortical cell death as evidenced by increased DNA condensation and fragmentation. In addition, the selective cortical apoptosis induced by PCP in vivo can be mimicked in vitro where it can also be demonstrated that cortical neurodegeneration negatively impacts synaptogenesis in the striatum, likely due to loss of corticostriatal afferent neurons and/or the blockade of NRs in the striatum. It seems probable that the dysfunction of striatal synaptogenesis contributes to the behavioral abnormalities observed following perinatal PCP administration.

GENE EXPRESSION CHANGES ASSOCIATED WITH ANESTHETIC-INDUCED NEURONAL CELL DEATH

Genomic data may be used to describe the gene expression changes induced by anesthetic agents including ketamine (66). To determine which of several candidate death genes were associated with anesthetic drug-induced apoptosis, gene microarray techniques were utilized. Neonatal rats (PND 7) were injected subcutaneously with either water or 40 mg/kg ketamine. Brain tissue was harvested one, two, and four hours after the injection to identify significant gene expression changes in the thalamus that might be related to acute ketamine exposure. Laser capture microdissection was used to collect approximately 500 cells from the thalamus. Microarray analyses identified 18 significant (fold change >1.5 and $p < 0.05$) gene expression changes in the brain one hour after treatment, 624 significant gene expression changes two hours after treatment, and 52 significant gene expression changes four hours after treatment. Of these, several genes were specifically associated with apoptosis (CYCS, ELL, PDCD8, PRKCB1, and RIPK1), and others associated with oxidative stress were found to be upregulated in the brain following a single injection of ketamine (67,68).

CONCLUSION

Although not fully developed, the working model for NR antagonist–induced neurodegeneration involves the modulation of normally occurring brain-sculpting mechanisms that control CNS development. Exposure of the developing brain to NR antagonists during critical developmental periods perturbs the NR system and results in enhanced neuronal cell death (69). Upregulation of the NR1 subunit by NR antagonists appears critical to the subsequent expression of cell death that includes a variety of intracellular components downstream. Iterative perturbations of the test system allowed refinement of the general model and reinforced the selection of specific pathways that explain the whole of the data to date.

Several data gaps/research needs have been identified as part of the systems toxicology approach. For example, proteomic and additional global genomic data are still needed. These datasets will be integrated into the model, discrepancies will be identified, and follow-on hypothesis-driven studies will be conducted. Thus, new data generated will be used to recast the model. Although many more studies will be necessary to achieve a quantitative, mathematical model, a general outline has been postulated and the test system has been selectively challenged in an iterative manner as defined by a systems toxicology approach. Precise developmental stage and dose-response experiments with the use of the phenotypic anchor neuronal cell death—and "omics" data remain to be completed. As these data become available the model will be improved allowing for a better understanding of the neurotoxicity associated with NR antagonists, some of which are commonly used pediatric anesthetic agents. Integration of these data with the adverse functional consequences now known to occur with suprathreshold exposures to these agents in rodents (70) and nonhuman primates (71) will form the basis for further refinement of the model and hopefully the development of rational therapeutic and preventative strategies.

REFERENCES

1. Bellinger DC. Children's cognitive health: the influence of environmental chemical exposures. Altern Ther Health Med 2007; 13(2):S140–S144.
2. Daston G, Faustman E, Ginsberg G, et al. A framework for assessing risks to children from exposure to environmental agents. Environ Health Perspect 2004; 112(2):238–256.
3. Ginsberg G, Slikker W Jr., Bruckner J, et al. Incorporating children's toxicokinetics into a risk framework. Environ Health Perspect 2004; 112(2):272–283.
4. Olney JW. New insights and new issues in developmental neurotoxicology. Neurotoxicology 2002; 23(6):659–668.
5. USEPA (U.S. Environmental Protection Agency). OPPTS 870.6300, Developmental Neurotoxicity Study, Health Effects Test Guidelines, EPA 712-C-98-239, Washington, DC, 1998. Available at: http://www.epa.gov/opptsfrs/publications/OPPTS_Harmonized/870_Health_Effects_Test_Guidelines/index.html.
6. USFDA Redbook 2000. Toxicological Principles for the Safety Assessment of Food Ingredients. IV.C.9.b. Guidelines for Developmental Toxicity Studies. Center for Food Safety and Applied Nutrition, U.S. Food and Drug Administration, College Park, MD, 2000. Available at: http://www.cfsan.fda.gov/~redbook/redivc9b.html.
7. OECD Test Guideline 426. OECD Guideline for Testing of Chemicals. Developmental Neurotoxicity Study. Organisation of Economic Co-operation and Development, Paris, France, 2007. Available at: http://miranda.sourceoecd.org/vl=5375569/cl=26/nw=1/rpsv/cw/vhosts/oecdjournals/1607310x/v1n4/contp1-1.htm.

8. USFDA Guidance for Industry: Nonclinical Safety Evaluation of Pediatric Drug Products. Center for Drug Evaluation and Research, U.S. Food and Drug Administration, Rockville, MD, 2006. Available at: http://www.fda.gov/cder/guidance/5671fnl.pdf.
9. Makris SL, Kim J, Ellis A, et al. Current and Future Needs for Developmental Toxicity Testing. Accepted, Birth Def Res Part B, Dev Repro Tox 2009.
10. Slotkin TA. Guidelines for developmental neurotoxicity and their impact on organophosphate pesticides: a personal view from an academic perspective. Neurotoxicology 2004; 25:631–640.
11. Auffray C, Imbeaud S, Roux-Rouquie M, et al. From functional genomics to systems biology: concepts and practices. C R Biol 2003; 326:879–892.
12. Waters M, Boorman G, Bushel P, et al. Systems toxicology and the Chemical Effects in Biological Systems (CEBS) knowledge base. EHP Toxicogenomics 2003; 111:15–28.
13. Noble D. The future: putting Humpty-Dumpty together again. Biochem Soc Trans 2003; 31:156–158.
14. Slikker W, Chang LW, eds. Handbook of Developmental Neurotoxicology. San Diego: Academic Press, 1998.
15. Slikker W Jr. Biomarkers of neurotoxicity: an overview. Biomed Environ Sci 1991; 4:192–196.
16. Anger WK. Worker exposures. In: Annau Z, ed. Neurobehavioral Toxicology. Baltimore: Johns Hopkins Press, 1986:331–347.
17. Ikonomidou C, Bosch F, Miksa M, et al. Blockade of NMDA receptors and apoptotic neurodegeneration in the developing brain. Science 1999; 283:70–74.
18. Scallet AC, Schmued LC, Slikker W Jr., et al. Developmental neurotoxicity of ketamine: Morphometric confirmation, exposure parameters, and multiple fluorescent labeling of apoptotic neurons. Toxicol Sci 2004; 80:1–7.
19. Miyamoto K, Nakanishi H, Moriguchi S, et al. Involvement of enhanced sensitivity of N-methyl-D-aspartate receptors in vulnerability of developing cortical neurons to methylmercury neurotoxicity. Brain Res 2001; 901:252–258.
20. Behar TN, Scott CA, Greene LC, et al. Glutamate acting at NMDA receptors stimulates embryonic cortical neuronal migration. J Neurosci 1999; 19:4449–4461.
21. Ikonomidou C, Bittigau P, Koch C, et al. Neurotransmitters and apoptosis in the developing brain. Biochem Pharmacol 2001; 62:401–405.
22. Pohl D, Bittigau P, Ishimaru MJ, et al. N-Methyl-D-aspartate antagonists and apoptotic cell death triggered by head trauma in developing rat brain. Proc Natl Acad Sci USA 1999; 96:2508–2513.
23. Hsu C, Hsieh YL, Yang RC, et al. Blockage of N-methyl-D-Aspartate receptors decreases testosterone levels and enhances postnatal neuronal apoptosis in the preoptic area of male rats. Neuroendocrinology 2000; 71:301–307.
24. Petry HM, Chen B, Yarbrough GI. Ethanol-induced apoptosis in the developing tree shrew brain. Program No 419.12. Abstract Viewer/Itinerary Planner. Washington, DC: Society for Neuroscience, 2003.
25. Llado J, Caldero J, Ribera J, et al. Opposing effects of excitatory amino acids on chick embryo spinal cord motoneurons: excitotoxic degeneration or prevention of programmed cell death. J Neurosci 1999; 19:10803–10812.
26. Terro F, Esclaire F, Yardin C, et al. N-methyl-D-aspartate receptor blockade enhances neuronal apoptosis induced by serum deprivation. Neurosci Lett 2000; 278:149–152.
27. Slikker W Jr., Xu Z, Wang C. Application of a systems biology approach to developmental neurotoxicology reproductive toxicology. J Reprod Toxicol 2005; 19:305–319.
28. Ideker T, Galitski T, Hood L. A new approach to decoding life: systems biology. Ann Rev Genomics Hum Genet 2001; 2:343–372.
29. Vaux DL, Korsmeyer SJ. Cell death in development. Cell 1999; 96:245–254.
30. Gold R, Hartung HP, Lassmann H. T-cell apoptosis in autoimmune diseases: termination of inflammation in the nervous system and other sites with specialized immune-defense mechanisms. Trends Neurosci 1997; 20:399–404.
31. Vila M, Przedborski S. Targeting programmed cell death in neurodegenerative disease. Nat Rev Neurosci 2003; 4:365–375.
32. Lorenzo HK, Susin SA, Penninger J, et al. Apoptosis inducing factor (AIF): a phylogenetically old, caspase-independent effector of cell death. Cell Death Differ 1999; 6:516–524.
33. Schultz DR, Harrington WJ Jr. Apoptosis: programmed cell death at a molecular level [Review]. Semin Arthritis Rheum 2003; 32(6):345–369.

34. Komuro H, Rakic P. Modulation of neuronal migration by NMDA receptors. Science 1993; 260:95–97.
35. Choi DW. Glutamate neurotoxicity and diseases of the nervous system. Neuron 1988; 1: 623–634.
36. Olney JW. Excitotoxic amino acids and neuropsychiatric disorders. Annu Rev Pharmacol Toxicol 1990; 30:47–71.
37. Ishii T, Moriyoshi K, Sugihara H, et al. Molecular characterization of the family of the N-methyl-D-aspartate receptor subunits. J Biol Chem 1993; 268:2836–2843.
38. Kutsuwada T, kashiwabuchi N, Mori H, et al. Molecular diversity of the NMDA receptor channel. Nature 1992; 358:36–41.
39. Monyer H, Sprengel R, Schoepfer R, et al. Heteromeric NMDA receptors: molecular and functional distinction of subtypes. Science 1992; 256:1217–1221.
40. Moriyoshi K, Masu M, Ishii T, et al. Molecular cloning and characterization of the rat NMDA receptor. Nature 1991; 354:31–37.
41. Zukin RS, Bennett MV. Alternatively spliced isoforms of the NMDAR1 receptor subunit. Trends Neurosci 1995; 18:306–313.
42. Durand GM, Gregor P, Zheng X, et al. Cloning of an apparent splice variant of the rat NMDA receptor NMDAR1 with altered sensitivity to polyamines and activators of protein kinase C. Proc Natl Acad Sci USA 1992; 89:9359–9363.
43. Paupard MC, Friedman LK, Zukin RS. Developmental regulation and cell-specific expression of NMDA receptor splice variants in rat hippocampus. Neuroscience 1997; 79(2):399–409.
44. Laurie DJ, Seeburg PH. Regional and developmental heterogeneity in splicing of the rat brain NMDAR1 mRNA. J Neurosci 1994; 14:3180–3194.
45. Ikonomidou C, Mosinger JL, Shahid Salles K, et al. Sensitivity of the developing rat brain to hypobaric/ischemic damage parallels sensitivity to N-methyl-aspartate neurotoxicity. J Neurosci 1989; 9:2809.
46. Wang C, Schowalter VM, Hillman GR, et al. Chronic phencyclidine increases NMDA receptor NR1 subunit mRNA in rat forebrain. J Neurosci Res 1999; 55:762–769.
47. Johnson KM, Phillips M, Wang C, et al. Chronic phencyclidine induces behavioral sensitization and apoptotic cell death in the olfactory and piriform cortex. J Neurosci Res 1998; 52:709–722.
48. Wang C, Kaufmann J, Sanchez-Rose M, et al. Mechanisms of N-methyl-D-aspartate-induced apoptosis in phencyclidine-treated cultured forebrain neurons. J Pharmacol Exp Ther 2000; 294:287–295.
49. Phillips M, Wang C, Johnson KM. Pharmacological characterization of locomotor sensitization induced by chronic phencyclidine administration. J Pharmacol Exp Ther 2001; 296:1–9.
50. Garthwaite G, Garthwaite J. Neurotoxicity of excitatory amino acid receptor agonists in rat cerebellar slices: dependence on calcium concentration. Neurosci Lett 1986; 66:93–198.
51. Choi DW. Ionic dependence of glutamate neurotoxicity. J Neurosci 1987; 7:369–379.
52. Malis CD, Bonventre JV. Mechanism for calcium potentiation of oxygen free radical injury to renal mitochondria. J Biol Chem 1985; 261:14201–14208.
53. Luetjens CM, Bui NT, Sengpiel B, et al. Delayed mitochondrial dysfunction in excitotoxic neuron death: cytochrome c release and secondary increase in superoxide production. J Neurosci 2000; 20:5715–5723.
54. Patel M, Day BJ, Crapo JD, et al. Requirement for superoxide in excitotoxic cell death. Neuron 1996; 16:345–355.
55. Salvemini D, Wang ZQ, Zweier JL, et al. A nonpeptidyl mimic of superoxide dismutase with therapeutic activity in rats. Science (Wash DC) 1999; 286:209–210.
56. McInnis J, Wang C, Anastasio N, et al. The role of superoxide and nuclear factor-κB signaling in N-Methyl-D-aspartate-induced necrosis and apoptosis. J Pharmacol Exp Ther 2002; 301:478–487.
57. Wang C, Anastasio N, Popov V, et al. Blockade of N-Methyl-D-aspartate receptors by phencyclidine causes the loss of corticostriatal neurons. Neuroscience 2004; 125:473–483.
58. Kamata H, Hirata H. Redox regulation of cellular signaling. Cell Signal 1999; 11:1–14.
59. Ko HW, Park KY, Kim H, et al. Ca^{2+} mediated activation of c-Jun N-terminal kinase and nuclear factor κB by NMDA in cortical cell cultures. J Neurochem 1998; 71:1390–1395.

60. Bowie A, O'Neill LA. Oxidative stress and nuclear factor κB activation. Biochem Pharmacol 2000; 59:13–23.

61. Tamatani M, Che YH, Matsuzaki H, et al. Tumor necrosis factor induced Bcl-2 and Bcl-x expression through NFκB activation in primary hippocampal neurons. J Biol Chem 1999; 274: 8531–8538.

62. Shou Y, Gunaseker PG, Borowitz JL, et al. Cyanide-induced apoptosis involves oxidative-stress-activated NF-KappaB in cortical neurons. Toxicol Appl Pharmacol 2000; 164:196–205.

63. Chen C, Edelstein LC, Gelinas C. The Rel/Nf-κB family directly activates expression of the apoptosis inhibitor Bcl-XL. Mol Cell Biol 2000; 20:2687–2695.

64. Wang C, McInnis J, West JB, et al. Blockade of phencyclidine-induced cortical apoptosis and deficits in prepulse inhibition by M40403, a superoxide dismutase mimetic. J Pharmacol Exp Ther 2003; 304:266–271.

65. Wang C, McInnis J, Ross-Sanchez M, et al. Long-term behavioral and neurodegenerative effects of perinatal phencyclidine administration: implications for schizophrenia. Neuroscience 2001; 107:535–550.

66. Wang C, Slikker W. Strategies and experimental models for evaluating anesthetics: effects on the developing nervous system. Anesth Analg 2008; 106(6): 1643–1658.

67. Wright LKM, Twaddle N, Branham W, et al. A single administration of ketamine produces an inflammatory response in the developing rat brain. Neurotoxicol Teratol 2005; 27:380.

68. Slikker W, Paule MG, Wright LKM, et al. Systems biology approaches for toxicology. J Appl Toxicol 2006; 27: 201–217.

69. Haberny KA, Paule MG, Scallet AC, et al. Ontogeny of the N-methyl-D-Aspartate (NMDA) receptor system and susceptibility to neurotoxicity. Toxicol Sci 2002; 68:9–17.

70. Jevtovic-Todorovic V, Hartman RE, Izumi Y, et al. 2003 Early exposure to common anesthetic agents causes widespread neurodegeneration in the developing rat brain and persistent learning deficits. J Neurosci 2003; 23: 876–882.

71. Paule MG, Li M, Zou X, et al. Early postnatal ketamine anesthesia causes persistent cognitive deficits in rhesus monkeys. Paper presented at the Society for Neuroscience Annual Meeting, 2009, Chicago, Illinois, # 413.7.

10 Developmental toxicology of the respiratory system

Raymond G. York and Robert M. Parker

ABNORMAL DEVELOPMENT OF THE RESPIRATORY SYSTEM

The lung is a remarkable organ in that it does not become functional until birth and then must be prepared to function adequately, or it would put the life of the organism at risk. From the moment of birth the lung provides the body with oxygen and eliminates carbon dioxide. Over the past few decades it has become increasingly apparent that both environmental and genetic factors operating during early life can induce persistent alterations in the respiratory system.

Prenatal lung development can be divided into a sequence of phases that takes place at specific times and involves specific events. As such, the timing and conditions of adverse exposure are critical because there are discrete windows of vulnerability in the developing respiratory system. Infant mortality in the United States has been steadily decreasing from 26.8 per 1000 live births in 1950, down to 7.1 per 100 in 1997 (1). Congenital malformations are one of the leading causes of infant mortality in the United States, accounting for approximately 20% of all infant deaths. Infant mortality from defects of the respiratory system was 0.25 per 1000 in the United States, in 1991–1997, with hypoplasia and dysplasia of the lungs accounting for 80% of all deaths in this category (1). Mortality from respiratory distress syndrome is also declining, from 2.3 deaths per 1000 live births in 1968, down to 0.6 per 100 in 1991 (2). This chapter will examine the normal and abnormal development of the respiratory system at the molecular, gene, cellular, and tissue level. The objectives of this chapter are to present current concepts of normal and abnormal lung development including transcription factors involved in respiratory system morphogenesis and regulating perinatal lung maturation; to integrate the growing body of evidence relating genetic factors to the structure and functions of the respiratory system; and to identify future directions for research into factors adversely influencing respiratory development.

INTRODUCTION AND BACKGROUND

The respiratory system consists of a number of anatomic regions: the extrathoracic region consisting of the anterior nose and the posterior nasal passages, larynx, and mouth; the tracheobronchial region consisting of the trachea, bronchi, bronchioles, and terminal bronchioles; and the alveolar-interstitial region consisting of the respiratory bronchioles, alveolar ducts, and sacs; and the interstitial connective tissue. The roles the perinatal lung plays in metabolism, immune defense, and endocrine function will not be addressed in this chapter. The mammalian fetus develops in an aquatic environment where the placenta provides not only respiratory functions but also gastrointestinal, hepatic, polyendocrine, and renal functions (3). Lung morphogenesis and development involve stereotypic airway branching with the differentiation of more than 40 different cell types, and a highly ordered airway branching system with 25,000 distinct terminations giving rise to more than 300 million alveoli (4). The lung is unique in that it grows while not fulfilling its postnatal function in utero, but it must immediately be ready to function as an efficient gas-exchanger at the moment of birth. Before birth, the human lung secretes a staggering 5 mL/kg/hr of fluid. At birth, this secretory function must be switched off, and the lung must become an absorptive organ (5).

Perinatal lung development can be divided into several distinct phases that are crucial to its future role as a gas-exchanger. Growth and development of the human respiratory system are not complete until approximately 18 to 20 years of age (6), with greater than 80% of alveoli arising postnally in the adult stage. In essence, lung development is a continuum from embryogenesis through early adulthood, and inter-ference with any of these phases can have an immediate effect on lung maturation or survival (7). Birth defects involving the respiratory system (mainly the lungs, trachea, and nose) are life-threatening but less common than those involving other major organs. Lung agenesis (failure of one or both lungs to develop) or hypoplasia (one or both lungs are abnormally small) are the most important of the respiratory system defects in terms of medical significance and prevalence. The overall rate of occurrence for respiratory system defects in the State of Illinois from 1995 to 1999 was 10.6 per 10,000 births. During that time the incidence of lung agenesis/hypoplasia was 3.0 per 10,000 live births (8). Another surveillance survey, the New South Wales Birth Defects Register, reported in 1994, an overall occurrence rate of respiratory system defects as 6 per 10,000 births, with defects of the nose, larynx/trachea/bronchus, lungs, and diaphragmatic hernia as 1, 2, 3, and 3 per 10,000 births, respectively (9).

LUNG MORPHOGENESIS

Lung development is divided into five distinct but overlapping phases on the basis of anatomic and histological characteristics: embryonic, glandular, canalicular, saccular, and alveolar (10,11) (Table 1). The embryonic phase encompasses the formation of the lung buds from the cranial foregut and three generations of branching to form the primordia of the two lungs, five lobes, and the bronchopulmonary segments. The embryonic phase leads to airway formation and occurs during the third to eighth week of gestation in humans (15). In the fourth week of gestation, the primordium of the human respiratory system appears as an outgrowth from the laryngotracheal groove that lies ventrally along the surface of the cranial foregut. Each terminal lobe of the endodermal growth, together with its surrounding splanchnic mesenchyme, constitutes a lung bud from which all the tissues of the corresponding lung and bronchial tree will be derived. The epithelium of the internal lining of the lungs, bronchi, trachea, and larynx is therefore endodermal in origin (16). Muscular and cartilaginous components of the respiratory system are derived from the splanchnopleuric mesoderm surround-ing the foregut. By the fifth week of gestation, after rapid endodermal branching, the trachea separates from the esophagus, and the left and right lung lobes with the main bronchi are established (10). By the seventh week of gestation, three generations of

TABLE 1   Phases of Lung Development (Approximate Ages)

	Embryonic	Glandular	Canalicular	Saccular	Alveolar
Mouse	GD 9–11.5	GD 11.5–16.5	GD 16.5–17.5	GD 17.5–PND5	PND5–28
Rat	GD 9–13	GD 13–18	GD 19–20	GD 21–PND7	PND7–PND21
Rabbit	GD 6–19	GD 19–24	GD 24–27	GD 27–28	GD 30–?
Sheep	–	GD–95	GD 95–120	GD 120–?	GD 120–?
Human	GW 3–7	GW 6–17	GW 16–26	GW 24–38	GW 38 to 8 yr

Abbreviations: GD, gestation day; GW, gestation week; PND, postnatal day.
Source: Adapted from Refs. 12–14,243,244.

branching form the primordium of the two lungs, five lobes (formed by the three right and two left bronchi), and the bronchopulmonary segments.

The glandular (also called the pseudoglandular) phase beginning at approximately the 6th week and lasts until the 16th to 17th week of gestation is the period of proliferation of bronchial branches (14 generations) forming terminal bronchioles with acinar tubules and buds. Vasculogenesis and innervation commence during this phase. A defining characteristic of this phase is the "pseudoglandular" appearance of lung tissue upon histological evaluation. This feature persists up to the 16th week of gestation. During this phase, the epithelial cells lining all branches of the bronchial tree are continuous cuboidal or low columnar (17). Mucous glands then appear in the trachea and soon begin to secrete mucus. Cartilaginous rings develop in the mesoderm during the 7th week of gestation, and cartilage reaches the lobar and segmental bronchi in the 11th and 12th weeks of gestation (18). This phase also includes the closure of the pleural and peritoneal cavities and the differentiation of the smooth muscle, and development of connective tissue, collagen, and elastin fibers (10).

The canalicular phase entails the further division into respiratory bronchioles with greater organization of the pulmonary vascular bed and increasing innervation. During this phase, the epithelium attenuates to form a flattened continuous layer that forms the alveolar epithelium in the adult. The cells of the conducting airways differentiate into diverse cell types. The trachea and bronchia become lined with pseudostratified epithelium consisting of goblet, basal, nonciliated secretory (Clara), and ciliated cells. The smaller conducting airways (bronchioles to alveoli) become lined with simple columnar epithelial cells composed primarily of ciliated and nonciliated (Clara) cells; in addition, a relatively rare cell type, the pulmonary neuroendocrine cells (PNEC), often clustered in a neuroendocrine body (NEB), first appear. The number and variety of cell types vary developmentally, usually along the proximal-distal axis of the lung, and are highly species specific. There is growth of capillaries adjacent to the respiratory bronchioles and the region becomes richly vascularized. The capillary bed comes into close apposition with the epithelial layer of the bronchioli establishing the air-blood barrier, and the blood supply increases (19). Acinar formation proceeds by further branching and lengthening of the terminal tubules. The production of surfactant follows the differentiation of the alveolar epithelial type I cells (AEC I) and type II cells (AEC II), which contain recognizable lamellar bodies from cuboidal epithelial cells (20).

The saccular-alveolar phase is characterized morphologically by a marked decrease in interstitial tissue and development of narrow and compact layers of lining cells. This phase is a particularly vulnerable time in mammalian development, and it marks the transition from a fluid-filled to an air-filled lung on which postnatal survival depends. Respiration becomes possible when some of the cells of the cuboidal type I epithelial cells of the respiratory bronchiole change, becoming increasingly squamous, providing close apposition between the respiratory epithelium and the pulmonary blood vessels. The peripheral lung tissues thin as the saccules dilate. The double-capillary structure in the saccular walls transforms into a single-capillary epithelial layer seen in the mature alveoli during the alveolar phase. Sufficient capillaries are now present to provide adequate gas exchange, permitting a prematurely delivered infant to survive. The future airspaces are filled with fetal pulmonary fluid that is rapidly resorbed by the blood and lymph capillaries. A small amount of the fetal pulmonary fluid in the airspaces is expelled via the trachea at the onset of labor and during vaginal delivery. At birth, pulmonary vascular resistance falls, pulmonary blood flow increases, lung fluid is resorbed, and surfactant lipids and proteins expressed by the type II cells are secreted into the peripheral saccules of the lung. The pulmonary surfactant reduces the surface tension and prevents alveolar collapse once the lung is filled with air. At birth, rat lungs, unlike humans, have no alveoli or alveolar ducts (21). Birth does not signal the end of lung development. There is a continuing complex process of lung growth after birth, permitting changing relationships of airway size, alveolar size, and surface area. Most alveolar development occurs after birth. The neonate has 50 million alveoli at birth and an additional 250 million develop during early childhood. Septation dramatically increases the alveolar surface area from 3 to 70 m^2 from birth through childhood (15).

TRANSCRIPTIONAL CONTROL OF LUNG MORPHOGENESIS

Detailed descriptions of the transcription factors involved with lung morphogenesis are beyond the scope of this chapter; however, several recent review articles on various aspects of lung morphogenesis are available (14,22–29). Detailed information on most transcription factors and genes discussed later can be found on the National Institutes of Health website called "Entrez Gene" (http://www.ncbi.nlm.nih.gov). Gene regulation and transcription factors involved in lung morphogenesis have been extensively studied in the mouse and therefore the transcription factors and the timing of the events described below will be primarily based on mouse data (Table 2).

During lung morphogenesis, multiple cell types are derived from the ectodermal, mesodermal, and endodermal cells. During the formation of the lung, multiple signaling centers are involved in the control of autocrine-paracrine signaling between cells that

TABLE 2 Transcription Factors Associated with Abnormalities in Lung Morphogenesis or Function

Branching defects
  N-MYC/MYCN, LKLF/KLF2, RDC1, CATENIN/CTNNB1, SOX 17, SMAD 1, SMAD2, SMAD3, SMAD4, SMAD5, SMAD7, ERM, TBX4/5, DICER, POD1/TCFZ1
Structure defects
  N-MYC/MYCN, TTF-1/TITF1, HOP/HOD, RAR, GLI2, GLI3, E2F1
Left-right symmetry defects
  FOXJ1, PITX2, ZIC3
Sacculation defects
  HIF2A, CREB1, CEBPA, HOXA5, CUTL1, PROP1, GR/NR3C1, GATA-6, SP-3, SRC1/TIF2/NCOA, p300/EP300, CBP/CREBBP, CARM1, MRG15/MORF4L1, SOX11, NF1, NFATC3(w/CNB1), E2F1
Alveolarization defects
  FOXA2, FOXF1, CUTL1, ERβ/ESR2, SMAD3
Vascularization defects
  HIF1A, HIF2A, FOXF1, FOXM1
Ciliogenesis defects
  FOXJ1
Differentiation defects
  FOXA1, TTF-1/TITF1, GATA-6, DICER
Differentiation defects (goblet cells)
  FOXA2, CATENIN/CTNNB1, SPDEF
Differentiation defects (basal cells)
  P63/TP73L
Differentiation defects (dendritic cells)
  RUNX3
Differentiation defects (pulmonary neuroendocrine cells)
  ASCL1/MASH1, GLI1, GFI1, SOX17, RBI
Differentiation defects (no pulmonary neuroendocrine cells)
  HES1
Differentiation defects (alveolar macrophages)
  PU.1/SPI1
Differentiation defects (submucosal glands)
  TCF/LEF

affect transcriptional processes regulating gene expression and cell differentiation (14). Lung morphogenesis depends on both tissue-selective and ubiquitous transcription factors and genes. The principles that govern the design and evolution of transcriptional networks operating during lung formation and function have just begun to be understood; however, *unique* transcription factors specifying lung formation have not yet been identified (14). To date, a number of transcription factors and their binding sites have been characterized and associated with the regulation of lung-specific genes during lung morphogenesis. This section provides a brief summary of the roles of a number of transcription factors influencing lung morphogenesis, with particular focus on developmental processes critical for lung morphogenesis and for lung function at the time of birth.

Transcription factors active in the endodermally derived cells of the developing lung tubules are required for normal lung morphogenesis and function, including TTF-1[a] (thyroid transcription factor 1), β-catenin, FOX (Forkhead orthologs), GATA (capable of binding to DNA sequence "GATA"), SOX (Sry-related HMG Box), and ETS (proteins capable of binding to the DNA-binding domain Ets) family members (14).

Prior to the formation of the thyroid, lungs, and stomach, transcription factors or markers characteristic of these organs are expressed along the foregut. Transcription factors controlling foregut growth and differentiation [e.g., FOXF1, GLI (zinc finger transcription factor family that mediates Hedgehog (*Hh*) signaling family members)] are also critical in lung morphogenesis. These include *Foxa2, Catnb, Sox17, Gata-6, Stat3* genes, and other transcription factors, expressed in the foregut well before the formation of the lung (30–33), while many of these early transcription factors are also involved in lung morphogenesis later in gestation and into the postnatal period (34). Deletion or partial deletion, over-expression or underexpression, or altered timing of expression of genes or transcription factors that regulate lung morphogenesis can result in malformations, growth retardation, functional deficits, or death. Drugs, chemicals, or physical factors that alter these complex temporal-spatial interactions may cause severe, even life-threatening, birth defects.

Transcription Factors During Embryonic and Pseudoglandular Periods

Fibroblast growth factor (FGF) signaling centers, located in the cardiac mesenchyme and surrounding splanchnic mesenchyme, signal the TFF-1-expressing cells. The lung primordium occurs in a region of the foregut with cells coexpressing TFF-1 and FOXA2, while the thyroid primordium occurs in a region of the foregut coexpressing TFF-1 and PAX-8 (paired box–containing genes) (35,36). The FGF's signal intensity and timing induce the formation of the lung primordium as the foregut forms (37). TTF-1 plays a critical role in the regulation of lung morphogenesis, epithelial cell differentiation, and the expression of genes on which perinatal respiratory adaption depends (14). The multiple functions of TTF-1 are highly dependent on its temporal-spatial expression, interactions with other transcription factors, and its responses to various external conditions, as occurs during injury after birth. TTF-1 directly binds and regulates the expression of genes selectively expressed in respiratory epithelial cells, including genes regulating surfactant synthesis and homeostasis [surfactant protein (*Sftp*) genes -a, -*b*, -*c*, and -*d*] (35,38). TTF-1 also influences vasculogenesis, host defense, fluid homeostasis, and pulmonary inflammation before birth (39).

The FOX transcription factors are expressed in the lung, including FOXA1, FOXA2, FOXF1, FOXJ1, FOXM1, and FOXP family members (40). Prior to lung formation FOXA1

---

[a]Capital letters are used to denote transcription factors (e.g., FOX), lowercase italicized letters are used to denote mouse genes [e.g., *shh* (the sonic hedgehog gene)], while capital italicized letters are used to denote human genes (e.g., *HASH1*).

and FOXA2 are expressed in the foregut endoderm. These transcription factors are expressed in an overlapping pattern with TTF-1 in respiratory cells later during lung morphogenesis and in the mature lung (41,42). FOXA1 and FOXA2 influence differentiation of respiratory epithelial cells, regulating genes important for lung function at birth (43–45). FOXA1/A2 regulates the production of SHH (sonic hedgehog) by endodermally derived cells. SHH is critical for differentiation of pulmonary mesenchymal precursors and is involved in the regulation of smooth muscle differentiation (14).

During lung morphogenesis, Wnt ligands activate multiple entities that cause uncoupling of β-catenin from the degradation pathway and its subsequent nuclear entry (46). The function of Wnt pathway in lung morphogenesis has recently been reviewed (47–49). The Wnt signaling and β-catenin pathway regulates branching morphogenesis, as well as airway epithelial differentiation.

Like FOXA2 and TTF-1, GATA-6 is expressed in respiratory epithelial cells throughout morphogenesis and is required for the survival of the endodermally derived progenitors that form the bronchiolar epithelium (32). Later in development, GATA-6 influences sacculation and alveolarization of the lung.

Defects Associated with Transcription Factors During Embryonic and Pseudoglandular Periods

Although TTF-1 is required for lung organogenesis, mice lacking TTF1 can form the upper trachea and mainstem bronchi (50,51). Lungs of TTF-1 knockout ($Titf-1^{-/-}$) homozygous mice have a tracheoesophageal fistula with intact main bronchial tubes and an absence of peripheral lung structures, which is consistent with TTF-1's role in regulating stereotypic lung arborization. TTF-1 deletions in the mouse also cause malformations of the forebrain, thyroid, and lung (50). Mutations in human $TTF-1$ gene have been associated with hypothyroidism and respiratory failure in infants (52). Deletion of transcription factors and signaling molecules, including SHH, SOX9, and RAR (retinoic acid receptor) and FGF receptors, disrupts tracheal cartilage formation in the mouse lung (53). Deletion of both $Foxa$ genes during the canalicular period severely disrupts branching morphogenesis, resulting in lung hypoplasia, cyst formation, and lack of epithelial and smooth muscle differentiation. $Foxa1^{-/-}$ homozygous mice survived after birth manifesting delays in lung maturation but no alterations in branching morphogenesis (54). Deletion of FOXA2 in respiratory epithelium also does not alter branching morphogenesis; however, it delays the maturation of the peripheral lung resulting in death shortly after birth. $Shh^{-/-}$ homozygous mice develop tracheoesophageal fistula with simple unbranched cyst-like lung sacs. Deletion of β-catenin is lethal in the early mouse embryonic period, well before lung development, while deletion or inhibition of β-catenin in respiratory cells impairs branching morphogenesis and inhibits peripheral airway cell differentiation (55,56). Expression of dominant-negative GATA-6 in mouse respiratory epithelium inhibited lung differentiation in late gestation and decreased expression of surfactant proteins and aquaporin-5 (57). Increased activity of GATA-6 perturbs pulmonary function and inhibits alveolarization (58,59).

Transcription Factors in the Mesenchyme During Branching Morphogenesis

The nonendodermal compartment of the lungs is formed from the splanchnic mesenchyme. This compartment includes cartilage, bronchial smooth muscle, arteries, veins, capillaries, lymphatics, and other stromal tissues. The transcriptional mechanisms that control differentiation and proliferation of these mesenchymally derived cell types are not well understood. The FOXF1, FOXM1B, HIF, GLI, and the HOX family members have been identified as playing important roles in the development of the pulmonary mesenchyme.

FOXF1 expression is limited to the more distal mesenchyme of the developing lung and the muscle layer of the bronchus. In its role regulating mesenchymal-epithelial interactions, FOXF1 reduces the expression of a number of genes critical for lung morphogenesis and function (60–62). FOXM1B activates genes involved with cell cycle progression, G_1/S and G_2/M, via Cdk-cyclin complexes and p300/CBP coactivators (63). FOXM1B activates cell cycle–promoting factors such as cyclin A2, cyclin E, cyclin B1, cyclin F, and Cdk1, while decreasing the levels of p21, a Cdk inhibitor (64). During lung development and repair, FOXM1B regulates genes essential for proliferation of alveolar and bronchial epithelial cells, as well as smooth muscle, extracellular matrix remodeling, and vasculogenesis.

Hypoxia-inducible factors (HIFs) are transcription factors that respond to physiological and pathophysiological factors related to hypoxia, and therefore play an important role in regulation of perinatal lung function and hypoxia-induced pulmonary vascular remodeling. The transcription targets of HIF-1α and role of HIF in pulmonary pathophysiology have been recently reviewed (65–67). The HIFs regulate the expression of genes involved in the control of angiogenesis, glucose metabolism, and cellular proliferation. HIF-1α is expressed at low levels in the fetal lung and in the bronchial epithelium of the adult (68,69) and is required for embryonic lung vascularization. During hypoxia, HIF-1α is induced in alveolar and bronchial epithelium, vascular endothelium, and in smooth muscle (70). HIF-2α mRNA is expressed in the endothelial cells and pulmonary mesenchyme at embryonic days 13.5 and 15.5 in mice (69). HIF-2α is expressed in the bronchial epithelial cells and type II cells increasing just around birth (around E18 in the mouse) and becoming abundant in the adult lung (71). Both HIF-1α and HIF-2α have critical roles in the regulation of perinatal lung function and hypoxia-induced pulmonary vascular remodeling.

GLI proteins are members of a zinc finger transcription factor family that mediates *Hh* signaling via the SHH pathway. GLI1 and GLI2 are activators of *Hh*-target genes while GLI3 is a repressor of the *Hh* gene but is also involved in target gene transcription (72). GLI1, GLI2, and GLI3 are expressed in an overlapping fashion in the lung mesenchyme. GLI1 is expressed in epithelium after E16.5 (73).

A subset of HOX genes (a family of regulatory homeobox genes expressed along the anteroposterior axis; HOXB3, -4, and -5) are expressed in the pulmonary mesenchyme with levels decreasing with advancing gestational age (74). Later HOXB3 and HOXB4 are expressed in tracheal, bronchial, and peripheral lung mesenchyme, while HOXB2 and HOXB5 are expressed in the peripheral lung tubular mesenchyme (75).

Defects Associated with Transcription Factors in the Mesenchyme During Branching

Deletion of FOXF1 before gastrulation is lethal while severe lung malformations were observed in *Foxf1*$^{+/-}$ heterozygous mice (60). Pulmonary vascular abnormalities were observed in *Foxm1*$^{-/-}$ homozygous mice (76). *Hif-1α*$^{-/-}$ homozygous mice die at midgestation due to defects in vasculogenesis (77). During hypoxic conditions, *Hif1α*$^{+/-}$ heterozygous mice were observed with pulmonary hypertension and pulmonary vascular remodeling (78), while *Hif2*$^{+/-}$ heterozygous mice were protected from pulmonary hypotension and right ventricular hypertrophy (79). *Hif2*$^{-/-}$ homozygous mice died after birth from decreased surfactant production and resultant atelectasis (68). *Gli2*$^{-/-}$ homozygous mice have foregut defects including lung hypoplasia and lobulation defects with stenosis of the trachea and esophagus. *Gli2*$^{-/-}$ and *Gli3*$^{-/-}$ homozygous mice lack esophagus, trachea, and lung (80). *Gli2*$^{-/-}$ homozygous and *Gli3*$^{+/-}$ heterozygous mice have esophageal atresia with tracheoesophageal fistula and severe lung hypoplasia (80). Deletion of GLI3 caused lung hypoplasia and altered lung shape (73). Although these

HOX genes are involved with tracheopulmonary development, only $Hoxa5^{-/-}$ homozygous mice had defects in trachea and lung structure with decreased expression of surfactant proteins, TTF-1, FOXA2, and N-Myc (81).

Transcription Factors During the Canalicular Phase

The transcription factors TTF-1, NF-1β, GATA-6, and other transcription factors, including members of the RB (retinoblastoma tumor suppressor protein), ETS, SOX, and FOX families play roles in cell-specific differentiation and gene expression in the conducting airways during the canalicular phase (E16.5–E17.5 in mice). Differentiation of distinct cell types along a cephalocaudal axis of the conducting airway is dependent on the distribution of the transcription factors, their concentration, activation, and interactions with other transcription factors.

Transcription factor p63, a tumor suppressor gene p53 homologue, is highly expressed in embryonic ectoderm and epithelial cells of the trachea and bronchi. Differentiation of the pseudostratified epithelium of conducting airways occurs prior to birth, and the basal cells located within this epithelium selectively express p63.

FOXJ1 is involved with the establishment of left-right symmetry in the early embryo and with ciliogenesis in the lung (82–84). Ciliogenesis of several cell types in the conducting airways is controlled by several genes, such as ezrin and calpastatin (85,86) that are regulated by FOXJ1.

The pulmonary neuroendocrine system consists of specialized airway epithelial cells often in close apposition with nerve fibers. These relatively rare PNEC are found as isolated cells or in clusters (NEB) (87). Several transcription factors have been shown to influence pulmonary neuroendocrine development including the human orthologs of *Mash1* (*HASH1*), HES1, GFI1 (growth factor independent 1), and pRB. The *HASH-1* gene is tightly involved in the formation of the neuronal network for autonomous control of ventilation in human and is expressed in normal neuroendocrine cells. GFI1 is a transcription factor that is coexpressed with *Mash1*/HASH1 in PNECs in the developing and mature lung.

SOX2, -9, -11, and -17 are expressed in the developing lung (88). SOX2 is expressed in ciliated cells in conducting airways (89), while SOX9 is expressed in the respiratory epithelial cells throughout lung morphogenesis (90). SOX17 expression is first detected in the mesenchyme around E10 to E11 and then later it is expressed in the epithelium of the conducting airways (E16–E17). SOX17 also interacts with β-catenin in the regulation of transcription of endodermal genes. Both SOX2 and SOX17 are abundantly expressed in ciliated cells of the perinatal lung. Their expression changes dramatically during repair of the conducting airways and redifferentiation of bronchiolar epithelium following injury (89).

Several members of the EST family [ERM family members (ezrin, radixin, and moesin), SPDEF (SAM pointed domain-containing ETS transcription factor), EST-1 (Ever Shorter Telomere 1), and PEA3 (Polyomavirus Enhancer Activator 3)] are expressed in the pulmonary mesenchyme and/or epithelial cells (91,92), and the spatial distribution of these transcription factors influence pulmonary cell differentiation. During lung morphogenesis, ERM and PEA3 are expressed in peripheral lung buds. ERM is coexpressed with SFTP-C and activates the *Sftpc* gene promoter in type II alveolar epithelial cells, while expression of the dominant-negative form of Erm inhibits differentiation of type II alveolar epithelial cells (92). SPDEF is expressed in epithelial cells of the extrapulmonary airways (93,94) and coactivates the expression of several genes involved in the conducting airways (14). Both ERM and SPDEF bind to TTF-1 and are involved either directly or synergistically with the regulation of transcriptional target genes (93).

Transcription Factors Associated with Defects During the Canalicular Phase

Deletion of p63 results in conducting airways with a simple columnar epithelium devoid of basal cells (95). FOXJ1 deletion causes a lack of cilia formation in the embryonic node, resulting in the failure to express proteins that control the site that influences lung lobulation. Although deletion of MASH1/HASH1 results in no detectable PNECs (96,97), progressive airway hyperplasia and metaplasia were caused by increased expression of MASH1/HASH1. $Gfi^{1-/-}$ homozygous mice had decreased numbers of PNECs and smaller NEBs indicating its role in the growth and maturation of PNEC/NEBs (98). $Sox9^{-/-}$ homozygous mice have multiple anomalies of the skeleton and cartilage, including severe tracheal cartilage malformations (99); however, deletion of SOX9 expression does not alter lung morphogenesis or function (90). $Sox11^{-/-}$ homozygous mice die at birth from cyanosis and have lung hypoplasia (100). Expression of SOX17 in the fetal lung disrupted peripheral lung formation, but when expressed in the postnatal lung, SOX17 induced focal hyperplasia and caused ectopic expression of proximal airway markers.

Transcription Factors During the Saccular-Alveolar Periods

Several transcription factors expressed by pre-type II and type II cells influence perinatal lung maturation, including TTF-1, FOXA2, NFATc3, C/EBPα, and the glucocorticoid receptor (GRα). These transcription factors regulate the expression of genes critical for postnatal respiration and regulating surfactant protein and lipid synthesis, fluid and solute transport, electrolyte homeostasis, vasculogenesis, and innate host defense. These transcription factors interact either directly, through complex-formation at transcriptional targets, or by influencing each other. FOXA2 directly regulates the TTF-1 promoter in vitro (101). Other transcription factors such as NF-1 (neurofibromatosis type 1), NFATc3 (nuclear factor of activated T cells), GATA-6, which partner with TTF-1 and other coactivators, play a role in the regulation of perinatal lung function, influencing or regulating type II cell function and the differentiation of type I alveolar epithelial cells.

The actions of glucocorticoids are mediated by the glucocorticoid receptor α and β (GRα, GRβ) present on pulmonary target cells (102,103). The effect of glucocorticoids on maturation of surfactant biosynthesis by fetal type II cells is dependent on mesenchymal cell paracrine signals inducing epithelial maturation (104,105). Stimulatory effects of glucocorticoids on surfactant lipid synthesis are associated with increased enzyme expression mediating lipid biosynthesis. Glucocorticoids have also been shown to influence the expression of epithelial Na^+ channels and Na^+-K^+-ATPase that mediate lung fluid resorption at birth.

FOXP1, -2, and -4 are expressed in respiratory epithelial cells and repress the activity of the CCSP (clara cell secretory protein) and SP-C (surfactant protein C) promoters (106,107). The repressor CtBP-1 (COOH-terminal binding protein 1) interacts with and represses the activity of FOXP1 and FOXP2 but not FOXP4 (108).

Transcription Factors Associated with Defects During the Saccular-Alveolar Periods

While mutation of TTF-1 (phosphorylation site mutation $Titf-1^{PM}$) or cell-selective deletion of FOXA2, C/EBPα, or calcineurin b1 (Cnb1) do not substantially alter branching morphogenesis in the early mouse lung, these mutations do impair pulmonary maturation during the saccular-alveolar phase, resulting in respiratory failure at birth (38,39,45,109,110). Although only subtle differences in lung differentiation and maturation were observed in these mutant mice, marked reductions in expression of the genes regulating surfactant homeostasis were observed. Mutation or deletion of TTF-1 or deletion of FOXA2 blocked C/EBPα and CNB1 expression in the peripheral epithelial cells prior to birth. Maturation of the peripheral saccules was retarded based on lack of type I and II epithelial cell differentiation and the absence of surfactant lipid synthesis (38,39,110). The lack of surfactant causes respiratory distress syndrome in preterm infants and is an important cause of morbidity and mortality in newborns. The importance of surfactant for perinatal survival is supported by the discovery that mutations in *Abca3*, *Sftpb*, and *Sftpc* genes cause lethal respiratory distress in mature infants after birth (111). Deletion of GRα in mice delays lung maturation causing respiratory distress at birth (112). FOXP1-deficient mice have cardiac effects but no alterations in lung morphogenesis suggesting that other FOXP family members may compensate for its loss in respiratory epithelial cells (113).

Transcription Factors During Alveolarization-Postnatal Lung Morphogenesis

GATA-6, TFF-1, ERβ, RARs, SMAD3, FOXA2, and FOXA1 have been shown to be involved with cellular processes controlling alveolarization-postnatal lung morphogenesis. Increased levels of either GATA-6 or TTF-1 in mouse respiratory epithelial cells resulted in disrupted postnatal alveolarization with airspace enlargement (59,114). ERβ-deficient mice have larger and fewer alveoli than alveoli observed in normal lungs (94). Dominant-negative expression of RARα in respiratory epithelial cells results in lungs with increased airspace size and decreased alveolar surface area (115). $Smad3^{-/-}$ homozygous mice had centrilobular emphysema associated with decreased tropoelastin and increased MMP-9 production (116). Mice with a deletion of FOXA2 in respiratory cells, $Foxf1^{+/-}$ heterozygous mice, and $RARβ^{-/-}$ homozygous mice, all had defects in alveolarization (44,60,117).

Other Transcription Factors Influencing Pulmonary Formation/Function

The research into lung morphogenesis has revealed numerous transcription factors that influence lung formation and function, and the number known is increasing at an accelerating pace. This list of transcription factors include STAT (signal transducer and activator of transcription) family, E2F1 and RB, SMADs and the TGF-β superfamily, BZIP (basic leucine Zipper DNA-binding motif) family, transcriptional coactivators, nuclear receptors, and homeodomain proteins.

STATs mediate signaling from multiple cytokines, growth factors, and other extracellular factors, and are important in lung inflammation and repair (118). E2F1 binds and regulates gene promoters involved with cell proliferation, differentiation, and development. RB associates with E2F1 and suppresses cell proliferation, differentiation, and development (119). RAR can convert E2F1 into a transcription suppressor causing inhibition of the growth of normal human bronchial epithelial cells (120). Many of the ligands of the TGF-β superfamily, including bone morphogenic proteins (BMPs), TGF-β1, -2, -3, and activin are expressed in the developing lung. The SMAD (mothers against DPP homolog 1) proteins are nuclear effectors of TGF-β superfamily ligands regulating their transcription. Recent reviews of the mechanisms of TGF-β-SMAD signaling are available (121,122). SMAD1 and -7 positively regulate pulmonary branching morphogenesis (123–125), while SMAD2, -3, and -4 negatively regulate branching morphogenesis (126). The BZIP transcriptions bind to sequence-specific DNA and activate or repress transcription (127). BZIP transcription factors including JunB, c-Jun, JunD, NRF2 (nuclear factor E2 p45-related factor 2), CREB (cAMP response element binding protein), C/EBP (CCAAT/enhancer binding protein), and ATF-2 (activating transcription factor-2) regulate gene expression in the developing lung.

The mammalian nuclear receptor superfamily is composed of more than 45 DNA-binding transcription factors whose nuclear transport and activity are modulated to either activate or suppress gene expression. Members of this superfamily include receptors (that bind lipid and prostaglandins as well as to DNA receptor sites) such as peroxisome proliferator activated receptors (PPARs) and liver X receptors (LXRs); the glucocorticoid receptors (GR); receptors for steroid hormones such as estrogen (ER); receptors for nonsteroid ligands such as RAR; and the vitamin D receptor (VDR) (128). Nuclear receptors are involved in the regulation of diverse aspects of lung morphogenesis, homeostasis, and inflammation.

Before transcription factors can bind to their target DNA elements in the regulatory regions of DNA directing RNA synthesis via RNA polymerase, at least two critical processes for activation must occur. First the tightly compacted chromatin structures consisting of DNA and histones must be opened. And second, the factor or complex of factors must interact with RNA polymerase to produce RNA. A number of coregulators involved with the opening of the chromatin structure and the interaction with transcription factors and RNA polymerase have been identified including factors that mediate ATP-dependent chromatin remodeling and histone modification. In partnership with transcription factors, RNA polymerase II–associating factors as well as cell-specific cofactors and negative cofactors function either as coactivators or corepressors (91). Many of these cofactors are recruited with TTF-1 to transcription sites on target genes. Many of these cofactors are involved in the development and function of the lung including four chromatin modifiers [p160, CBP/p300 (CREB binding protein and p300) (129), CARM1 (coactivator-associated arginine methyltransferase 1) (130), and MRG15 (MORF-related gene on chromosome 15) (131)] that are required for normal lung morphogenesis.

Homeodomain proteins bind to specific transcriptional control regions of target genes and perform fundamental roles in morphogenesis, influencing segmentation organogenesis patterning and cell fate determination. The homeodomain proteins are classified into six distinct classes: HOX, Extended HOX, NK, Paired, LIM, POU (Pit-Oct-Unc family), and an atypical class (132). On the basis of studies with knockout mice, six homeodomain transcription factors [HOXA5, TTF-1, PITX2 homeobox, CUTL1 (human gene locus for CDP named cut-like 1), HOP (homeobox protein), and Prophet of PIT1 (PROP1)] have been shown to have important roles in lung morphogenesis and function (14).

COMMON ANOMALIES OF THE RESPIRATORY SYSTEM

Although rare in comparison with other types of malformations, congenital lung malformations lead to considerable morbidity and mortality. The purpose of this section is to provide brief descriptions of the more common congenital anomalies of the respiratory system.

- Tracheomalacia (133) is softening of the tracheal wall due to an abnormality of the cartilaginous ring and hypotonia of the myoelastic elements. With this condition, the trachea and other major airways are unable to maintain airway patency. The primary cause of this defect is congenital immaturity of the tracheal cartilage.
- Tracheal stenosis (134) is characterized by the narrowing of the tracheal rings of cartilage and is usually associated with anomalies of the pulmonary artery.
- Tracheal bronchus (135) occurs when the upper right bronchus originates from the trachea rather than distal to the carina (where the trachea splits into the two primary bronchi), being either simply displaced or supernumerary. Supernumerary bronchi may end blindly into lung tissue (apical lungs or lobes).

- Bronchial atresia (136) results from focal obliteration of a proximal segmental bronchus that lacks communication with the central airways. The distal bronchi fill with mucus and form a bronchocele. This condition generally produces no symptoms.
- Bronchogenic cysts (137) are thought to originate from the primitive ventral foregut, and develop as a result of abnormal budding during the embryonic phase and are usually asymptomatic.
- Pulmonary underdevelopment (138) are classified into three groups:
  1. Agenesis, when bronchus and lung are absent
  2. Aplasia, when rudimentary bronchus is present without lung tissue
  3. Hypoplasia, when the bronchus and lung are reduced
  The etiology of the lung agenesis and aplasia are genetic, teratogenic, or mechanical and occurs during the embryonic period. Lung hypoplasia or small lungs are usually associated with premature deliveries, while profound lung hypoplasia is usually associated with congenital diaphragmatic hernias and currently is unsalvageable.
- Scimitar syndrome (136), also known as hypogenetic lung syndrome, consists of hypoplasia of the right lung, cardiovascular anomalies, abnormal diaphragm, hemivertebrae, and anomalies of the genitourinary tract.
- Cystic adenomatoid (139,140) is a multicystic mass of pulmonary tissue that causes respiratory distress in neonates and infants, including tachypnea, retractions, and cyanosis.
- Lobar emphysema (141,142) is the overdistention of one or two lobes of the lung. It results from a check-valve mechanism at the bronchial level that causes progressive hyperinflation of the lung by allowing more air to enter the involved area on inspiration than leaves on expiration. It is often associated with abnormal cartilaginous ring development or other cardiovascular anomalies such as a large pulmonary artery.
- Pulmonary isomerism (143) is an anomaly of the number of lung lobes. In the common variety of pulmonary isomerism, the right lung has two lobes and the left lung has three lobes. This anomaly may be associated with situs inversus, asplenia, polysplenia, and/or anomalous pulmonary drainage.
- Azygous lobe (143) is a malformation of the right upper lobe caused by an aberrant azygous vein located in the pleural mesentery.
- Polyalveolar lobe (143) is when the numbers of alveoli are increased to more than three times the normal. When extra lung fluid is retained, respiratory distress may occur in the first days of life.
- Hamartomas (144) are lung nodules that contain cartilage, respiratory epithelium, and collagen. They may be in the lung tissue or the bronchial lumen and can cause airway obstruction.
- Pulmonary sequestration (145) is an aberrant or accessory lung tissue that has no normal airway or vascular connections and accounts for 6% of all respiratory congenital malformations. Intralobar sequestration is contained within the lung and has a visceral pleural covering; extralobar sequestration is a mass of abnormal lung tissue that develops late in gestation and is surrounded by its own separate pleura. Approximately 95% of extrapulmonary cases are left sided, with boys affected four times more than girls.
- Tracheoesophageal fistula (146) is an abnormal communication between the esophagus and trachea that can be traced back to the incomplete separation of the primitive foregut into trachea and esophagus and is often associated with gastrointestinal and other malformations.

- Lung cysts are fluid-filled membraneous sacs that arise from any of the parenchymal tissues of the lung. They can cause symptoms if they enlarge and occupy substantial space (147).
- Vascular anomalies (148) of the respiratory system include compression and displacement of the trachea resulting from a malformed aorta arch complex resulting in the formation of a ring encircling both the trachea and esophagus; it includes the classic double aortic arch.
- Diaphragmatic hernia (149) is a defect or hole in the diaphragm that allows the abdominal contents to move into the chest cavity thereby impeding proper lung formation and/or expansion. Newborns often have severe respiratory distress, and pulmonary hypoplasia is directly related to the presence of abdominal organs in the chest cavity.
- Alveolar capillary dysplasia (150), a fatal condition, is caused by reduction of the distal arteriolar blood supply, misalignment of the pulmonary veins, and increased connective tissue between the alveolar epithelium and the capillary endothelium.
- Pulmonary lymphangiectasis (151,152) is a disorder in which the normal pulmonary lymphatics are dilated. It may be associated with congenital heart disease in which the pulmonary venous pressure is elevated. Pulmonary lymphangiectasis can also be observed with lymphangiomatosis, in which proliferation of the lymphatic tissue and channels occurs.
- Respiratory distress syndrome (153), previously called hyaline membrane disease, is caused by pulmonary immaturity and surfactant deficiency; the incidence and severity is inversely related to gestational age. It can also result from a genetic problem with the production of surfactant-associated proteins.

RESPIRATORY SYSTEM DEVELOPMENTAL TOXICANTS AND CONDITIONS

Developmental anomalies of the respiratory system have many underlying etiologies or pathogenic mechanisms including chromosomal aberrations, single-gene (Mendelian) disorders, and sequences (multiple defects that are related to a single problem in morphogenesis). These include inherited genetic conditions, maternal disorders, biochemical and physiological perturbations, insults from endogenous substances such as hormones and vitamins, environmental toxicants, therapeutic agents, drugs of abuse, and miscellaneous procedures (Table 3). Morphological defects in the lung include vascular abnormalities, tracheal and bronchial abnormalities, hypoplasia or agenesis of one or both the lobes of the lung, and hamartomatous malformations (229). Developmental biochemical defects can include imperfections in pulmonary surfactants structure or production, active transpulmonary ion transport system, certain hormones and enzymes, growth factors, and collagen and elastic fiber formations (230). Physiological factors such as fetal breathing movements, lung expansion, apoptosis, airway resistance, birthing, and lung compliance (231,232). Depending on the severity of the physical or chemical insult to the developing pulmonary system, there is a tendency toward retardation of lung maturation, leading to hypoplasia and atelectasis, which reduces alveolarization, causing perturbations in the surfactant system and poor gas exchange.

Normal pulmonary vascular development is critical to normal development of the lung. There are four main features by which vascular development may be assessed, and these include the branching pattern of the developing vasculature, the number of arteries, the structure of the wall of developing arteries, and the arterial size. A criterion to distinguish between maladaptation, maldevelopment, and underdevelopment of the abnormal pulmonary vasculature has been developed (233). Lung vascular growth

TABLE 3 Reported Respiratory Developmental Toxicants and Conditions

Environmental agents
Nitrofen (diphenyl ether herbicide) (154–156); O,O,S-TMP (trimethylphosphorothioate) (157,158); cadmium (159,160); lead (161), ozone (162,163); hyperoxia (164, 165); hypoxia (166); nitrogen dioxide (167); high-boiling coal liquid (168); semicarbazide (169); biphenyl carboxylic acid (170,171); amitraz (formamidine insecticide) (172); perfluoroalkyl acids (e.g., PFOS, PFOA) (173,174); monocrotaline (pyrrolizidine alkaloid) (175); cis-hydroxyproline (176); aviglycine (aminoethoxyvinylglycine) (177); mycotoxins (178)

Genetic conditions
Trisomy 13,18,21; Pallister-Hall, Pfeiffer, Smith-Lemli-Opitz, Pena-Shokeir, Matthew-Wood, Larsen-like, myasthenia gravis, thymic hypoplasia, pterygium, hydrolethalus, tight skin, lung agenesis, neonatal osseous dysplasia, osteosclerotic, Jeune, Meelel, and Kartagener syndromes (179)

Maternal/fetal disorders
Alloxan-induced diabetes (180,181); malnutrition (165,182); zinc deficiency (183–185); copper deficiency (185–188)

Therapeutic agents
β-Adrenergic agents (e.g., propranolol) (189,190); chlorphentermine (e.g., phentermine analog) (191); indomethacin (192), betamethasone (prolonged use) (193); bisdiamine (dichloroacetamide) (194–197); bromodeoxyuridine (170,198,199); chlorambucil (200); methylsalicylate (200); dihydrotestosterone (201); hydrocortisone (202); dexamethasone (203); isoxsuprine (vasodilator) (204,205); Fumagillin (fungal antibiotic) (206); terbutaline (203,207); thalidomide (206); limbitrol (chlordiazepoxide + amitriptyline) (208); Su-5416 (antiangiogenic agent) (206); SB-210661 (benzofuranyl urea derivative) (170,171); DW-116 (fluoroquinolone antibacterial) (209); CI-921 (anilinoacridine compound) (210); Aceon (ACE inhibitor) (211); telmisartan (angiotensin AT1 receptor blocker) (212)

Endogenous compounds
Glucocorticoids (213,214); thyroid hormones (T3,T4) (215–217); cAMP (cyclic adenosine monophosphate) (218,219); retinoids (166); epinephrine (220)

Miscellaneous
Radiation exposure (221), persistent patent ductus arteriosus (222,223); ligated ductus arteriosus (224); mechanical injury (barotraumas, volutrauma) (225,226); thyroidectomy (227); high altitude (228)

---

involves two basic processes: vasculogenesis, the formation of blood vessels from endothelial cells within the mesenchyme, and angiogenesis, the formation of new blood vessels from sprouts of preexisting vessels. Since angiogenesis is necessary for normal alveolarization during a critical period of lung development, it is not surprising that antiangiogenic agents such as thalidomide, Fumagillin, and Su-5416 decrease alveolarization causing lung hypoplasia.

The lungs of a newborn with congenital diaphragmatic hernia, a defect or hole in the diaphragm that allows the abdominal contents to move into the chest cavity, have significantly reduced number of airways and vascular generations in the lungs (234). Affected newborns have severe respiratory distress because of the pulmonary hypoplasia. Several extracellular matrix components are expressed in the basement membrane during different phases of lung development and maturation. One such molecule, laminin-beta1, which is integral to lung branching morphogenesis, is upregulated in lung maturation and downregulated in hypoplastic lungs (235). Adverse stimuli such as hypoxia, hyperoxia, or glucocorticoid treatment can also disrupt alveolarization leading lung hypoplasia.

Gene targeting technology has been used to eliminate or knockout the genes, to create null mutant mice. The absence of specific gene(s) involved in respiratory system development helps shed light on their roles in the respiratory system morphogenesis and differentiation, and/or biochemical and functional maturation. One can identify the specific anomalies and/or fetal disease antenatally by an armamentarium of X ray, CT and MRI scans, sonograms, and amniocentesis tools (229).

Miscellaneous causes of abnormal lung development include poor mechanical ventilation, such as barotrama and volutrauma, high and low O_2 exposure, radiation exposure, and high-altitude exposure during the perinatal period (236). Persistent patent ductus arteriosus causes bronchopulmonary dysplasia to preterm infants (225). Exposure to environmental agents like perfluoroalkyl acids, nitrofen, and metals such as cadmium and lead can lead to developmental anomalies from respiratory distress syndrome to diaphragmatic hernias (154,173).

Maternal disorders such as malnutrition, involving amino acid, mineral, and/or vitamin deficiencies, diseases such as diabetes mellitus and vaginal/placental infections, and loss of amniotic fluids can cause pulmonary hypoplasia and respiratory distress syndrome (237–239). Women take over 700 prescribed generic drugs ($\sim$1500 trade names) and even more over-the-counter drugs and recreational drugs including nicotine, alcohol, and drugs of abuse, during pregnancy or while breast-feeding (240). For many drugs there is scant information on use during pregnancy and lactation, and their contraindications and FDA pregnancy categories are often confusing and limited in helping with prescribing decisions. Some of the classes of therapeutic agents that are contraindicated during pregnancy are angiogenesis-converting enzyme (ACE) inhibitors, β-adrenergic drugs, antiangiogenic agents, glucocorticoids, cytotoxic agents, and certain combination drugs (189,202).

FUTURE DIRECTIONS

Genetic factors strongly affect susceptibility to congenital anomalies and also influence disease-related quantitative traits. Identifying the relevant genes is difficult because each causal gene only makes a small contribution to the overall inheritable genotype. Through the advances of the human genome project and International HapMap Project (241), millions of single-nucleotide polymorphisms (SNPs) are now available in a public database for genome-wide association studies of common anomalies and complex syndromes, as well as high throughput genotyping capabilities (242). Genetic association studies are limited because only a small number of genes can be studied at a time, but offer a potentially powerful approach for mapping causal genes with modest effects. Taking advantage of the SNP, high throughput genotyping, and DNA specimen and clinical database availability, a genome-wide scan of candidate genes could be used to identify those genes that contribute to the development of acute lung disease in newborns. Genes identified as contributors to lung anomalies in animal models can guide selection of candidate genes for use in human association studies. An overlap of significant genes identified in different animal models indicates both the commonality of processes likely to be affected in lung anomalies as well as the complexity of the processes involved. For instance, the identification of genes involved in the pathogenesis of respiratory distress syndrome would provide an important step toward the study of mechanisms that, if deranged, may increase susceptibility to respiratory distress syndrome. Such studies could provide insight into why only a portion of prematurely born infants develop respiratory distress syndrome and may also help identify gene targets for therapeutic intervention.

REFERENCES

1. Hoyert DL, Kochanek KD, and Murphy SL. Deaths: final data for 1997. Nat Vital Stat Rep 1999; 47(19):1–105.
2. Singh GK, Yu SM. Infant mortality in the United States: trends, differentials, and projections, 1950 through 2010. Am J Pub Health 1995; 85:957–964.
3. Longo LD. Respiratory gas exchange in the placenta. In: Handbook of Physiology, Section 3, The Respiratory System. Vol. 4. Bethesda, MD: American Physiological Society, 1987.

4. Boyden EA. Development and growth of the airways. In: Hoden WA, ed. Development of the Lung. New York: M. Decker, 1977:3–35.
5. Bush. An update in pediatric lung disease 2006. Respir Crit Care Med 2007; 175:532–540.
6. Levitsky MG. Effects of aging on the respiratory system. Physiologist 1984; 27:102–107.
7. Stocks J. Developmental physiology and methodology. Am J Resp Crit Care Med 1995; 151(suppl):515–517.
8. Illinois Department of Public Health. Adverse Pregnancy Outcome Reporting System. November, 2001.
9. NSW Public Health Bulletin 1996; 7(7):67–69.
10. Farrell PM. Morphological aspects of lung maturation. In: Farrell PM, ed. Lung Development: Biological and Clinical Perspectives. New York: Academic Press, 1982:13–24.
11. Ballard PL. Hormones and lung maturation. Monogr Endocrinol 1986; 28:1–354.
12. Lau C, Kavlock R. Functional toxicity in the developing heart, lung, and kidney. In: Kimmel CA, Bulke-Sam J, eds. Developmental Toxicology. 2nd ed. New York: Raven Press, Ltd, 1994:119–188.
13. Zoetis T, Hurtt M. Species comparison of lung development. Birth Defects Res 2003; 68(pt B):121–124.
14. Maeda Y, Dave V, Whitsett JA. Transcriptional control of lung morphogenesis. Physiol Rev 2007; 87:219–244.
15. O'Rahilly R. Early human development and the chief sources of information on staged human embryos. Eur J Obstet Gynecol Repr Biol 1979; 9:273–280.
16. Low FN, Sampaio MM. The pulmonary alveolar epithelium as an entodermal derivative. Anat Rec 1957; 127:51–63.
17. Bucher U, Reid L. Development of the intrasegmental bronchial tree: the pattern of branching and development of cartilage at various stages of intrauterine life. Thorax 1961; 16:207–223.
18. Emery J, ed. The Anatomy of the Developing Lung. London: William Heinemann, 1969, 1–223.
19. Thurlbeck WM. Postnatal growth and development of the lung. Am Rev Resp Dis 1978; 111:803–844.
20. Sadler TW. Langman's Medical Embryology. 8th ed. New York: Lippincott Williams & Wilkins, 2000:260–269.
21. Engel S. The structure of the respiratory tissue in the newly born. Acta Anat 1955; 19:353–365.
22. Badri KR, Zhou Y, Schuger L. Embryological origin of the airway smooth muscle. Proc Am Thorac Soc 2008; 5:4–10.
23. Warburton D, Zhao J, Berderich MA, et al. Molecular embryology of the lung: then, now, and in the future. Am J Physiol 1999; 276:L697–L704.
24. Warburton D, Schwarz M, Tefft D, et al. The molecular basis of lung morphogenesis. Mech Dev 2000; 92:55–81.
25. Warburton D, Bellusci S, Del Moral P-M, et al. Growth factor signaling in lung morphogenetic centers: automaticity, stereotypy and symmetry. Respir Res 2003; 4:5–22.
26. Warburton D, Bellusci S. The molecular genetics of lung morphogenesis and injury repair. Paediatr Respir Rev 2004; 5(suppl A):S283–S287.
27. Hislop AA. Airway and blood vessel interaction during lung development. J Anat 2002; 201:325–334.
28. Cardoso WV. Lung morphogenesis revisited: old facts, current ideas. Dev Dyn 2000; 219:121–130.
29. Cardoso WV, Lu J. Regulation of early lung morphogenesis: questions, facts and controversies. Development 2006; 133:1611–1624.
30. Huelsken J, Vogel R, Brinkmann V, et al. Requirement for beta-catenin in anterior-posterior axis formation in mice. J Cell Biol 2000; 148:567–578.
31. Kanai-Azuma M, Kanai Y, Gad JM, et al. Deletion of definitive gut endoderm in Sox17-null mutant mice. Development 2002; 129:2367–2379.
32. Morrisey EE, Tang Z, Sigrist K, et al. GATA6 regulates HNF4 and is required for differentiation of visceral endoderm in the mouse embryo. Genes Dev 1998; 12:3579–3590.

33. Takada K, Noguchi K, Shi W, et al. Targeted disruption of the mouse Stat3 gene leads to early embryonic lethality. Proc Natl Acad Sci U S A 1997; 94:3801–3804.

34. Karet KS. Regulatory phases of early liver development: paradigms of organogenesis. Nat Rev Genet 2002; 3:499–512.

35. Bohinski RJ, Di Lauro R, Whitsett JA. The lung-specific surfactant protein B gene promoter is a target for thyroid transcription factor 1 and hepatocyte nuclear factor 3, indicating common factors for organ-specific gene expression along the foregut axis. Mol Cell Biol 1994; 14:5671–5681.

36. Di Palma T, Nitsch R, Mascia A, et al. The paired domain-containing factor Pax8 and the homeodomain-containing factor TTF-1 directly interact and synergistically activate transcription. J Biol Chem 2003; 278:3395–3402.

37. Seris AE, Doherty S, Parvatiyar P, et al. Different thresholds of fibroblast growth factors pattern the ventral foregut into liver and lung. Development 2005; 132:35–47.

38. Davé V, Childs T, Xu Y, et al. Calcineurin/Nfat signaling is required for the perinatal lung maturation and function. J Clin Invest 2006; 116:2597–2609.

39. DeFelice M, Silberschmidt D, Di Lauro R, et al. TTF-1 phosphorylation is required for peripheral lung morphogenesis, perinatal survival, tissue-specific gene expression. J Biol Chem 2003; 278:35574–35583.

40. Costa RH, Kalinichenko VV, Lim L. Transcription factors in mouse lung development and function. Am J Physiol Lung Cell Mol Physiol 2001; 280:L823–L838.

41. Besnard V, Wert SE, Whitsett JA. Immunohistochemical localization of Foxa1 and Foxa2 in mouse embryos and adult tissues. Gene Expr Patterns 2004; 5:193–208.

42. Zhao L, Lim L, Costa RH, et al. Thyroid transcription factor-1, hepatocyte nuclear factor-3beta, surfactant protein B, C, and Clara cell secretory protein in developing mouse. J Histochem Cytochem 1996; 44:1183–1193.

43. Wan H, Dingle S, Xu Y, et al. Compensatory roles of Foxa1 and Foxa2 during lung morphogenesis. J Biol Chem 2005; 280:13809–13816.

44. Wan H, Kaestner KH, Ang SL, et al. Foxa2 regulates alveolarization and goblet cell hyperplasia. Development 2004; 131:953–964.

45. Wan H, Xu Y, Ikegami M, et al. Foxa2 is required for transition to air breathing at birth. Proc Natl Acad Sci U S A 2004; 101:14449–14454.

46. Nelson WJ, Nusse R. Convergence of Wnt, beta-catenin, cadherin pathways. Science 2004; 303:1483–1487.

47. Pongracz JE, Stockley RA. Wnt signaling in lung development and diseases. Respir Res 2006; 7:15.

48. Shannon JM, Hyatt BA. Epithelial-mesenchymal interactions in the developing lung. Annu Rev Physiol 2004; 66:625–645.

49. Warburton D, Bellusci S, De Langhe S, et al. Molecular mechanisms of early lung specification and branching morphogenesis. Pediatr Res 2005; 57:26R–37R.

50. Kimura S, Hara Y, Pineau T, et al. The T/ebp null mouse: thyroid-specific enhancer-binding protein is essential for the organogenesis of the thyroid, lung, ventral forebrain, pituitary. Genes Dev 1996; 10:60–69.

51. Minoo P, Su G, Drum H, et al. Defects in tracheoesophageal and lung morphogenesis in Nkx2.1(−/−) mouse embryos. Dev Biol 1999; 209:60–71.

52. Devriendt K, Vanhole C, Matthijs G, et al. Deletion of thyroid transcription factor-1 gene in infant with neonatal thyroid dysfunction and respiratory failure. N Eng J Med 1998; 338:1317–1318.

53. Davis S, Bove KE, Wells TR, et al. Tracheal cartilaginous sleeve. Pediatr Pathol 1992; 12:349–364.

54. Besnard V, Wert SE, Kaestner KH, et al. Stage-specific regulation of respiratory epithelial cell differentiation by Foxa1. Am J Physiol Lung Cell Mol Physiol 2005; 289:L750–L759.

55. Mucenski ML, Wert SE, Nation JM, et al. β-Catenin is required for specification of proximal/distal cell fate during lung morphogenesis. J Biol Chem 2003; 278:40231–40238.

56. Shu W, Guttentag S, Wang Z, et al. Wnt/beta-catenin signaling acts upstream of N-myc, BMP4, FGF signaling to regulate proximal-distal patterning in the lung. Dev Biol 2005; 283:226–239.

57. Liu C, Morrisey EE, Whitsett JA. GATA-6 is required for maturation of the lung in late gestation. Am J Physiol Lung Cell Mol Physiol 2002; 283:L468–L475.

58. Koutsourakis M, Keijzer R, Visser P, et al. Branching and differentiation defects in pulmonary epithelium with elevated Gata6 expression. Mech Dev 2001; 105:105–114.
59. Liu C, Ikegami M, Stahlman MT, et al. Inhibition of alveolarization and altered pulmonary mechanics in mice expressing GATA-6. Am J Physiol Lung Cell Mol Physiol 2003; 285:L1246–L1254.
60. Kalinichenko VV, Lim L, Stolz DB, et al. Defects in pulmonary vasculature and perinatal lung hemorrhage in mice heterozygous null for the Forkhead Box f1 transcription factor. Dev Biol 2001; 235:489–506.
61. Lim L, Kalinichenko VV, Whitsett JA, et al. Fusion of the lung lobes and vessels in mouse embryos heterozygous for the forkhead box f1 targeted allele. Am J Physiol Lung Cell Mol Physiol 2002; 282:L1012–L1022.
62. Mahlapuu M, Enerback S, Carlsson P. Haploinsufficiency of the forkhead gene Foxf1, a target for sonic hedgehog signaling, causes lung and foregut malformations. Development 2001; 128:2397–2406.
63. Costa RH. FoxM1 dances with mitosis. Nat Cell Biol 2005; 7:108–110.
64. Kalinichenko VV, Gusarova GA, Tan Y, et al. Ubiquitous expression of the forkhead box M1B transgene accelerates proliferation of distinct pulmonary cell types following lung injury. J Biol Chem 2003; 278:37888–37894.
65. Haddad JJ. Oxygen-sensing mechanisms and the regulation of redox-responsive transcription factors in development and pathophysiology. Respir Res 2002; 3:26.
66. Ruas JL, Poellinger L. Hypoxia-dependent activation of HIF into a transcriptional regulator. Semin Cell Dev Biol 2005; 16:514–522.
67. Semenza GL. Pulmonary vascular responses to chronic hypoxia mediated by hypoxia-inducible factor 1. Proc Am Thorac Soc 2005; 2:68–70.
68. Compernolle V, Brusselmans K, Acker T, et al. Loss of HIF-2alpha and inhibition of VEGF impair fetal lung maturation, whereas treatment with VEGF prevents fetal respiratory distress in premature mice. Nat Med 2002; 8:702–710.
69. Jain S, Maltepe E, Lu MM, et al. Expression of ARNT, ARNT2, HIF-1 alpha, HIF2 alpha and Ah receptor mRNAs in the developing mouse. Mech Dev 1998; 73:117–123.
70. Yu AY, Frid MG, Shimoda LA, et al. Temporal, spatial, oxygen-regulated expression of hypoxia-inducible factor-1 in the lung. Am J Physiol Lung Cell Mol Physiol 1998; 275:L818–L826.
71. Ema M, Taya S, Yokotani N, et al. A novel bHLH-PAS factor with close sequence similarities to hypoxia inducible factor 1alpha regulates the VEGF expression and is potentially involved in lung and vascular development. Proc Natl Acad Sci U S A 1997; 94:4273–4278.
72. Kasper M, Regl G, Frischauf AM, et al. GLI transcription factors: mediators of oncogenic Hedgehog signaling. Eur J Cancer 2006; 42:437–445.
73. Gridley JC, Bellusci S, Perkins D, et al. Evidence for the involvement of the Gli gene family in embryonic mouse lung development. Dev Biol 1977; 188:337–348.
74. Bogue CW, Gross I, Vasavada H, et al. Identification of Hox genes in newborn lung and effects of gestational age and retinoic acid on their expression. Am J Physiol Lung Cell Mol Biol 1994; 266:L448–L454.
75. Bogue CW, Lou LJ, Vasavada H, et al. Expression of Hoxb genes in the developing mouse foregut and lungs. Am J Respir Cell Mol Biol 1996; 15:163–171.
76. Kim IM, Ramakrishna S, Gusarova GA, et al. The forkhead box m1 transcription factoris essential for embryonic development of pulmonary vasculature. J Biol Chem 2005; 280:22278–22286.
77. Ryan HE, Lo J, Johnson RS. HIF-1 alpha is required for solid tumor formation and embryonic vascularization. EMBO J 1998; 17:3005–3015.
78. Yu AY, Shimoda LA, Iyer NV, et al. Impaired physiological responses to chronic hypoxia in mice partially deficient for hypoxia-inducible factor 1alpha. J Clin Invest 1999; 103:691–696.
79. Brusselmans K, Compernole V, Tjwa M, et al. Heterozygous deficiency of hypoxia-inducible factor-2alpha protects mice against pulmonary hypotension and right ventricular dysfunction during prolonged hypoxia. J Clin Invest 2003; 111:1519–1527.
80. Motoyama J, Lui J, Mo R, et al. Essential function of Gli2 and Gli3 in the formation of the lung, trachea and oesophagus. Nat Genet 1998; 20:54–57.

81. Aubin J, Lemieux M, Tremblay M, et al. Early postnatal lethality in Hoxa-5 mutant mice is attributable to respiratory tract defects. Dev Biol 1997; 192:432–445.
82. Tichelaar JW, Lim L, Costa RH, et al. HNF-3/forkhead homologue-4 influences lung morphogenesis and respiratory epithelial cell differentiation in vivo. Dev Biol 1999; 213:405–417.
83. Tichelaar JW, Wert SE, Costa RH, et al. HNF-3/forkhead homologue-4 (HFH-4) is expressed in ciliated epithelial cells of the developing mouse. J Histochem Cytochem 1999; 47:823–832.
84. You L, Huang T, Richer EJ, et al. Role of the f-box factor foxj1 in differentiation of ciliated airway epithelial cells. Am J Physiol Lung Cell Mol Physiol 2004; 286:L650–L657.
85. Huang T, You Y, Spoor MS, et al. Foxj1 is required for apical localization of ezrin in airway epithelial cells. J Cell Sci 2003; 116:4935–4945.
86. Gomperts BN, Gong-Cooper X, Hackett BP. Foxj1 regulates basal body anchoring to the cytoskeleton of ciliated pulmonary epithelial cells. J Cell Sci 2004; 117:1329–1337.
87. Van Lommel A, Bolle T, Fannes W, et al. The pulmonary neuroendocrine system: the past decade. Arch Histol Cytol 1999; 62:1–16.
88. Ishii Y, Rex M, Scotting PJ, et al. Region-specific expression of chicken Sox2 in the developing gut and lung epithelium: regulation by epithelial-mesenchymal interactions. Dev Dyn 1998; 213:464–475.
89. Park KS, Wells JM, Zorn AM, et al. Transdifferentiation of ciliated cells during repair of respiratory epithelium. Am J Respir Cell Mol Biol 2006; 34:151–157.
90. Lazzaro D, Price M, de Felice M, et al. The transcription factor TTF-1 is expressed at the onset of thyroid and lung morphogenesis and in restricted regions of the foetal lung. Development 1991; 113:1093–1104.
91. Roder RG. Transcriptional regulation and the role of diverse coactivators in animal cells. FEBS Lett 2005; 579:909–915.
92. Liu Y, Jiang H, Crawford HC, et al. Role for the ETS domain transcription factors Pea3/Erm in mouse lung development. Dev Biol 2003; 261:10–24.
93. Lin S, Perl AK, Shannon JM. Erm/Thyroid transcription factor 1 interactions modulate surfactant protein C transcription. J Biol Chem 2006; 281:16716–16726.
94. Patrone C, Cassel TN, Pettersson K, et al. Regulation of postnatal lung development and homeostasis by estrogen receptor beta. Mol Cell Biol 2003; 23:8542–8552.
95. Daniely Y, Liao G, Dixon D, et al. Critical role of p63 in the development of a normal esophageal and tracheobronchial epithelium. Am J Physiol Cell Physiol 2004; 287: C171–C181.
96. Borges M, Linnoila RI, van de Velde HJ, et al. An achaete-scute homologue essential for neuroendocrine differentiation in the lung. Nature 1997; 386:852–855.
97. Ito T, Udaka N, Yazawa T, et al. Basic helix-loop-helix transcription factors regulate the neuroendocrine differentiation of fetal mouse pulmonary epithelium. Development 2000; 127:3913–3921.
98. Kazanjian A, Wallis D, Au N, et al. Growth factor independence-1 is expressed in primary human neuroendocrine lung carcinomas and mediates the differentiation of murine pulmonary neuroendocrine cells. Cancer Res 2004; 64:6874–6882.
99. Mori-Akiyama Y, Akiyama H, Rowitch DH, et al. Sox9 is required for the determination of chondrogenic cell lineage in the cranial neural crest. Proc Natl Acad Sci U S A 2003; 100:9360–9365.
100. Sock E, Rettig SD, Enderich J, et al. Gene targeting reveals a widespread role for the high-mobility group transcription factor Sox11 in tissue remodeling. Mol Cell Biol 2004; 24:6635–6644.
101. Ikeda K, Shaw-White JR, Wert SE, et al. Hepatocyte nuclear factor 3 activates transcription for thyroid transcription factor 1 in respiratory epithelial cells. Mol Cell Biol 1996; 16:3626–3636.
102. Pujols L, Mullol J, Perez M, et al. Expression of the human glucocorticoid receptor alpha and beta isoforms in human respiratory epithelial cells and their regulation by dexamethasone. Am J Respir Cell Mol Biol 2001; 24:49–57.
103. Korn SH, Wouters EF, Wesseling G, et al. In vitro and in vivo modulation of alpha- and beta-glucocorticoid-receptor mRNA in human bronchial epithelium. Am J Respir Crit Care Med 1997; 155:1117–1122.

104. Batenburg JJ, Effring RH. Pre-translational regulation by glucocorticoid of fatty acid and phosphatidylcholine synthesis in type II cells from the fetal rat lung. FEBS Lett 1992; 307:164–168.

105. Smith BT, Post M. Fibroblast-pneumonocyte factor. Am J Physiol Lung Cell Mol Physiol 1989; 257:L174–L178.

106. Shu W, Yang H, Zhang L, et al. Characterization of a new subfamily of winged-helix/ forkhead (Fox) genes that are expressed in the lung and act as transcriptional repressors. J Biol Chem 2001; 276:27488–27497.

107. Lu MM, Li S, Yang H, et al. Foxp4: a novel member of the Foxp subfamily of winged-helix genes co-expressed with Foxp1 and Foxp2 in pulmonary and gut tissues. Gene Expr Patterns 2002; 2:233–228.

108. Li S, Weidenfeld J, Morrisey EE. Transcriptional and DNA binding activity of the Foxp1/2/4 family is modulated by heterotypic and homotypic protein interactions. Mol Cell Biol 2004; 24:809–822.

109. Basseres DS, Levantini E, Ji H, et al. Respiratory failure due to differentiation arrest and expansion of alveolar cells following lung-specific loss of the transcription factor C/EBPalpha in mice. Mol Cell Biol 2006; 26:1109–1123.

110. Martis PC, Whitsett JA, Xu Y, et al. C/EBPalpha is required for lung maturation at birth. Development 2006; 133:1155–1164.

111. Whitsett JA, Weaver TE. Hydrophobia surfactant proteins in lung function and disease. N Eng J Med 2002; 347:2141–2148.

112. Cole TJ, Blendy JA, Monaghan AP, et al. Targeted disruption of the glucocorticoid receptor gene blocks adrenergic chromaffin cell development and severely retards lung maturation. Genes Dev 1995; 9:1608–1621.

113. Wang B, Weidenfeld J, Lu MM et al. Foxp1 regulates cardiac outflow tract, endocardial cushion morphogenesis and myocyte proliferation and maturation. Development 2004; 131:4477–4487.

114. Wert SE, Dey CR, Blair PA, et al. Increased expression of thyroid transcription factor-1 (TFF-1) in respiratory epithelial cells inhibits alveolarization and causes pulmonary inflammation. Dev Biol 2002; 242:75–87.

115. Yang L, Nalter A, Yan C. Overexpression of dominant negative retinoic acid receptor alpha causes severe alveolar abnormality in transgenic neonatal lungs. Endocrinology 2003; 144:3004–3011.

116. Chen H, Sun J, Buckley S, et al. Abnormal mouse lung alveolarization caused by Smad3 deficiency is a developmental antecedent of centrilobular emphysema. Am J Physiol Lung Cell Mol Physiol 2005; 288:L683–L691.

117. Snyder JM, Jenkins-Moore M, Jackson SK, et al. Alveolarization in retinoic acid receptor-beta-deficient mice. Pediatr Res 2005; 57:384–391.

118. Chen W, Daines MO, Khurana K, et al. Turning off signal transducer and activator of transcription (STAT): the negative regulation of STAT signaling. J Allergy Clin Immunol 2004; 114:476–490.

119. Frolov MV, Dyson NJ. Molecular mechanisms of E2F-dependent activation and pRB-mediated repression. J Cell Sci 2004; 117:2173–2181.

120. Lee HY, Dohi DF, Kim YH, et al. All-trans retinoic acid converts E2F into a transcriptional suppressor and inhibits the growth of normal human bronchial epithelial cells through a retinoic acid receptor-dependent signaling pathway. J Clin Invest 1998; 101:1012–1019.

121. Goumans MJ, Mummery C. Functional analysis of the TGFbeta receptor/Smad pathway through gene ablation in mice. Int J Dev Biol 2000; 44:253–265.

122. Ten Dijke P, Hill CS. New insights into TGF-beta-Smad signaling. Trends Biochem Sci 2004; 29:265–273.

123. Chen C, Chen H, Sun J, et al. Smad1 expression and function during mouse embryonic lung branching morphogenesis. Am J Physiol Lung Cell Mol Physiol 2005; 288:L1033–L1039.

124. Zhao J, Crowe DL, Castillo C, et al. Smad7 is a TGF-beta-inducible attenuator of Smad2/3-mediated inhibition of embryonic lung morphogenesis. Mech Dev 2000; 93: 71–81.

125. Zhao J, Shi W, Chen H, et al. Smad7 and Smad6 differentially modulate transforming growth factor beta-induced inhibition of embryonic lung morphogenesis. J Biol Chem 2000; 275:23992–23997.

126. Zhao J, Lee M, Smith S, et al. Abrogation of Smad3 and Smad2 or of Smad4 gene expression positively regulates murine embryonic lung branching morphogenesis in culture. Dev Biol 1998; 194:82–195.
127. Vinson C, Myakishev M, Acharya A, et al. Classification of human B-ZIP proteins based on dimerization properties. Mol Cell Biol 2002; 22:6321–6335.
128. Perissi V. Rosenfeld MG. Controlling nuclear receptors: the circular logic of cofactor cycles. Nat Rev Mol Cell Biol 2005; 6:542–554.
129. Yao TP, Oh SP, Fuchs M, et al. Gene dosage dependent embryonic development and proliferation defects in mice lacking the transcriptional integrator p300. Cell 1998; 93:361–372.
130. Yadav N, Lee J, Kim J, et al. Specific protein methylation defects and gene expression perturbations in coactivator-associated arginine methyltransferase 1-deficient mice. Proc Natl Acad Sci U S A 2003; 100:6464–6468.
131. Tominaga K, Kirtane B, Jackson JG, et al. MRG15 regulates embryonic development and cell proliferation. Mol Cell Biol 2005; 25:2924–2937.
132. Banerjee-Basu S, Sink DW, Baxevanis AD. The homeodomain resource: sequences, structures, DNA binding sites and genomic information. Nucleic Acids Res 2001; 29:291–293.
133. Berrocal T, Madreid C, Novo S, et al. Congenital anomalies of the tracheobronchial tree, lung, and mediastinum: embryology, radiology, and pathology. Radiographics 2003; 24:e17.
134. Benjamin B. Tracheomalacia in infants and children. Ann Otol Rhinol Laryngol 1984; 93:438–442.
135. Doolittle AM, Mair EA. Tracheal bronchus: classification, endoscopic analysis, and airway management. Otolaryngol Head Neck Surg 2002; 126:240–243.
136. Keslar P, Newman B, Oh KS. Radiographic manifestation of anomalies of the lung. Radiol Clin North Am 1991; 29:255–270.
137. St-Georges R, Deslauriers J, Duranceau A, et al. Clinical spectrum of bronchogenic cysts of the mediastinum and lung in the adult. Ann Thor Surg 1991; 52:6–13.
138. DiFiore JW, Fauza DO, Slavin R, et al. Experimental fetal tracheal ligation reverses the structural and physiological effects of pulmonary hypoplasia in congenital diaphragmatic hernia. J Pediatr Surg 1994; 29(2):248–256.
139. Morelli L, Piscioli I, Licci S, et al. Pulmonary congenital cystic adenomatoid malformation, type I, presenting as a single cyst of the middle lobe in an adult: case report. Diagn Pathol 2007; 2:17–20.
140. Lujan M, Bosque M, Mirapeix RM, et al. Late-onset congenital cystic adenomatoid malformation of the lung. Embryology, clinical symptomatology, diagnostic procedures, therapeutic approach and clinical follow-up. Respiration 2002; 69(2):148–154.
141. Lincoln JC, Stark J, Subramanian S, et al. Congenital lobar emphysema. Ann Surg 1971; 173(1):55–62.
142. Frenckner B, Freyschuss U. Pulmonary function after lobectomy for congenital lobar emphysema and congenital cystic adenomatoid malformation: a follow-up study. Scand J Thorac Cardiovasc Surg 1982; 16(3):293–298.
143. Hubbard AM, States LJ. Fetal magnetic resonance imaging. Top Magn Reson Imaging 2001; 12(2):93–103.
144. Cosio BG, Villena V, Echave-Sustaeta J. Endobronchial hamartoma. Chest 2002; 122(1):202–205.
145. Bratu I, Flageole H, Chen MF, et al. The multiple facets of pulmonary sequestration. J Pediatr Surg 2001; 36(5):784–790.
146. Karnak I, Senocak ME, Ciftci AO, et al. Congenital lobar emphysema: diagnosis and therapeutic considerations. J Pediatr Surg 1999; 34:1347–1351.
147. Nuchtern JG, Harberg FJ. Congenital lung cysts. Semin Pediatr Surg 1994; 3(4):233–243.
148. Berdon WE, Baker DH. Vascular anomalies in the infant lung: rings, slings, and other things. Semin Roentgenol 1972; 7:39–47.
149. Clugston RD, Greer JJ. Diaphragm development and congenital diaphragmatic hernia. Semin Pediatr Surg 2007; 16(2):94–100.
150. Hubbard AM, Crombleholme TM. Anomalies and malformations affecting the fetal/neonatal chest. Semin Roentgenol 1998; 33(2):117–125.

151. Antonetti M, Manuck TA, Schramm C, et al. Congenital pulmonary lymphangiectasia: a case report of thoracic duct agenesis. Pediatr Pulmonol 2001; 32(2):184–186.
152. Lee SY, Yang SR, Lee KR. Congenital pulmonary lymphangiectasia with chylothorax. Asian Cardiovasc Thorac Ann 2002; 10(1):76–77.
153. Hallman M, Glumoff V, Ramet M. Surfactant in respiratory distress syndrome and lung injury. Comp Biochem Physiol A Mol Integr Physiol 2002; 129(1):287–294.
154. Manson J. Mechanism of nitrofen teratogenesis. Environ Health Persp 1986; 70:137–147.
155. Lau C, Cameron AM, Irsula O, et al. Teratogenic effect of nitrofen on cellular and functional maturation of the rat lung. Toxicol Appl Pharmacol 1988; 95:412–422.
156. Stone LC, Manson JM. Effects of the herbicide 2, 4-dichlorophenyl-p-nitrophenyl ether (TOK): effects on the lung maturation of rat fetus. Toxicology 1981; 20:195–207.
157. Koizumi A, Montalbo M, Nguyen O, et al. Neonatal death and lung injury in rats caused by intrauterine exposure to O,O,S-trimethylphosphorothioate. Arch Toxicol 1988; 61:378–386.
158. Koizumi A, Sageshima M, Wada Y, et al. Immature alveolar/blood barrier and low disaturated phosphatidycholine in fetal lung after intrauterine exposure to O,O,S-trimethylphosphorothioate. Arch Toxicol 1989; 63:331–335.
159. Daston GP. Toxic effects of cadmium on the developing rat lung: II. Glycogen and phospholipid metabolism. J Toxicol Environ Health 1982; 9:51–61.
160. Daston GP, Grabowski CT. Toxic effects of cadmium on the developing rat lung: I. Altered pulmonary surfactant and the induction of respiratory distress syndrome. J Toxicol Environ Health 1979; 5:973–983.
161. Stevenson AJ, Kacew S, Singhai RL. Influence of lead on hepatic, renal and pulmonary nucleic acid, polyamine, and cyclic adenosine 3', 5'-monophosphate metabolism in neonatal rats. Toxicol Appl Pharmacol 1977; 40:161–170.
162. Mariassy AT, Abraham WM, Phipps RJ, et al. Effect of ozone on the postnatal development of lamb Mucociliary apparatus. J Appl Physiol 1990; 68:2504–2510.
163. Sherwin RP, Richters V. Effect of 0.3 ppm ozone exposure on type II cells and alveolar walls of newborn mice: an image-analysis quantitation. J Toxicol Environ Health 1985; 16-535–546.
164. Yam J, Roberts RJ. Oxygen-induced lung injury in the newborn piglet. Early Hum Dev 1980; 4:411–424.
165. Sosenko IR, Frank L. Guinea pig lung development: antioxidant enzymes and premature survival in high O_2. Am J Physiol 1987; 252:693–698.
166. Massaro GD, Massaro D. Formation of pulmonary alveoli and gas-exchange surface area: Quantitation and regulation. Annu Rev Physiol 1996; 58:73–92.
167. Freeman G, Juhos LT, Furiosi NJ, et al. Delayed maturation of the rat lung in an environment containing nitrogen dioxide. Am Rev Respir Dis 1974; 10:754–759.
168. Springer DL, Hackett PL, Miller RA, et al. Lung development and postnatal survival for rats exposed in utero to a high-boiling coal liquid. J Appl Toxicol 1986; 6:129–133.
169. De la Fuente M, Hernanz A, Alia M. Effect of semicarbazide on the perinatal development of the rat: changes in DNA, RNA and protein content. Methods Find Exp Clin Pharmacol 1983; 5:287–297.
170. Babiuk RP, Greer JJ. Diaphragm defects occur in a CDH hernia model independently of myogenesis and lung formation. Am J Lung Cell Mol Physiol 2002; 283:1310–1314.
171. Mey J, Babiuk RP, Clugston R, et al. Retinal dehydrogenase-2 is inhibited by compounds that induce congenital diaphragmatic hernias in rodents. Am J Path 2003; 162:673–679.
172. Kim JC, Shin JY, Yang YS, et al. Evaluation of developmental toxicity of Amitraz in Sprague-Dawley rats. Arch Environ Contam Toxicol 2007; 52:137–144.
173. Lau C, Butenhoff JL, Rogers JM. The developmental toxicity of perfluoroalkyl acids and their derivatives. Tox Appl Pharm 2004; 198:231–241.
174. Lau C, Thibodeaux JR, Hanson RG, et al. Exposure to perfluorooctane sulfonate during pregnancy in rat and mouse. II: Postnatal evaluations. Toxicol Sci 2003; 74:382–392.
175. Todd L, Mullen M, Oiley PM, et al. Pulmonary toxicity of monocrotaline differs at critical periods of lung development. Pediatr Res 1985; 19:731–737.
176. King GM, Adamson YR. Effects of cis-hydroxyproline on type II Cell development in fetal rat lung. Exp Lung Res 1987; 12:347–362.
177. Notices. Fed Regist 2005; 70(85):23162–23167.

178. Etzel RA, Montana E, Sorenson WG, et al. Acute pulmonary hemorrhage in infants associated with exposure to Stachybotrys atra and other fungi. Arch Pediatr Adolesc Med 1998; 152:757–762.
179. Taeusch HW, Ballard RA, Gleason CA, et al. Avery's Diseases of the Newborn. 8th ed. Philadelphia: Elsevier Saunders, 2005:1633.
180. Bose CL, Manne DN, D'Ercole AJ, et al. Delayed fetal pulmonary maturation in a rabbit model of the diabetic pregnancy. J Clin Invest 1980; 66:220–226.
181. Sosenko IRS, Lawson EE, Demottaz E, et al. Functional delay in lung maturation in fetuses of diabetic rabbit. J Appl Physiol 1980; 48:643–647.
182. Fariday EE. Effect of maternal malnutrition on surface activity of fetal lungs of rats. J Appl Physiol 1975; 39:535–540.
183. Daston GP. Fetal zinc deficiency as a mechanism for cadmium induced toxicity to the developing rat lung and pulmonary surfactant. Toxicology 1982; 24:55–63.
184. Vojnik C, Hurly LS. Abnormal prenatal lung development resulting from maternal zinc deficiency in rats. J Nutr 1977; 107:862–872.
185. Hurley LS. Teratogenic aspects of manganese, zinc and copper nutrition. Physiol Rev 1981; 61:249–295.
186. Kilburn KH, Hess RA. Neonatal death and pulmonary dysplasia due to d-penicillamine in the rat. Teratology 1982; 26:1–9.
187. Maki JM, Sormunen R, Lippo S, et al. Lysyl oxidase is essential for normal development and function of the respiratory system and for the integrity of elastic and collagen fibers in various issues. Am J Pathol 2005; 167:927–936.
188. Keen CL, Uriu-Hare JY, Hawks S. Effects of copper deficiency on prenatal development and pregnancy outcome. Am J Clin Nutr Suppl 1998; 67:1003S–1011S.
189. Kudlacz EM, Navarro HA, Eylers JP, et al. Prenatal exposure to propranolol via continuous maternal infusion: effects on physiological and biochemical processes mediated by beta adrenergic receptors in fetal and neonatal rat lung. J Pharmacol Exp Ther 1990; 252:42–50.
190. Petit KP, Nielsen HC. Chronic in utero beta blockade alters fetal lung development. Dev Pharmacol Ther 1992; 19:131–140.
191. Thoma-Laurie D, Walker ER, Reasor MJ. Neonatal toxicity in rats following in utero exposure to chlorphentermine or phentermine. Toxicology 1982; 24:85–94.
192. Harker LC, Kirkpatrick SE, Friedman WF, et al. Effects of indomethacin on fetal rat lungs: A possible cause of persistent fetal circulation (PFC). Pediatr Res 1981; 15:147–151.
193. Johnson JW, Mitzner W, Beck JC, et al. Long-term effects of betamethasone on fetal development. Am J Obstet Gynecol 1981; 141:1053–1064.
194. Kilburn KH, Hess RA, Lesser M, et al. Perinatal death and respiratory apparatus dysgenesis due to a bis (dichloroacetyl) diamine. Teratology 1982; 26:155–162.
195. Tasaka H, Takenaka H, Okamato N, et al. Diaphragmatic hernia induced in rat fetuses by administration of bisdiamine. Cong Anom 2008; 32:347–355.
196. Hashimoto R, Inouye M, Murata Y. Pathogenesis of congenital diaphragmatic hernia induced by transplacental infusion of bisdiamine into rats. Cong Anom 1998; 38:143–152.
197. Hashimoto R, Inouye M, Murata Y. Hypoplastic lung observed in rat with chemical-induced congenital diaphragmatic hernia. A preliminary report. Environ Med 1999; 43:66–68.
198. Bannigan JG, Cottell DC. Development of the lung in mice with bromodeoxyuridine-induced cleft palate. Teratology 1991; 44:165–176.
199. Nagai A, Matsumiya H, Yasui S, et al. Administration of bromodeoxyuridine in early postnatal rats results in lung changes at maturity. Exp Lung Res 1993; 19:203–219.
200. Kavlock RJ, Chernoff N, Rogers E. An analysis of fetotoxicity using biochemical endpoints of organ differentiation. Teratology 1982; 26:183–194.
201. Dammann CEL, Ramadurai SM, McCants DD, et al. Androgen regulation of signaling pathways in late fetal mouse lung development. Endrocrinology 2000; 141:2923–2929.
202. Carson SH, Taeusch HW Jr., Avery ME. Inhibition of lung division after hydrocortisone injection into fetal rabbits. J Appl Physio 1973; 34:660–663.
203. Kudlacz EM, Navarro HA, Kavlock RJ, et al. Regulation of postnatal beta-adrenergic receptor/adenylate cyclase development by prenatal agonist stimulation and steroids: alterations in rat kidney and lung after exposure to terbutaline or dexamethasone. J Dev Physiol 1990; 14:273–281.

204. Kanjanapone V, Hartig-Beecken I, Epstein MF. Effect of isoxsuprine on fetal lung surfactant in rabbits. Pediatr Res 1980; 14:278–281.
205. Kero P, Hirvonen T, Valimaki I. Prenatal and postnatal isoxsurprine in respiratory distress syndrome. Lancet 1973; 2:198.
206. Jakkula M, Le Cras T, Gebb S, et al. Inhibition of angiogenesis decreases alveolarization in the developing rat lung. Am J Physiol Lung Cell Mol Physiol 2000; 279:L600–L607.
207. Bergman B, Hedner T. Antepartum administration of terbutaline and the incidence of hyaline membrane disease in preterm infants. Acta Obstet Gynecol Scand 1978; 57:217–221.
208. Beyer BK, Guram MS, Geber WF. Incidence and potentiation of external and internal fetal anomalies resulting from chlordiazepoxide and amitriptyline alone and in combination. Teratology 2005; 30:39–45.
209. Kim J-C, Yun H-I, Shin H-C, et al. Embryo lethality and teratogenicity of a new fluoroquinolone antibacterial DW-116 in rats. Arch Toxicol 2000; 74:120–124.
210. Henck JW, Brown SL, Anderson JA. Developmental toxicity of CI-921, an anilinoacridine antitumor agent. Toxicol Sci 1992; 18:211–220.
211. Waknine Y. FDA Safety Changes: Aceon, Metadate CD, Reyataz. Medscape Medical News, November 12, 2008:1–3.
212. Micardis (elmisartan) Product Monograph, Boehringer Ingelheim, June 19, 2008:1–27.
213. Liggins GC. Premature delivery of foetal lambs infused with glucocorticoids. J Endocrinol 1969; 45:515–523.
214. Wang NS, Kotas RV, Avery ME, et al. Accelerated appearance of osmiophilic bodies in fetal lungs following steroid injection. J Appl Physiol 1971; 30:362–365.
215. Redding RA, Pereira C. Thyroid function in respiratory distress syndrome (RDS) of the newborn. Pediatrics 1974; 54:423–428.
216. Cuestas RA, Lindall A, Engel RR. Low thyroid hormones and respiratory-distress syndrome in the newborn. N Engl J Med 1976; 295:297–302.
217. Volpe MV, Nielsen HC, Archavachotikul K, et al. Thyroid hormone affects distal airway formation during the late pseudoglandular period of mouse lung development. Mol Genet Metab 2003; 80:242–254.
218. Xorbet Aj, Flax P, Alston C, et al. Effect of aminophyllin and dexamethasone on secretion of pulmonary surfactant in fetal rabbits. Pediatr Res 1978; 12:797–799.
219. Hadjigeorgiou E, Kitsiou S, Psaroudakis A, et al. Antepartum aminophylline treatment for prevention of respiratory distress syndrome in premature infants. Am J Obstet Gynecol 1979; 135:257–260.
220. Lawson EE, Brown ER, Torday JS, et al. The effect of epinephrine on tracheal fluid flow surfactant efflux in fetal sheep. Am Rev Respir Dis 1978; 118:1023–1026.
221. Penny DP, Shapiro DL, Rubin P, et al. Effects of radiation on the mouse lung and potential induction of radiation pneumonitis. Virchows Arch B 1981; 37:327–336.
222. Verma, R. Respiratory distress syndrome of the newborn infant. Obst Gyn Survey 1995; 50:542–555.
223. Whitsett JA, Stahlman MT. Impact of advances in physiology, biochemistry, and molecular biology on pulmonary disease in neonates. Am J Respir Crit Care Med 1988; 157:67–71.
224. Wild LM, Nickerson PA, Morin FC III. Ligating the ducts arteriosus before birth remodels the pulmonary vasculature of the lamb. Pediatr Res 1989; 25:251–257.
225. Jobe, A. The new BPD: an arrest of lung development. Ped Res 1999; 46:641–645.
226. Pierce RA, Albertine KH, Starcher BC, et al. Chronic lung injury in preterm lambs: disordered pulmonary elastin deposition. Am J Physiol Lung Mol Physiol 1997; 272:452–460.
227. Erenberg A, Rhodes ML, Weinstein MM, et al. The effects of fetal thyroidectomy on ovine fetal lung maturation. Pediat Res 1979; 13:230–235.
228. Scherrer U, Turini P, Thalmann S, et al. Pulmonary hypertension in high-altitude dwellers: novel mechanism, unsuspected predisposing factors. In: Roach R, Wagner PD, Hackett P, eds. Hypoxia and Exercise. Springer US, 2007:277–291.
229. Shanmugam G, MacArthur K, Pollock JC. Congenital lung malformations-antenatal and postnatal evaluations and management. Euro J Cardiothor Surg 2005; 27:45–52.
230. Reiser K, McCormick R, Rucker RB. Enzymatic and nonenzymatic cross-linking of collagen and elastin. FASEB J 1992; 6:2439–2449.

231. Harding R. Foetal pulmonary development: role of respiratory movements. Equine Vet J Suppl 1997; 24:32–39.
232. Deutsh GH, Pinar H. Prenatal lung development. In: Voelkel NF, MacFee W, eds. Chronic Obstructive Lung Diseases. London: BC Decker, 2002:1–14.
233. Geggel RL, Reid LM. The structure basis of PPHN. Clin Perinatol 1984; 2:525–549.
234. Areechhon W, Reid L. Hypoplasia of the lungs with congenital diaphragmatic hernia. Br Med J 1963; 1:230–233.
235. Cheng ZH, Cilley RE, Miller SA, et al. Laminin may play a role in altered epithelial morphogenesis and delayed maturation of murine hypoplastic lungs. Am J Respir Crit Care Med 2000; 161:A563.
236. Jobe AH, Ikegami M. Antenatal infection/inflammation and postnatal lung maturation and injury. Respir Res 2001; 2:27–32.
237. Murthy EK, Pavlic-Renar I, Metelko Z. Diabetes and pregnancy. Diabetologia Croatica 2002; 31:131–144.
238. McArdle HJ, Ashworth CJ. Micronutrients in fetal growth and development. Br Med Bul 1999; 55:499–510.
239. Perlman M, Williams J, Hirsch M. Neonatal pulmonary hypoplasia after prolonged leakage of amniotic fluid. Arch Dis Childhood 1976; 51:349–353.
240. Blackburn S. Drugs for pregnant and lactating women. J Peri Neonatal Nursing 2004; 18:72–73.
241. Gibbs JR and Singleton A. Application of genome-wide single nucleotide polymorphism typing: simple associations and beyond. Nature 2003, 426:789–796.
242. Hirschhorn JN and Daly MJ. Genome-wide association studies for common diseases and complex traits. Nat Rev Genet 2005, 6:95–108.
243. Willet KE, McMenamin P, Pinkerton KE, et al. Lung morphometry and collagen and elastin content: changes during normal development and after prenatal hormone exposure in sheep. Pediatr Res 1999; 45(5):615–625.
244. Pinkerton KE, Joad JP. The mammalian respiratory system and critical windows of exposure for children's health. Environ Health Perspect 2000; 108(suppl 3):457–462.

11 Developmental toxicity of the kidney

Gregg D. Cappon and Mark E. Hurtt

INTRODUCTION

Human renal development involves two basic processes, the morphological development of kidney cells and structure followed by the acquisition of function. The anatomic formation of the human kidney occurs exclusively in utero. However, the acquisition of adult-like function encompasses in utero and postnatal development, beginning concurrent with formation of the nephron and continuing with rapid attainment of mature functional capability occurring after birth. The processes occurring during morphological and functional development are quite different, but both are susceptible to toxic injury. The extended period of functional kidney development presents the potential for differing vulnerability to toxic insult compared to the adult kidney.

Developmental toxicity to the kidney can be manifested by abnormal structural development and/or abnormal functional development. Traditionally, evaluations of developmental toxicity have focused on the adverse effects of in utero exposure. However, recent concerns regarding the effects of environmental chemicals on children along with an increased focus on the safety of drugs in pediatric patients have increased the scrutiny on the potential for drugs and chemicals to impact postnatal development. Further heightening the concern of potential adverse effects of xenobiotics on development is the continuing emergence of information supporting the concept of fetal origins of adult disease and the potential role that adverse kidney development may have as a mechanism for adult disease. The objective of this chapter is to provide an overview of the potential manifestations of toxicity to the kidney during pre- and postnatal development and to highlight the potential role of the developing kidney in programming of adult disease. The purpose of this chapter is not to catalog specific agents that produce developmental toxicity of the kidney as those reviews have been previously provided (1,2). Instead, this chapter will focus on manifestations of developmental kidney toxicity with a focus on age-related sensitivity to various manifestations of toxicity.

KIDNEY STRUCTURE, FUNCTION, AND DEVELOPMENT

Kidney Structure and Function

To understand the potential age-associated manifestations of toxicity to the developing kidney, it is necessary to have an understanding of mature kidney structure and function and the developmental biology of the kidney. Kidney structure and function has been presented in detail elsewhere; therefore, only a cursory review will be provided here (3,4). Mammals have two kidneys each of which receives its blood supply from a single renal artery that originates from the aorta and drains into a single renal vein that connects to the inferior vena cava. Normally, urine empties from each kidney into a single ureter and then into the bladder. The basic structural unit of the kidney is the nephron that may be considered in three portions, the vascular element, the glomerulus, and the tubular element. The kidney itself may be readily subdivided into three anatomic areas, the cortex, medulla, and papilla. The renal cortex constitutes the major portion of the kidney and contains the glomerulus, a complex specialized capillary bed that functions to create a virtually protein and cell-free ultrafiltrate that passes into the tubular portion of the nephron. The volume and composition of the glomerular filtrate is progressively altered as fluid passes through each of the tubular segments, ultimately resulting in urine formation.

The kidney functions to regulate the internal body environment by three general mechanisms, glomerular filtration, tubular reabsorption, and tubular secretion. The movement of water and solute across the glomerular capillary wall to form an ultra-filtrate of plasma is called glomerular filtration; the rate at which this occurs is called the glomerular filtration rate (GFR). The movement of a substance from tubular fluid back into the plasma is called tubular reabsorption and results in retention of that substance within the body. The movement of a substance from plasma into the tubular fluid is called tubular secretion and results in excretion of that substance into the urine.

The kidneys are structurally similar between species, with several exceptions (5). Probably the most notable structural difference between species is in the papilla. Most small mammals, such as rat, mouse, and rabbit, are unipapillate (have a single papilla in each kidney), while larger mammals, including humans, are multipapillate. Regardless of whether they are unipapillate or multipapillate, urine exits from the tip of the papilla into the renal pelvis in all mammals.

Kidney Development

Kidney anatomical and functional development, including molecular genetics, has been reviewed in detail elsewhere and will not be reproduced in detail here (6–10). In the most simple terms though, the development of the kidney consists of three stages, the aglomerular pronephros; the mesonephros, which represents beginning of formation for the functional kidney; and the metanephros. In most mammals, including humans and rodents, the pronephros and mesonephros appear early in gestation and undergo regression. In the human, the pronephros appears at 3 weeks of gestation and regresses by gestation week 5. The pronephros is nonfunctional, but the pronephric duct becomes the mesonephric duct, which subsequently evolves into the ureteric bud. The meso-nephros develops at 3 to 4 weeks of gestation and degenerates between gestation weeks 5 and 12. Subsequently, the metanephros develops from the interaction between the ureteric bud and the surrounding mesenchyme to form the basic structures of the functioning nephron. In humans, development of nephrons starts at about 5 weeks of gestation. The period between gestation weeks 18 and 32 is a critical time point in renal development as it is at this time that nephrogenesis reaches its peak (11). Nephrons and kidney vasculature develop during the same time frame and are complete by about week 35 of gestation. Thus, the anatomic formation of the kidney occurs exclusively in utero for humans.

Functional development of the kidney is a relatively prolonged process, with some functions of the kidney developing early in gestation while others do not mature until at least 1 year after birth. Urine production in the human kidney is known to begin around 10 to 12 weeks of gestation even though the placenta is primarily responsible for salt and water homeostasis throughout gestation. Renal blood flow (RBF) and GFR progressively increase during the second half of gestation and achieve full-term levels between gestation weeks 32 and 35. There is a maturational increase in glucose transport of the fetal kidneys during gestation, but other tubular functions such as bicarbonate reabsorption and acid production remain low in the fetus. The newborn human infant is incapable of excreting concentrated urine at birth and this function reaches maturity by the first year of life, underscoring the remarkable growth and functional changes that begin at birth. The most notable change is an increase in RBF and a subsequent increase in GFR. Tubular functions mature over the first several years, urine pH is achieved by first week of postnatal age, and maximum urinary acidification can be achieved as early as one to two months of postnatal age. Adult values for RBF,

TABLE 1   Age at Completion of Nephrogenesis

Species	Age at completion of nephrogenesis
Man	Before birth (35 weeks' gestation)
Baboon	Before birth (25 weeks' gestation)
Rat	Postnatal weeks 4–6
Mouse	Before birth
Sheep	Before birth
Dog	Postnatal week 2
Guinea pig	Before birth
Pig	Postnatal week 3

Source: From Refs. 16–19.

TABLE 2   Maturation of Selected Kidney Functions in Human and Rat

	Human	Rat
GFR	1–2 yr of age (16)	Postnatal week 6 (20)
Concentrating ability	Unable to excrete concentrated urine at birth, matures by 1 yr (16)	Unable to excrete concentrated urine at birth, matures by ~PND 11 (21)
Acid-base equilibrium	Little control at birth, matures by 2 mo (22)	Increase dramatically between term fetus and PND 17 (23)
Urine volume control	Present at 3 days after birth and increases over first weeks of life (22)	Low at birth, but develops rapidly during first week of life (24,25)

GFR, urea clearance, maximum tubular excretory capacity, and maximum renal concentrating capacity are achieved by one to two years of age.

It is important to note that the kidney's response to the environmental changes brought about by birth, and its success at maintaining homeostasis are heavily influenced by the gestational age at delivery. Intrauterine growth retardation can permanently reduce the number of functioning nephrons that may lead to renal failure at birth, and one study indicated that the formation of new nephrons does not necessarily continue during postnatal development in cases of extremely low birth weight infants (12–14). Therefore, preterm infants may be at significantly higher risk for renal toxicity than term infants, and this sensitivity may be present throughout the patient's lifetime (15).

Comparative Development in Experimental Species

The comparative development of the kidney has been previously reviewed (8). In general, mammalian kidneys follow similar developmental pathways; however, the time frame with regard to birth varies between species. The end of the anatomical development of the kidney is marked by completion of nephrogenesis that occurs prior to birth in humans, monkeys, mice, sheep, and guinea pigs, and after birth in rats, dogs, and pigs (Table 1).

With the exception of acid base equilibrium and control of urine volume that develop postnatally in all species, differences in functional maturation of the kidney relative to birth are noted among different species. For example, concentrating ability develops postnatally in humans, rats, rabbits, and sheep; but develops prenatally in dogs and guinea pigs. The postnatal development of selected kidney functions in rat and human is presented in Table 2.

MANIFESTATIONS OF KIDNEY DEVELOPMENTAL TOXICITY IN LABORATORY ANIMALS

Kidney development is continuum of a series of events that proceed in carefully timed sequence. Perturbations in any of the normal developmental process can potentially alter subsequent morphogenesis and result in congenital malformations. Adverse kidney development arising from in utero xenobiotic exposure is not limited to congenital malformations, but may also manifest in functional defects. While malformations are associated with dysmorphogenesis resulting in structural or anatomical alteration at birth typically associated with teratogenesis, functional deficits refer to the failure of the developing organ system to acquire normal physiological function and represent another class of developmental toxicity, which may be just as deleterious to the organism as malformations. While most of the focus for functional toxicity has been on nervous system and behavior, kidney function has also been shown to be altered by in utero exposure. Often times, the kidney functional deficit is accompanied by an anatomical correlate; however, though rarely experimentally confirmed, the lack of dysmorphogenesis may not always reflect that the kidney can achieve mature functional capability (26,27).

Abnormal Structural Development of the Kidney

The urinary tract has been shown to be a sensitive target for xenobiotics (28) and comprehensive lists of toxicants affecting kidney development have been published (1,2). In humans, congenital abnormalities of the kidney and urinary tract are relatively common, reportedly occurring in 3% to 4% of live births, accounting for almost one-third of all congenital malformations (6). Some of the developmental abnormalities of the kidney are asymptomatic and inconsequential, but others manifest only in later life and may be important causes of morbidity in older children and adults (29).

The use of animal models to evaluate potential developmental toxicity of xenobiotics is a linchpin of safety evaluation for pharmaceuticals and environmental agents. The most commonly used animal models for investigation of potential dysmorphogenesis are the rat and rabbit. Table 3 provides an overview of the common structural anomalies noted in rat and rabbit developmental toxicity studies.

In general, the kidney morphological abnormalities in experimental models arising from dysmorphogenesis are similar to those noted in humans and can be grouped as renal agenesis (absent kidney development), renal dysgenesis (maldevelopment affecting size, shape, or structure), and anomalies in shape and position (e.g., ectopic kidney, renal fusion) (32).

Abnormal Functional Development

The functional integrity of the kidney is vital to maintain total body homeostasis, as the kidney plays a primary role in excretion of metabolic wastes and in regulation of extracellular fluid volume, electrolyte composition, and acid-base balance. An insult to

TABLE 3 Kidney Abnormalities in Common Laboratory Animals

Observation	Classification
Absent kidney	Malformation
Absent renal papilla	Malformation
Dilated renal pelvis	Variation
Color of kidney (e.g., discolored, pale, hemorrhagic)	General observation
Misshapen kidney	Variation
Fused kidney	Malformation
Size of kidney (small, enlarged)	Malformation
Malpositioned kidney	Malformation
Cyst on kidney	General observation
Supernumerary kidney	Malformation

Source: Terminology and classification of findings used by the Pfizer Developmental and Reproductive Toxicology group, in agreement with published industry standards (30,31).

the developing kidney could disrupt any of these processes and thereby have profound effects on the entire organism. While the nervous system and behavior have been the primary focus of investigations for impaired function due to in utero and postnatal exposure, the kidney has been reported among the most sensitive organ systems when biochemical endpoints are included in the criteria for determining developmental toxicity (33).

Assessment of renal function following developmental exposure can be accomplished using the same techniques that are used for evaluation of renal function in adult animals. In adult animals, the standard battery of noninvasive tests includes measurements of urine volume and osmolality, pH, and urinary composition (e.g., electrolytes, glucose, protein). GFR can be measured directly by determining creatinine or insulin clearance. However, in juvenile animal studies these evaluations of kidney function are complicated due to the ability of the kidney to compensate for developmental insult resulting in a damaged kidney (34). Therefore, most investigators have found it necessary to challenge the capacity of the kidney prior to making functional assessment in juvenile animals (2). Commonly used approaches to evaluate kidney function in juvenile animals involve manipulation of fluid volume either through liquid deprivation (test for renal concentrating ability) or by administration of a water bolus followed by determination of the minimal osmotic concentration that can be excreted.

While functional measures for the assessment of developmental toxicity of the kidney may provide certain insights into toxicological process, in general neither physiological nor biochemical measurements have been shown to have major advantages in sensitivity in detection of developmental insults than traditional morphological examinations (2). This may be in part due to testing paradigms that favor evaluation of morphological effects and the resulting relative paucity of data on functional effects. One promising area for future research incorporating functional evaluations may be to better understand the long-term consequences of minor structural alterations in development.

SENSITIVE WINDOWS FOR DEVELOPMENTAL TOXICITY OF THE KIDNEY

Because of the rapid changes occurring during development, the principle of critical windows of sensitivity based on the developmental stage of the conceptus is a primary consideration for developmental toxicologists. Consistent with this idea, manifestation of developmental toxicity to the kidney induced by xenobiotics can vary depending on the developmental period at which exposure occurs. In general, the period for greatest sensitivity to dysmorphogenesis will encompass the period of rapid cell division and tissue differentiation.

In Utero Exposure

Since structural development of the kidney in humans is completed prior to birth, xenobiotic-induced dysmorphogenesis of the kidney only occurs following in utero exposure. However, because nephrogenesis is not completed until PND 8 in rats, postnatal exposure in rats could potentially impact structural development. On the basis of animal studies, this sensitive window for induction of dysmorphogenesis can be further refined. In utero exposure can also cause abnormal functional development, and the window of sensitivity to functional effects overlaps with the period for induction of frank structural alterations and extends the window of sensitivity for developmental toxicity.

A classic example of sensitive windows of exposure for impairing kidney development is the case of angiotensin-converting enzyme (ACE) inhibitors (for review see

Refs. 2, 8, and 35). ACE is a primary component of the renin-angiotensin system that regulates a variety of body functions, including blood pressure and extracellular fluid volume. ACE catalyzes the conversion of angiotensin I to angiotensin II, which in turn, acts as a vasoconstrictor. ACE inhibitors were developed to lower blood pressure by interrupting this system. The role of ACE inhibitors in infant renal failure has been well documented, and potential harmful effects of the ACE inhibitors in humans include fetal skull abnormalities, intrauterine growth restriction, fetal and neonatal renal dysfunction and failure, oligohydramnios, pulmonary hypoplasia, respiratory distress syndrome, and fetal and neonatal hypotension (35–41). The pathogenic mechanisms of renal dysfunction in affected fetuses and neonates are not fully understood. Histologically, there is tubular dysgenesis and acute tubular necrosis, and recent evidence indicates that renin-angiotensin system inactivity causes chronic low perfusion pressure of the fetal kidney and subsequent renal lesions and early anuria (42).

Rat studies illustrate the importance of the timing of exposure to ACE inhibitors as a critical factor in the development of altered renal morphology. The rat is susceptible to altered renal morphology mainly during the last 5 days of pregnancy and the first two weeks of life (16,43,44). Treatment of newborn rats with an ACE inhibitor for the first 12 days of postnatal life resulted in marked renal abnormalities such as few and immature glomeruli, distorted and dilated tubules, few and short thick arterioles (16,43). Abnormalities in kidney structure produced by postnatal treatment with the ACE inhibitor enalapril correlated with alterations in urine osmolality, GFR, and effective renal plasma flow (45,46). In contrast to the adverse effects on kidney function when treatment initiated late in pregnancy or in early postnatal life, treatment with the ACE inhibitors enalapril and losartan when initiated on postnatal days 14 and 21, respectively, caused no changes in renal morphology (38,44,46). Similarly, exposure of rats to the ACE inhibitors rentiapril and quinapril during gestational periods for traditional teratology studies (gestation day 17 and 20, respectively) did not produce adverse kidney morphology (47,48). Instead, it appears that the time frame in which rats are susceptible to renal injury induced by ACE inhibitors correlates with the critical period for nephrogenesis and marked tubular growth and differentiation. Further, since the human kidney is more mature than the rat kidney at birth, the adverse effects of renin-angiotensin blockers on human kidneys are more likely to occur when exposure occurs during the last few weeks of pregnancy rather than during postnatal development, and in fact ACE inhibitors are recommended for use to manage hypertension in pediatric patients from one year of age (49).

Pediatric Exposure

Children are increasingly exposed to a variety of potentially nephrotoxic agents, sometimes from very young ages. Because structural development of the kidney occurs prior to birth, the focus for postnatal developmental toxicity is increased sensitivity to functional toxicity, based on the developmental stage of the kidney in pediatric populations. In many cases, the developing kidney appears to be more resistant than the adult to toxic agents. Examples of experimental acute nephrotoxicants for which younger animals are less sensitive than adults include sodium dichromate, uranyl nitrate, and mercuric chloride (50–52). This age-related decrease in sensitivity in experimental animals has also been shown following seven days of treatment with gentamicin in newborn animals (53). This decrease in sensitivity corresponded with studies that showed lower renal concentrations of gentamicin and other aminoglycosides in younger animals as compared to adult animals (54–56). The finding with gentamicin is consistent with animal studies with cisplatin where nephrotoxicity was less in young rats and was more evident as the rats grew older. Similar to what was observed with gentamicin-increased sensitivity to cisplatin-induced renal toxicity as animals matured corresponded with increased renal cisplatin concentration (57,58). These findings suggest that a mechanism underlying reduced renal toxicity to some nephrotoxicants in young animals may be the result of reduced perfusion of superficial cortical nephrons in the developing kidney, such that these nephrons receive a lower dose of the agent. The increased renal toxicity of these agents as animals mature is consistent with the postnatal increase in RBF and corresponding increase in GFRs.

It would be wrong to assume that the developing kidney is always going to be less sensitive to toxic insult than the adult kidney. For example, the difference in functional maturation of the kidney is recognized to have major implications on the efficacy and safety of loop diuretics in pediatric patients. The loop diuretics furosemide and bumetanide are frequently employed in the pediatric population for the management of fluid overload. The compounds act mainly by inhibiting sodium reabsorption in the nephron at the thick ascending limb of Henle's loop. Pharmacokinetic differences between adults and infants include a reduced clearance and prolonged half-life, which may cause accumulation of these agents to potentially toxic levels in pediatric patients if dosing intervals are not adjusted (59).

Another compound that has received scrutiny due to concern over reports of enhanced renal toxicity in children is the cancer chemotherapeutic ifosfamide. Ifosfamide is an alkylating agent that has been incorporated into frontline therapy for a number of malignant pediatric tumors. Several preliminary studies suggested young age (less than 5 years of age) as a risk factor for ifosfamide-induced nephrotoxicity, although this conclusion has been debated (60–62). A hypothesized mechanism for the increased childhood sensitivity to ifosfamide involves greater exposure to chloroacetaldehyde, the nephrotoxic metabolite of ifosfamide (63). It has been proposed that maturation of renal CYP P450 expression and activity necessary for formation of chloroacetaldehyde may be related to increased sensitivity in young children, but insufficient information on the ontogeny of renal CYP 450 is available to support this hypothesis (63). However, when the nephrotoxicity of chloroacetaldehyde was compared in isolated human pediatric and adult renal tubules, the sensitivity of pediatric tubules to the toxic effects of chloroacetaldehyde and the rate of their chloroacetaldehyde uptake were not statistically different from those found in adult tubules. Therefore, it appears that the lack of increased susceptibility of pediatric tubules to chloroacetaldehyde toxicity does not support a theory of increased risk for ifosfamide-induced nephrotoxicity in children relative to adults (64).

KIDNEY DEVELOPMENT AND THE FETAL PROGRAMMING OF ADULT DISEASE

Kidney structure can be altered by in utero exposure to xenobiotics, and the function of the mature kidney can be impaired by developmental exposure, in some cases without obvious structural deficiencies. Emerging evidence also suggests that alterations in renal development may be an underlying mechanism for programming of adult disease. The concept of fetal origin of adult diseases or the Barker hypothesis was first proposed on the basis of epidemiologic associations between high rates of infant death and high incidence of death due to ischemic heart disease and coronary artery disease in adult life (65). Subsequent epidemiological research has found links between low birth weight and later adult disease (66,67). These studies suggest that environmental forces impacting intrauterine growth and development may program the organism for subsequent hypertension, cardiovascular, and renal disease in adult lie. The primary measure of adverse intrauterine environment has been birth weight. However, while low birth weight due to fetal malnutrition is almost surely a risk factor, the importance of low birth weight in the genesis of hypertension and subsequent renal and cardiovascular disease is uncertain given that low birth weight can only account for a minor proportion of adults with hypertension (68). These observations have led to the idea that manifestations of poor intrauterine environment other than birth weight might be better predictors for adult hypertension. Accumulating evidence indicates that kidney development may be one such indicator.

The potential for the involvement of kidney development in fetal origins of adult disease was initially proposed on the basis of associations observed in low birth weight infants between retardation of renal development (as demonstrated by reduced nephron number) and blood pressure later in life (69) and has been reviewed elsewhere (70,71). On the basis of these observations, it was suggested that glomerular hyperfiltration resulting from reduced nephron number could stimulate physical and cellular factors leading to systemic hypertension and progressive deterioration of renal function. In this scheme, maternal undernutrition could lead to permanent structural alterations within the kidneys that contribute to a propensity for adult cardiovascular disease.

Support for a link between low birth weight and reduced nephron number has been provided by several studies (72–74). In fact, the number of nephrons in human autopsy samples was shown to range from 227,327 to 1,825,380 with a strong correlation with birth weight (75). Thus, a glomerular origin of hypertension may extend across the normal range of birth weights and not be confined to low birth weight individuals. Furthermore, a study showing that kidneys of individuals with a history of hypertension contained about only half the number of nephrons as those with no history of hypertension provide support for the hypothesis that reduced nephron number, whether due to developmental or other factors, is associated with hypertension (76).

Animal studies also support the hypothesis that reduced nephron number may be the underlying factor in the susceptibility of low birth weight infants to hypertension (70,71,77). A series of animal models investigating underlying mechanisms of programming of adult disease has been developed, and a feature of many of these animal models is alteration in renal development (78,79). For example, intrauterine growth retardation induced by uterine ligation resulted in offspring with significant reductions in glomeruli number. Studies of uteroplacental insufficiency have also shown reduced nephron numbers in rat, rabbit, and pigs, and in these studies the decrease in nephron number correlated to a decrease in GFR (80–83). A model of maternal protein restriction throughout pregnancy has also been shown to cause persistent deficit in nephron number with a corresponding decrease in GFR (78). While many of these models caused intrauterine growth retardation, studies showed that early in utero exposure of rat fetuses to β-lactam antibiotics could produce lower nephron numbers with minimal effect on birth weight (84). Also, early in utero exposure of fetuses to hyperglycemia, a model not commonly associated with fetal growth retardation, cause oligonephronia with fetuses exposed to high glucose concentrations having 10% to 35% fewer nephrons (85). In the rat, prenatal dexamethasone for two-day periods in late gestation resulted in a reduction in glomerular number with corresponding hypertension (86,87). Similarly, in utero exposure to dexamethasone resulted in reduced nephron number and increased glomerular volume in adult animals (88). The timing at which treatment with synthetic glucocorticoids that is most effective in inducing subsequent hypertension in rats and sheep corresponds with the timing for initiation of nephrogenesis in the metanephric kidney in each species. This suggests that a sensitive period for hypertensive programming may correspond with formation of the metanephric kidney during early gestation. Interestingly, given the role of ACE inhibitors in causing fetotoxicity during late gestation, some preliminary experimental evidence has led to a hypothesis that perturbation of the renin-angiotensin system during early gestation be a factor in programming of postnatal hypertension (70,77).

In summary, there are many studies in which a low nephron number has been induced by manipulations of pregnant animals that negatively affected the intrauterine environment. When nephron number is compromised during kidney development, functional changes occur that can lead to hypertension and/or renal disease. The nephron deficit appears to be determined before birth, and not to be made up for after birth. In some of these models, the decrease in nephron number occurs in the absence of any effect on birth weight. This suggests that the health implications of poor intrauterine environment may be much greater than predicted solely by the incidence of babies with low birth weight.

CONCLUSION

The development of the kidney is a relatively prolonged process, initiating early in gestation and continuing through early postnatal development. Developmental toxicity of the kidney can be expressed by both morphological alterations and by impaired function depending on the stage of kidney development at the time of insult. Evidence indicates that the extended developmental timelines of the kidney present multiple sensitive windows for developmental toxicity. Also, an increasing amount of evidence indicates that perturbations of the intrauterine environment can program adult disease. Altered renal development, especially reduced nephron number, can be influenced by a number of environmental factors that adversely affect the intrauterine environment. It has been suggested that the reduction in nephron number and subsequent impairment of kidney function early in development may have implications for development of cardiovascular disease in adults. This is an important area for future investigation.

REFERENCES

1. McCormack KM, Hook JB, Gibson JE. Developmental anomalies of the kidney: a review of normal and aberrant renal development. In: Hook JB, ed. Toxicology of the Kidney. New York: Raven Press, 1981:227–250.
2. Lau C, Kavlock RJ. Functional Toxicity in the developing heart, lung, and kidney. In: Kimmel CA, Buelke-Sam J, eds. Developmental Toxicology. 2nd ed. New York: Raven Press, Ltd., 1994:119–188.
3. Sands JM, Verlander JW. Anatomy and physiology of the kidneys. In: Tarloff JB, Lash LH, eds. Toxicology of the Kidney. 3rd ed. Boca Raton: CRC Press, 2005.
4. Schnellmann RG. Toxic responses of the kidney. In: Klassen CD, ed. Casarett and Doull's Toxicology: The Basic Science of Poisons. 6th ed. New York: McGraw-Hill, 2001:491–514.
5. Sellers RS, Khan NM. Age, sex, and species differences in nephrotoxic response. In: Hook JB, Robin S, Goldstein AY, eds. Toxicology of the Kidney. 3rd ed. Boca Raton: CRC Press, 2005:1059–1097.
6. Carlson BM. Urogenital system. In: Copland B, ed. Human Embryology and Developmental Biology. 2nd ed. St Louis: Mosby, 1999:359–397.
7. Saxen L. The developing kidney in toxicity tests. In: Bourdeau P, Somers E, Richarson GM, et al, eds. Short-Term Toxicity Tests for Non-Genotoxic Effects. Chichester: John Wiley, 1990:135–153.
8. Zoetis T, Hurtt ME. Species comparison of anatomical and functional renal development. Birth Defects Res B Dev Reprod Toxicol 2003; 68:111–120.
9. Chevalier RL. Renal function in the fetus, neonate, and child. In: Wein AJ, Kavoussi LR, Novick AC, et al., eds. Campbell-Walsh Urology. 9th ed. Vol 4. Philadelphia: Saunders 2007.
10. Glassberg KI. Normal and abnormal development of the kidney: a clinician's interpretation of current knowledge. J Urol 2002; 167:2339–2350; discussion 2350–2331.

11. Gasser B, Mauss Y, Ghnassia JP, et al. A quantitative study of normal nephrogenesis in the human fetus: its implication in the natural history of kidney changes due to low obstructive uropathies. Fetal Diagn Ther. 1993; 8:371–384.
12. Steele BT, Paes B, Towell ME, et al. Fetal renal failure associated with intrauterine growth retardation. Am J Obstet Gynecol 1988; 159:1200–1202.
13. Merlet-Benichou C. Influence of fetal environment on kidney development. Int J Dev Biol 1999; 43:453–456.
14. Rodriguez MM, Gomez AH, Abitbol CL, et al. Histomorphometric analysis of postnatal glomerulogenesis in extremely preterm infants. Pediatr Dev Pathol 2004; 7:17–25.
15. Chevalier RL. Developmental renal physiology of the low birth weight pre-term newborn. J Urol 1996; 156:714–719.
16. Gomez RA, Sequeira Lopez ML, Fernandez L, et al. The maturing kidney: development and susceptibility. Ren Fail 1999; 21:283–291.
17. Gubhaju L, Black MJ. The baboon as a good model for studies of human kidney development. Pediatr Res 2005; 58:505–509.
18. Kleinman LI. Developmental renal physiology. Physiologist 1982; 25:104–110.
19. Fouser L, Avner ED. Normal and abnormal nephrogenesis. Am J Kidney Dis 1993; 21:64–70.
20. Horster M. Nephron function and perinatal homeostasis. Ann Rech Vet 1977; 8:468–482.
21. Kavlock RJ, Gray JA. Evaluation of renal function in neonatal rats. Biol Neonate 1982; 41:279–288.
22. Walker DG. Functional differentiation of the kidney. In: Barnes AC, ed. Intra-Uterine Development. Lea & Febiger: Philadelphia, 1968:245–252.
23. Schwartz GJ, Olson J, Kittelberger AM, et al. Postnatal development of carbonic anhydrase IV expression in rabbit kidney. Am J Physiol 1999; 276:F510–F520.
24. Falk G. Maturation of renal function in infant rats. Am J Physiol 1955; 181:157–170.
25. Heller H. The response of newborn rats to administration of water by the stomach. J Physiol 1947; 106:245–255.
26. Daston GP, Rehnberg BF, Carver B, et al. Functional teratogens of the rat kidney: I. Colchicine, dinoseb, and methyl salicylate. FundApplToxicol 1988; 11:381–400.
27. Daston GP, Rehnberg BF, Carver B, et al. Functional teratogens of the rat kidney: II. Nitrofen and ethylenethiourea. Fund Appl Toxicol 1988; 11:401–415.
28. Wilson JG, Warkany J. Malformations in the genito-urinary tract induced by vitamin A deficiency in the rat. Am J Anat 1948; 83:357–407.
29. Guay-Woodford LM. Hereditary nephropathies and abnormalities of the urinary tract. In: Goldman L, Ausiello DA, eds. Cecil Medicine. Philadelphia: Saunders, 2007.
30. Wise LD, Beck SL, Beltrame D, et al. Terminology of developmental anomalies in common laboratory mammals (version 1). Teratology 1997; 55:249–292.
31. Solecki R, Bergmann B, Bürgin H, et al. Harmonization of rat fetal external and visceral terminology and classification: Report of the Fourth Workshop on the Terminology in Developmental Toxicology, Berlin, 18–20 April 2002. Repro Tox 2002; 17:625–637.
32. Elder JS. Congenital anomalies and dysgenesis of the kidneys. In: Kliegman RM, Behrman RE, Jenson HB, et al., eds. Nelson Textbook of Pediatrics. 18th ed. Philadelphia: Saunders, 2007.
33. Kavlock RJ, Chernoff N, Rogers E, et al. An analysis of fetotoxicity using biochemical endpoints of organ differentiation. Teratology 1982; 26:183–194.
34. Kavlock RJ, Daston GP. Detection of renal dysfunction in neonatal rats: methodologies and applications. Prog Clin Biol Res 1983; 140:339–356.
35. Buttar HS. An overview of the influence of ACE inhibitors on fetal-placental circulation and perinatal development. Mol Cell Biochem 1997; 176:61–71.
36. Hanssens M, Keirse MJ, Vankelecom F, et al. Fetal and neonatal effects of treatment with angiotensin-converting enzyme inhibitors in pregnancy. Obstet Gynecol 1991; 78:128–135.
37. Shotan A, Widerhorn J, Hurst A, et al. Risks of angiotensin-converting enzyme inhibition during pregnancy: experimental and clinical evidence, potential mechanisms, and recommendations for use. Am J Med 1994; 96:451–456.
38. Sedman AB, Kershaw DB, Bunchman TE. Recognition and management of angiotensin converting enzyme inhibitor fetopathy. Pediatr Nephrol 1995; 9:382–385.
39. Laube GF, Kemper MJ, Schubiger G, et al. Angiotensin-converting enzyme inhibitor fetopathy: long-term outcome. Arch Dis Child Fetal Neonatal Ed 2007; 92:F402–F403.

40. Hegde AU, Parekji S, Ali US, et al. Angiotensin converting enzyme inhibitor fetopathy. Indian Pediatr 1999; 36:79–82.
41. Pryde PG, Sedman AB, Nugent CE, et al. Angiotensin-converting enzyme inhibitor fetopathy. J Am Soc Nephrol 1993; 3:1575–1582.
42. Gribouval O, Gonzales M, Neuhaus T, et al. Mutations in genes in the renin-angiotensin system are associated with autosomal recessive renal tubular dysgenesis. Nat Genet 2005; 37:964–968.
43. Lasaitiene D, Chen Y, Guron G, et al. Perturbed medullary tubulogenesis in neonatal rat exposed to renin-angiotensin system inhibition. Nephrol Dial Transplant 2003; 18: 2534–2541.
44. Hilgers KF, Norwood VF, Gomez RA. Angiotensin's role in renal development. Semin Nephrol 1997; 17:492–501.
45. Guron G, Adams MA, Sundelin B, et al. Neonatal angiotensin-converting enzyme inhibition in the rat induces persistent abnormalities in renal function and histology. Hypertension 1997; 29:91–97.
46. Guron G, Marcussen N, Nilsson A, et al. Postnatal time frame for renal vulnerability to enalapril in rats. J Am Soc Nephrol 1999; 10:1550–1560.
47. Dostal LA, Kim SN, Schardein JL, et al. Fertility and perinatal/postnatal studies in rats with the angiotensin-converting enzyme inhibitor, quinapril. Fundam Appl Toxicol 1991; 17:684–695.
48. Cozens DD, Barton SJ, Clark R, et al. Reproductive toxicity studies of rentiapril. Arzneimittelforschung 1987; 37:164–169.
49. Mitsnefes MM. Hypertension in children and adolescents. Pediatr Clin N Am 2006; 53:493–512.
50. Daston GP, Kavlock RJ, Rogers EH, et al. Toxicity of mercuric chloride to the developing rat kidney I Postnatal ontogeny of renal sensitivity. Toxicol Appl Pharmacol 1983; 71: 24–41.
51. Appenroth D, Gambaryan S, Friese KH, et al. Influence of metyrapone and phenobarbital on sodium dichromate nephrotoxicity in developing rats. J Appl Toxicol 1990; 10:227–232.
52. Pelayo JC, Andrews PM, Coffey AK, et al. The influence of age on acute renal toxicity of uranyl nitrate in the dog. Pediatr Res 1983; 17:985–992.
53. Klein J, Koren G, MacLeod SM. Comparison of methods for prediction of nephrotoxicity during development. Dev Pharmacol Ther 1992; 19:80–89.
54. Marre R, Tarara N, Louton T, et al. Age-dependent nephrotoxicity and the pharmacokinetics of gentamicin in rats. Eur J Pediatr 1980; 133:25–29.
55. Lelievre-Pegorier M, Sakly R, Meulemans A, et al. Kinetics of gentamicin in plasma of nonpregnant, pregnant, and fetal guinea pigs and its distribution in fetal tissues. Antimicrob Agents Chemother 1985; 28:565–569.
56. Provoost AP, Adejuyigbe O, Wolff ED. Nephrotoxicity of aminoglycosides in young and adult rats. Pediatr Res 1985; 19:1191–1196.
57. Ali BH, Al-Moundhri M, Tageldin M, et al. Ontogenic aspects of cisplatin-induced nephrotoxicity in rats. Food Chem Toxicol 2008; 46:3355–3359.
58. Jongejan HT, Provoost AP, Wolff ED, et al. Nephrotoxicity of cis-platin comparing young and adult rats. Pediatr Res 1986; 20:9–14.
59. Eades SK, Christensen ML. The clinical pharmacology of loop diuretics in the pediatric patient. Pediatr Nephrol 1998; 12:603–616.
60. Aleksa K, Woodland C, Koren G. Young age and the risk for ifosfamide-induced nephrotoxicity: a critical review of two opposing studies. Pediatr Nephrol. 2001; 16: 1153–1158.
61. Ashraf MS, Brady J, Breatnach F, et al. Ifosfamide nephrotoxicity in paediatric cancer patients. Eur J Pediatr 1994; 153:90–94.
62. Loebstein R, Koren G. Ifosfamide-induced nephrotoxicity in children: Critical review of predictive risk factors. Pediatrics 1998; 101:e8.
63. Aleksa K, Matsell D, Krausz K, et al. Cytochrome P450 3A and 2B6 in the developing kidney: implications for ifosfamide nephrotoxicity. Pediatr Nephrol 2005; 20:872–885.
64. Dubourg L, Taniere P, Cochat P, et al. Toxicity of chloroacetaldehyde is similar in adult and pediatric kidney tubules. Pediatr Nephrol 2002; 17:97–103.
65. Barker DJ. The fetal and infant origins of adult disease. BMJ 1990; 301:1111.

66. Curhan GC, Chertow GM, Willett WC, et al. Birth weight and adult hypertension and obesity in women. Circulation 1996; 94:1310–1315.
67. Barker DJ, Hales CN, Fall CH, et al. Type 2 (non-insulin-dependent) diabetes mellitus, hypertension and hyperlipidaemia (syndrome X): relation to reduced fetal growth. Diabetologia 1993; 36:62–67.
68. Rostand SG. Oligonephronia, primary hypertension and renal disease: 'is the child father to the man?' Nephrol Dial Transplant 2003; 18:1434–1438.
69. Brenner BM, Chertow GM. Congenital oligonephropathy and the etiology of adult hypertension and progressive renal injury. Am J Kidney Dis 1994; 23:171–175.
70. McMillen IC, Robinson JS. Developmental Origins of the Metabolic Syndrome: Prediction, Plasticity, and Programming. Physiol Rev 2005; 85:571–633.
71. Moritz KM, Singh RR, Probyn ME, et al. Developmental programming of a reduced nephron endowment: more than just a baby's birth weight. Am J Physiol Renal Physiol 2009; 296:F1–F9.
72. Hinchliffe SA, Howard CV, Lynch MR, et al. Renal developmental arrest in sudden infant death syndrome. Pediatr Pathol 1993; 13:333–343.
73. Hinchliffe SA, Lynch MR, Sargent PH, et al. The effect of intrauterine growth retardation on the development of renal nephrons. Br J Obstet Gynaecol 1992; 99:296–301.
74. Manalich R, Reyes L, Herrera M, et al. Relationship between weight at birth and the number and size of renal glomeruli in humans: a histomorphometric study. Kidney Int 2000; 58:770–773.
75. Hughson M, Farris AB III, Douglas-Denton R, et al. Glomerular number and size in autopsy kidneys: the relationship to birth weight. Kidney Int 2003; 63:2113–2122.
76. Keller G, Zimmer G, Mall G, et al. Nephron number in patients with primary hypertension. N Engl J Med 2003; 348:101–108.
77. Moritz KM, Dodic M, Wintour EM. Kidney development and the fetal programming of adult disease. Bioessays 2003; 25:212–220.
78. Woods LL. Maternal nutrition and predisposition to later kidney disease. Curr Drug Targets 2007; 8:906–913.
79. Armitage JA, Khan IY, Taylor PD, et al. Developmental programming of the metabolic syndrome by maternal nutritional imbalance: how strong is the evidence from experimental models in mammals? J Physiol 2004; 561:355–377.
80. Merlet-Benichou C, Gilbert T, Muffat-Joly M, et al. Intrauterine growth retardation leads to a permanent nephron deficit in the rat. Pediatr Nephrol 1994; 8:175–180.
81. Bassan H, Trejo LL, Kariv N, et al. Experimental intrauterine growth retardation alters renal development. Pediatr Nephrol 2000; 15:192–195.
82. Bauer R, Walter B, Bauer K, et al. Intrauterine growth restriction reduces nephron number and renal excretory function in newborn piglets. Acta Physiol Scand 2002; 176:83–90.
83. Pham TD, MacLennan NK, Chiu CT, et al. Uteroplacental insufficiency increases apoptosis and alters p53 gene methylation in the full-term IUGR rat kidney. Am J Physiol Regul Integr Comp Physiol 2003; 285:R962–R970.
84. Nathanson S, Moreau E, Merlet-Benichou C, et al. In utero and in vitro exposure to beta-lactams impair kidney development in the rat. J Am Soc Nephrol 2000; 11:874–884.
85. Amri K, Freund N, Vilar J, et al. Adverse effects of hyperglycemia on kidney development in rats: in vivo and in vitro studies. Diabetes 1999; 48:2240–2245.
86. Ortiz LA, Quan A, Weinberg A, et al. Effect of prenatal dexamethasone on rat renal development. Kidney Int 2001; 59:1663–1669.
87. Ortiz LA, Quan A, Zarzar F, et al. Prenatal dexamethasone programs hypertension and renal injury in the rat. Hypertension 2003; 41:328–334.
88. Wintour EM, Moritz KM, Johnson K, et al. Reduced nephron number in adult sheep, hypertensive as a result of prenatal glucocorticoid treatment. J Physiol 2003; 549: 929–935.

12 Developmental toxicology of the liver

Kartik Shankar and Harihara M. Mehendale

INTRODUCTION

The liver is strategically positioned between the digestive system and the systemic circulation and is designed to act as a "first-pass" for both nutrients and xenobiotics alike. The adult liver is a heterogeneous tissue with hepatocytes as the principal functional cell type accounting for approximately 80% of the parenchymal volume (1). Hepatocytes are polarized epithelial cells that carry out a multitude of metabolic functions and possess both endocrine and exocrine properties. Further, hepatocytes exhibit a regioselective gene expression phenotype and associated functional properties, depending on their acinar location. Biliary epithelial cells, Kupffer cells (resident liver macrophages), sinusoidal endothelial cells, stellate cells (Ito cells), and pit cells (liver-specific natural killer cells) make up the remainder of the liver. While the vast majority of research has examined both the biology of and toxicity to hepatocytes, nonhepatocyte cells are also known to play critical roles in maintaining liver function and influencing the extent of injury and repair. The liver plays a major role in normal energy metabolism, including glycogen storage, maintaining systemic glucose levels via gluconeogenesis, very-low density lipoprotein secretion, plasma protein synthesis, hormone production, and decomposition of red blood cells. In addition, the liver produces bile that is essential for the absorption of dietary lipids from the intestines and excretion of endogenous compounds and xenobiotics.

Venous blood enters the liver through the portal veins draining blood circulation from the intestines and the stomach. Hence, the liver is generally the primary site for detoxification of toxic endo- and exobiotics or metabolism of nutrients (vitamins, macronutrients viz. carbohydrate, fat, protein), xenobiotics (drugs, environmental chemicals, bioactive phytochemicals), and microbial waste. To meet the needs of functioning as a sentinel, the liver is equipped with a high metabolic capacity with numerous phase I and II xenobiotics metabolizing enzymes. These enzymes localized in the membranes of the smooth endoplasmic reticulum, mitochondria, or cytosol, carry out an array of oxidation, hydrolysis, and conjugation reactions that are aimed at making xenobiotics less toxic, more water soluble, and eventually excreted via the urine or secreted into bile to be excreted via feces. The profile of these enzymes expressed by the liver are influenced by several factors, including developmental stage, diet, gender, concurrent disease, and other environmental factors (such as drugs and toxicants). This, in turn, may significantly alter the ability of the liver to maintain or perform its functions of detoxification and metabolism (2).

Ironically, being the first site of contact with xenobiotics and possessing a high metabolic capacity make the liver a frequent target of toxicity. Mechanisms of hepatotoxicity have been a topic of intense investigation for more than four decades and these mechanisms are well understood for many hepatotoxicants. However, these are beyond the scope of this chapter. In adults, drug-induced hepatotoxicity represents approximately 6% of all adverse reactions and the most frequent cause of postmarketing withdrawal of medications. Although over 300 drugs in current use have been implicated to cause liver injury, the majority of these remain non-dose-dependent hepatotoxins, with less than 20 having a known mechanism of toxicity. This has fueled considerable effort in identifying early biomarkers for hepatotoxicity in adults and understanding underlying mechanisms of toxicity. In addition, a large number of chemicals are intrinsically hepatotoxic and cause dose-dependent liver injury. Carbon tetrachloride (CCl_4), chloroform, dimethylnitrosamine, trinitrotoluene, aflatoxin B1, and phalloidin, a mushroom poison, are well-known examples. However, a number of diverse chemical compounds are known to target the liver specifically during development. We will provide a brief overview of the known literature later in this chapter. The dynamic processes of hepatocyte death, proliferation, and replacement are critical in protecting the liver from many harmful effects of xenobiotics, which may threaten hepatic survival and function. Further insults during development may lead to long-term programming of hepatic physiology, which may be revealed only in later life. This aspect of developmental programming sometimes referred to as the "developmental origins of health and disease" or the Barker hypothesis focuses on critical targets that are susceptible to programming in early life (endocrine axes such as the hypothalamic-pituitary-adrenal and growth hormone signaling, estrogen and androgen signaling, etc., and epigenetic regulation of nuclear receptors) maybe a critical target of early developmental toxicity in the liver.

EMBRYONIC LIVER DEVELOPMENT

A basic understanding of liver development is essential to appreciate the complexity of the processes and identify potential targets of toxicants that can modulate development of a healthy liver. Much of what is known about early liver development has come from studying mouse models, since these are amenable to precise genetic manipulations. However, the basic underlying events are highly conserved in rat, human, and other models of liver development. The major liver cells, hepatocytes and bile duct cells, originate from a common precursor, the hepatoblast, that derives from the definitive endoderm (2–4). In the mouse, the definitive endoderm emerges from the primitive streak at embryonic day 7.0 (ED 7), displacing the embryonic endoderm of the yolk sac. By ED 8.0, the ventral wall of the foregut endoderm is positioned adjacent to the developing heart (5,6). This is critically important in the development of the liver, as signals from the developing heart are the first signals to induce changes in the neighboring endoderm (6,7). The hepatic diverticulum (or hollow outgrowth) is lined by the endodermal cells and invades upward and forward into the *septum transversum*, a mass of mesoderm between the vitelline duct and the pericardial cavity (5,8). Fate-mapping experiments have revealed that the liver arises from the lateral domains of the endoderm in the developing ventral foregut (7,9). In response to specific signals from the cardiac mesenchyme around ED 8.5 (in the mouse), the median and lateral domains of the foregut endoderm come together and form the hepatic endoderm. The future pancreas is also specified in the adjacent caudal area of the foregut endoderm. These events correspond to about three weeks of human gestation.

Both the signals from the cardiac mesenchyme and septum transversum, and the resulting signaling cascades in the primordial foregut are under intense investigation in basic developmental biology. The use of genetic knockdown and genetic lineage marking studies have shown that fibroblast growth factor (FGF) from the cardiac mesenchyme and bone morphogenetic protein (BMP) from the septum transversum mesenchymal cells coordinately induce the liver program (3,4,6,10–12). Mitogen-activated protein kinase (MAPK) signaling is activated in the endodermal hepatic progenitors in response to FGF (13). These cells specified to become the liver in the embryo form the earliest hepatoblasts, which are bipotential progenitors that give rise to both hepatocytes and cholangiocytes. Hepatoblasts express the hepatocyte-specific genes such as albumin and transthyretin (14). The stimulation by FGF on hepatic specification is only transient, and later acts as a stimulus for expansion of the newly specified progenitor cell populations. This is influenced by a variety of signals including Wnts (15–17). As the hepatoblasts proliferate, the endodermal cells transition into a pseudostratified columnar shape, a process controlled by the *homeobox* transcription factor *Hhex* (18). Following this, the hepatic cells proliferate into the surrounding stroma and begin organogenesis. The transcription factors *prox1* and *OC-1* and *-2* regulate the interaction of the hepatoblasts with the stromal cells by controlling the expression of cell adhesion proteins (19,20).

In concert with the invasion of the budding liver into the stroma, vasculogenesis and angiogenesis occur through interactions of the developing endoderm and endothelial cell precursors (21). The fetal liver serves as a transient site for hematopoiesis in amniotes and hence the development of the liver vasculature is necessary for fetal viability. As mentioned previously, the hepatic bud emerges around ED 8.5. The septum transversum of mesenchymal origin has been suggested to contain vasculogenic endothelial cells (21). Studies using CD31 (platelet endothelial cell adhesion molecule, PECAM) and Flk-1 (VEGFR2, receptor for VEGF) as markers of angioblasts and embryonic endothelial cells, respectively, revealed that cells positive for both markers are detected interceding between the thickening hepatic epithelium and the septum transversum as early as ED 8.5 to ED 9.0. These studies indicate that prior to blood vessel formation and function, angioblasts or endothelial cells physically interact with the nascent hepatic cells, preceding liver bud formation (21). In mice lacking *flk-1*, where no angioblasts or endothelial cells develop, following normal thickening of the hepatic endoderm, hepatic cells fail to proliferate into the surrounding septum transversum. Hence, interactions with the angioblasts and/or early endothelial cells are essential for liver bud growth prior to vascular development and recruitment of hematopoetic cells (21). Further, after hepatic bud formation, interactions with endothelial cells promote hepatic morphogenesis. In human liver development, angioblasts are present by five weeks of gestation and form into sinusoids, the template on which the three-dimensional growth of hepatic cords occurs. The liver cell plates are initially three to five cells in thickness and gradually transform into plates of one cell thickness (2).

In mid-gestation the hepatoblast population undergoes approximately 8 doublings (in the rat) and expands around 80-fold in volume (8). Several transcription factors are critical in this process. Genetic studies have shown that Xbp1, FoxM1b, c-jun, and HNF-4α are critical for this process of expansion and hepatic maturation (8,22–24). In addition, β-catenin, a key downstream effector of Wnt signaling, has been shown to play an important role in achieving this expansion. Liver-specific deletion of β-catenin results in smaller livers, and forced overexpression results in increased liver size (15,25). Similarly, loss of hepatocyte growth factor or its receptor c-Met results in severe hypoplasia of the liver (26,27). Wnt/β-catenin signaling is also important for regulating biliary epithelial differentiation (15,28). The development of liver structure and function progresses well into early postnatal life, through lactation, and postweaning. Studies have shown that the hepatic gene expression is maintained, in large part, by active transcription (29). Not surprisingly, a large number of important transcription factors orchestrate the expression of the complete set of genes representative of hepatocyte function. HNF-1α and β, C/EBP-α, FoxA, and HNF-4α control hepatocyte differentiation and function (30–34). While the precise roles of these factors are just being elucidated using molecular genetic techniques, no information is currently available on the effects of specific toxicants during very early stages on liver development and maturation. In utero, the fetus is supplied with a continuous flow of high-carbohydrate, low-fat, and high–amino acid nutrients through the placenta. Following birth, the newborn is fed at intervals mainly with milk, a relatively high-fat diet. At weaning, again the diet changes to a more high-carbohydrate and low-fat diet. The liver has to adapt to these dietary changes through regulation of carbohydrate, protein, and fat metabolism. This also allows functional zonation of the liver to occur, with Zone 1 (periportal) hepatocytes mainly carrying out gluconeogenesis, β-oxidation, cholesterol biosynthesis, and bile secretion, whereas Zone 3 (centrilobular) hepatocytes carry out glycolysis, lipogenesis, ketogenesis, and phase II metabolism.

SELECTED CLASSES OF TOXICANTS CAUSING DEVELOPMENTAL LIVER TOXICITY
Mycotoxins

Mycotoxins are biologically active, small molecular weight (MW ~ 700) secondary metabolites produced by fungi, including mushrooms, yeast, and molds. Mycotoxins exert potent toxic effects in both animals and people (35,36). Mycotoxins can appear in the food chain as a result of fungal infection of crops, either by being eaten directly by humans, or indirectly by being used as livestock feed. Mycotoxins are generally resistant to decomposition or being broken down during digestion, so they remain in contaminated food grain and are carried up the food chain in meat and dairy products. Exposure to high-temperature treatments such as cooking and freezing also does not destroy mycotoxins (35). Broadly, aflatoxins, ochratoxin, citrinin, ergot, and *Fusarium* toxins are the most well studied (36). Aflatoxins are produced by two species of *Aspergillus* fungi, *A. flavus* and *A. parasiticus*. Among the four forms of aflatoxin (B1, B2, G1, and G2), aflatoxin B1 is the most toxic and is a potent carcinogen. Aflatoxins are acutely toxic, immunosuppressive, mutagenic, teratogenic, and carcinogenic (37,38). The primary target organ is the liver. Aflatoxin B1 is converted in the liver of a number of species, including rats, to its hydroxylated derivative alflatoxin M1 (36). Aflatoxin B1 is a potent hepatocarcinogen in experimental animals, but tumors may also occur in the colon and kidneys. In rats, tumors develop after the administration of 0.2 μg/day or a single dose of 0.5 mg of aflatoxin B1. Aflatoxins produce liver tumors in mice, rats, fish, ducks, marmosets, tree shrews, and monkeys after administration by several routes, including orally (38). The developmental toxicity has been studied in various animals. Aflatoxin B1 has been reported to be teratogenic and/or embryotoxic in rats (39), mice (40), hamsters (41), chick embryos (42), tadpoles (43), and Japanese medaka eggs (44). Gestational exposure to aflatoxin B1 also induced skeletal malformations, intrauterine growth retardations, cleft palate, and impaired locomotor, learning, and early avoidance response development in rats (45).

Lifetime exposure to aflatoxin B1 at 2 ppm from conception onward leads to development of malignant hemorrhagic liver tumors, histologically diagnosed as hemangiosarcoma (46). Other investigators have reported that exposure to aflatoxin B1 (or its metabolites) limited to gestation also causes liver pathology in the offspring (47). Offspring exposed to aflatoxin B1 in utero developed inflammatory, hyperplastic, and neoplastic changes in the liver. Hepatic and biliary cysts and cholangiocarcinomas and hepatocellular carcinomas were observed in the offspring (47). In other studies, using thin layer chromatography of offspring liver homogenates, investigators confirmed the transfer of toxic aflatoxin B1 metabolites from pregnant rats to their litters (48). Extensive biliary and hepatic cystic lesions were observed only in young animals. Multifocal hepatic necrosis, bile ductular proliferation, areas of altered hepatocytes, neoplastic nodules, and hepatocellular carcinoma were observed in both adult and newborn animals (48). These studies suggest that aflatoxin B1 has significant transplacental toxicity in animal models.

Aflatoxin M1 is the primary metabolite of hepatic aflatoxin B1 metabolism secreted into milk. Contamination into milk can occur following ingestion of contaminated feed, and aflatoxin M1 levels in dairy milk are required to be below 0.5 ppb by the FDA. The conversion rate of aflatoxin B1 to M1 in mammals, including humans, varies between 0.5% to 6% of the ingested dose (49). Previous studies have indicated that, in one-day-old ducklings, aflatoxin M1 is less toxic than aflatoxin B1. However, studies in rats have shown similar toxicity of aflatoxin B1 and M1 (50). Hence, aflatoxin M1 is categorized as a Group 2B carcinogen by the International Agency for Research on Cancer (*possibly carcinogenic to humans*). No adequate epidemiological studies exist on the dose-response relationships between the intake of aflatoxin M1, exposure to hepatitis B or C virus, and liver cancer. Using the Frog Embryo Teratogenesis Assay–*Xenopus* (FETAX), Vismara et al., showed that the metabolite aflatoxin M1 was not embryotoxic up to doses of 256 µg/L (51). Since aflatoxin M1 is the primary metabolite in milk, the toxicity to the offspring liver during lactation via aflatoxin exposure through milk is considered low.

Other mycotoxins such as deoxynivalenol (vomitoxin) display a quite different profile of toxicity. Rats fed deoxynivalenol-supplemented diets throughout gestation and lactation showed no clinical signs of toxicity except lower birth weights. On mating, significantly lower reproductive success was noted. In Swiss Webster mice, deoxynivalenol treatment to dams from gestation days 8 to 11 caused 100% resorptions at doses higher than 10 mg/kg/day and 80% resorptions at 5 mg/kg/day. Rats challenged with deoxynivalenol at doses between 0 and 5 mg/kg orally from gestation days 6 to 19 showed increased fetal resorptions and skeletal malformations at 2.5 and 5 mg/kg. No adverse effects were observed at 0.5 and 1 mg/kg doses. Similarly, intragastric gavage of deoxynivalenol (at doses ranging from 0.2 to 10 mg/kg) to rats through 7 to 15 days of gestation revealed skeletal malformations with a no-observed-effect-level of 0.2 mg/kg/day. There is no evidence of specific developmental liver toxicity following deoxynivalenol in rodents. In a recent study, Tiemann et al. examined whether deoxynivalenol exposure to pigs during gestation altered IGF/IGFBP components in maternal or fetal liver (52). Since the insulin-like growth factor (IGF) axis is highly responsive to nutritional status, these investigators examined expression of IGF-I, -II, their receptors, and mRNA expression of IGF binding proteins (IGFBP)-1,-2, and -3. The hepatic transcript levels of all aforementioned genes were not affected by feeding pregnant sows for 35 days at different stages of gestation with diets naturally contaminated with deoxynivalenol.

Ochratoxin is a mycotoxin produced by *Penicillium* and *Aspergillus* species that exists in three secondary metabolite forms, A, B, and C. Ochratoxin D is a nonchlorinated form of ochratoxin A, and ochratoxin C is an ethyl ester form of ochratoxin A. Ochratoxin A, which is the most common and most toxic form, is a developmental neurotoxicant with exposure to it during gestation causing a dose-dependent increase in cerebral necrosis, malformation, and fetal death (36,53). While ochratoxin crosses the placenta (54), no effects of ochratoxin on the developing liver have been reported even though high levels of DNA adducts were observed in the fetus. There is an apparent gender difference in the development of DNA adducts with greater adducts being found in the kidneys of male progeny, while female progeny contained exclusively liver DNA adducts (55). Another mycotoxin, citrinin, also causes increased fetal mortality in mice and rats when given at high doses. However, embryotoxicity was not associated with malformations. In addition, citrinin treatment in rats during gestation caused significant maternal hepatocytic degeneration, karyomegaly (enlarged nucleus), and marginally condensed chromatin (56). However, histopathological changes in the fetal liver were not investigated and no data exist to make a conclusive description of changes.

Fumonisin B1 is the most prevalent of the fumonisins and produced by the fungus *Fusarium moniliforme* and other *Fusarium* species. It is stable at food processing temperatures and light. Fumonisin B1 is the most common fungal metabolite associated with corn (57). There is inadequate evidence in humans for the carcinogenicity of toxins derived from *F. moniliforme* and there is limited evidence in experimental animals for the carcinogenicity of fumonisin B1. It is therefore classified as "possibly carcinogenic to humans." In vivo studies on the toxicity of fumonisin B1 indicate that liver and kidneys are the main target organs. Fumonisins are known to be poorly absorbed, not metabolized, and rapidly eliminated (58). In rats, fumonisin B1 is widely distributed, but small amounts accumulate in the liver and kidneys. Chronic exposure to fumonisin B1 in the diet or given intraperitoneally to rats results in nephrotoxicity, associated with apoptosis of tubular epithelial cells in the outer medulla (59). In both rats and mice, short- and long-term treatments with fumonisin B1 result in hepatotoxicity. Reduced liver weights, increased hepatocyte apoptosis, and increased serum alanine amino-transferase have been reported (60,61). Toxicity is greater in females compared with males. Treatment with fumonisin B1 to pregnant rats (with doses ranging from 0 to 15 mg/kg/day) from gestation days 3 through 16, revealed maternal toxicity (liver and kidney) but very little fetal toxicity. Fetal body weights, at gestation day 17 were not altered, expect for fetal weights in day 20 fetuses at the highest dose (15 mg/kg/day) were decreased. However, no dose-related effects on skeletal and soft tissue malfor-mation were observed (62,63). No studies have reported specific developmental liver toxicity following fumonisin B1 during gestation.

Arsenic

The toxicity of metals and arsenicals in particular has been recognized for millennia. The chemistry of arsenic is complex as it may occur as trivalent chloride or oxides, or as pentavalent forms of arsenic pentoxide, arsenic acid, or arsenates. Inorganic arsenic can enter the environment through anthropogenic sources that include smelting of copper, zinc, and lead and the manufacture of glass, pesticides, herbicides, and other chemicals (64). The major source of occupational exposure is during manufacture of chemicals involving arsenicals. Worldwide, millions of people in developing countries are exposed to arsenic from emissions, mining, industrial, or pesticide use, and drinking contaminated well water. Arsenic in the bedrock or soil is easily dissolved in the surrounding ground water, resulting in high levels of arsenic in drinking water. Southeast Asia and Bangladesh and surrounding countries are particularly affected, where a large percentage of drinking water has levels far exceeding the WHO guide-lines and the EPA's maximum contaminant level (10 ppb). Furthermore, use of arsenic-contaminated groundwater for irrigation can lead to secondary contamination of food (64).

The toxicity of arsenic has been widely studied mainly in adults. Chronic arsenic poisoning or arsenicosis is characterized by the classical dermal stigmata, together with internal pathology such as liver and lung injury. Arsenic is a potent human carcinogen causing cancer of the bladder, lung, skin, and possibly liver, kidney, and prostate in humans (65,66). In addition, multiple reports indicate relationships between arsenic exposure and noncancer toxicities including type 2 diabetes, skin diseases, immune dysfunction, chronic cough, and nervous system toxicity (67). In a recent review, Liu and Waalkes have extensively summarized the hepatic affects of arsenic exposure (65). Data show that human exposure to inorganic arsenic has resulted in elevated serum transaminases, hepatomegaly, hepatoportal sclerosis, hepatic fibrosis, and cirrhosis (65). A series of published epidemiological studies of a population in southwest Taiwan also suggests an association between environmental arsenic exposure and increased risk of developing liver cancers (68,69). Hepatocarcinogenicity of trimethylarsine oxide given via drinking water for up to two years was examined in F344 rats. A dose-dependent induction of hepatocellular adenoma occurred, and liver tumor multiplicity increased up to 2.5-fold (11). Another form of methylated arsenic, dimethylarsenic acid (DMA) when given to F344 rats in the drinking water, dose-dependently promoted tumors in the liver and other organs, which were initiated using a multicarcinogen cocktail (70). Similarly, evidence for the propensity of monoarsenious acid to promote carcinogen-initiated tumors also exists. Exposure of nontumorigenic hepatic epithelial cells to arsenic in culture induces malignant transformation, with the transformed cells capable of forming tumors in nude mice (71). Several mechanisms for arsenic carcinogenesis have been suggested and have been reviewed extensively. Primarily, increased oxidative DNA damage, impaired DNA repair, epigenetic alterations, modulation of hormone signaling, and resistance to apoptosis have been documented (65).

In the past few years, it has become increasingly evident that gestational exposure to arsenic causes adverse affects on growth, development, and adult risk to liver cancer. Studies in both humans and animals indicate that arsenic crosses the placenta and enters the fetal circulation (72–74). In addition, other evidence also suggests that arsenic may accumulate in the placenta at levels higher than in the mother. In an elegant series of studies, short-term exposure to inorganic arsenic in utero to mice produced a variety of internal tumors in the offspring in adulthood (75–77). These included tumors of the liver, lung, ovary, and adrenal in adulthood following transplacental exposure to arsenic. Gestation is a period of high sensitivity to chemical carcinogenesis in rodents and probably humans. In this model, a short (10-day) exposure to arsenic was sufficient to be a complete carcinogen in the offspring. Newborn mice transplacentally exposed to arsenic showed changes in expression of genes involved in the metabolism of endo- and exobiotics (CYP2A4, 3A25, 2F2, 7B1, HSD17-β dehydrogenase) and IGF signaling (IGF-1, IGFR2, and IGFBP-1 and -3). These changes were associated with loss of methylation of GC-rich DNA promoter regions of genes in newborn mice livers (78). An important target of epigenetic methylation changes was estrogen receptor-α (ER-α). This was accompanied with a "feminized" gene expression profile in the liver (79). Other studies have found hypermethylation of tumor suppressor genes including p53 and p16 (80). Hence, alterations in DNA methylation status are an important mechanism in aberrant gene expression and hepatocarcinogenesis following gestational arsenic exposure.

Perfluoroalkyl Acids

The perfluoroalkyl acids are a family of synthetic fluorocarbons consisting of a perfluorinated carbon tail and an acidic functional moiety, usually a carboxylate or a sulfonate. These compounds have a variety of industrial and consumer applications due to their strong surface tension–reducing properties. However, due to the carbon-fluorine bond, these compounds are resistant to environmental and metabolic degradation, consequently resulting in exceptionally long elimination half-lives (81). Perfluorooctane sulfonate (PFOS) is a sulfonated perfluoroalkane with a large repertoire of applications and is widespread in the environment. The presence of PFOS has been documented in both human and wildlife populations. PFOS is readily absorbed but poorly metabolized and cleared, thus producing long half-lives in rats (7.5 days), monkeys (200 days), and humans (estimated at 8.7 years). The rates of elimination and body burden accumulation, which appear to be dependent on the carbon chain length and functional moieties, are important determinants of toxicity (82,83). In addition, there is considerable species difference in elimination half-lives. In adult animal models, hepatic toxicity and interference with thyroid hormone status have been noted. Subchronic exposure to PFOS resulted in reduction of body weight, liver hypertrophy, and decreased serum cholesterol and triglycerides. In addition, PFOS has been shown to cause peroxisome proliferation in rats and mice. Cellular mechanisms include alterations in mitochondrial bioenergetics and cell-cell communications through gap junctions and alterations in hepatic fatty acid binding protein expression. While much of what is known about mechanisms of toxicity of PFOS is from studies in adult animals as noted above, newer studies are beginning to address developmental liver toxicity due to PFOS.

Recent studies have evaluated the developmental toxicity of PFOS in rats and mice (83,84). PFOS was given via oral gavage at doses ranging from 1 to 10 mg/kg (for rats) and 1 to 20 mg/kg (for mice) during gestation. Exposure to PFOS during gestation led to decreased weight gains and accumulation of PFOS in the maternal livers to four times the concentration in the serum (83). Although PFOS did not decrease implantation sites per dam, the percentage of live fetuses were decreased at the highest dose. Furthermore, increased relative liver weights and the incidence of birth defects including cleft palate, anasarca, ventricular septal defect, and enlargement of the right atrium of the heart were observed in both rats and mice. However, even at the lower doses, a significant increase in neonatal mortality was observed in the offspring. The morbidity and mortality appeared to be related to the body burden of PFOS. While neonates at the high doses were born alive, they survived only for a few hours and cross-fostering did not improve survival of these neonates (82). The findings suggest that PFOS exposure may alter postnatal organ and/or endocrine development. Indeed, hypothyroxinemia was observed as early as postnatal day 2, and serum T4 levels were suppressed in all dose groups. In addition, relative liver weights were increased at the 2 mg/kg dose (in rats) and 5 mg/kg (in mice) prior to and at postnatal day 21. However, these differences were normalized after weaning. Histological examination of fetal livers exposed to PFOS during gestation revealed eosinophilic granularity characteristic of peroxisome proliferation (84).

Rosen et al. (84) have suggested that the liver is a likely target for PFOS, based on hepatomegaly observed with PFOS administration and the activation of peroxisome proliferator–activated receptor α (PPAR-α), which occurred in fetal and adult liver. Global gene expression profiling has been employed to evaluate transcriptional changes induced by PFOS in the fetal mouse liver (84,85). The transcriptional profile observed following in utero exposure to PFOS resembles that of PFOS and PFOA in adult rats and is consistent with PPAR-α activation. PFOS has been shown to be a weak PPAR-α agonist in transactivation assays and PPAR-α dependent changes in transcriptional profiles occur in the absence of changes in PPAR-α mRNA. The expression profile is similar to that observed following known PPAR-α ligands, clofibrate, gemfibrozil, and trichloroethylene (85). PFOS exposure stimulates increases in fatty acid transporters and enzymes involved in catabolic activation, it also increases genes regulating fatty acid biosynthesis in the fetal liver. This has been proposed to lead to futile metabolic cycling leading to energy expenditure and wasting (85). Hence, collectively the evidence points to a transcriptional basis of hepatic pathology regulated by PPAR-α, associated with gestational exposure to PFOS. Indices of functional liver pathology (bilirubin, transaminases, gamma-glutamyl transpeptidase (γ-GGT), biliary secretion, etc.) in offspring with gestational exposure to PFOS are not available.

LIVER AS A TARGET FOR LONG-TERM DEVELOPMENTAL PROGRAMMING

Developmental toxicity has traditionally been identified by impaired growth, development, or morphological alterations identifiable at or immediately after birth. However, in the last decade or so it has become clear that susceptibility to chronic diseases can be programmed in utero and early postnatal life. In a series of epidemiological studies, Barker and colleagues at the University of Southampton discovered an inverse relationship between birth weight and risk of mortality due to coronary heart disease (86). Mortality due to coronary artery disease was almost twice more likely in those at the lower end of the weight distribution at one year of age compared with men who had higher body weights at one year. Despite early criticism about the early observations, it is now unequivocally accepted that early influences play critical roles in shaping the risk of adult diseases and many large cohorts have reproduced the original findings, reviewed extensively in several excellent articles (87–89). This concept that many metabolic and cardiovascular (CV) diseases are programmed in utero either by maternal health and nutrition or environmental clues has gained prominence and is often referred to as the "Fetal Origin of Adult Diseases" or the Barker Hypothesis. Over the last two decades, epidemiological studies have extended the initial associations between birth weight and later CV disease to include associations between early growth patterns and an increased risk for hypertension, insulin resistance, type 2 diabetes, and obesity in later life. Further, the concept has now become an important paradigm in establishing developmental programming of carcinogenesis.

In large epidemiological cohorts, being small for gestational age or born prematurely (generally <32 weeks of gestation) has been used as a surrogate for impaired fetal growth and hence fetal programming. Experimentally manipulating birth weights in a variety of animal models (rats, mice, sheep, pigs) also recapitulates the developmental programming phenotype. These are generally achieved by restricting calories or protein macronutrient intake during pregnancy or via other surgical manipulations that cause intrauterine growth retardation (uterine artery ligation). Exposure to excess glucocorticoids or inflammatory cytokines during gestation also leads to altered fetal programming. These manipulations have also been shown to increase risk for a variety of diseases for the offspring in later life, including shorter life span, hypertension, insulin resistance, obesity, and neurodevelopmental defects (88). Experimentally challenging the offspring by changing the postnatal environment (high-fat diet challenge, increased neonatal nutrition, etc.) further exacerbates the development of disease processes.

Since there are several examples of physiological insults that do not lead to overt fetal/embryo mortality, but indeed lead to greater susceptibility to adult-onset diseases, it is conceivable that exposure to certain environmental chemicals may do the same. Indeed several environmental chemicals, especially those with the ability to interfere with or mimic endogenous hormone activities (like those of estrogens, androgens, and corticosteroids) have been evaluated, and this topic has been recently reviewed (90–93). Several "environmental estrogens" such as bisphenol A, nonylphenol, diethylstilbestrol, phthalate esters, and organotins have shown to affect systems regulating adipocyte differentiation, insulin secretion/signaling, or energy balance, thereby modifying risk of obesity and certainly the continuum of metabolic diseases (type 2 diabetes, atherosclerosis, obesity, and hypertension). Relatively much less is known about "programming of hepatic physiology" following exposure to these chemicals and is likely to play a significant role in later life. Since this area of research is rapidly expanding, in the following section we will highlight a few examples where developmental exposure is known to affect adult liver physiology in the offspring.

Organotins

Organotin compounds or stannanes are chemical compounds based on tin with hydrocarbon substituents. Organotins are a diverse group of widely distributed environmental chemicals that have pleiotropic toxic effects. Trialkylorganotins, such as tributyltin chloride and bistriphenyltin oxide have adverse effects on invertebrate and vertebrate endocrine systems. Organotins persist in the environment in aquatic dietary sources such as fish and shellfish. Organotins have also been used as antifungal agents in wood preservation, water systems, plastics, and textiles. Exposure may also occur to humans from these sources (94).

Organotins gained notoriety initially as endocrine disruptors that caused imposex (the induction of male sexual features) in mollusks, at least in part due to the alterations in sex hormone production via changes in aromatase activity (which converts testosterone to estradiol) (95). Exposure to tributyltin also leads to masculinization of certain species of fish, but evidence for alterations in male to female ratios in vertebrates is not strong. Toxicities to the brain, liver, and immune function are primary effects of tributyltin exposure. Recent evidence has also indicated that in vivo tributyltin may mediate its action through mechanisms other than modulation of steroid metabolism. Foremost among these is through transcriptional regulation of genes via regulation of transcription factors. In the gastropod *Thais clavigara* the retinoid X receptor (RXR) homolog is activated by tributyltin (95). RXR, a member of the nuclear receptor superfamily, contains two signature domains: DNA-binding domain and ligand-binding domain (LBD). RXR is a ligand-dependent transcription factor and is endogenously activated by 9-*cis*-retinoic acid. RXR plays an important role in fundamental biological processes such as reproduction, cellular differentiation, adipogenesis, bone development, hematopoesis, and pattern formation during embryogenesis. Besides acting as a homodimer, RXR plays a central role in regulating the activity of other nuclear hormone receptors by acting as a partner for heterodimers. RXR forms a functional heterodimer with retinoic acid receptor (RAR), PPAR-α, -γ, and -δ, thyroid hormone receptor, vitamin D receptor, and many other nuclear receptors. Among these, PPAR-γ is of notable importance, since it is a master regulator of adipogenesis (differentiation and maintenance of the adipocyte phenotype). Ectopic expression of PPAR-γ2 in fibroblasts and nonadipose cells is sufficient for the induction of multitude of adipose-specific genes (96). In addition, expression of PPAR-γ1 in livers of mice via adenoviral expression results in development of fatty liver and expression of several adipocyte factors including adipisin, adiponectin, aP2, CD36, and other adipose genes (97). Recently, Grun et al. utilized a GAL4 reporter assay to screen for the ability of tributyltin and bis(triphenyltin) oxide to activate the ligand binding domain of RXR-α, PPAR-α, and PPAR-γ among other nuclear receptors (98). Their studies revealed that both compounds were potent agonistic ligands for RXR and PPAR-γ with EC_{50} values in the nanomolar range. Consistent with these findings, tributyltin enhanced both hormonally induced and basal adipogenesis in 3T3-L1 preadipocytes. Further, acute treatment of mice with tributyltin increased expression of PPAR-γ target gene, C/EBP-β, by almost threefold over controls, along with other lipogenic genes, fatty acid synthase, and acetyl CoA carboxylase. Most strikingly, in utero exposure to tributyltin from gestation day 12 to 18 in mice led to disorganization of hepatic architecture at birth and significantly increased lipid accumulation in the liver as assessed by Oil Red O staining. Tributyltin-exposed offspring developed increased adiposity (20% greater epididymal fat pad weights over controls) in adulthood without changes in body weight (98). These findings demonstrate that organotins cause developmental toxicity associated with endocrine disruption via nuclear receptor transcriptional regulation.

Ethanol

Despite recognition of the harmful effects of ethanol on the growing fetus for nearly three decades, ethanol continues to be the most common teratogen ingested during pregnancy (99). One of every 29 women who know they are pregnant reports ethanol consumption (100). The toxic effects of in utero ethanol exposure are manifested by a constellation of physical, behavioral, and cognitive abnormalities commonly referred to as fetal alcohol spectrum disorder (FASD) or alcohol-related birth defects (ARBD). In addition to mental retardation, in utero ethanol exposure results in increased rates of miscarriage, reduced birth weight, growth retardation, and teratogenic effects. While the teratogenic effects of ethanol have been extensively studied and reviewed, newer research is uncovering more subtle effects of programming following intrauterine ethanol exposure (101). The liver of the developing offspring appears to be a central target for the long-term programming by ethanol. In this section, we will specifically summarize the evidence wherein exposure to ethanol leads to long-term developmental toxicity to the liver in the offspring.

The liver is central to triglyceride, lipid, and metabolic homeostasis. The adjustments to fuel utilization in the postprandial state and during periods of fasting are orchestrated in large part by the liver through control of hepatic hormones and through regulation of gluconeogenesis. These processes are tightly regulated by hormonal signals including insulin, glucocorticoids, glucagon, and thyroid hormone. Dysregulation of these signaling pathways results in the development of metabolic disorders such as insulin resistance, type 2 diabetes, and hyperlipidemia. Primarily, hepatic gluconeogenesis is an important process that maintains systemic glycemic control, a process that goes awry in type 2 diabetes. Binding of insulin to its receptor (insulin receptor, IR) leads to the autophosphorylation of the β-subunits and the tyrosine phosphorylation of proteins called insulin receptor substrates (IRS). IRS activates phosphoinositide 3-kinase (PI3K) through its SH2 domain, thus increasing the intracellular concentration of the secondary lipid messenger, PIP2. This, in turn, activates Akt/PKB (protein kinase B) via Phosphoinositide-dependent kinase 1 (PDK-1) (102). Akt controls a variety of cellular functions via phosphorylation of downstream targets including GSK3 and FoxO1. Insulin also exerts signaling via Ras/Raf/MAPK and other pathways, but effects of insulin on metabolism are mainly mediated via the insulin-IRS/PI3K/AKT pathway (102). Inhibition of the key enzyme controlling hepatic gluconeogeneis, phosphoenolpyruvate carboxykinase (PEPCK), by insulin occurs transcriptionally via a complex mechanism that involves FoxO1 and PPAR-γ coactivator-1α (PGC-1α). However, several factors that impinge on Akt signaling, such as the phosphatase, PTEN, and tribbles homolog TRB3, can also regulate this process indirectly via Akt signaling (103,104).

Offspring of rats exposed to ethanol during gestation at levels that do not cause frank teratogenicity, develop hypertriglyceridemia and insulin resistance in adulthood. Both basal and insulin-stimulated glucose uptake in skeletal muscle is impaired by gestational ethanol (105,106). Consistent with these findings, Chen and Nyomba reported that offspring exposed to ethanol during gestation developed growth retardation and insulin resistance in later life (107,108). This was associated with decreased expression of Glut4 in the skeletal muscles and greater circulating concentrations of the hormone resistin, known to antagonize insulin action (108). Further, insulin-stimulated phosphorylation of IRS-1, association of the p85 subunit of PI3K, and downstream activation of Akt and GSK3β were all lower in skeletal muscles from offspring exposed to ethanol during gestation (109). Hepatic insulin signaling is also impaired in the ethanol-exposed offspring. Activation of hepatic phosphoinositide-dependent kinase 1 (PDK-1), Akt (at Ser473 and Thr308), and PKCχ was reduced following insulin. These impairments in insulin signaling are associated with increased expression and acetylation of PTEN and TRB3, along with increased histone deacetylase (HDAC1) activities (110). The functional consequences of these hepatic alterations are increased gluconeogenesis and reduced insulin signaling, contributing to increased risk of type 2 diabetes in the offspring exposed to ethanol.

In conclusion, a number of chemical insults that are not overtly toxic to the embryo do result in subtle changes in hepatic development, perhaps secondary to other endocrine changes systemically. This in turn sets the stage for adverse outcomes in later life.

THE DEVELOPING LIVER AND RESILIENCY TO ACUTE HEPATOTOXIC INSULT

It is generally accepted that younger populations are at greater risk of toxic effects from exposure to a variety of insults. Although this is certainly true for several developmentally sensitive organ systems (neurodevelopment, skeletal formation, etc.), neonate and postnatally developing livers (in rats) are resistant to a number of acute hepatotoxic chemicals. Previous studies have demonstrated that neonates and young rats are resilient to a number of chemicals such as CCl₄ (111,112), allyl alcohol (113), galactosamine (114), and acetaminophen (115). Furthermore, young rats are refractory to a lethal combination of chlordecone (CD) and CCl₄ (116). Dietary exposure to a nontoxic dose of CD (10 ppm in diet for 15 days) is known to potentiate hepatotoxicity and lethality of a low dose of CCl₄ (100 μL/kg) in male (117) and female (118) adult rats. Studies have revealed that inhibition of liver cell division and repair, thereby permitting progression of liver injury after exposure to the combination of CD and CCl₄ is central to the remarkable potentiation of toxicity in adults (119). Studies have shown that mechanisms relating to bioactivation to toxic species are relatively unchanged in younger animals and insufficient to explain their remarkable resiliency (112,116,120). Using colchicine antimitosis to block cell division, the resiliency of young rats to either single toxicants or mixtures has been shown to stem from higher plasticity in hepatocellular regeneration and efficient tissue repair mechanisms in younger animals (120). The liver has an extraordinary ability to regenerate in response to cell loss by physical damage, toxic injury, or infections. The final outcome of hepatotoxicity depends not only on how much injury is initiated but also on mechanisms that govern progression of injury and on the adequacy of compensatory tissue repair (119,121). It should be recalled that newly dividing/divided cells overexpress endogenous inhibitors of death proteins such as calpains and phospholipases (122,123).

For these reasons, a three-stage model has evolved to study the dynamic interaction of liver injury and compensatory tissue repair (124). During stage I, toxic chemicals or their metabolites following bioactivation inflict cellular injury via covalent binding, lipid peroxidation, altering redox status, or other mechanisms. In stage II, tissue injury progresses due to mechanisms other than those involving the parent compound or metabolites, but due to enzymatic degradation of neighboring healthy cells via proteases, phospholipases, secondary inflammation, or "run-away" oxidative damage (125). In stage III, tissue repair occurs promptly and sufficiently to overcome injury, and to restore cellular/organ structure and function. If tissue repair is delayed and/or inhibited, injury progresses in an unrestrained manner leading to organ failure and animal death.

Recent studies have led to the concept that leakage of hydrolytic degradative enzymes from damaged or necrosed hepatocytes destroys the neighboring healthy or partially damaged hepatocytes. Initially the relevance of these findings was established with the role of calpain in expanding liver injury initiated by CCl₄ in rats and acetaminophen in mice (121–123). Calpains are a family of cytoplasmic proteases that are ubiquitously expressed and require Ca²⁺ for their activation. After being released into the Ca²⁺-rich extracellular environment by dying hepatocytes, activated calpain attacks the plasma membrane proteins (e.g., receptors, transporters, and cytoskeletal proteins) on neighboring hepatocytes, causing leakage of cellular contents, necrotic injury, and death of the neighboring cells, thus progressing injury (125). Similarly, Napirei et al. reported deoxyribonuclease-I (Dnase-I) leaking from necrosed hepatocytes contributes to the progression of injury by acetaminophen (126). Another important family of degradative enzymes expressed in hepatocytes and macrophages noted for producing injury is secretory phospholipase A2 (sPLA2). sPLA2 hydrolyzes the ester bond at the *sn*-2 position of glycerolipids in the presence of Ca^{2+} resulting in the release of arachidonic acid. The further fate of arachidonic acid is determined by the action of cyclooxygenases-1 and -2 (COX-1,-2) and 5-lipooxygenase (5-LO) enzymes. Under necroinflammatory conditions, such as drug/toxicant-induced injury, COX-2 levels are induced leading to conversion of arachidonic acid into hepatoprotective prostaglandins (PGE_2 and PGJ_2) diverting metabolism away from the 5-LO pathway. Recent studies suggest that following acute injury, insufficient COX-2 may have detrimental consequences leading to progression of liver injury via sPLA2 (123,127). Hence, COX-2 plays a protective role in mitigating liver injury and the inhibition of hepatic COX-2 following the widely used class of anti-inflammatory drugs may inadvertently be sensitizing the liver to toxic insult. While all of the aforementioned processes are presumably subject to developmental regulation, little research has been focused on identifying these processes in the developing liver. For example, differences in COX-2 enzyme expression remain unknown. However, it is known that newly divided cells and cells from newborn animals overexpress endogenous inhibitors such as calpastatin and annexins, and are consequently resistant to the action of these lytic enzymes (128). Further, the processes of cell proliferation are already ongoing in newborn and young animals that are protectively "primed" to respond to toxic injury. Hence, it is intuitive to understand the resiliency of younger animals to hepatotoxicity when tissue repair is taken into consideration. These recent findings also explained the earlier observation that in the absence of new cells (antimitotic action of colchicine), liver injury was accelerated and progressed (129,130). The final outcome of toxicity hence depends on the dynamic and opposing interplay between bioactivation-based liver injury and tissue repair response.

In conclusion, the liver is a target of both acute and chronic developmental toxicity during various periods of sensitivity throughout gestation and early life. The remarkable capacity of the liver to regenerate in postnatal life may serve as an important safeguard in protecting the liver from acute damage and failure.

REFERENCES

1. Wallace AD, Meyer SA. Hepatotoxicity. In: Smart RC, Hodgson E, eds. Molecular and Biochemical Toxicology. New York: John Wiley, 2008:671–692.
2. Jonas MM, Perez-Atayde AR. Liver Disease in infancy and childhood. In: Schiff ER, Sorrell MF, Schiff L, et al., eds. Schiff's Diseases of the Liver. Philadelphia: Lippincott Williams & Wilkins, 2006:1305–1350.
3. Gualdi R, Bossard P, Zheng M, et al. Hepatic specification of the gut endoderm in vitro: cell signaling and transcriptional control. Genes Dev 1996; 10(13):1670–1682.
4. Deutsch G, Jung J, Zheng M, et al. A bipotential precursor population for pancreas and liver within the embryonic endoderm. Development 2001; 128(6):871–881.
5. Zaret KS. Molecular genetics of early liver development. Annu Rev Physiol 1996; 58:231–251.
6. Jung J, Zheng M, Goldfarb M, et al. Initiation of mammalian liver development from endoderm by fibroblast growth factors. Science 1999; 284(5422):1998–2003.
7. Tremblay KD, Zaret KS. Distinct populations of endoderm cells converge to generate the embryonic liver bud and ventral foregut tissues. Dev Biol 2005; 280(1):87–99.
8. Zhao R, Duncan SA. Embryonic development of the liver. Hepatology 2005; 41(5):956–967.
9. Chalmers AD, Slack JM. The Xenopus tadpole gut: fate maps and morphogenetic movements. Development 2000; 127(2):381–392.
10. Rossi JM, Dunn NR, Hogan BL, et al. Distinct mesodermal signals, including BMPs from the septum transversum mesenchyme, are required in combination for hepatogenesis from the endoderm. Genes Dev 2001; 15(15):1998–2009.
11. Shin D, Shin CH, Tucker J, et al. Bmp and Fgf signaling are essential for liver specification in zebrafish. Development 2007; 134(11):2041–2050.
12. Zaret KS, Grompe M. Generation and regeneration of cells of the liver and pancreas. Science 2008; 322(5907):1490–1494.

13. Calmont A, Wandzioch E, Tremblay KD, et al. An FGF response pathway that mediates hepatic gene induction in embryonic endoderm cells. Dev Cell 2006; 11(3): 339–348.

14. Zaret KS. Genetic programming of liver and pancreas progenitors: lessons for stem-cell differentiation. Nat Rev Genet 2008; 9(5):329–340.

15. Monga SP, Monga HK, Tan X, et al. Beta-catenin antisense studies in embryonic liver cultures: role in proliferation, apoptosis, and lineage specification. Gastroenterology 2003; 124(1):202–216.

16. Tan X, Yuan Y, Zeng G, et al. Beta-catenin deletion in hepatoblasts disrupts hepatic morphogenesis and survival during mouse development. Hepatology 2008; 47(5):1667–1679.

17. Ober EA, Verkade H, Field HA, et al. Mesodermal Wnt2b signalling positively regulates liver specification. Nature 2006; 442(7103):688–691.

18. Bort R, Signore M, Tremblay K, et al. Hex homeobox gene controls the transition of the endoderm to a pseudostratified, cell emergent epithelium for liver bud development. Dev Biol 2006; 290(1):44–56.

19. Sosa-Pineda B, Wigle JT, Oliver G. Hepatocyte migration during liver development requires Prox1. Nat Genet 2000; 25(3):254–255.

20. Margagliotti S, Clotman F, Pierreux CE, et al. The Onecut transcription factors HNF-6/OC-1 and OC-2 regulate early liver expansion by controlling hepatoblast migration. Dev Biol 2007; 311(2):579–589.

21. Matsumoto K, Yoshitomi H, Rossant J, et al. Liver organogenesis promoted by endothelial cells prior to vascular function. Science 2001; 294(5542):559–563.

22. Duncan SA. Mechanisms controlling early development of the liver. Mech Dev 2003; 120(1):19–33.

23. Costa RH, Kalinichenko VV, Holterman AX, et al. Transcription factors in liver development, differentiation, and regeneration. Hepatology 2003; 38(6):1331–1347.

24. Parviz F, Matullo C, Garrison WD, et al. Hepatocyte nuclear factor 4alpha controls the development of a hepatic epithelium and liver morphogenesis. Nat Genet 2003; 34(3): 292–296.

25. Suksaweang S, Lin CM, Jiang TX, et al. Morphogenesis of chicken liver: identification of localized growth zones and the role of beta-catenin/Wnt in size regulation. Dev Biol 2004; 266(1):109–122.

26. Bladt F, Riethmacher D, Isenmann S, et al. Essential role for the c-met receptor in the migration of myogenic precursor cells into the limb bud. Nature 1995; 376(6543): 768–771.

27. Schmidt C, Bladt F, Goedecke S, et al. Scatter factor/hepatocyte growth factor is essential for liver development. Nature 1995; 373(6516):699–702.

28. Hussain SZ, Sneddon T, Tan X, et al. Wnt impacts growth and differentiation in ex vivo liver development. Exp Cell Res 2004; 292(1):157–169.

29. Derman E, Krauter K, Walling L, et al. Transcriptional control in the production of liver-specific mRNAs. Cell 1981; 23(3):731–739.

30. Kaestner KH, Hiemisch H, Schutz G. Targeted disruption of the gene encoding hepatocyte nuclear factor 3gamma results in reduced transcription of hepatocyte-specific genes. Mol Cell Biol 1998; 18(7):4245–4251.

31. Zaret KS. Regulatory phases of early liver development: paradigms of organogenesis. Nat Rev Genet 2002; 3(7):499–512.

32. Wang ND, Finegold MJ, Bradley A, et al. Impaired energy homeostasis in C/EBP alpha knockout mice. Science 1995; 269(5227):1108–1112.

33. Hayhurst GP, Lee YH, Lambert G, et al. Hepatocyte nuclear factor 4alpha (nuclear receptor 2A1) is essential for maintenance of hepatic gene expression and lipid homeostasis. Mol Cell Biol 2001; 21(4):1393–1403.

34. Odom DT, Zizlsperger N, Gordon DB, et al. Control of pancreas and liver gene expression by HNF transcription factors. Science 2004; 303(5662):1378–1381.

35. Fink-Gremmels J. Mycotoxins: their implications for human and animal health. Vet Q 1999; 21(4):115–120.

36. Peraica M, Radic B, Lucic A, et al. Toxic effects of mycotoxins in humans. Bull World Health Organ 1999; 77(9):754–766.

37. Hayes WA. Aflatoxins. In: Hayes WA, ed. Mycotoxin Teratogenecity and Mutagenicity. Boca Raton: CRC Press, 1981:44–47.

38. Hood DR, Szczech MG. Teratogenicity of fungal toxins and fungal-produced anti-microbial agents. In: Keeler RF, Tu TA, eds. Handbook of Natural Toxins Plant and Fungal Toxins. New York: Marcel Dekker, 1983:201–235.
39. Elegbe RA, Amure BO, Adekunle AA. Induction of implantation by aflatoxin B1 in the rat. Acta Embryol Exp (Palermo) 1974; 0(2):199–201.
40. Arora RG, Frolen H, Nilsson A. Interference of mycotoxins with prenatal development of the mouse. I. Influence of aflatoxin B1, ochratoxin A and zearalenone. Acta Vet Scand 1981; 22(3–4):524–534.
41. DiPaolo JA, Elis J, Erwin H. Teratogenic response by hamsters, rats and mice to aflatoxin B1. Nature 1967; 215(5101):638–639.
42. Bassir O, Adekunle A. Teratogenic action of aflatoxin B1, palmotoxin B0 and palmo-toxin G0 on the chick embryo. J Pathol 1970; 102(1):49–51.
43. Gabor M, Puscariu F, Deac C. [Teratogenic action of aflatoxins in tadpoles]. Arch Roum Pathol Exp Microbiol 1973; 32(2):269–275.
44. Llewellyn GC, Stephenson GA, Hofman JW. Aflatoxin B1 induced toxicity and teratogenicity in japanese Medaka eggs (Oryzias latipes). Toxicon 1977; 15(6):582–587.
45. Kihara T, Matsuo T, Sakamoto M, et al. Effects of prenatal aflatoxin B1 exposure on behaviors of rat offspring. Toxicol Sci 2000; 53(2):392–399.
46. Ward JM, Sontag JM, Weisburger EK, et al. Effect of lifetime exposure to aflatoxin b1 in rats. J Natl Cancer Inst 1975; 55(1):107–113.
47. Grice HC, Moodie CA, Smith DC. The carcinogenic potential of aflatoxin or its metabolites in rats from dams fed aflatoxin pre- and postpartum. Cancer Res 1973; 33(2):262–268.
48. Naidu NR, Sehgal S, Bhaskar KV, et al. Cystic disease of the liver following prenatal and perinatal exposure to aflatoxin B1 in rats. J Gastroenterol Hepatol 1991; 6(4):359–362.
49. Neal GE, Eaton DL, Judah DJ, et al. Metabolism and toxicity of aflatoxins M1 and B1 in human-derived in vitro systems. Toxicol Appl Pharmacol 1998; 151(1):152–158.
50. Pong RS, Wogan GN. Toxicity and biochemical and fine structural effects of synthetic aflatoxins M 1 and B 1 in rat liver. J Natl Cancer Inst 1971; 47(3):585–592.
51. Vismara C, Di Muzio A, Tarca S, et al. Aflatoxin M1 effects on Xenopus laevis development. Birth Defects Res B Dev Reprod Toxicol 2006; 77(3):234–237.
52. Tiemann U, Brussow KP, Danicke S, et al. Feeding of pregnant sows with mycotoxin-contaminated diets and their non-effect on foetal and maternal hepatic transcription of genes of the insulin-like growth factor system. Food Addit Contam 2008; 18:1–9.
53. Szczech GM, Hood RD. Brain necrosis in mouse fetuses transplacentally exposed to the mycotoxin ochratoxin A. Toxicol Appl Pharmacol 1981; 57(1):127–137.
54. Appelgren LE, Arora RG. Distribution of 14C-labelled ochratoxin A in pregnant mice. Food Chem Toxicol 1983; 21(5):563–568.
55. Petkova-Bocharova T, Stoichev II, Chernozemsky IN, Castegnaro M, Pfohl-Leszkowicz A. Formation of DNA adducts in tissues of mouse progeny through transplacental contamination and/or lactation after administration of a single dose of ochratoxin A to the pregnant mother. Environ Mol Mutagen 1998; 32(2):155–162.
56. Singh ND, Sharma AK, Dwivedi P, et al. Citrinin and endosulfan induced maternal toxicity in pregnant Wistar rats: pathomorphological study. J Appl Toxicol 2007; 27(6):589–601.
57. Stockmann-Juvala H, Savolainen K. A review of the toxic effects and mechanisms of action of fumonisin B1. Hum Exp Toxicol 2008; 27(11):799–809.
58. Martinez-Larranaga MR, Anadon A, Diaz MJ, et al. Toxicokinetics and oral bioavail-ability of fumonisin B1. Vet Hum Toxicol 1999; 41(6):357–362.
59. Voss KA, Riley RT, Norred WP, et al. An overview of rodent toxicities: liver and kidney effects of fumonisins and Fusarium moniliforme. Environ Health Perspect 2001; 109(suppl 2):259–266.
60. Bondy G, Suzuki C, Barker M, Armstrong C, et al. Toxicity of fumonisin B1 administered intraperitoneally to male Sprague-Dawley rats. Food Chem Toxicol 1995; 33(8):653–665.
61. Bondy GS, Suzuki CA, Fernie SM, et al. Toxicity of fumonisin B1 to B6C3F1 mice: a 14-day gavage study. Food Chem Toxicol 1997; 35(10–11):981–989.

62. Collins TF, Shackelford ME, Sprando RL, et al. Effects of fumonisin B1 in pregnant rats. Food Chem Toxicol 1998; 36(5):397–408.
63. Collins TF, Sprando RL, Black TN, et al. Effects of fumonisin B1 in pregnant rats. Part 2. Food Chem Toxicol 1998; 36(8):673–685.
64. Vahter ME. Interactions between arsenic-induced toxicity and nutrition in early life. J Nutr 2007; 137(12):2798–2804.
65. Liu J, Waalkes MP. Liver is a target of arsenic carcinogenesis. Toxicol Sci 2008; 105(1): 24–32.
66. IARC (International Agency for Research on Cancer). Monographs on the evaluation of carcinogenic risk of chemicals to humans. Some Metals and Metallic Compounds. Lyon: IARC Press, 2004:39–141.
67. WHO. Arsenic and arsenic compounds. International Programme on Chemical Safety. Geneva: World Health Organization, 2001.
68. Chen CJ, Chuang YC, Lin TM, et al. Malignant neoplasms among residents of a blackfoot disease-endemic area in Taiwan: high-arsenic artesian well water and cancers. Cancer Res 1985; 45(11 pt 2):5895–5899.
69. Chen JG, Chen YG, Zhou YS, et al. A follow-up study of mortality among the arseniasis patients exposed to indoor combustion of high arsenic coal in Southwest Guizhou Autonomous Prefecture, China. Int Arch Occup Environ Health 2007; 81(1):9–17.
70. Nishikawa T, Wanibuchi H, Ogawa M, et al. Promoting effects of monomethylarsonic acid, dimethylarsinic acid and trimethylarsine oxide on induction of rat liver preneoplastic glutathione S-transferase placental form positive foci: a possible reactive oxygen species mechanism. Int J Cancer 2002; 100(2):136–139.
71. Zhao CQ, Young MR, Diwan BA, et al. Association of arsenic-induced malignant transformation with DNA hypomethylation and aberrant gene expression. Proc Natl Acad Sci U S A 1997; 94(20):10907–10912.
72. Hood RD, Vedel GC, Zaworotko MJ, et al. Uptake, distribution, and metabolism of trivalent arsenic in the pregnant mouse. J Toxicol Environ Health 1988; 25(4):423–434.
73. Devesa V, Adair BM, Liu J, et al. Arsenicals in maternal and fetal mouse tissues after gestational exposure to arsenite. Toxicology 2006; 224(1–2):147–155.
74. Concha G, Vogler G, Lezcano D, et al. Exposure to inorganic arsenic metabolites during early human development. Toxicol Sci 1998; 44(2):185–190.
75. Waalkes MP, Liu J, Diwan BA. Transplacental arsenic carcinogenesis in mice. Toxicol Appl Pharmacol 2007; 222(3):271–280.
76. Waalkes MP, Ward JM, Diwan BA. Induction of tumors of the liver, lung, ovary and adrenal in adult mice after brief maternal gestational exposure to inorganic arsenic: promotional effects of postnatal phorbol ester exposure on hepatic and pulmonary, but not dermal cancers. Carcinogenesis 2004; 25(1):133–141.
77. Waalkes MP, Ward JM, Liu J, et al. Transplacental carcinogenicity of inorganic arsenic in the drinking water: induction of hepatic, ovarian, pulmonary, and adrenal tumors in mice. Toxicol Appl Pharmacol 2003; 186(1):7–17.
78. Xie Y, Liu J, Benbrahim-Tallaa L, et al. Aberrant DNA methylation and gene expression in livers of newborn mice transplacentally exposed to a hepatocarcinogenic dose of inorganic arsenic. Toxicology 2007; 236(1–2):7–15.
79. Liu J, Xie Y, Ducharme DM, et al. Global gene expression associated with hepatocarcinogenesis in adult male mice induced by in utero arsenic exposure. Environ Health Perspect 2006; 114(3):404–411.
80. Chanda S, Dasgupta UB, Guhamazumder D, et al. DNA hypermethylation of promoter of gene p53 and p16 in arsenic-exposed people with and without malignancy. Toxicol Sci 2006; 89(2):431–437.
81. Das KP, Grey BE, Zehr RD, et al. Effects of perfluorobutyrate exposure during pregnancy in the mouse. Toxicol Sci 2008; 105(1):173–181.
82. Lau C, Thibodeaux JR, Hanson RG, et al. Exposure to perfluorooctane sulfonate during pregnancy in rat and mouse. II: postnatal evaluation. Toxicol Sci 2003; 74(2):382–392.
83. Thibodeaux JR, Hanson RG, Rogers JM, et al. Exposure to perfluorooctane sulfonate during pregnancy in rat and mouse. I: maternal and prenatal evaluations. Toxicol Sci 2003; 74(2):369–381.

84. Rosen MB, Schmid JE, Das KP, et al. Gene expression profiling in the liver and lung of perfluorooctane sulfonate-exposed mouse fetuses: comparison to changes induced by exposure to perfluorooctanoic acid. Reprod Toxicol 2009; 27(3–4):278–288.

85. Bjork JA, Lau C, Chang SC, et al. Perfluorooctane sulfonate-induced changes in fetal rat liver gene expression. Toxicology 2008; 251(1–3):8–20.

86. Barker DJ, Winter PD, Osmond C, et al. Weight in infancy and death from ischaemic heart disease. Lancet 1989; 2(8663):577–580.

87. Gluckman PD, Hanson MA. Living with the past: evolution, development, and patterns of disease. Science 2004; 305(5691):1733–1736.

88. McMillen IC, Robinson JS. Developmental origins of the metabolic syndrome: prediction, plasticity, and programming. Physiol Rev 2005; 85(2):571–633.

89. Couzin J. Quirks of fetal environment felt decades later. Science 2002; 296(5576):2167–2169.

90. Heindel JJ, vom Saal FS. Role of nutrition and environmental endocrine disrupting chemicals during the perinatal period on the aetiology of obesity. Mol Cell Endocrinol 2009; 304(1–2):90–96.

91. Newbold RR, Padilla-Banks E, Jefferson WN. Environmental estrogens and obesity. Mol Cell Endocrinol 2009; 304(1–2):84–89.

92. Baillie-Hamilton PF. Chemical toxins: a hypothesis to explain the global obesity epidemic. J Altern Complement Med 2002; 8(2):185–192.

93. Heindel JJ. Endocrine disruptors and the obesity epidemic. Toxicol Sci 2003; 76(2):247–249.

94. Appel KE. Organotin compounds: toxicokinetic aspects. Drug Metab Rev 2004; 36(3–4):763–786.

95. Nishikawa J, Mamiya S, Kanayama T, et al. Involvement of the retinoid X receptor in the development of imposex caused by organotins in gastropods. Environ Sci Technol 2004; 38(23):6271–6276.

96. Tontonoz P, Hu E, Spiegelman BM. Stimulation of adipogenesis in fibroblasts by PPAR gamma 2, a lipid-activated transcription factor. Cell 1994; 79(7):1147–1156.

97. Yu S, Matsusue K, Kashireddy P, et al. Adipocyte-specific gene expression and adipogenic steatosis in the mouse liver due to peroxisome proliferator-activated receptor gamma1 (PPARgamma1) overexpression. J Biol Chem 2003; 278(1):498–505.

98. Grun F, Watanabe H, Zamanian Z, et al. Endocrine-disrupting organotin compounds are potent inducers of adipogenesis in vertebrates. Mol Endocrinol 2006; 20(9):2141–2155.

99. Randall CL. Alcohol and pregnancy: highlights from three decades of research. J Stud Alcohol 2001; 62(5):554–561.

100. Eustace LW, Kang DH, Coombs D. Fetal alcohol syndrome: a growing concern for health care professionals. J Obstet Gynecol Neonatal Nurs 2003; 32(2):215–221.

101. Shankar K, Ronis MJ, Badger TM. Effects of pregnancy and nutritional status on alcohol metabolism. Alcohol Res Health 2007; 30(1):55–59.

102. Taniguchi CM, Emanuelli B, Kahn CR. Critical nodes in signalling pathways: insights into insulin action. Nat Rev Mol Cell Biol 2006; 7(2):85–96.

103. Du K, Herzig S, Kulkarni RN, et al. TRB3: a tribbles homolog that inhibits Akt/PKB activation by insulin in liver. Science 2003; 300(5625):1574–1577.

104. Koo SH, Satoh H, Herzig S, et al. PGC-1 promotes insulin resistance in liver through PPAR-alpha-dependent induction of TRB-3. Nat Med 2004; 10(5):530–534.

105. Elton CW, Pennington JS, Lynch SA, et al. Insulin resistance in adult rat offspring associated with maternal dietary fat and alcohol consumption. J Endocrinol 2002; 173(1):63–71.

106. Pennington JS, Shuvaeva TI, Pennington SN. Maternal dietary ethanol consumption is associated with hypertriglyceridemia in adult rat offspring. Alcohol Clin Exp Res 2002; 26(6):848–855.

107. Chen L, Nyomba BL. Effects of prenatal alcohol exposure on glucose tolerance in the rat offspring. Metabolism 2003; 52(4):454–462.

108. Chen L, Nyomba BL. Glucose intolerance and resistin expression in rat offspring exposed to ethanol in utero: modulation by postnatal high-fat diet. Endocrinology 2003; 144(2):500–508.

109. Chen L, Yao XH, Nyomba BL. In vivo insulin signaling through PI3-kinase is impaired in skeletal muscle of adult rat offspring exposed to ethanol in utero. J Appl Physiol 2005; 99(2):528–534.

110. Yao XH, Nyomba BL. Hepatic insulin resistance induced by prenatal alcohol exposure is associated with reduced PTEN and TRB3 acetylation in adult rat offspring. Am J Physiol Regul Integr Comp Physiol 2008; 294(6):R1797–R1806.

111. Cagen SZ, Klaassen CD. Hepatotoxicity of carbon tetrachloride in developing rats. Toxicol Appl Pharmacol 1979; 50(2):347–354.

112. Dalu A, Warbritton A, Bucci TJ, et al. Age-related susceptibility to chlordecone-potentiated carbon tetrachloride hepatotoxicity and lethality is due to hepatic quiescence. Pediatr Res 1995; 38(2):140–148.

113. Rikans LE. Influence of aging on chemically induced hepatotoxicity: role of age-related changes in metabolism. Drug Metab Rev 1989; 20(1):87–110.

114. Abdul-Hussain SK, Mehendale HM. Studies on the age-dependent effects of galactosamine in primary rat hepatocyte cultures. Toxicol Appl Pharmacol 1991; 107(3):504–513.

115. Green MD, Shires TK, Fischer LJ. Hepatotoxicity of acetaminophen in neonatal and young rats. I. Age-related changes in susceptibility. Toxicol Appl Pharmacol 1984; 74 (1):116–124.

116. Cai Z, Mehendale HM. Resiliency to amplification of carbon tetrachloride hepatotoxicity by chlordecone during postnatal development in rats. Pediatr Res 1993; 33(3):225–232.

117. Klingensmith JS, Mehendale HM. Potentiation of CCl4 lethality by chlordecone. Toxicol Lett 1982; 11(1–2):149–154.

118. Agarwal AK, Mehendale HM. Potentiation of CCl4 hepatotoxicity and lethality by chlordecone in female rats. Toxicology 1983; 26(3–4):231–242.

119. Mehendale HM. Amplified interactive toxicity of chemicals at nontoxic levels: mechanistic considerations and implications to public health. Environ Health Perspect 1994; 102(suppl 9):139–149.

120. Dalu A, Mehendale HM. Efficient tissue repair underlies the resiliency of postnatally developing rats to chlordecone + CCl4 hepatotoxicity. Toxicology 1996; 111(1–3):29–42.

121. Mehendale HM. Tissue repair: an important determinant of final outcome of toxicant-induced injury. Toxicol Pathol 2005; 33(1):41–51.

122. Limaye PB, Apte UM, Shankar K, et al. Calpain released from dying hepatocytes mediates progression of acute liver injury induced by model hepatotoxicants. Toxicol Appl Pharmacol 2003; 191(3):211–226.

123. Bhave VS, Donthamsetty S, Latendresse JR, et al. Secretory phospholipase A2 mediates progression of acute liver injury in the absence of sufficient cyclooxygenase-2. Toxicol Appl Pharmacol 2008; 228(2):225–238.

124. Wang T, Shankar K, Ronis MJ, et al. Mechanisms and outcomes of drug- and toxicant-induced liver toxicity in diabetes. Crit Rev Toxicol 2007; 37(5):413–459.

125. Mehendale HM, Limaye PB. Calpain: a death protein that mediates progression of liver injury. Trends Pharmacol Sci 2005; 26(5):232–236.

126. Napirei M, Basnakian AG, Apostolov EO, et al. Deoxyribonuclease 1 aggravates acetaminophen-induced liver necrosis in male CD-1 mice. Hepatology 2006; 43(2): 297–305.

127. Bhave VS, Donthamsetty S, Latendresse JR, et al. Inhibition of cyclooxygenase-2 aggravates secretory phospholipase A2-mediated progression of acute liver injury. Toxicol Appl Pharmacol 2008; 228(2):239–246.

128. Limaye PB, Bhave VS, Palkar PS, et al. Upregulation of calpastatin in regenerating and developing rat liver: role in resistance against hepatotoxicity. Hepatology 2006; 44(2): 379–388.

129. Mangipudy RS, Rao PS, Mehendale HM. Effect of an antimitotic agent colchicine on thioacetamide hepatotoxicity. Environ Health Perspect 1996; 104(7):744–749.

130. Kulkarni SG, Warbritton A, Bucci TJ, et al. Antimitotic intervention with colchicine alters the outcome of o-DCB-induced hepatotoxicity in Fischer 344 rats. Toxicology 1997; 120(2):79–88.

Cardiovascular development and malformation

John M. DeSesso and Arthi G. Venkat

INTRODUCTION

The importance of the cardiovascular system is manifested by its early appearance in the embryo and its status as the first system to function. The adult heart is ventrally located in the center of the chest with its apex to the left side and its base facing the midline. The ventricles, which do most of the pumping, are located anteriorly, and the atria, which accept blood flow from the veins, are located posteriorly. It is important to understand the maturation and growth of these structures in three dimensions. The development of the cardiovascular system has been described in detail in textbooks of embryology (1–5) and developmental biology (6), as well as books devoted solely to cardiac development (7). The information in this chapter is summarized from those sources and, for the sake of brevity, is limited to the development of the heart and great vessels/aortic arches.

The sequence of events during formation of the heart is the same among mammals regardless of the lengths of gestation, although the timing of specific events differs. The times of various events when mentioned in this chapter relate to human cardiac development although most have been described by sequence rather than by developmental age. Table 1 presents a summary of important gestational milestones in cardiac development in rats, mice, rabbits, monkeys, and humans. To ascertain the timing of events in heart development for these and various other experimental animals, the reader is referred to more detailed accounts of comparative gestational milestones (8,9).

Adult Configuration of the Heart and Vascular System

To understand the embryogenesis of the heart, it is helpful to appreciate the adult configuration of the heart and chest. This section provides a cursory overview of the cardiac anatomy. More detailed accounts can be found in textbooks of anatomy (10–12). The heart occupies the midline, anterior portion of the chest, just beneath the sternum. If the heart is visualized as a four-sided pyramid, its base is at the right edge of the sternum and faces toward the right arm; its apex lies in the medial portion of the left side of the chest and ends at the fourth intercostal space (just below the left nipple in males). Thus, most of the heart is found in the middle portion of the chest (Fig. 1).

Internally, the heart comprises four chambers: two smaller thin-walled atria that exhibit smooth luminal surfaces, and two larger muscular ventricles that have trabeculated surfaces (Fig. 2). Blood from venous returns enters the atria prior to filling the ventricles; blood is pumped out of the heart by way of the outflow vessels (aorta for the left ventricle and pulmonary artery for the right ventricle). The atria are located superiorly to the ventricles in the heart's posterior region. They are separated from each other by the thin interatrial septum, which presents with the fossa ovalis (a depression on the right atrial surface). The ventricles make up the anterior aspect of the heart. The wall separating the ventricles is the interventricular septum, which is composed of two portions: a thick inferiorly located muscular section and a small membranous region located near the atrioventricular orifices. To keep blood flowing in one direction, there are valves that separate each atrium from its respective ventricle.

TABLE 1   Comparative Milestones in Cardiovascular Development (in Gestational Days)

Milestone	Rat	Mouse	Rabbit	Monkey	Human
One-cell zygote	1	1	1	1	1
Blastocyst	5	4	5.5	5–8	5
Implantation	5.5–6	4.5–5	7–7.5	9	6–7.5
Cardiogenic plate	7.3–8.5	8.3	8.3	–	19–20
First contractions	9.5	8	8.5	–	21–24
Aortic arch 1	10	8.5	9.25	21–23	22
Sinus venosus	10.5	–	–	24–26	24–27
Cardiac looping	10	8.5	9.5	25	25–27
Circulation begins	10–10.5	8.5	–	–	26
Aortic arch 2	10.5–11	9.2	9.5	21–23	30
Aortic arch 3	11	10	9.75	27–28	28–31
Aortic arch 4 (AA 1 regressing)	11.8–13	10.5	11	28–29	31–34
Endocardial cushions	12–12.4	9.8	11	–	26–28
Aortic arch 6	12.2	10.75	11.75	29–30	32–35
Interventricular septum begins	12.2–13	9–10.75	12	–	29–35
Septum primum	12–12.5	11–11.5	13	–	28–37
Endocardial cushions fused	12.5	11	14	–	35–40
Spiral septum begins	–	11.5	–	–	32–35
Ostium secundum	13.25	11	12.75	–	40
Septum secundum	14	–	13	–	40
Foramen ovale	14	11.5	14	34	41–44
Interventricular septum complete	15[a]	13	16.5	–	43–46
Spiral septum complete	15.5	13–14	16.5	36	35–46

[a]Septum may not close completely until perinatal period; see discussion at page 235.

These atrioventricular valves are attached to a connective tissue frame that separates the atria from the ventricles.

The walls of the heart comprise three layers: the smooth innermost layer that lines the walls of the chambers facing the blood is the endocardium; the thick middle layer or myocardium that is made up of the cardiac muscle, connective tissue, and blood vessels; and the external smooth layer of serous secreting cells or epicardium, which lines the outer surface of the heart and faces the lumen of the pericardial cavity.

The heart is the pump for the two circulations of the body: (*i*) the systemic circulation to the body is supplied by blood from the left atrium and ventricle and (*ii*) the pulmonary circulation wherein the lungs receive blood from the right atrium and ventricle and return oxygenated blood to the left side of the heart for distribution to the body. Under normal conditions, the two circulations do not mix in the adult heart.

EMBRYONIC DEVELOPMENT
Early Stages
In mammals, creation of the zygote occurs through fertilization in the ampullary region of the uterine tube. In most mammalian species, the zygote undergoes rapid mitotic cleavage divisions during the next five to six days, transforming itself from a single cell to a cluster of smaller cells called a morula. As the number of cells in the morula increases, a central cavity appears and the organism is termed a blastocyst, which enters the maternal uterine lumen, adheres to the uterine epithelium, and implants into the uterine wall. The anatomy of the blastocyst comprises an outer shell of cells, the trophoblast (which gives rise to the placenta and fetal membranes), the much smaller inner cell mass (which gives rise to the embryo proper), and a fluid-filled blastocyst

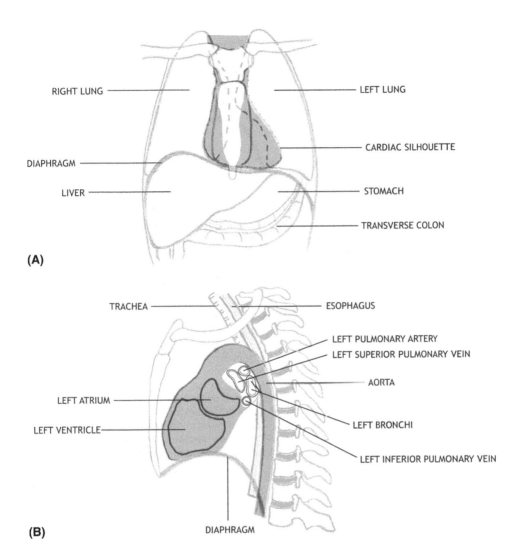

FIGURE 1 Position of the heart in the adult human. The heart occupies the central region of chest with its apex directed toward the fourth intercostal space on the left when viewed from the anterior aspect (**A**). When viewed laterally, the heart is positioned immediate deep to the sternum in the anterior portion of the chest (**B**); the aorta is positioned posteriorly, near the vertebral column.

cavity (Fig. 3). The inner cell mass quickly divides into a bilayered disk of cells that unequally partitions the blastocyst cavity into a smaller amniotic cavity and a larger primitive yolk sac cavity. One layer of the inner cell mass (epiblast) is associated with the developing amniotic cavity, which will give rise to the entire embryo proper; the other (hypoblast) is associated with the developing yolk sac cavity.

The zygote, morula, and blastocyst obtain their nutrition from diffusion. But as the embryo gets larger, diffusion is not efficient. During this early phase, transfer of nutrients into the zygote is governed by Fick's law of diffusion, which maintains that

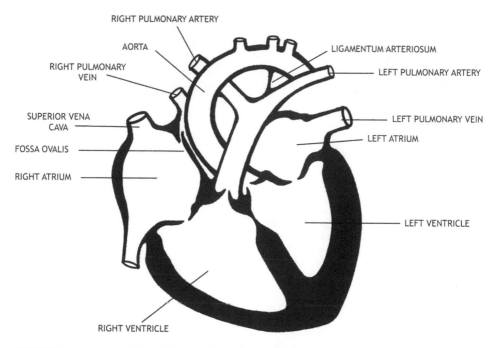

FIGURE 2 Anatomy of the adult heart. Note the superiorly located, thin-walled atria that lead to their respective muscular ventricles by means of the atrioventricular valves. The right atrium receives deoxygenated blood from the superior and inferior venae cavae; the right ventricular outflow is the pulmonary artery, which divides into right and left arteries under the arch of the aorta and supplies the respective lungs. Oxygenated blood returns to the left atrium via the pulmonary veins; the left outflow tract is the aorta, which is the distributor of arterial blood to the body. Note the fossa ovalis in the interatrial wall of the right atrium, and the interventricular septum that comprises a large inferior muscular septum and a small membranous portion that is associated with the origin of the two major outflow tracts.

the rate of transfer is proportional to the surface area available between two entities and the efficiency of the exchange mechanism (14). Fick's law predicts that diffusion becomes increasingly inefficient when the diameter of the zygote exceeds 0.2 mm. Thus, if a zygote is to grow larger than the maximum diameter of efficiency, a system or a set of mechanisms must develop that will improve the exchange of nutrients and metabolic wastes between zygotes and the maternal organism. In mammals, the manner in which physiologic exchange is improved typically involves the apposition of embryonal to maternal tissues in the form of a placenta (15). Placentae can be classified by the tissues that participate in the apposition. In rodents and primates, the chorion is apposed to the endometrium and is vascularized by blood vessels from the allantois. The resulting placenta is classified as chorioallantoic (16).

Placenta

Before the vascular system develops, the zygote implants into the endometrium and develops a syncytium from invasive, multinucleated trophoblastic cells. Inside the syncytium, spaces termed lacunae develop. The embryo burrows into the uterine wall,

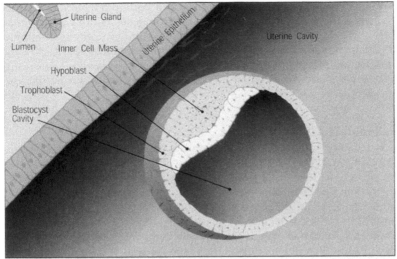

Human Blastocyst: gestational day 6

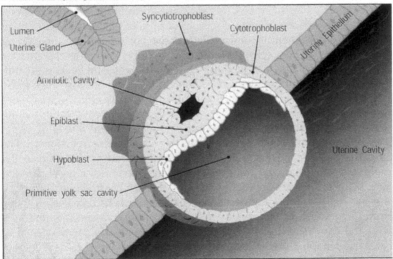

Human Blastocyst: gestational day 8

FIGURE 3 Diagrams that portray a blastocyst. The upper panel depicts a blastocyst floating freely in the lumen of the uterus. The single layer of cells that composes the outer shell is the trophoblast, which will give rise to the placenta and most of the fetal membranes, while the group of cells internal to the shell is the inner cell mass that will develop into the embryo proper. The layer of inner cell mass that is adjacent to the blastocyst cavity is the hypoblast. The lower panel shows the relationships of various parts of the implanting blastocyst when the layer of inner cell mass above the hypoblast has differentiated into epiblast. Together, the epiblast and hypoblast make up the bilaminar embryonic disc. The space appearing within the inner cell mass is the amniotic cavity. *Source*: From Ref. 13.

and the edge of the syncytiotrophoblast erodes into some of the maternal blood vessels. These little lacunar spaces are somewhat interconnected. At 13 to 14 days post fertilization, enough erosion will have occurred for maternal blood to flow from the maternal sinusoids into the lacunae, eventually filling them. Differential pressure levels in the lacunae will cause blood to wash through the lacunae and back into maternal veins in the uterine wall. This is uteroplacental circulation—it is composed entirely of maternal blood, beginning about 14 days after fertilization. It is more efficient than simple diffusion, but it will not be sufficient to support the development of the rapidly growing embryo. High-resolution ultrasound suggests that the rate of flow is very low during the first 10 weeks of gestation; the extent of uteroplacental circulation during the first trimester remains a point of controversy.

Early Steps in Cardiovascular Development

The embryonic circulatory system first arises from the extraembryonic mesenchyme, where blood islands form in the walls of the yolk sac. Cells in the yolk sac wall clump together to form clusters with an outer continuous surface and some cells sloughed to the middle of the cluster. The latter cells become nucleated blood cells, while the cells on the outside become the lining of the blood vessels. Gradually, the clusters merge with each other forming the rudiment of a vascular tree. This type of vessel formation is vasculogenesis (the differentiation of cells from mesenchyme into both blood cells and endothelial cells). In addition to the first blood islands in the wall of the yolk sac, others soon develop in the connecting stalk, as well as in the walls of the allantois. These are all extraembryonic sites. Shortly thereafter, blood islands also form in the embryo proper.

In the embryo proper, as cells ingress through the primitive streak, some of them receive signals that determine them to ultimately become portions of the developing heart and circulatory system. These cells are located in a region known as the anterior heart field. Fate mapping has shown that those cells in the anterior heart field that ingress near the cranial end of the primitive streak develop into the outflow tract; those that ingress through mid-streak develop into ventricle; and those that ingress more caudally become atrium. It is thought that as the cells ingress at different locations along the primitive streak, they migrate through areas of different concentrations of signal molecules such as retinoic acid, which primes them to receive inductive signals. The source of the inductive signals appears to be a specialized area of hypoblast near the caudal end of the prechordal plate, the anterior visceral endoderm (for a more detailed treatment, see Ref. 17). The cells of the anterior heart field assemble in the presumptive cephalic region of the embryo. A horseshoe-shaped mass of mesoderm coalesces above the anterior visceral endoderm at the cranial end of the oropharyngeal membrane, a structure found at the cranial end of the notochord that eventually breaks down to provide access to the developing digestive tube. This horseshoe-shaped mass is known as the cardiogenic plate. It is located subjacent and anterior to the neural plate at the cephalic end of the embryo. It will eventually develop a lumen, as described in the following text, and will contribute to much of the heart (Fig. 4).

Formation of the Cardiac Tube

During early development, the embryo can be envisaged as a disk of cells that is essentially a two-dimensional structure. During the third week of gestation, the embryo begins to assume a three-dimensional shape by undergoing a complex series of changes known as folding. The cardiogenic plate is located in the splanchnopleuric mesoderm (i.e., mesenchyme related to the wall of the gut) and is anterior to the presumptive brain before the embryo undergoes cephalocaudal and lateral folding. Folding takes place simultaneously along two axes of the embryo.

1. Cephalocaudal folding brings the cardiovascular tube, which began anterior to the brain, to a position inferior to the head. In the adult configuration, nerves that

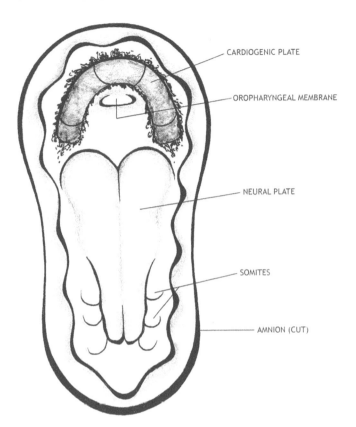

CARDIOGENIC PLATE

OROPHARYNGEAL MEMBRANE

NEURAL PLATE

SOMITES

AMNION (CUT)

FIGURE 4 Dorsal view of an embryo with the amnion removed depicting the horseshoe-shaped cardiogenic plate. The cardiogenic plate forms from mesodermal tissue that lies anterior to both the neural plate (the forerunner of the brain) and the prechordal plate, and lies beneath the ectoderm.

stream downward from the brain stem through the neck to innervate the heart are a result of this folding.

2. Lateral folding brings the endocardial tubes together near the midline and they fuse. The result when viewed from the ventral aspect is an "X"-shaped structure, with single midline, ventral endocardial tube that diverges into two laterally extending vessels at the cranial and caudal ends (Fig. 5A). The caudal vessels will bring blood to the heart (the presumptive venous return), whereas the cranial extensions will carry blood away from the heart (presumptive arterial outflow). The first contractions of heart are observed at, or shortly after, the time the endocardial tubes fuse. This occurs at approximately 23 days in humans.

Regions of the Developing Cardiac Tube

The cardiac tube proper, which is now in the midline and located ventrally, is anchored at two points (where the cranial and caudal vessels diverge), but most of the midline tube is free-floating in an open area called the coelom (the precursor of the pericardial cavity). As the cardiac tube continues to grow, five regions are described (Fig. 5B).

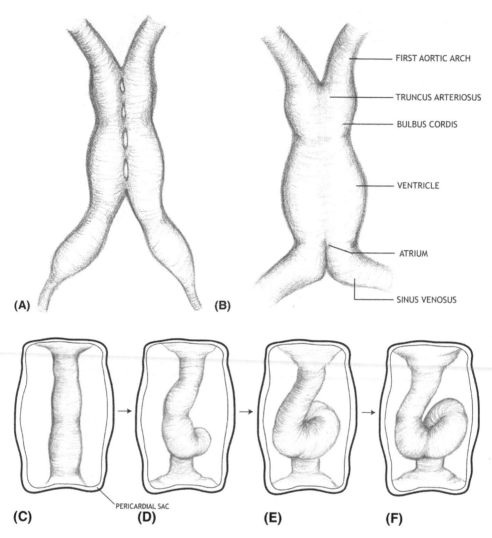

FIRST AORTIC ARCH

TRUNCUS ARTERIOSUS

BULBUS CORDIS

VENTRICLE

ATRIUM

SINUS VENOSUS

(A)   (B)

PERICARDIAL SAC

(C)   (D)   (E)   (F)

FIGURE 5 Ventral views of the early development of the heart tube. The endocardial tubes fuse to from a midline cardiac tube (**A**) that lengthens in such a manner that the tube outgrows the confines of the nascent pericardial sac. The five regions of the early heart tube are labeled (**B**). To remain within the confines of the pericardial sac the heart tube bends in a process called cardiac looping (**C–F**) that foreshadows the adult configuration of the heart.

If one follows the flow of blood through the developing cardiovascular tube, the regions are:

- Sinus venosus—where veins from the body wall (cardinal system), yolk sac (vitelline veins), and placenta (allantoic/umbilical veins) join to enter the heart;
- Atrium—the chamber that accepts blood from right and left sinus venarum (gives rise to the auricles of the atria in the adult heart);
- Ventricle—the main pumping portion of the heart tube (gives rise to the left ventricle in the adult heart);

- Bulbus cordis—a dilatation of the tube near the point where the cardiac tube bends (proximal region gives rise to the right ventricle);
- Truncus arteriosus—the outflow tract of the cardiac tube (which, together with the distal region of bulbus cordis, will divide into the proximal portions of the aorta and pulmonary artery).

Cardiac Looping and Early Chamber Organization

The cardiac tube grows at a faster rate than the rate of expansion of the coelom. Consequently, as the tube gets longer, it is forced to bend and twist to remain within the confines of the pericardial sac. This process is called "cardiac looping." The first signal of asymmetry in the developing heart tube is the shifting of transcription factor HAND-1 to the left side of the caudal region of the heart tube followed by the expression of a second transcription factor (HAND-2) in the region of the primitive right ventricle. As cardiac looping continues, the sinus venosus (the inflow tract involved with the atrium) shifts cephalically, moving posterior to the ventricle. As it does so, the atria move up and to the right, assuming a position posterior and slightly superior to the ventricles. Simultaneously, the apex of the loop moves such that it ultimately points toward the left, which puts the apex of the heart on the left side (Fig. 5C). As cardiac looping concludes, it is possible to discern the locations of the four presumptive chambers of the adult heart, but the chambers are not yet separated from each other.

It is important to recognize that the description of cardiac looping (and all of heart development) provided in this chapter is greatly simplified. A more complete discussion of cardiac looping can be found in Manner (18) and details of the molecular control of this process are discussed elsewhere (4,5,17).

Ventricle Separation

Bulbus cordis and truncus arteriosus are sequential chambers in the cardiac tube. As this combined structure folds back on the atria during cardiac looping, it foreshadows a four-chambered heart (Fig. 6A). When seen in a dissected view of the developing heart, a small ripple appears in the floor of the ventricle at the juncture between the bulbus cordis and the primitive ventricle. That ripple is slightly to the left of the base of the bulbus cordis and marks the site of the early interventricular septum. This will develop into the muscular interventricular septum. It enlarges as the presumptive ventricles expand, accreting into the septum some of the tissue from the medial walls of the growing ventricles (Fig. 7).

The membranous portion of the interventricular septum is formed in association with the spiral septum and the division of the outflow tract. The distal portion of bulbus cordis is the beginning of that outflow tract, which will ultimately become the aorta and the pulmonary trunk and will be discussed later. Blood flows from the sinus venosus to the atrium, through the atrioventricular opening (which is shaped like a dumbbell or an "H") into the ventricle (Fig. 6B). The orifice will be pinched off in the midline (crossbar of the "H" or bar of the dumbbell) to form two orifices, one entering each presumptive ventricle. The blood exits the ventricular region through the outflow tract that is formed from the distal bulbus cordis and the truncus arteriosus. The pinching off appears visually simple, but it involves a well-orchestrated and tightly timed series of complex developmental events including the growth and fusion of endocardial cushions and the septum primum, which will be discussed in following sections, as well as the incorporation of the proximal portion of the pulmonary vein (20). Simultaneously, the chambers expand, the inner curvature of the heart is remodeled, and the position of the presumptive right ventricle and outflow tract rotate toward their adult positions.

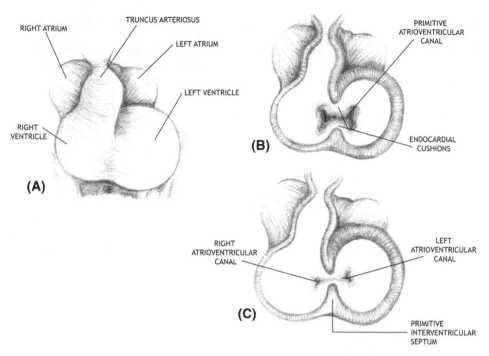

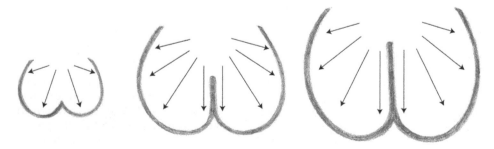

FIGURE 6 Diagrams that portray the early development of the interventricular septum and atrioventricular orifices. A ventral view of the fully looped cardiac tube is depicted in **A**. Panels **B** and **C** portray dissected views as development progresses. In **B**, the primitive atrioventricular canal appears as a foreshortened capital "H." The tissues that compress the lumen of the atrioventricular lumen in **B** and **C** are endocardial cushions. The ridge that appears on the floor of the presumptive ventricular wall and grows upward is the nascent muscular interventricular septum. As the interventricular septum and endocardial cushions grow, the middle region of the primitive atrioventricular canal is obliterated dividing the lumen into separate passages for the right and left sides of the heart the right and left atrioventricular canals.

FIGURE 7 Growth of the muscular portion of the interventricular septum occurs by accretion as the two ventricles expand. Note that if growth of the septum is limited to this single mode of growth, the septum will not completely partition the two ventricles. The remaining superior portion of the interventricular septum (membranous septum) develops in concert with the growth of the spiral septum. *Source*: From Ref. 19.

Atrial Separation

The position of the bulbus cordis/truncus arteriosus in relation to the primitive atrium is a landmark that denotes the location where a septum will develop that separates the primitive atrium internally into the right and left atria (Fig. 8). The septation of the atria is more complex than that of the ventricles because there are two septa required to separate the atria and the necessity for a right-to-left shunt of blood as the embryo develops the two vascular circuits: the systemic circuit that distributes blood to the body and the pulmonary circuit that caries blood to and from to the lungs.

In the plane of the location where the bulbus cordis/truncus arteriosus appear to rest on the roof of the primitive atrium, an indentation in the roof of the atria marks the site where septum primum begins to grow from the posterior wall. As it grows, the inferior free edge of septum primum outlines an orifice (ostium primum) that connects the two presumptive atria and allows blood to move from the right to the left. With continued development, the septum primum grows with the consequence that the ostium primum gets smaller and blood flows less freely. Note that there is not much blood entering the left atrium from the venous system in the embryo because the lungs are not functioning to oxygenate the blood, and, therefore, blood returning from the lung is only the small amount of blood that had provided oxygen and nutrients to the growing lung bud tissue.

As the septum primum lengthens and the ostium primum constricts, blood has more difficulty passing from the right atrium to the left atrium. Most of the blood returning from the body travels via the inferior vena cava, which enters the inferior aspect of the right atrium. The trajectory of the blood from the inferior vena cava points toward the superior aspect of the septum primum. This directed flow encourages cell death in that area of the septum primum where a new opening, ostium secundum, breaks through the superior region of the septum primum. As the ostium secundum enlarges, the ostium primum closes completely when the inferior edge of septum primum fuses with the floor of the atrium.

Initially, blood can pass in either direction through the ostia. Shortly after the septum primum fuses with the floor of the atrium, a second more substantial septum, the septum secundum, starts to form just to the right of the septum primum in the right atrium. The septum secundum is a sickle-shaped structure that descends from the anterior wall of the right atrium. It grows inferiorly far enough to cover the ostium secundum, but it does not grow all the way to the atrial floor. Because it does not attach to the atrial floor, an aperture remains that allows blood to pass to the left atrium. The septum secundum is a secure rigid structure, unlike the septum primum that is rather flimsy. Upon completion, the combined structure allows blood to pass through the opening bounded by the incomplete free edge of the septum secundum where it pushes the septum primum open, and enters the left atrium. This right-to-left shunt acts as a one-way valve that allows blood to travel in only one direction. It is called the foramen ovale.

The volume of the heart grows rapidly. To handle the blood flow entering the heart, the atria expand by incorporating the proximal portions of the veins that adjoin the atria. Thus, the majority of the smooth walls of the right atrium is derived from the right sinus venosus; the left sinus venous is less conspicuous and becomes the coronary sinus (the site for venous drainage for the heart itself). In the left atrium, the proximal walls of the original, single pulmonary vein are intussuscepted to the extent that there are eventually four entry points into the chamber as more and more of the vessel is absorbed into the left atrial wall.

Spiral Septum and Membranous Portion of the Interventricular Septum

Throughout the length of the truncus arteriosus and the distal one-half to two-thirds of the bulbus cordis, a pair of truncal ridges that run lengthwise in the direction of the

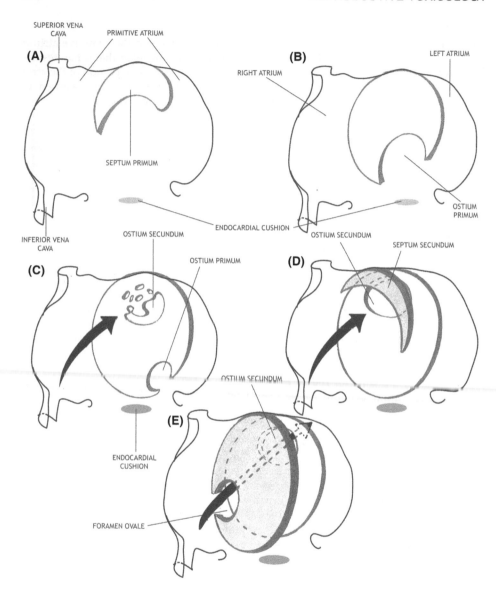

FIGURE 8 Diagrams that depict the internal development of the atria with special attention to the interatrial septum. Shortly after cardiac looping is completed (**A**), a ridge appears in the caudal region of the cranial aspect of the primitive atrium. The ridge quickly develops into a thin septum primum that grows inferoanteriorly, dividing the primitive atrium into a left and right atrium (**B**). The opening at the free edge of septum primum allows blood to pass from the right side (where blood returns to the heart from the body and the placenta); this passage is termed the ostium primum (**C**). Most of the blood returning to the heart enters the presumptive right by means of the inferior vena cava. The blood from that vessel is directed high against the wall of the septum primum (**C**). As the ostium primum closes due to growth of the septum primum, a second opening (ostium secundum) appears in the septum primum. Simultaneously, a second more substantial septum secundum appears in the ventral cranial wall of the right atrium (**D**). The septum secundum grows toward the dorsal wall of the right atrium. The septum secundum does not form completely, and the oblique canal formed by the free edge of septum secundum together with the ostium secundum is the foramen ovale (**E**).

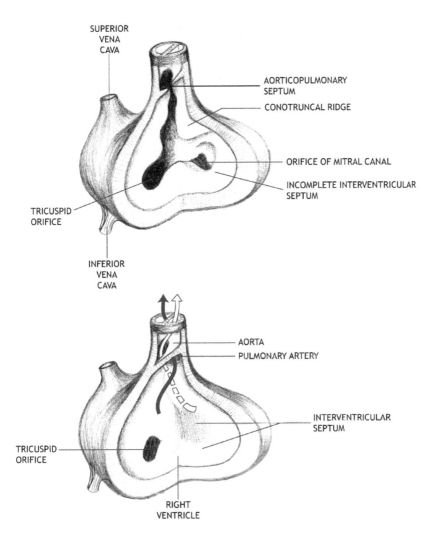

FIGURE 9 Division of the outflow tract by development of the spiral septum. The spiral septum is formed by the growth and merging of the conotruncal ridges in the wall of the truncus arteriosus. The ridges are invaded by migrating neural crest cells from the hindbrain region. The inferior edge of the spiral septum fuses with the endocardial cushions in the region surrounding the atrioventricular canals. The most inferior part of the spiral septum contributes to the membranous portion of the interventricular septum.

bloodflow develops in the walls on opposite sides. The truncal ridges extend into the lumen toward each other, where they will eventually meet to form a septum that partitions the outflow tract (Fig. 9). The orientation of the truncal ridges shifts as one progresses along the outflow tract such that the truncal ridges appear to turn in a clockwise direction as one progresses in the direction of bloodflow. This corkscrew type of arrangement is the basis for the naming of the spiral septum. The spiral septum is the progenitor of the aorticopulmonary septum in the configuration of the adult outflow tract (the wall that is shared by and separates the aorta and pulmonary artery) and is a significant contributor to the membranous portion of the interventricular septum (which joins the muscular interventricular septum and completes the separation of the ventricles). While the completion of the membranous interventricular septum appears to be grossly complete in the embryos of most short-gestation species used in safety testing, there is evidence that the septum may remain patent until several days after birth in rats (21–23).

In contrast to the septation that divides the chambers of the heart, the development of the spiral septum requires a supplement of extracardiac tissue to succeed. This extracardiac tissue is supplied by neural crest cells that emigrate from the hindbrain of the developing neural tube. The neural crest cells migrate into the third and fourth pharyngeal arches and enter the developing outflow tract of the heart where they take up residence in, and contribute to, the truncal ridges that give rise to the spiral septum. In the absence or perturbations of neural crest cell migration, the spiral septum fails to form properly.

Atrioventricular Valves

The description of cardiac development up to this point has been described in terms of a tube, the walls of which are rather thin. In reality, the walls of the cardiac tube are relatively thick. In early embryos, the heart tube has three regions: (*i*) an inner surface, the endocardium, that faces the lumen; (*ii*) an outer surface, the epimyocardium, that faces the presumptive pericardial cavity; and (*iii*) an area between these surfaces that is filled with a gelatinous material called cardiac jelly. Cardiac jelly is a well-hydrated colloid that acts as a space holder that allows the cardiac tube to bend during looping without kinking or obstructing the lumen. In the region where the orifices between the atria and ventricles develop, the cardiac jelly takes on a special role. In this territory, it is known as the endocardial cushions, which surround the presumptive atrioventricular canals. This tissue will give rise to both the valves between the atria and ventricles (mitral and tricuspid) and the supporting structure for the valves (cardiac skeleton). The cells in the region will be mesenchyme cells and will not be contractile.

As development progresses, the cells of the epimyocardium secrete particles called adherons. Adherons, which are visible with a light microscope, contain TGF-β, fibronectins, proteoglycans, as well as a variety of inductive signals. One of the inductive signals stimulates the cells in endocardium to migrate into the cardiac jelly where they differentiate into the mesenchymal cells of the endocardial cushions that participate in making the tissue that separates the atria from the ventricles and supports the atrioventricular valves. As the cells proliferate and occupy more space, the cardiac jelly disappears.

The atrioventricular valves are formed from mesenchymal cells of the endocardial cushions that overlie the walls of the ventricles, forming a cap of cells (Fig. 10). With continued expansion of the ventricles, the territory beneath the mesenchymal cap is vacated in such a way that the medial edge of the mesenchymal cap becomes a free-standing valve leaflet that is attached by cords (chordae tendineae) to muscular tissue (papillary muscles) near the base of the ventricular lumen. As these valves become competent, the flow of blood through the heart becomes increasingly unidirectional.

Tissue Differentiation—Cardiomyocytes and Conduction System

As mentioned earlier, shortly after the endocardial tubes fuse, the walls of the tube begin to twitch due to random contractions of the differentiating cardiomyocytes. Any fluid within the lumen would slosh back and forth in no particular direction. As the heart undergoes cardiac looping, significant changes occur in the walls of the heart. The walls, which were occupied by cardiac jelly, become populated by additional nascent

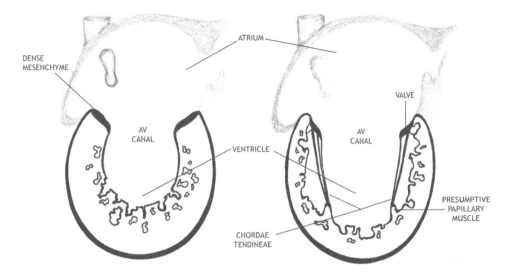

FIGURE 10 Atrioventricular valve formation. As the ventricles expand, dense mesenchymal tissue appears near the orifice of the atrioventricular canal. With continued expansion of the ventricles, the tissue under the dense mesenchyme is hollowed out leaving space immediately beneath the presumptive valves, but maintaining a connection between the free edge of the valves and cord-like strands (the chordae tendineae) that connect to elevations in the floor of the ventricles (the presumptive papillary muscles). The papillary muscles and chordae tendineae prevent the prolapse of the valves into the atrium during contraction of the ventricle.

cardiomyocytes and these cells begin to contract rhythmically in a peristaltic manner. In addition, as the four chambers appear and the circulations separate, the contractions are coordinated such that blood will leave both atria at the same time, and shortly thereafter both ventricles will simultaneously contract. Despite the fact that young cardiomyocytes exhibit differentiated characteristics (e.g., contractile filaments), they retain the ability to undergo mitosis and to continue differentiation. As the heart develops and its pulsation rate matures, there appears simultaneously a conduction system that coordinates the contractions of the muscle cells. The conduction system, which includes elements like the atrioventricular (AV) node, Purkinje fibers, and bundle of His, differentiate from contractile cardiomyocytes (*not* neural crest cells) that sequester excess amounts of glycogen and are located immediately subjacent to the endocardium on the interventricular septum and the atrioventricular floor (24). It is believed that hemodynamic flow through the developing heart causes certain areas of the endocardium to undergo shear stress and to secrete the cytokine endothelin-1, which induces the differentiation of the conducting system (24).

Embryonic Arterial System
The arterial system of the embryo develops simultaneously with the heart. At the time the cardiogenic plate is forming, blood islands appear on both sides of the developing nervous system. The blood islands coalesce to form the primitive dorsal aortae, which eventually connect to the developing heart. As the embryo undergoes cephalocaudal folding and the heart attains its position in the ventral region of the embryo, the distal portions of the aortae are pulled ventrally as well. Because the caudal portions of the aorta are located dorsal to the developing gut tube and they retain their association with

the dorsal embryonic structures (neural tube, notochord, somites), the aortae in the cephalic region of the embryo bend ventrally and assume a C-shaped configuration. The largest portions of the aortae (which are located in the thorax and abdomen) are located dorsal to the gut tube and are termed "dorsal aortae." Eventually, the dorsal aortae fuse into a single, definitive dorsal aorta. The portion that is located ventral to the pharynx (the portion of the gut tube in the head and neck regions) is known variously as the ventral aorta or aortic sac and is a transient structure that is incorporated into the truncus arteriosus. The connections between these two vessels, which traverse the tissues on either side of the pharynx, are aortic arches. As the head and neck continue to grow, additional pairs of aortic arches appear such that eventually there will be five pairs of aortic arches in mammals. Each pair of aortic arches is associated with a pair of mesenchymal bars (pharyngeal arches) that are associated with the development of the jaws and neck.

The five definitive pairs of aortic arches in mammals are numbered I–VI (Fig. 11; aortic arch V is very transient and is not depicted). They develop in a cephalocaudal direction such that aortic arch I develops first, followed by aortic arch II. However, as

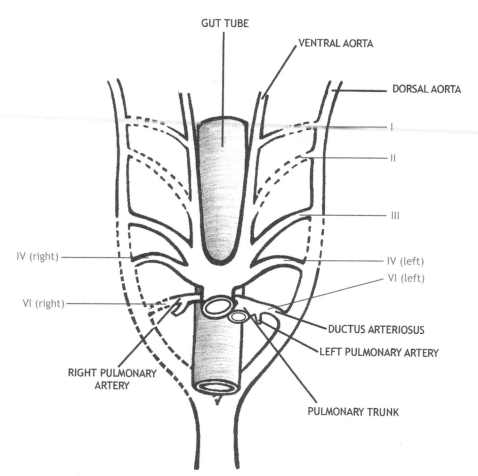

FIGURE 11 Diagram of the aortae and aortic arch system. Obliterated sections are depicted with broken lines. Table 2 provides a list of the adult derivatives of the aortae and aortic arches.

TABLE 2   Derivatives of the Embryonic Arterial System

Embryonic vessel	Right side derivative	Left side derivative
Aortic arch I	Maxillary artery	Maxillary artery
Aortic arch II	Hyoid artery	Hyoid artery
	Stapedial artery (disappears)	Stapedial artery (disappears)
Aortic arch III	**Proximal arch**: common carotid artery	**Proximal arch**: common carotid artery
	Distal arch: internal carotid artery	**Distal arch**: internal carotid artery
Aortic arch IV	Proximal portion of right subclavian artery	Arch of the aorta (medial portion)
Aortic arch VI	Right pulmonary artery	Left pulmonary artery
		Ductus arteriosus
Ventral aorta		
• Cranial to arch III	• External carotid artery	• External carotid artery
• Between arch III and IV	• Common carotid artery	• Common carotid artery
• Between arch IV and VI	• Brachiocephalic artery	• Ascending aorta
Dorsal aorta		
• Cranial to arch III	• Internal carotid artery	• Internal carotid artery
• Between arch III and IV	• Disappears with growth of neck	• Disappears with growth of neck
• Between arch IV and VI	• Central right subclavian artery	• Descending aorta
• Caudal to arch VI	• Disappears	• Descending aorta

aortic arch III appears, aortic arch I begins to disappear as an entity. By the time aortic arch IV is developing, aortic arch I is no longer present. The changes in the vasculature are partly driven by development of the neck, which appears as the embryonic face lifts off the chest and the head assumes an erect position.

The aortic arches contribute greatly to the vasculature of the upper chest, neck, and head. Table 2 summarizes the derivatives of the aortic arches. While all of the aortic arches are important, the two caudalmost arches (aortic arches IV and VI) are the only ones to undergo asymmetric development; these two arches play integral roles in the development of the heart. Aortic arch IV is the vascular trunk that delivers blood to the body. On the left side, it develops into the arch of the aorta; on the right side it becomes the right subclavian artery, which delivers blood to the right arm and the right neck and head. Aortic arch VI delivers blood to the lungs. On the right side, the distal portion of aortic arch VI disappears and the remainder becomes the right pulmonary artery. On the left side, the proximal portion of the sixth arch becomes the left pulmonary artery; the distal portion of the sixth aortic arch remains as the ductus arteriosus in the embryo/fetus. The ductus arteriosus is the third right-to-left shunt, which allows blood being sent by means of the pulmonary artery to bypass the nonfunctioning embryonic lungs. After birth, the ductus arteriosus closes and becomes the ligamentum arteriosum.

Circulatory Changes at Birth

In an embryo/fetus, blood is oxygenated, nutrient fortified, and cleansed at the placenta. Blood from the placenta enters the embryo/fetus via the umbilical vein that travels in the umbilical cord. The umbilical vein travels on the inner surface of the abdominal wall in a midline fascial structure (falciform ligament) to the liver where it enters a preferential channel (ductus venosus) that bypasses most of the liver and conveys the blood to the inferior vena cava, and thence to the right atrium. Blood from the right atrium would be destined for the lungs, but because the fetal lungs are not functioning to oxygenate blood, it is necessary to move the majority of the blood efficiently from the presumptive pulmonary circulation (the right side) to the systemic circulation on the left side. In the atrium, this is accomplished through the foramen ovale, which acts like a flutter valve that opens when pressure from the blood delivered by the inferior vena cava pushes through to the left atrium. Blood returning from the upper portion of the body via the superior vena cava enters the right atrium, passes to the right ventricle, and exits toward the lungs by means of the pulmonary artery. Most of the blood in the pulmonary artery passes through the ductus arteriosus to join the aorta rather than following the pulmonary arteries to the lungs. Blood exits the fetus en route to the placenta by means of the umbilical arteries (branches of the external iliac arteries) that course toward the midline umbilicus and exit into the umbilical cord.

At birth, a series of dramatic changes occur in the circulation of the neonate. Not only do the lungs begin to function with the requirement to fill the previously unused pulmonary circulation, but also the right-to-left vascular shunts must be closed. As the vasculature of the lungs opens, the pressure in the right atrium drops so that it is lower than the pressure in the left atrium with the consequence that the flutter valve at the foramen ovale no longer conducts blood to the left atrium. Eventually, over the course of the first year of life in most people, the septum primum fuses with the septum secundum to form the interatrial wall. In rats, fusion occurs more quickly and is reported to have occurred by the third day of postnatal life (23). The location of the foramen ovale is marked by a thumb print–shaped depression on the interatrial wall of the right atrium: the fossa ovalis. In approximately 27% of healthy people, the fusion of the septa does not take place (25). In these people, elevated pressure in the right atrium can reopen the right-to-left shunt. This condition is known as "probe patent foramen ovale." Interestingly, this condition has been associated with a higher than expected incidence of migraine headaches and stroke (26), as well as a greater sensitivity to altitude sickness and decompression illness among SCUBA divers (27).

The other major right-to-left shunt, the ductus arteriosus, constricts such that it closes within the first 15 to 24 hours after birth, thereby closing the last major right-to-left shunt. Simultaneously, the vessels within the neonate that carried blood to and from the placenta also undergo constriction and closure. The umbilical vein becomes the round ligament that is found in the free edge of the falciform ligament in the abdomen; the ductus venosus constricts to form ligamentum vensosum in the liver. The umbilical arteries contract and become the cord-like medial umbilical ligaments on the anterior abdominal wall. It should be noted that the umbilical cord should not be tied off immediately after birth, but should remain connected to both the neonate and the placenta until the cord stops pulsating. The pulsation is a consequence of blood that was within the placenta entering the neonate. This is an important phenomenon because the blood that was in the placenta helps to provide blood volume for the previously greatly underperfused pulmonary circulation.

CARDIOVASCULAR MALFORMATIONS

As a group, cardiovascular malformations are the most frequent-encountered birth defects in humans, with moderate to severe forms affecting nearly 1% of newborns (28,29,30). In some cases, cardiovascular malformations are minor or treatable, but they are often fatal or have a marked adverse impact on one's quality or length of life. While certain genetic conditions (e.g., Down syndrome) and exposure to some drugs [e.g., isotretinoin (Accutane®)] have been associated with cardiovascular malformations, the underlying causes of most cardiovascular malformation have not been determined.

TABLE 3 Incidences and Proportions of Cardiac Malformations in Human Neonates

Malformation	Incidence per 10^6 live births	Percentage of cardiac defects
Total cardiac malformations	9596	100
Ventricular septal defect	3750	39
Atrial septal defect	941	10
Persistent ductus arteriosus	799	8
Pulmonary stenosis	729	7.6
Tetralogy of Fallot	421	4.5
Coarctation of aorta	409	4
Aortic stenosis	401	4
Atrioventricular septal defect	348	3.6
Transposition of great vessels	315	3.3

Source: From Ref. 28.

Consequently, the relative roles of genetics and environmental agents as causal influences in cardiovascular malformation are unknown. There are at least 35 recognized types of cardiovascular malformations, the most prevalent of which are listed and briefly described in Table 3 (examples of common malformations depicted in Fig. 12).

Inspection of Table 3 reveals that the overwhelming majority of the most commonly observed cardiac malformations is the result of faulty septation. Because of the complexity of the processes involved, this should not be surprising. The most common cardiac malformation observed in babies is ventricular septal defect. The defect usually involves the membranous interventricular septum. Interestingly, between 85% and 90% of these defects close spontaneously by the time of the first birthday (29). Atrial septal defects account for 10% of cardiac malformations. This usually involves failure of foramen ovale to close, leaving a patency between the right and left atria (often due to a short septum primum). This is not the same condition as probe patent foramen ovale (which is not a malformation), in which the foramen is functionally closed, but the septum primum and septum secundum have not fused. A high percentage of defects (~15.4%) occurs because of problems with the spiral septum. When the spiral septum does not form, there is a persistent truncus (resulting in a mixture of oxygenated and unoxygenated blood in the outflow tract). If the spiral septum deviates to one side, conditions involving unequal sizes of the outflow tract (pulmonary stenosis, aortic stenosis, tetralogy of Fallot) result. Two malformations noted in the table occur if changes that usually occur at birth in the ductus arteriosus go awry. Ductus arteriosus has smooth muscular tissue that contracts at birth in response to a combination of increased oxygen content and prostaglandins in the blood. If the ductal tissue fails to respond, the ductus does not close (resulting in persistent ductus arteriosus), allowing oxygenated and deoxygenated blood to mix prior to distribution to the body. In some cases, ductal tissue extends into the wall of the aorta. In these cases, the extraductal tissue responds appropriately to the signals for contraction and causes the lumen of the aorta to be narrowed near the point where the ligamentum arteriosum attaches to the aorta resulting in coarctation of the aorta.

Cardiovascular malformations result from disruptions in any of the developmental processes involved in the formation of the heart. Major categories of developmental processes that contribute to heart development were described by Clark (31,32) and are presented later. Because of the complexity of cardiac embryology, it is overly simplistic to categorize any cardiac malformation as being due to interruption of only one of the morphogenetic processes. Nevertheless, it is instructive to think about how disruption of these processes can lead to malformations of the heart.

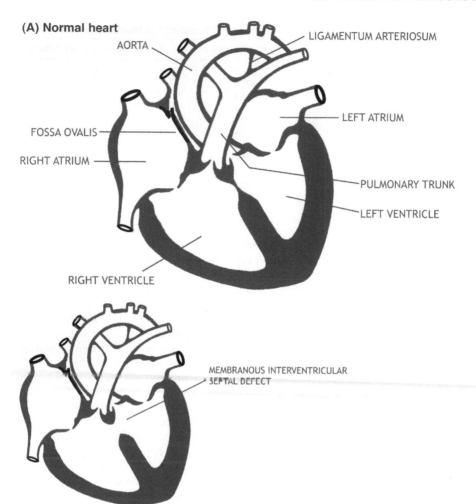

(A) Normal heart

AORTA

LIGAMENTUM ARTERIOSUM

FOSSA OVALIS

RIGHT ATRIUM

LEFT ATRIUM

PULMONARY TRUNK

LEFT VENTRICLE

RIGHT VENTRICLE

MEMBRANOUS INTERVENTRICULAR SEPTAL DEFECT

(B) Interventricular septal defect

FIGURE 12 Illustrations of the anatomy of the normal heart (**A**) and the organization of hearts that are afflicted with several common cardiac malformations. Ventricular septal defects (**B**) are the most common congenital heart defects and usually occur in the membranous portion of the septum. Atrial septal defects (**C**) often involve failures of growth of either septum primum or septum secundum such that foramen ovale remains open after birth. Malformations involving the spiral septum (**D–F**) make up a high percentage of malformations. These include absence of the spiral septum resulting in persistent truncus arteriosus (**D**). When the septum spirals the wrong way, the result is transposition of the great vessels (**E**), in which blood from the aorta is sent back to the lungs and blood from the pulmonary artery is distributed back to the body. This malformation is fatal unless it is accompanied by an interventricular septal defect. If the spiral septum is shifted so that it disproportionately divides the outflow tract, conditions of pulmonary or aortic stenosis result. In some cases, a combination of malformations occur such as tetralogy of Fallot (**F**) in which the heart exhibits an overriding aorta and pulmonary stenosis (from a displaced spiral septum), interventricular septal defect, and hypertrophy of the right ventricle. Two malformations are involved with inappropriate closure of the ductus arteriosus. If the ductus fails to close (**G**), there is a persistent ductus arteriosus that allows for mixing of oxygenated and unoxygenated blood prior to distribution of blood to the body. In the other malformation, closure of the ductus is associated with a narrowing of the lumen of the aorta, known as coarctation of the aorta (**H**).

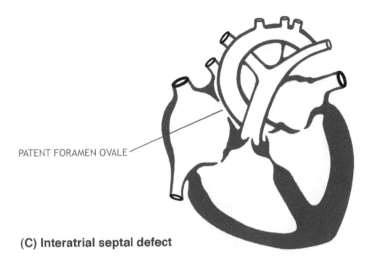

PATENT FORAMEN OVALE

(C) Interatrial septal defect

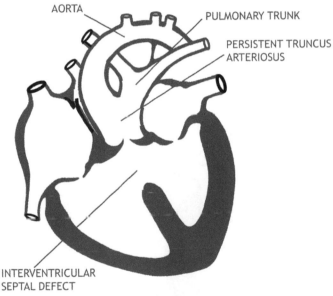

AORTA

PULMONARY TRUNK

PERSISTENT TRUNCUS ARTERIOSUS

INTERVENTRICULAR SEPTAL DEFECT

(D) Persistent truncus arteriosus

FIGURE 12   (*Continued*)

Cellular Migration, Particularly of the Neural Crest Cells

Neural crest cell migration contributes cells that participate in division of the outflow tract. Disturbance of this process leads to malformations that include subarterial ventricular septal defects (type I VSDs), tetralogy of Fallot, persistent truncus, and transposition of the great vessels (31).

Cardiac Hemodynamics

As blood flows through the developing heart, the differential pressure on the various areas of the chamber walls contributes to alterations in the shape of the chamber.

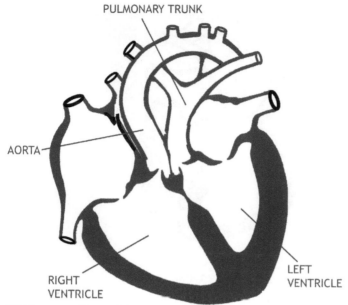

(E) Transposition of the great vessels

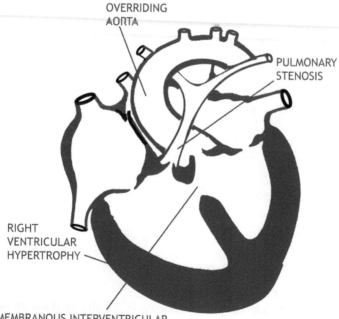

(F) Tetralogy of Fallot

FIGURE 12 (*Continued*)

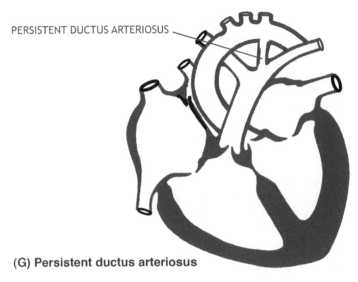

PERSISTENT DUCTUS ARTERIOSUS

(G) Persistent ductus arteriosus

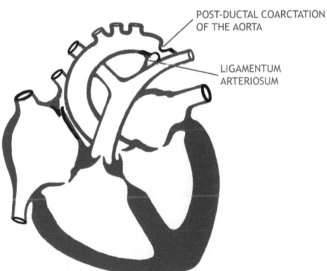

POST-DUCTAL COARCTATION
OF THE AORTA

LIGAMENTUM
ARTERIOSUM

(H) Coarctation of the aorta

FIGURE 12 (*Continued*)

Deviant cardiac hemodynamics can lead to atypical distention of the cardiac chambers and valves, which can modify their shape and function (33). Malformations that have been associated with abnormal hemodynamics include hypoplastic left heart syndrome, coarctation of the aorta, and perimembranous (type II) VSD (32).

Cell Death
Programmed cell death transforms the developing heart by removing tissue, which is significant in the development of cardiac valves, the trabeculae carneae of the

ventricular wall, and timely appearance of shunts between the developing right and left hearts (34). Excessive cell death can lead to aberrant septal perforations; insufficient cell death contributes to Ebstein's anomaly, a condition wherein the tricuspid valve does not detach from the ventricular wall (32).

Extracellular Matrix Function
Cardiac jelly plays a major role in the development of endocardial cushions that surround the atrioventricular orifice and the outflow tract of the great arteries (31). Atrioventricular septal defects can occur when there is insufficient extracellular matrix to allow formation of fully functional cardiac cushions (32).

Targeted Growth
Targeted growth processes contribute to the proper formation of many cardiac structures. Abnormal targeted growth can lead to several disorders, including those related to abnormal venous return that results from the faulty incorporation of the common pulmonary vein into the left atrium (32).

Establishment of Visceral Situs and Cardiac Looping
Visceral situs establishes the right and left sides of the body. Aberrant signals involved in establishing visceral situs can result in cardiac looping defects wherein ventricular inversion or reversed right-left position of organs occurs (32).

CONCLUSION
Cardiovascular development is elegantly controlled and implemented. In mammals, the progression of developmental events follows a nearly identical course regardless of species, although the timing of events differs on the basis of the total duration of the gestational period. The rudiments of the cardiovascular system appear early in gestation, and the heart is the first organ system to begin functioning. During its earliest phase of development, blood islands appear within the embryo in the anterior heart field; these coalesce to form a horseshoe-shaped cardiogenic plate that cavitates to produce the cardiac tube. As the embryo assumes a three-dimensional shape by undergoing cephalocaudal and lateral folding, the cardiac tube assumes a ventral midline position. The developing heart is contained within the nascent pericardial cavity and, with continued growth, it lengthens and undergoes cardiac looping. Cardiac looping results in an anatomic arrangement that prefigures the four chambers of the adult heart. Thereafter, septa appear that (*i*) divide the atria and ventricles from each other and, in conjunction with the development of the neural crest-enhanced spiral septum, partitions the outflow tract and (*ii*) separate the vascular system into a systemic and a pulmonary circulation. Because the lungs do not function until birth, a system of right-to-left shunts allows the circulatory system to function with the placenta serving as the major organ for provision of gases and nutrients to the fetus (and removes carbon dioxide and wastes from the fetal blood) until birth. The final anatomic changes occur shortly after birth as the foramen ovale and ductus arteriosus close. A variety of morphogenetic processes are involved in the development of the heart, and perturbations of any of them can result in cardiac malformations. The incidence and description of the most common malformations were previously described.

REFERENCES
1. Hamilton WJ, Mossman HW. Hamilton, Boyd and Mossman's Human Embryology. 4th ed. Baltimore: Williams and Wilkins, 1972.
2. Moore KL, Persaud TVN. The Developing Human: Clinically Oriented Embryology. 7th ed. Philadelphia: W.B. Saunders, 2008.

3. Sadler TW. Langman's Medical Embryology. 10th ed. Baltimore: Lippincott Williams & Wilkins, 2006.
4. Schoenwolf GC, Bleyl SB, Brauer PR, et al. Larsen's Human Embryology. 4th ed. New York: Churchill Livingstone, 2008.
5. Carlson BM. Human Embryology and Developmental Biology. 4th ed. Philadelphia: Elsevier Mosby, 2009.
6. Gilbert SF. Developmental Biology. 8th ed. Sunderland, MA: Sinauer Associates, 2006.
7. Harvey RP, Rosenthal N. Heart Development. San Diego: Academic Press, 1999.
8. Sissman NJ. Developmental landmarks in cardiac morphogenesis: comparative chronology. Am J Cardiol 1970; 25:141–148.
9. DeSesso JM. Comparative features of vertebrate embryology. In: Hood RD, ed. Chapter 6 in Developmental and Reproductive Toxicology: A Practical Approach. 2nd ed. Boca Raton, FL: CRC Press, Inc., 2006:147–197.
10. Drake RL, Vogl W, Mitchell AWM. Gray's Anatomy for Students. Philadelphia: Elsevier Churchill Livingstone, 2005.
11. Moore KL, Dalley AF. Clinically Oriented Anatomy. 5th ed. Philadelphia: Lippincott Williams and Wilkins, 2006.
12. Warwick R, Williams PL. Gray's Anatomy 35th British Edition. Philadelphia: W. B. Saunders, 1973.
13. DeSesso JM, Scialli AR, Holson JF. Apparent lability of neural tube closure in laboratory animals and humans. Am J Med Genet 1999; 87:143–162.
14. Lodish H, Baltimore D, Berk A, et al. Molecular Cell Biology. 3rd ed. New York: W. H. Freeman and Company, 1995:635–636.
15. Mossman HW. Comparative morphogenesis of the foetal membranes and accessory uterine structures. Contrib Embryol Carneg Inst 1937; 26:129–246.
16. Ramsey EM. The Placenta, Human and Animal. New York: Praeger, 1982.
17. Brand T. Molecular networks in cardiac development. In: Unsicker K, Krieglstein K, eds. Cell Signaling and Growth Factors in Development. Weinheim: Wiley-VCH Verlag GmbH, 2006:841–908.
18. Manner J. The anatomy of cardiac looping: a step towards the understanding of the morphogenesis of several forms of congenital cardiac malformations. Clin Anat 22:21–35.
19. Langman J. Medical Embryology. 4th ed. Baltimore: Williams and Wilkins, 1981.
20. Webb S, Brown NA, Anderson RH. Formation of the atrioventricular septal structures in the normal mouse. Circ Res 1998; 82:645–656.
21. Solomon HW, Wier PJ, Fish CJ, et al. Spontaneous and induced alterations in the cardiac membranous ventricular septum of fetal, weanling, and adult rats. Teratology 1997; 55:185–194.
22. Fleeman TL, Cappon GD, Hurtt ME. Postnatal Closure of Membranous Ventricular Septal Defects in Sprague-Dawley Rat Pups after Maternal Exposure with Trimethadione. Birth Defects Res B Dev Reprod Toxicol 2004; 71:185–190.
23. Momma K, Ito T, Ando M. In situ morphology of the foramen ovale in the fetal and neonatal rat. Pediatr Res 1992; 32:669–672.
24. Pennisi DJ, Rentschler S, Gourdie RG, et al. Induction and patterning of the cardiac conduction system. Int J Dev Biol 2002; 46:765–775.
25. Hagan PT, Scholz DG, Edwards WD. Incidence and size of patent foramen ovale during the first 10 decades of life: an autopsy study of 965 normal hearts. Mayo Clin Proc 1984; 59:17–20.
26. Desai AJ, Fuller CJ, Jesurum JT, et al. Patent foramen ovale and cerebrovascular diseases. Nat Clin Pract Cardiocasc Med 2006; 3:446–455.
27. Allemann Y, Hutter D, Lipp E, et al. Patent foramen ovale and high-altitude pulmonary edema. JAMA 2006; 296:2954–2958.
28. American Heart Association. Congenital heart defects in children factsheet. Available at: http://www.americanheart.org/presenter.jhtml?identifier=12012, 2005.
29. Hoffman JIE, Kaplan S. The incidence of congenital heart disease. J Am Coll Cardiol 2002; 39:1890–1900.
30. American Heart Association. Congenital cardiovascular defects statistics. Available at: http://www.americanheart.org/presenter.jhtml?identifier=4576, 2005.

31. Clark EB. Mechanisms in the pathogenesis of congenital cardiac malformations. In: Pierpont ME, Moller JH, eds. Genetics of Cardiovascular Disease. Boston: Martinus Nijhoff Publishing, 1986:3–11.

32. Clark EB. Growth, morphogenesis and function: the dynamics of heart development. In: Moller JH, Neal WA, Lock JE, eds. Fetal, Neonatal and Infant Heart Disease. New York: Appleton-Century-Crofts, 1992:1–17.

33. Clark EB, Hu N. Hemodynamics of the developing cardiovascular system. Ann N Y Acad Sci 1990; 588:41–47.

34. Pexieder T. Cell death in the morphogenesis and teratogenesis of the heart. Adv Anat Embryol Cell Biol 1975; 52:1–100.

14 Male reproductive toxicity

Robert W. Kapp, Jr.

INTRODUCTION

The biological function of the male reproduction system is to produce male sex hormones and healthy male gametes and to provide a system and route for the gametes to be transported to the female gamete at the appropriate time and in adequate condition such that fertilization and development of an offspring can occur. This extraordinary feat is accomplished through a series of highly coordinated and complex processes that ultimately produce spermatozoa and sex hormones that are vital to the development, regulation, and integration of male reproductive function. Simplistically, the human male reproductive system includes those organs that operate cooperatively toward the reproductive process. The coordination of reproduction is ultimately controlled by the hypothalamic-pituitary-testicular axis (HPT axis), with the hypothalamus neuroendocrine liaison to and from the central nervous system (CNS). The reproductive system of the male mammal is composed of the HPT axis; the external organs, which include the penis, scrotum, and urethra; and the internal organs, which include the testis, vas deferens, epididymis (caput, corpus, and cauda), seminal vesicles, coagulating glands, prostate, bulbourethral glands, and the ejaculatory ducts. The male sex organs work together to produce and release semen into the reproductive system of the female during sexual intercourse. Chapter 2 explains in detail the normal development of the male reproductive system.

As described by Lipshultz and Howards (1) in their pivotal book titled *Infertility in the Male*, exposure to toxicants that may affect the male reproduction system can occur over a lifetime including exposure in utero during embryofetal development, during infancy prior to the initiation of spermatogenesis, during adolescence at the outset of spermatogenesis, and during the life span of the adult as spermatogenesis matures. Interestingly, while testicular function wanes in later life, the process of spermatogenesis in males is continuous from puberty until death. Hence, there are a vast number of sites and life stages that, if altered, could effectively disrupt the male reproductive development and/or process. This chapter concentrates on damage to the adult male reproductive system.

Background

In a broad sense, toxicological damage to the adult male reproductive system includes toxicants that affect spermatogenesis and/or the physical delivery of spermatozoa to the ovulating female. These include those materials that inhibit cell division or differentiation such as toxicants that affect the Sertoli cells (which nourish and support the developing spermatogonia within the seminiferous tubules of the testes). Other toxicants can have a direct effect on the DNA of the stem cells or developing spermatogonia and can cause death or genetic mutation within the spermatogonial DNA. Other toxicants can cause toxic damage by inhibiting the metabolism of the spermatozoa, which are highly dependent on access to energy to get to their target oocytes. Other toxicants can interfere with the production or release of testosterone from the Leydig cells (outside of the seminiferous tubules), which results in effects on spermatogenesis, less nutrition to the spermatocytes, and less libido, which can ultimately result in reproductive impairment.

Reproductive toxicity is evident given the fact that infertility in humans is estimated to be 10% to 15% with approximately one-third of the cases due to deficiencies in the male, one-third due to the female, with the remaining one-third, the causes are unknown (2,3). It has been shown that genetic aberrations as well as age, use of tobacco, drugs, and alcohol contribute heavily to the decrease in quality of semen, quantity of sperm, and presumably reproductive fertility (4–6). However, as much as 30% of infertility cannot be clearly traced to any specific cause and may be associated with environmental chemical exposure (7). Sharpe and his colleagues (8,9) proposed that the decrease in semen quality could be due to the effects of environmental estrogens (endocrine disruptors) on the developing testes. However, Safe (10) suggests that the endocrine disruptor theory should be viewed with some skepticism. Safe states the proposal that endocrine-active environmental contaminants causing wildlife concerns are also contributing to the increased incidence of male reproductive problems is not supported by most research studies. One report indicated that that men born after 1970 manifested sperm counts 25% lower than men born prior to 1959 (11). Olsen et al. (12) revisited the data using different statistical models and found an increase in average sperm numbers. On the contrary, Sikka and Wang (13) indicate that newer approaches to detection of effects in sperm measuring Y chromosome deletions have strengthened the hypothesis that indicate sperm decline may be a result of environmental toxins. The arguments for and against the significance of endocrine disruption and the effects on male reproduction are still present today.

Our understanding of the general causes of decrease in male infertility remains a daunting task because there can be changes in sperm production and/or content that are asymptomatic and there are numerous nontoxicological factors at play that would be difficult to quantify. Complicating this is the fact that the male reproductive system is a complex organ that includes the testis, which contains germ cells, Sertoli cells, and Leydig cells, a variety of accessory organs, all in proximity, and all are regulated by various endocrine and cytokine, that is, paracrine and autocrine, mechanisms. Elucidation of toxicity in the male reproductive system involves examination of events or exposures that interfere with either sperm cell production or sperm cell delivery.

Assessing Male Reproductive Toxicity

There are several general methods of assessing the effects of toxicants in male reproductive systems. One can examine toxicants in in vivo animal studies, in in vitro cellular laboratory studies, clinical studies, or epidemiological studies. Animal studies and cellular studies provide the most flexibility, and protocols that are acceptable for submission for industrial and pharmacological purposes have been worked out by the US Food and Drug Administration (FDA), US Environmental Protection Agency (EPA), Organization for Economic Cooperation and Development (OECD), International Conference on Harmonization (ICH), etc. (These protocols are detailed in chaps. 5 and 6.)

Clinical reproductive performance in human males has been evaluated historically by analysis of the following:

1. Reproductive history including infertility, birth defects, spontaneous abortions, prenatal mortality, age, smoking, and alcohol use, etc.
2. Sperm analysis including sperm count, motility, morphology, Y body. There is wide variability in normal semen characteristics within the male population (14).
3. Hormone evaluations including luteinizing hormone (LH), follicle-stimulating hormone (FSH), and testosterone. Here also there is a wide variability in spermatological changes that can be noted within the normal hormonal concentration ranges. Testosterone concentration ranges are very broad, and Leydig cells tend to compensate for damage by increased production of testosterone.

4. Testicular histopathology. While this gives reliable information, unfortunately, this is a highly invasive procedure and not practically available for use in man.

One can also examine male toxicity by various classification methods such as direct versus indirect toxicants. Indirect toxicants include those that must be activated metabolically to another form before they can produce effects as well as those, such as thyroid hormone, that may affect changes in the organism, which eventually lead to alterations in the testes. Another method of classifying testicular toxicity is by cell type. This method examines the specificity of toxicant actions on various cell types due to the characteristic affects of cells and/or their metabolic potential.

This chapter briefly reviews male reproductive toxicants that can be classified by indirect versus direct mechanisms and those that affect the various cell types found within the male reproductive system. While this results in some redundancy, it is hoped that it will provide a clearer picture of the mechanisms of male reproductive toxicity.

DIRECT AND INDIRECT MODES OF MALE REPRODUCTIVE TOXICITY

Direct-acting toxicants are generally reactive by themselves or may be structurally similar to endogenous molecular communication chemicals. If they are similar, they may interfere with signaling pathways by inappropriately activating or deactivating the receptors involved. They may affect most any cell type or location within the reproductive system.

Direct-Acting Reproductive Toxicants

Some examples of compounds that are structurally similar to signaling chemicals include steroids, cimetidine, diethylstilbestrol, galactose, and azathioprine 6-mercaptopurine. Those which are reactive but do not resemble existing chemical messengers include alkylating agents cadmium, boron, lead, and mercury (15).

Lead is one of the best-known toxicants that can damage sperm directly by interaction with DNA. Lead has many targets within the reproductive system. Studies indicate that lead can also produce interference with the HPT axis (16,17). Lead has been shown to reduce the amount of excreted sulfonated steroids and can reduce the level of testosterone and sperm concentration (18). Studies have also shown that lead can cause disruption of Sertoli cell (19) and prostate secretory functions (20) as well as decreases in libido and increases in sperm abnormalities (21).

Cadmium directly affects the endocrine vasculature that surrounds the testes and epididymis. Sperm must be kept at a proper temperature, that is, 3°F lower than the body temperature, or they will succumb to heat. Extensive testicular vascularization that is designed to remove the excess heat is highly susceptible to direct exposure to cadmium. The damage to the testicular vasculature subsequent to exposure to cadmium reduces the perfusion rate and hence the ability to ameliorate the higher temperature, which can destroy the proximate sperm cells (22). Cadmium can also directly injure the testes by inducing morphological changes in Sertoli cells (23,24). Although the lesions from cadmium exposure are primarily vascular in nature, which can cause sperm cell death, cadmium exposure can also cause Leydig cell atrophy, tumors, and decreased androgen production (25).

Ionizing radiation and alkylating agents used in chemotherapy such as methotrexate, adriamycin, chlorambucil (Leukeran), docetaxol (Taxotere), paclitaxil (Taxol), melphalan (Alkeran), busulfan, vincristine, and vinblastine can directly damage DNA or interfere with concomitant protein respiration or function during critical phases of spermatogenesis. DNA changes can lead to cell death or mutation. Ethylene oxide—a commonly used gas sterilant used for medical devices and in some foodstuffs also has direct actions on cellular biomolecules called protamines, which bind to the DNA in late-stage spermatogenesis and are very susceptible (22).

Indirect-Acting Reproductive Toxicants

Some chemicals must be metabolically activated to produce reproductive toxicity. The testis have microsomal monooxygenase and epoxide hydrases and transferases that metabolize many xenobiotic compounds to toxicants that can and frequently do generate intermediates that are capable of affecting the male reproductive system. Some examples of compounds that are indirect-acting toxicants include diethylstilbestrol, ethanol, chlorcyclizine, dibromochloropropane, polycyclic aromatic hydrocarbons (PAHs), and cyclophosphamide. Some reactive products can also mimic endogenous molecules and ultimately disrupt homeostasis. Some of these types of chemicals include halogenated hydrocarbons, anticonvulsants, and salicylazosulfapyridine (15).

The antineoplastic drugs cyclophosphamide and ifosfamide (Holoxan) are converted by mixed function oxidase enzymes in the liver to active alkylating metabolites, 4-hydroxycyclophophosphamide, aldophosphamide, acrolein, and phosphoramide mustard, with the first-step transformation to hydroxylated intermediates by the cytochrome P450 system. With the solvent n-hexane, the toxicity is not due to hexane, but rather to one of its metabolites, hexane-2, 5-dione. This material reacts with the amino group of the side chain of lysine residues in proteins around the sperm DNA, causing cross-linking and a loss of protein function (26). Phthalates, which are used as plasticizers, are ubiquitous in the environment and have been shown to affect male reproduction after they have been metabolized by interference with testicular testosterone biosynthesis (27). Dibromochloropropane (DBCP) is believed to cause damage to the seminiferous tubules, which affect spermatogonial cells and/or Sertoli cells indirectly through the inhibition of oxidative phosphorylation (28). PAHs are also well known as reproductive toxicants and not only can cause reproductive toxicity indirectly via conversion to active metabolites (29) but also have direct toxic effects on Sertoli cells in vitro via apoptosis (30). Other examples of indirect-acting male reproductive toxicants include acrylamide whose active metabolite is glycidamide (31) and vinyl chloride, which is a breakdown intermediate from common dry cleaning solvents (22).

Theophylline (THP) and 1,3-dinitrobenzene (1,3-DNB) are believed to induce infertility indirectly by incapacitating the nurturing Sertoli cells and subsequently causing germ cell apoptosis in the testicular seminiferous epithelium via alterations to the genes within the ubiquitin-protease pathway as well as the mitochondrial pathway (32,33).

TOXICITY BY CELL TYPE

Cell-type toxicity examines the specificity of toxicant actions on various cell types within the male reproductive system due to the characteristic effects of cells and/or their metabolic potential. For some toxicants, there is differential sensitivity between somatic and meiotic cells. There are several cell types that can be examined in the male reproductive system. Perturbations at almost any point in the regulation of these complex processes controlling successful reproduction could theoretically impact the system including the HPT axis, the testes that includes (i) the spermatogenic compartment (containing the seminiferous tubules and Sertoli cells), (ii) the steroidogenic compartment (containing interstitial component containing Leydig cells), (iii) the blood-testis barrier (BTB), and (iv) the endothelial vasculature, and lastly, the accessory glands.

Hypothalamic-Pituitary-Testicular Axis

The male reproductive system is managed by the CNS and more specifically at the neuroendocrine system, which is made up of several interdependent parts. Specifically, the hypothalamus, pituitary gland, and testes work together with astonishingly elegant coordination to assure the proper development and regulation of several of the body's systems including the reproductive and immune systems. Because of the close coordination and control of these three elements, they are frequently described as a single system (HPT axis). Fluctuations in the hormones of one gland can cause changes in the hormones produced by other glands and can have both local and widespread effects throughout the body. This axis controls not only reproductive development and function but also the reproductive aging process.

Hypothalamic neuroendocrine neurons secrete gonadotrophin-releasing hormone (GnRH), also known as luteinizing hormone–releasing hormone (LHRH), which travels to the anterior portion of the pituitary through the hypophyseal portal system at the median eminence. This, in turn, carries the GnRH to the pituitary gland that contains the gonadotrope cells, where GnRH activates its own specific receptor, gonadotropin-releasing hormone receptor (GnRHR) (34). At the pituitary, the GnRH stimulates the secretion of the gonadotropins LH and FSH. LH is occasionally referred to as interstitial cell–stimulating hormone (ICSH) in males (35). These hormone releases are controlled by the size and frequency of the GnRH pulses from the hypothalamus. In males, the GnRH pulses are at a constant frequency (36). LH stimulates the interstitial Leydig cells located in the testes to produce testosterone, and FSH stimulates the maturation of germ cells and induces Sertoli cells to secrete inhibin and stimulates the formation of Sertoli-Sertoli tight junctions (zonula occludens). FSH also enhances the production of androgen-binding protein (ABP) by the Sertoli cells of the testes by binding to FSH receptors on their basolateral membranes, which is critical for the initiation of spermatogenesis. Recent research has identified the possibility of an additional level of control—a neuro-steroid axis—which helps the pituitary cortex to regulate the hypothalamic production of GnRH (37). The formation of GnRH, LH, and FSH is similar to that present in the female; however, the effects are significantly different in males. In the male, LH binds to the Leydig cells, causing them to secrete testosterone. FSH stimulates maturation of seminiferous tubules and the process of spermatogenesis. Testosterone is required for normal spermatogenesis and its presence inhibits the production of GnRH by the hypothalamus. FSH also stimulates Sertoli cells to release ABP that binds specifically to testosterone, dihydrotestosterone, and 17-β-estradiol. Binding testosterone and dihydrotestosterone renders these hormones less hydrophilic permitting them to concentrate within the seminiferous tubules, which promote spermatogenesis in the seminiferous tubules and spermatid maturation within the epididymis.

Activin and inhibin are two closely related protein complexes that have opposing biological effects. Activin is produced by the testes and the pituitary and participates in a number of activities including androgen synthesis by enhancing LH action and spermatogenesis. Activin also enhances FSH biosynthesis and secretion, and participates in a number of activities including the regulation of cell proliferation, differentiation, apoptosis, metabolism, homeostasis, immune response, wound repair, and endocrine function (38,39). Conversely, inhibin is produced by Sertoli cells and downregulates FSH production and GnRH release from the hypothalamus (40). Further, the Sertoli cells also produce small amounts of estrogen. After puberty, these hormone levels remain relatively constant.

Toxicity at the HPT Axis

As described above, the HPT axis is a delicately balanced feedback system, which has many sites that can modulate the overall system. At each one of the points in the negative feedback hierarchy previously described, toxicants can affect changes to the HPT axis, subsequently decreasing the trophic effects to the accessory sex organs from the enhanced or diminished secretion of testosterone. There is evidence that suggests a variety of chemicals such as heavy metals, pesticides, and polychlorinated biphenyls can impair reproduction by disruption at the hypothalamic level (41–43). Studies with pesticides suggest some organic compounds can disrupt neuroendocrine systems by impairing the neurotransmitter systems that control GnRH and gonadotropin secretion (44). Neuroendocrine neurons have nerve endings that contain monoamines such as norepinephrine, dopamine, and serotonin, which affect their activity. Reserpine, chlorpromazine, and monoamine oxidase (MAO) inhibitors are also known to modify the activities of brain monoamines that, in turn, affect gonadotropins (35). Bartsch et al. (45) suggested that growth factor secretions and cellular interactions may be susceptible to toxicants. Alcohol (ethanol) exposure has been shown to be associated with low levels of GnRH and pituitary LH in male rats (46). It appears that alcohol could deregulate the series of complex mechanisms that produce impulses external to the hypothalamus that activate the GnRH pulse generator, which subsequently results in diminished LH release. Gonadotoxins can act on the neuroendocrine processes in the brain or they could act directly on the testes. Cooper et al. (47) proposes that hepatic and/or renal toxicants that alter the transformation of endogenous sex steroids could interfere with the pituitary feedback balance. Cunha et al. (48) suggested that androgen-directed differentiation and maintenance of male sex organs are influenced by the effects of 5α-dihydrotestosterone [which is produced from testosterone in target organs by 5α-reductase enzyme (49)] and androgen receptor function in stromal cells. Stromal cells produce growth factors, which, in turn, influence the growth of epithelium. Basal cells within the epithelium could also be targets of toxicants of accessory sex glands, and basal cells appear to be involved in the regulation of epithelial cells since they express receptors for epidermal growth factor (EGF) and insulin-like growth factor (IGF-I) (50). Goserelin is a synthetic analogue of the naturally occurring LHRH. It rapidly binds to the LHRH receptor cells in the pituitary gland thus leading to an initial increase in production of LH resulting in an initial increase in the production of corresponding sex hormones. Goserelin is used to treat hormone-sensitive prostate cancer in men and breast cancer in pre-perimenopausal women and some benign gynecological disorders such as endometriosis, uterine fibroids, as well as endometrial thinning (51).

Alcohol may also impair the function of protein kinase C, which is a key enzyme in LH production (52). Also suggested is that ethanol may reduce the number of active binding sites rendering LH less potent. Alcohol does not appear to affect FSH synthesis. Hence, ethanol appears to interfere with male reproduction by changing endocrine regulation and ultimately Leydig cell secretion of testosterone. Further studies have shown that propanolol, cannabis, carbon disulfide, and some phthalates can also disrupt the HPT axis. It has also been shown that carbon disulfide and chlordecone can cause decreased sexual drive by limiting androgen production, which affects the CNS neuroendocrine system and ultimately libido (22).

Lead at blood levels >40 μg/dL suppresses serum testosterone, intratesticular testosterone, and spermatogenesis by disrupting the signals between the hypothalamus and the pituitary (53). It is postulated that alterations in the catecholaminergic neurotransmitter system are involved in these signal disruptions (54). Cadmium exposure during adulthood increases norepinephrine in the posterior hypothalamus and decreases neurotransmitter content in both the anterior and mediobasal hypothalamus. The data suggest that cadmium exerts an age-dependent effect on the HPT axis that may be related to selective cadmium accumulation (55).

Testes

In mammals, the testes are contained within extensions of the abdomen called the scrotum. The scrotum is positioned outside of the pelvic cavity to maintain a temperature of about 35°C—approximately 2° cooler than normal body temperature of 37°C

(56). This lower temperature is optimum for spermatogenesis. In healthy adult human males, average testicular volume is 18 cm^3/testis, with normal size ranging from 12 to 30 cm^3 (57). Testes are components of both the reproductive system and the endocrine system. Both of these functions of the testes—sperm development and endocrine secretion—are under control of LH and FSH, which are produced by the anterior pituitary. Hence, the testes not only produce spermatozoa, they produce male sex steroids called androgens—the most common and abundant of which is testosterone. Androgens stimulate or control the development and maintenance of masculine characteristics by binding to androgen receptors. This includes the activity of the accessory male sex organs and development of male secondary and tertiary sex characteristics. However, androgens also function as paracrine hormones, which are required, along with FSH, by the Sertoli cells to support normal sperm production throughout adulthood.

The testis is roughly compartmentalized into spermatogenic (containing the seminiferous tubules and Sertoli cells) and steroidogenic (containing Leydig cells) areas. Contained in the testes are very finely coiled tubes called seminiferous tubules. The tubules are lined with a layer of cells (gametes) present from puberty into old age, which develop into sperm cells (spermatozoa). In the seminiferous tubules, primary germ cells develop into spermatogonia, spermatocytes, spermatids, and subsequently spermatozoa through the process of spermatogenesis by a series of mitotic and meiotic cell divisions, which is a coordinated effort, including the HPT axis, Sertoli cells, Leydig cells, germ cells, and the vascular endothelium. The developing spermatogonia proliferate, become primary spermatocytes, and travel through the seminiferous tubules to the rete testis located in the mediastinum testis. In this structure, fluid is reabsorbed and the sperm are concentrated. Subsequently, the maturing spermatocytes are transported to the efferent ducts that are unilaminar and are composed of columnar ciliated and nonciliated (absorptive) cells. The ciliated cells ensure further homogeneous absorption of water from the testicular fluid to concentrate the spermatocytes. The efferent ducts lead to the epididymis, where the developing gametes mature and are stored until ejaculation. The maturation process of the sperm involves the acquisition of a tail and subsequent motility. At the outset of the process, the original stem cells are located at the outer edge of the tubules. As they change into haploid spermatozoa, they migrate toward the lumen of the tubules where they are eventually transported to the epididymis and then out of the body as ejaculate.

In the human male, spermatogenesis takes just over nine weeks to produce mature spermatozoa from a stem cell. Clermont (58) showed spermatogenesis occurs in six distinctive stages in humans. The Sertoli cell provides the proper environment and it serves as an androgen receptor and site of synthesis for ABP during spermatogenesis (59). Upon signaling, primarily by LH, the Leydig cells secrete androgenic steroids, which, along with autocrine and paracrine cytokines and growth factors such as transferrin, ABP, inhibin, activin, proopiomelancortin derivatives, neuropeptides, interleukin factors, IL-1 and IL-6, tumor necrosis factor (TNF), interferon gamma (IFN-γ), leukemia inhibitory factor (LIF), and stem cell factor (SCF), mediate spermatogenesis (35,60–62). This process is extremely complicated and we are only beginning to unravel the complex interplay among the chemical regulators that guide the process of spermatogenesis.

Toxicity at the Testis
Clearly, there are numerous obvious targets for toxicants in the scheme of testis and male endocrine regulation. Spermatogenesis results in the production of 100 million or more sperm per day. This provides a vast array of rapidly dividing energy-dependent

cells that are susceptible to various insults, which can result in reduced overall spermatozoa concentrations, increased numbers of morphologically defective spermatozoa, as
well as changes in androgenic hormone secretions with the ultimate outcome of impaired
fertility. With the close proximity of the testicular spermatogenic and steroidogenic
compartments, these chemical insults can occur individually or in combination with direct
or indirect events—both of which can be equally disruptive. Direct-acting toxicants can be
identified as reactive chemicals or chemicals that are structurally similar to endogenous
cellular signaling chemicals that disrupt the communication pathway. Many indirect-
acting toxicants must first be modified via metabolic bioactivation within the cells or tissue
or may be as a result of alterations on other systems that could affect a change within the
testes. Interestingly, detrimental reactive metabolites are often created from the host's
metabolic defenses in eliminating xenobiotics from the system. Other indirect toxicants
can affect a change in other ways. Data suggest that the thyroid hormone directly affects
the development of prepubertal testes and the regulation of FSH-R and ABP gene
expression in Sertoli cells as well as the LH-R mRNA levels in Leydig cells, which may
lead to modulating the effect of gonadotropins on testes function (63).

Spermatogenic Compartment Cells (Seminiferous Tubules)

Spermatogonia (germ cells)   These cells are diploid gametes that ultimately develop into
haploid spermatozoa through spermatogenesis. Mammalian spermatogonial stem cells,
sometimes called stem cells, are a small population of adult tissue-specific stem cells
present in the testis. Formation of the spermatogonial stem cell population early in life
and differentiation in postpubertal individuals of spermatogonial stem cells to spermatozoa are responsible for continual production of sperm in the testis.

   Germ cell toxicity   Germ cells are completely dependent on proper coordination
of their milieu and all the concomitant processes within the testes to survive—hence
their response to significant disturbances is usually death (64). Each stage of germ cell
has its own sensitivity to various chemical insults with the earlier stages being generally
more sensitive; however, this can vary depending on the chemical and timing of the
insult. Germ cell division is rapid and presents a significant target for alkylating agents.
A number of toxicants that affect germ cells are cancer chemotherapeutic agents, which
are alkylating agents such as busulfan, chlorambucil, cyclophosphamide, nitrogen
mustard, doxorubicin, corticosteroids, methotrexate, procarbazine, vincristine, and
vinblastine (65). Industrial chemicals that can damage rapidly dividing cells include
acrylamide and ethylene oxide; PAHs and ethylene dibromide are also able to directly
interact with DNA (22). Ionizing radiation and ultraviolet and X rays can also directly
affect dividing germ cells resulting in mutated and/or lethal damage. Early studies on
YY fluorescent sperm revealed increases in aneuploidy in humans exposed to radiation,
Adriamycin®, Flagyl®, and DBCP (14,66). Testicular biopsies of men exposed to DBCP
show that the site of action is the spermatogonia (67). Smoking (68), caffeine, and
alcohol (69) have all been shown to increase the incidence of aneuploidy in germ cells
by using fluorescent in situ hybridization (FISH) technique. These alkylating chemicals
and physical agents can cause alterations in sperm chromosome number or structure, or
by causing molecular gene mutations in sperm that can be transmitted to offspring.
Increases in aneuploidy can result in increased infertility since most of these abnormal
germ cells die before progressing to the final spermatozoa stage. Dixon and Lee (70)
showed that some alkylating agents result in cells being unable to make normal DNA
repairs on DNA that has been damaged. Chromosomal damage has also been shown in
chromosomal analyses developed by Working (71). Lead has been linked to chromosome aberrations such as chromosome and chromatid breaks (72). It has been demonstrated that increased levels of reactive oxygen species (ROS) can cause significant

damage to sperm DNA, which can result in male infertility. Extensive studies have been conducted and reviewed that establish a link between oxidative stress and human sperm pathology (73). The most direct evidence suggesting ROS could cause sperm damage was first demonstrated by Kodama et al. (74) where they showed that infertile men had increased levels of 8-hydroxydeoxyguanosine (8OHdG) when compared with those not exposed to ROS. 8OHdG is an oxidative DNA adduct that is considered to be a specific biomarker of oxidative DNA damage (75). DNA strand breaks are associated with damaged sperm in both sperm cell gel electrophoresis (SCGE) or the comet assay, and the terminal deoxynucleotidyl transferase nick end-labeling (TUNEL) assay provides yet additional evidence of germ cell damage (75). Additional studies appear to show that ROS are implicated in other aspects of male infertility such as altered sperm membrane function and impaired metabolism (76,77).

Sertoli cells These are cells that nourish, support, and protect developing germ cells from the walls of the seminiferous tubules. Sertoli cells are bound to each other through tight junctions and form a ring between the basal membrane of the tubule and its lumen. They act as a physicochemical barrier that prevents the entrance of substances into the tubule lumen, as well as retaining the luminal fluid, thereby ensuring an appropriate environment for gamete development and differentiation. Sertoli cells also act as phagocytes, consuming the residual cytoplasm during spermiogenesis (the last stage of sperm formation) and are the primary secretory cells in the seminiferous tubules. Sertoli cells secrete tissue plasminogen activator (TPA), inhibin, transferrin, and ABP in addition to other proteases. ABP is a glycoprotein that binds specifically to testosterone, dihydrotestosterone (DHT), and 17-β-estradiol enabling spermatogenesis in the seminiferous tubules and sperm maturation in the epididymides. It has been shown that Sertoli cells are also necessary to maintain germ cell glutathione metabolism, to stimulate both RNA and DNA synthesis in germ cells, and to induce appropriate germ cell surface antigens by way of Sertoli cell paracrine secretions (78).

Translocation of germ cells from the base to the lumen of the seminiferous tubules occurs by conformational changes in the lateral margins of the Sertoli cells. In response to testosterone secreted by the Leydig cells and to FSH secreted by the pituitary gland, Sertoli cells produce a large variety of chemical messengers that stimulate spermatozoan proliferation and differentiation (79). Since Sertoli cells can only support a limited number of germ cells through development, some germ cells must be eliminated via apoptosis. This regulated process of cell death occurs in two phases and is critical for the normal development of spermatozoa in the adult testis (80–83).

Sertoli cell toxicity Sertoli cells not only provide support for germ cells, they also have specialized tight junctions (see section "The Blood-Testis Barrier") that isolate the germ cells from somatic cells and from many chemicals that might be circulating throughout the testes. Because of their location and function, Sertoli cell damage is usually accompanied by disruption of the structural and/or metabolic support of germ cell development. Sertoli cell damage is usually accompanied by inter/intracellular vacuolization of smooth endoplasmic reticulum and/or swelling of the basal cell cytoplasm (64,84). The resulting effects include germ cell disruption, that is, premature exfoliation, spermatid retention, germ cell degeneration, as well as decreased testicular weight and degeneration of seminiferous tubules (85). Sertoli cell toxicants frequently have a general indirect influence on various stages of germ cells because their support of the germ cells may be impaired and the BTB may be physically disrupted, thus allowing access of potential toxicants to the germ cells. Chemical insults usually make the Sertoli cell dysfunctional rather than killing it outright. Sertoli cell damage generally results in prolonged spermatozoa cell reduction or even irreversible testicular atrophy

(86). This may be in part due to the fact that there is limited replacement of Sertoli cells with little or no cell division in these cells in the adult male. The Sertoli cells also provide various cytokines that may assist in initiation of the spermatogenic cycle as well as in response to bacterial infection of the testis (22,87).

Effects of toxicants in Sertoli cells can be evaluated in animal models by examination of changes in morphology, biochemistry, FSH responsiveness, Sertoli cell enzymes, and secretion such as ABP, inhibin and lactate and pyruvate production, and microtubule formation (88).

It has been shown that n-hexane and methyl n-butyl ketone are reproductive toxicants that interfere with microtubule formation, which is an important part of the Sertoli cell cytoskeleton, due to their common metabolite 2, 5-hexanedione (89). Carbendazim (CBZ) inhibits Sertoli cell microtubules similarly to colchicine by decreasing the rate and stability of microtubule assembly by binding to the heterodimer (90).

Phthalate esters such as di-n-butyl phthalate (DBP), mono-(2-ethylhexyl) phthalate (MEHP), one of the active metabolites of di (2-ethylhexyl) phthalate (DEHP), and di-n-pentylphthalate (DPP) also can affect Sertoli cells by causing a breakdown of the attachments between Sertoli cells and germ cells (85). Cytochrome P450 bioactivation of tri-o-cresyl phosphate (TOCP) to saligenin cyclic-o-tolyl phosphate (SCOTP) in the Leydig cells can result in impaired Sertoli cells and ultimately impaired or cessation of spermatogenesis (91).

1,3-DNB and its structurally related analogues also appear to disrupt Sertoli cell function. Recent study results by Lee et al. (92) suggest that 1, 3-DNB suppresses Sertoli cell proliferation, increases necrosis and apoptotic cell death, and modulates the cell cycle distribution during G2/M (interphase gap 2 and cell division mitosis phase) cell cycle arrest.

The Blood-Testis Barrier

The male gonad possesses a specialized biological barrier that is referred to as the blood - testis barrier (BTB). This is a physical barrier between the blood vessels and the seminiferous tubules within the testes. It is formed with tight connections between the Sertoli cells at the lumen of seminiferous tubules and the lumen of interstitial capillaries. Data show that molecules less than 3.6 Å in size can get through the BTB and, based on lipid solubility, larger molecules are restricted from passage (88). Spermatogonia are located in indentation pockets of the basal compartment of the seminiferous tubules, which is insulated from the interstitial space and tubular lumen by a basement membrane and tight junctions between the Sertoli cells. Further differentiated germ cells such as primary and secondary spermatocytes and spermatids are located in the adluminal compartment. Each germ cell remains in intimate contact with Sertoli cells, which provide them with nutrients and growth factors (93). The BTB has several functions such as (i) confers cellular polarity, (ii) maintains an immunological barrier, and (iii) restricts the paracellular diffusion of molecules (94).

However, the BTB must be restructured at stage VIII to IX of the seminiferous epithelial cycle to facilitate the transit of primary preleptotene spermatocytes across the BTB, while differentiating into leptotene and zygotene spermatocytes. Studies have shown that the movement of germ cells across the seminiferous tubules is dependent on the Sertoli cells, regulated by testosterone and cytokines initiating the cascade of events that changes the cytoskeleton to permit such movement (78,95). Without this timely movement of developing primary spermatocytes across the BTB from the basal to the adluminal compartments at stages VIII to IX of the epithelial cycle, spermatogenesis is disrupted, leading to infertility (96). Several molecules have been shown to regulate Sertoli cell–tight junction dynamics. These include transforming growth factor b3

(TGFb3), occludin, protein kinase A, protein kinase C, and signaling pathways such as the TGFb3/p38 mitogen-activated protein kinase pathway (97). The function of the BTB may be to prevent an autoimmune reaction. Mature sperm arise long after immune tolerance is established in infancy. Therefore, since sperm are antigenically different from *self* tissue, a male animal can and sometimes will react immunologically to his own sperm, making antibodies against the sperm. In experimental animals, active immunization with testicular tissue or adoptive transfer of T lymphocytes causes autoimmune orchitis. In human males, infection and inflammation of the reproductive tract including the testes are widely accepted as important etiological factors of infertility (98). Thus, the BTB may reduce the likelihood that sperm proteins will induce an immune response, reducing fertility. The components of the complex process of permitting the passage of the selected cells and not compromising the immunological barrier function remain largely unknown (96,99).

BTB toxicity   The BTB is relatively resistant to some toxicants that are known to cause spermatogenic damage. However, DBCP, exposure to antineoplastic platinum-derived drugs such as cisplatin and carboplatin, vitamin A deficiency, and cytochalasin D treatment will disrupt the tight junctions of the Sertoli cells (70,85,88,100).

In addition, the endothelial vasculature that surrounds the testes has attributes that make it somewhat vulnerable to chemical exposures. The characteristics are described by Chubb (101) and Sundaram and Witorsch (88) and include (*i*) low blood flow and oxygen uptake to seminiferous tubules and (*ii*) confinement of the blood supply primarily to the interstitial cells.

Steroidogenic Compartment Cells (Interstitial Cells)

Leydig cells   These are cells found adjacent to the seminiferous tubules in the testicles. Their primary function is to secrete testosterone; they have round vesicular nuclei and an eosinophilic cytoplasm that releases androgens. The primary androgens they secrete include testosterone, androstenedione, and dehydroepiandrosterone (DHEA), when stimulated by the pituitary hormone ICSH. ICSH (aka LH) increases cholesterol desmolase activity leading to testosterone synthesis and secretion by Leydig cells. FSH from the anterior pituitary increases the response of Leydig cells to ICSH by increasing the number of ICSH receptors expressed on Leydig cells (88,102). It is further hypothesized that there remains a strong correlation between Leydig cell structure and steroidogenic function. The release of precursor cholesterol located in the Leydig cell membrane, which is destined to become converted to testosterone in the mitochondrion, is triggered by the release of sterol carrier protein 2 (SCP2) via ICSH stimulation. The cholesterol is redirected to a P450 enzyme that converts it to pregnenolone. The pregnenolone is subsequently metabolized to testosterone (103). Leydig cells also secrete a variety of other critical peptides including proopiomelanocortin (PMOC)-derived peptides, enkephalins, dynorphins, corticotrophin-releasing factor (CRF), oxytocin, and renin-angiotensins (103).

Leydig cell toxicity   Leydig cell damage is characterized by changes in testosterone levels due to changes in the Leydig cells. Decreases in testosterone result in degeneration of spermatids in certain developmental stages. Testosterone is also critical in the support of male sexual behavior, in maintaining the reproductive tract per se as well as in areas other than sexual performance such as secondary and tertiary male characteristics, that is, body hair, voice changes, and muscle mass as well as hematopoiesis and the immune system (104).

While the toxic effects of testosterone are important, it is difficult to tease out which effects are primary toxic effects and which are secondary responses due to hormonal deregulation from a variety of different controllers in the milieu. In addition, some of the toxicants that specifically affect Leydig cells can also cause responses in other tissue types further confounding the results. Historically, Leydig cell damage is measured via monitoring androgen levels, mating behavior, and accessory sex organ weights (88).

Ethane-1,2-dimethanesulfonate (EDS) affects androgen production in Leydig cells by interfering with the steroid synthesis process. Twenty-four hours post EDS exposure, Leydig cells show significant signs of apoptosis. Biochemical studies have shown that on exposure to EDS, the biosynthesis of testosterone is compromised between the cyclic adenosine monophosphate activation of protein kinase A and the cholesterol side chain cleavage enzyme (P450$_{scc}$) (105,106). Steroidogenesis acute regulatory protein (StAR) regulates the transport of cholesterol across mitochondrial membranes. Studies suggest that EDS impaired mitochondrial electrochemical potential (δ-psi), resulting in an increase in available StAR into the mitochondria (107). Given the research on Leydig cells to date, Teerds and Rijntjes (106) speculate that EDS affects Leydig cells by impairing the δ-psi and StAR functioning, which leads to changes in intracellular glutathione levels and ultimately a decrease in testosterone biosynthesis.

2,3,7,8-Tetrachlorodibenzo-p-dioxin (TCDD) has been shown to cause effects on Leydig cells as well as the seminal vesicle, the prostate, and epididymis (108). Studies suggest that 7,12-dimethylbenz[a]anthracene (DMBA), a reactive metabolite of TCDD, induces decreases in Leydig cell androgen concentrations and steroidogenic activity by inhibiting cytochrome P4501B1 (CYP1B1) expression levels (109).

Both Sertoli and Leydig cells are sensitive to phthalate esters. DEHP not only affects Sertoli cells, but it has been shown that effects of DEHP on Leydig cell steroidogenesis are influenced by the stage of development at exposure and may occur through modulation of testosterone biosynthesis pathway and serum LH (aka ICSH) levels (110).

As noted previously, alcohol affects many organs within the body. Studies have shown alcohol appears to activate specific intracellular death-related pathways within Leydig cells. The bax-dependant caspase-3 (a class of cysteine proteases), known to mediate a crucial stage of the apoptotic process, is expressed and promotes cell death activation and induction of apoptosis in Leydig cells (111).

Marihuana has been implicated in damage to the male reproductive system (112). The primary active ingredients include delta-9-trans-tetrahydrocannabinol, cannabidiol, and cannabinol. Reduced prostate and seminal vesicle weights, as well as altered testicular function, have been partially explained by the effect of marijuana in lowering serum testosterone needed for proper function and support. This lowered testosterone production by the Leydig cells is thought to be one of the primary effects; however, some of the weight changes may be due to a direct action of marijuana, and perhaps some of the other nonpsychoactive cannabinoids in marijuana, on the tissues themselves. Since other effects have been noted, it remains difficult to determine the primary site of action (88,113).

Endothelial Vasculature
Testicular vasculature controls the microcirculation and fluid dynamics within the testis. It is somewhat unique since the narrow testicular artery has high resistance resulting in low capillary pressure in the testis relative to the rate of oxygen uptake (114). The mechanisms maintaining the high basal permeability are not well understood; however, the rat testicular endothelium contains few vesicles (115) and perhaps the permeability may be caused by the few open interendothelial cell junctions observed in capillaries and postcapillary venules (116). It has been shown that the endothelial cell proliferation rate is higher than in other stationary organs (117). Leydig cells and possibly others secrete vascular endothelial cell growth factor (VEGF), and its receptor VEGF-R2 is expressed on testicular blood vessels (117). Studies have shown that testicular VEGF secretion is increased by hormonal stimulation of Leydig cells and that VEGF, via the VEGF-R2,

regulates endothelial cell proliferation in the rat testis (118). It appears that Leydig cells are critical to this regulation by altering the restriction of transport of LH and cellular nutrient, and altering the flow of testosterone secretions. This results in changes to testicular blood flow, formation of interstitial fluid, and microvessel permeability of the endothelial vasculature as well as endothelial cell proliferation (117,119). Testicular interstitial fluid surrounds the seminiferous tubules and other cells of the interstitial tissue and mediates the endocrine interactions between the testicular cells and circulation as well as paracrine interactions between testicular cells (120).

Endothelial vasculature toxicity   Vascular damage that produces even mild restriction in the vascular endothelium blood flow that surrounds the testes can reduce access of the germ cells to oxygen and nutrients (118,121). Severe reduction to oxygen and nutrients can produce germ cell death as well as necrosis of both the Leydig and Sertoli cells. Blood circulation functions to remove toxicants as well as to provide nutrients, hence anything that impedes the flow could cause toxicity at any point in the male reproductive system. Chubb (101) reported that lead, cobalt, and cadmium readily disrupt the endothelial vasculature of the testes. Cadmium severely and selectively affects the circulation in the testes, which ultimately causes Leydig cell damage and seminiferous tubule necrosis in the rat; however, this damage is not seen in the dog. Cadmium is also reported to cause changes in the BTB by fragmenting the microfilament bundles in the Sertoli tight junctions (122,123). More recent studies in mice have shown that expression of the metal ion transporter ZIP8 located at Slc39a8 gene—which enhances cadmium uptake 10-fold—is instrumental in the degree of severity of cadmium toxicity in the endothelial vasculature of the testes; however, the elucidation of the specific mechanism and how it relates to humans have yet to be determined (124).

ACCESSORY SEX ORGANS
From the efferent ducts, the developing gametes migrate to the epididymis, which is a tightly coiled tube about 5 m in length and connects the efferent ducts from each testicle to the vas deferens. The epididymis is divided into three generalized regions: (*i*) the head (Caput), which receives the diluted spermatozoa from the efferent ducts, (*ii*) the body (Corpus), where the spermatozoa begin to mature by gaining motility, and (*iii*) the tail (Cauda), which contains thicker myoepithelium and absorbs fluid to continue to concentrate sperm and stores them (125). When the sperm leave the testes, they are immature and incapable of fertilization. Spermiogenesis is the process by which spermatids develop into spermatozoa. During this process they gain a tail for motility, an acrosome (acrosomal cap) to facilitate in penetration of the oocyte, and additional mitochondria to provide energy to complete the fertilization. As the immature sperm move through the epididymis, they are subjected to a changing environment in the caput and corpus sections, which results in sperm maturation. The cauda section of the epididymis is primarily a storage site. The spermatozoa move from the epididymis into the vas deferens, which connect the left and right epididymis to the ejaculatory ducts. During transit through the duct system, that is, ejaculation, the smooth muscle in the walls of the vas deferens contracts in a radially symmetrical pattern of contraction, which propagates in a wave down the muscular tube propelling the sperm forward (peristalsis). The sperm is transferred from the vas deferens into the urethra, collecting secretions from the male accessory sex glands such as the seminal vesicles, prostate gland, and the bulbourethral (Cowper's) glands, which form the bulk of semen. Fluids are forcibly expelled from the accessory glands producing the seminal plasma. The seminal vesicles produce a viscous fluid containing high levels of fructose, prostaglandins, and various clotting proteins, which empties into the ejaculatory duct (126). The prostate produces an alkaline fluid that

helps neutralize the highly acidic female vaginal secretions and empties into urethra effectively closing off urine flow during ejaculation. Cowper's glands secrete an alkaline mucus-like fluid into the urethra that not only lubricates the urethra but also helps to neutralize the female vaginal secretions (127).

Accessory Sex Organ Toxicity

Toxicants can affect the accessory sex organs by altering concentration of hormones and growth factors that maintain the accessory organs. As previously noted, disruptions to the HPT axis can result in changes that would subsequently decrease the trophic effects to the accessory sex organs by enhancing or diminishing the secretion of testosterone. Specifically, effects can occur at the conversion of testosterone to 5α-dihydrotestosterone by 5α-reductase in the accessory organs. There is evidence that binding of testosterone or 5α-dihydrotestosterone to androgen receptors and the induction of various genes are susceptible to toxicants (45,128,129). For example, the blockage of normal androgen receptor function by a toxicant could produce an increase or decrease in gene product depending on the effect (130,131).

Methylmecury has been shown to cause epididymal atrophy in mice (132) while lead studies have shown a reduction in epididymal epithelial cell height (133). Exposures to industrial chemicals such as DBCP (134), 1,3-DNB (135), EDS (136), nitrobenzene (137), and tricresyl phosphate (138) cause effects on epididymal weights but no changes in testosterone levels. Carbendazim produces multiple effects on the testes as well as epididymal weights (139). α-Chlorohydrin can cause damage to the epididymis as well as sulfasalizine, which affects the sperm maturation process in the epididymis (140).

Boric acid affects prostate weight and the caput and corpus (but not the cauda) sections of the epididymis (111). Studies show that short-term treatment with cyclophosphamide affects spermatozoa in transit through the caput and corpus sections of the epididymis, while the spermatozoa in the caudal epididymis are resistant (142). Gellert et al. (143) showed that hexachlorophene causes fibrosis of the prostate, which ultimately results in impaired ejaculation in rats.

Chronic exposure of rats to lithium resulted in a decreased epithelial cell height in seminal vesicles (144). Seethalakshmi et al. (145) showed that high doses of cyclosporine A decreased prostate, seminal vesicle, and epididymal weights.

Studies have shown that 2-bromopropane can cause direct effects on spermatogonia such as decreased sperm counts and increased sperm abnormalities. It is suggested that the accompanying low accessory sex organ weights are secondary effects due to reduced levels of testosterone (146).

Oxidative stress characterized by ROS is also associated with male infertility via loss of sperm motility (147) and chromatin damage (148). It has been shown that both the epididymis (149) and seminal vesicles (150) shield the spermatozoa from oxidative stress by secreting antioxidants. Aging causes not only an increase in overall oxidative damage but also a decrease in the ability to respond to stress with the release of antioxidants (151). Studies have shown that significant age-related decreases in antioxidant enzyme activity were noted in the cauda epididymis and seminal vesicles, which ultimately result in an increase of impaired spermatozoa (152).

CONCLUSIONS

The development and functioning of the male reproductive system involves complex interrelated actions at many levels including molecular, organelle (e.g., mitochondria), cellular, tissue, organ, and systemic. The reproductive system is responsible for the production of male sex hormones, functional haploid gametes (sperm), for providing a system and route for the sperm to be delivered to the female for successful fertilization for offspring development. Development and functions are controlled and affected by other systems, including the neuroendocrine system. The reproductive system itself is composed of internal structures—HPT axis, testes, epididymides, and accessory sex organs [seminal vesicles, coagulating glands, prostate, Cowper's (bulbourethral) glands] and ejaculatory ducts—and external organs—the scrotum, penis, and urethra. Human infertility rates are relatively high (10–15%) and rising, with the cause(s) of at least one-third of the cases currently unknown. Male reproductive toxicity may be due to genetic aberrations (inherited or de novo), age related, from exposure to drugs of abuse, pharmaceuticals, environmental chemicals (pollutants), and now also endocrine disruptors. The major effects of reproductive toxicants are on spermatogenesis or the physical delivery of the sperm to the receptive female. This chapter has presented examples of the many reproductive toxicant classes and chemicals and their mechanism(s) of action when known. The various sensitivities and vulnerabilities, as presented (based on life stage, location, target cell, and type of insult), are balanced by the redundancies and protective roles of the various components (also presented) to this complex and vital system.

REFERENCES

1. Lipshultz LI, Howards SS. Evaluation of the subfertile man. In: Lipschultz LI, Howards SS, eds. Infertility in the Male. New York: Churchill Livingstone, 1983:187–206.
2. WHO. Towards more objectivity in diagnosis and management of male infertility. Int J Androl 1997; 7(suppl):1–53.
3. de Kretser DM, Baker HWG. Infertility in men: recent advances and continuing controversies. J Clin Endocrinol Metab 1999; 84:3443–3450.
4. Wyrobeck AJ, Eskenazi B, Young S, et al. Advancing age has differential effects on DNA damage, chromatin integrity, gene mutations, and aneuploidies in sperm. Proc Natl Acad Sci U S A 2006; 103(25):9601–9606.
5. Wyrobeck AJ, Gordon LA, Burkhart JG, et al. An evaluation of human sperm as indicators of chemically induced alterations of spermatogenic function. Mutat Res 1983; 115:73–148.
6. Eskenazi B, Wyrobek AJ, Sloter E, et al. The association of age and semen quality in healthy men. Hum Reprod 2003; 18:447–454.
7. Carlson E, Giwercman A, Keiding N, et al. Evidence for decreasing quality of semen during the past 50 years. Br Med J 1992; 305:609–613.
8. Sharpe RM. Endocrinology and paracrinology of the testes. In: Lamb JC, Foster PMD, eds. Physiology and Toxicology of Male Reproduction. New York: Academic Press, 1988:177–184.
9. Sharpe RM, Fisher JS, Millar MM, et al. Gestational and lactational exposure of rats to xenoestrogens results in reduced testicular size and sperm production. Environ Health Perspect 1995; 103(12):1136–1143.
10. Safe S. Endocrine disruptors and human health: is there a problem. Toxicolo 2004; 205:3–10.
11. Brake A, Krause W. Decreasing quality of semen. Br Med J 1992; 305:1498.
12. Olsen GW, Bodmer KM, Ramlow JM, et al. Have sperm count been reduced 50% in 50 years? A statistical model re-visited. Fertil Steril 1995; 63:887–893.
13. Sikka SC, Wang R. Endocrine disruptors and estrogenic effects on male reproductive axis. Asian J Andrology 2008; 10:134–145.
14. Wyrobeck AJ, Gordon LA, Burkhart JG, et al. An evaluation of the mouse-sperm morphology test and other sperm tests in non-human animals. Mutat Res 1983; 115:1–72.
15. Mattison DR, Thomford PJ. The mechanisms of action of reproductive toxicants. Toxicol Pathol 1989; 17(2):364–376.
16. Sokol R. Lead exposure and its effects on the reproductive system. In: Golub M, ed. Metals, Fertility, and Reproductive Toxicity. Boca Raton, FL: Taylor and Francis; 2005:117–154.
17. de Queiroz EKR, Waissmann W. Occupational exposure and effects on the male reproductive system. Cadernos de Saúde Pública, Rio de Janeiro 2006; 22(3):485–493. Available at: http://www.scielosp.org/pdf/csp/v22n3/03.pdf.

18. Apostoli PL, Romeo E, Peroni A, et al. Steroid hormone sulphation in lead workers. Br J Ind Med 1989; 46:204–208.
19. McGregor AJ, Mason HJ. Chronic occupational lead exposure and testicular endocrine function. Human Exp Toxicol 1990; 9:371–376.
20. Alloway BJ. Heavy metals in soils. New York: John Wiley Inc, 1990.
21. Lancranjan I, Popescu HI, Gavănescu O, et al. Reproductive ability of workmen occupationally exposed to lead. Arch Environ Health 1975; 30(8):396–401.
22. DeMott RP, Borgert CJ. Reproductive toxicology. In: Williams PL, James RC, Roberts SM, eds. Principles of Toxicology. 2nd ed. New York: John Wiley & Sons, Inc., 2000:211–238.
23. Boscolo P, Sacchettoni-Logroscino G, Ranelletti FO, et al. Effects of long-term cadmium exposure on the testis of rabbits: ultrastructural study. Toxicol Lett 1985; 24:145–149.
24. Jequier, AM. Physical agents, drugs and toxins in the causation of male infertility. In: Jequier AM, ed. Male Infertility: A Guide for the Clinician. Edinburgh: Blackwell Science Ltd., 2000:314–331.
25. Waissmann W. Endocrinopatologia associada ao trabalho. In: Mendes R, organizador, eds. Patologia do trabalho. São Paulo: Editora Atheneu, 2003:1093–1138.
26. Goel, SK, Rao, GS, Pandya, PK. Hepatotoxic effects elicited by n-hexane or n-heptane. J Appl Toxicol 1988; 8(2):81–84.
27. Martino-Andrade JA, Chahoud I. Reproductive toxicity of phthalate esters. Mol Nutr Food Res 2010; 54:148–157.
28. Stein KE, Brown TM. Reproductive toxicology and teratology. In: Principles of Toxicology. 2nd ed. Boca Raton: Taylor & Francis Group, LLC, 2006:126.
29. Jeng HA, Liang Y. Alteration of sperm quality and hormone levels by polycyclic aromatic hydrocarbons on airborne particulate particles. J Environ Sci Health Part A 2008; 43:675–681.
30. Raychoudhury SS, Kubinski D. Polycyclic aromatic hydrocarbon-induced cytotoxicity in cultured rat sertoli cells involves differential apoptotic response. Environ Health Perspect 2003; 111(1):33–38.
31. Besaratinia A, Pfeifer GP. Genotoxicity of acrylamide and glycidamide. J Nat Cancer Inst 2004; 96(13):1023–1029.
32. Tengowski MW, Feng D, Sutovsky M, et al. Differential Expression of genes encoding constitutive and inducible 20S proteasomal core subunits in the testis and epididymis of theophylline- or 1,3-dinitrobenzene-exposed rats. Biol Reprod 2007; 76(1):149–163.
33. Muguruma M, Yamazaki M, Okamura M, et al. Molecular mechanism on the testicular toxicity of 1,3-dinitrobenzene in Sprague-Dawley rats: preliminary study. Arch Toxicol 2005; 79:729–736.
34. Charlton H. Hypothalamic control of anterior pituitary function: a history (review). J Neuroendocrinol 2008; 20(6):641–646.
35. Thomas MJ, Thomas JA. Toxic responses of the reproductive system. In: Klaassen CD, ed. Casarett & Doull's Toxicology: The Basic Science of Poisons. New York: MeCgraw-Hill, 2001:673–710.
36. Dungan HM, Clifton DK, Steiner RA. Minireview: kisspeptin neurons as central processors in the regulation of gonadotropin-releasing hormone secretion. Endocrinology 2006; 147(3):1154–1158.
37. Vadakkadath Meethal S, Liu T, Chan H, et al. Identification of a regulatory loop for the synthesis of neurosteroids: A StAR-dependent mechanism involving HPG axis receptors. J Neurochem 2009; 110:1014–1027.
38. Chen YG, Wang Q, Lin SL, et al. Activin signaling and its role in regulation of cell proliferation, apoptosis, and carcinogenesis. Exp Biol Med (Maywood) 2006; 231(5): 534–544.
39. Sulyok S, Wankell M, Alzheimer C, et al. Activin: an important regulator of wound repair, fibrosis, and neuroprotection. Mol Cell Endocrinol 2004; 225(1–2):127–132.
40. Skinner M, McLachlan R, Bremner W. Stimulation of Sertoli cell inhibin secretion by the testicular paracrine factor PModS. Mol Cell Endocrinol 1989; 66(2):239–249.
41. Cooper RL, Goldman JM, Rehnberg GL. Pituitary function following treatment with productive toxins. Environ Health Perspect 1986; 70:177–184.
42. Cooper RL, Goldman JM, Stoker TE. Neuroendocrine and reproductive effects of contemporary-use pesticides. Toxicol Industrial Health 1999; 15:26–36.

43. Khan IA, Thomas P. Disruption of neuroendocrine control of luteinizing hormone secretion by Aroclor 1254 involves inhibition of hypothalamic tryptophan hydroxylase activity. Biol Reprod 2001; 64:955–964.
44. Kapp, Jr. RW, Thomas JA. Toxicology of the Endocrine system. In: Ballantyne B, Marrs TC, Syversen T, eds. General and Applied Toxicology, third edition. John Wiley & Sons, Ltd., 2009: 1525–1559.
45. Bartsch W, Klein H, Schiemann U, et al. Enzymes of androgen formation and degradation in the human prostate. Ann N Y Acad Sci 1990; 595:53–66.
46. Salonen I, Pakarinen P, Huhtaniemi. I. Effect of chronic ethanol diet in expression of gonadotropin genes in the male rat. J Pharmacol Exp Ther 1992; 260:463–467.
47. Cooper RL, Goldman JM, Tyrey L. The hypothalamus and pituitary as targets for reproductive toxicants. In: Korach KS, ed. Reproductive and Developmental Toxicology. New York: Marcel Dekker, 1998:195–210.
48. Cunha GR, Alarid ET, Tumer T, et al. Normal and abnormal development of the male urogenital system: role of androgens, mesenchymal-epithelial interactions, and growth factors. J Androl 1992; 13(6):465–475.
49. Marks LS. 5α-Reductase: history and clinical importance. Rev Urol 2004; 6(suppl 9): S11–S21.
50. Fiorelli G, De Bellis A, Longo A, et al. Growth factors in the human prostate. J Steroid Biochem Mol Biol 1991; 40(1–3):199–205.
51. Kotake T, Usami M, Akaza H, et al. Goserelin acetate with or without antiandrogen or estrogen in the treatment of patients with advanced prostate cancer: a multicenter, randomized, controlled trial in Japan. Jpn J Clin Oncol 1999; 29(11):562–570.
52. Emanuele MA, Halloran MM, Uddin S, et al. Effects of alcohol on the neuroendocrine control of reproduction. In: Zakhari S, ed. Alcohol and the Endocrine System: NIAAA Research Monograph No. 23. Bethesda, MD: National Institute on Alcohol Abuse and Alcoholism, 1993:89–116.
53. Sokol RZ, Berman N, Okuda H, et al. Effects of lead exposure on GnRH and LH secretion in male rats: response to castration and p-tyrosine (AMPT) challenge. Reprod Toxicol 1998; 12(3):347–355.
54. Goyer RA. Lead toxicity: current concerns. Environ Health Perspect 1993; 100:177–187.
55. Lafuente A, Marquez N, Perez-Lorenzo M, et al. Cadmium effects on hypothalamus-pituitary-testicular axis in male rats. Exp Biol Med 2001; 226(6):605–611.
56. Vash PD, Engels TM, Kandeel FR, et al. Scrotal cooling increases rectal temperatures in man. Exp Biol Med 2002; 227(2):105–107.
57. Behre HM, Yeung CH, Holstein AF, et al. Diagnosis of male infertility and hypogonadism. In: Nieschlag SMA, Nieschlag E, Behre H, eds. Andrology: Male Reproductive Health and Dysfunction. 2nd ed. Heidelberg: Springer, 2000:89–124.
58. Clermont Y. The cycle of the seminiferous epithelium in man. Am J Anat 1963; 112:35–51.
59. Herbert, DC, Supakar PC, Roy AK. Male reproduction. In: Witorsch RJ, ed. Reproductive Toxicology. 2nd ed. New York: Raven Press, 1995:1–21.
60. Skinner MK, Norton JN, Mullaney BP, et al. Cell-cell interactions and the regulation of testis function. Ann NY Acad Sci 1991; 637:354–363.
61. Huleihel M, Lunenfeld E. Regulation of spermatogenesis by paracrine/autocrine factors. Asian J Androl 2004; 6(3):259–268.
62. Xiong X, Wang A, Liu G, et al. Effects of p,p'-dichlorodiphenyldichloroethylene on the expressions of transferrin and androgen-binding protein in rat Sertoli cells. Environ Res 2006; 101(3):334–339.
63. Rao JN, Liang JY, Chakraborti P, et al. Effect of thyroid hormone on the development and gene expression of hormone receptors in rat testes in vivo. J Endocrinolg Investig 2003; 26(5):435–443.
64. Creasy DM. Evaluation of testicular toxicity in safety evaluation studies: the appropriate use of spermatogenic staging. Toxicol Pathol 1997; 25(2):119–131.
65. Chapman RM. Gonadal injury resulting from chemotherapy. Am J Int Med 1983; 4:149–161.
66. Kapp RW. Detection of aneuploidy in human sperm. Environ Health Perspect 1979; 31:27–31.

67. Potashnik G, Abeliovich D. Chromosomal analysis and health status of children conceived to men during or following dibromochloropropane-induced spermatogenic suppression. Andrologia 1985; 17:291–296.
68. Shi Q, Ko E, Barclay L, et al. Cigarette smoking and aneuploidy in human sperm. Mol Reprod Dev 2001; 59:417–421.
69. Robbins WA, Vine MF, Truany KY, et al. Use of fluorescence in situ hybridization (FISH) to assess effects of smoking, caffeine, and alcohol and aneuploidy load in sperm of healthy men. Environ Mol Mutagens 1997; 30:175–183.
70. Dixon RL, Lee IP. Pharmacolokinetic and adaptation factors in testicular toxicity. Fed Proc 1980; 39:66–72.
71. Working PK. Germ cell genotoxicity: methods for assessment of DNA damage and mutagenesis. In: Working PK, ed. Toxicology of the Male and Female Reproductive Systems. New York: Hemisphere Publishing, 1989:231–255.
72. Thomas JA, Brogan WC. Some actions of lead on the sperm and on the male reproductive system. Am J Ind Med 1983; 4:127–134.
73. Griveau JF, Le Lannou D. Reactive oxygen species and human spermatozoa: physiology and pathology. Int J Androl 1997; 20:61–69.
74. Kodama H, Yamagichi R, Fukada J, et al. Increased oxidative deoxyribonucleic acid damage in the spermatozoa of infertile male patients. Fertil Steril 1997; 68: 519–524.
75. Ong C-N, Shen H-M, Chia S-E. Biomarkers for male reproductive health hazards: are they available? Toxicol Lett 2002; 134:17–30.
76. Mazzilli F, Rossi T, Marchesini M, et al. Superoxide anion in human semen related to seminal parameters and clinical aspects. Fertil Steril 1994; 62:862–868.
77. Sikka SC. Role of oxidative stress and antioxidants in andrology and assisted reproductive technology. J Androl 2004; 25(1):5–18.
78. Mruk DD, Cheng CY. Sertoli cell proteins in testicular paracriny. In: Jégou B, Pineau C, Saez J, eds. Testis, Epididymis and Technologies in the Year 2000. Berlin: Springer-Verlag, 2000:197–228.
79. Hall J. Unit XIV Endocrine and reproduction. In: Hall J, ed. Guyton and Hall Physiology Review. Philadelphia: Saunders, 2006:219–246.
80. de Franca LR, Ghosh S, Ye SJ, et al. Surface and surface-to-volume relationships of the Sertoli cell during the cycle of the seminiferous epithelium in the rat. Biol Reprod 1993; 49:1215–1228.
81. Rodriguez I, Ody C, Araki K, et al. An early and massive wave of germinal cell apoptosis is required for the development of functional spermatogenesis. EMBO J 1997; 16:2262–2270.
82. Bartke A. Apoptosis of male germ cells, a generalized or a cell type–specific phenomenon? Endocrinology 1995; 136:3–4.
83. Boekelheide K, Fleming SL, Johnson KJ, et al. Role of sertoli cells in injury-associated testicular germ cell apoptosis. Proc Soc Exp Biol Med 2000; 225(2):105–115.
84. Chapin RE. Morphologic evaluation of seminiferous epithelium of the testis. In: Lamb JC, Foster PMD, eds. Physiology and Toxicology of Male Reproduction. San Diego: Academic Press, 1988:155–177.
85. Foster PMD. The sertoli cell. In: Sicalli AR, Clegg ED, eds. Reversibility in Testicular Toxicity Assessment. Boca Raton: CRC Press, 1992:57–86.
86. Hild SA, Reel JR, Larner JM, et al. Disruption of spermatogenesis and sertoli cell structure and function by the indenoppyridine CBD-4022 in rats. Biol Reprod 2001; 65:1771–1779.
87. Winnall WR, Muir JA, Hedger MP. Differential responses of epithelial Sertoli cells of the rat testis to Toll-Like receptor 2 and 4 ligands: implications for studies of testicular inflammation using bacterial lipopolysaccharides. Innate Immunity 2009, doi:10.1177/1753425909354764; (First published online December 18, 2009).
88. Sundaram K, Witorsch RJ. Toxic effects on the testes. In: Witorsch RJ, ed. Reproductive Toxicology. 2nd ed. New York: Raven Press, 1995:99–121.
89. Boekelheide K, Fleming SL, Allio T, et al. 2,5-hexanedione-induced testicular injury. Annu Rev Pharmacol Toxicol 2003; 43:125–147.
90. Boekelheide K, Johnson KJ, Richburg JH. Sertoli Cell Biology. San Diego: Elsevier Academic Press, 2005.

91. Chapin RE, Phelps JL, Somkuti SG, et al. The interaction of Sertoli and Leydig cells in the testicular toxicity of tri-o-cresyl phosphate. Toxicol Appl Pharmacol 1990; 104 (3):483–495.
92. Lee WS, Yoon H-J, Oh JH, et al. 1,3-Dinitrobenzene induces apoptosis in TM4 mouse Sertoli cells: involvement of the c-Jun N-terminal kinase (JNK) MAPK pathway. Toxicol Lett 2009; 189:145–151.
93. Ogawa T, Kita K, Kubota Y. Proliferation of spermatogonial stem cells and spermatogenesis in vitro. Reprod Med Biol 2006; 5:169–174.
94. Li MWM, Mruk DD, Lee WM, et al. Disruption of the blood-testis barrier integrity by bisphenol A in vitro: is this a suitable model for studying blood-testis barrier dynamics? Int J Biochem Cell Biol 2009; 41:2302–2314.
95. Xia W, Mruk DD, Lee WM, et al. Differential Interactions between Transforming Growth Factor-β3/TβR1, TAB1, and CD2AP Disrupt Blood-Testis Barrier and Sertoli-Germ Cell Adhesion. J Appl Biol Chem 2006; 281:16799–16813.
96. Yan HHN, Mruk DD, Lee WM, et al. Blood-testis barrier dynamics are regulated by testosterone and cytokines via their differential effects on the kinetics of protein endocytosis and recycling in Sertoli cells. FASEB J 2008; 22:1945–1959.
97. Lui WL, Mruk D, Lee WM, et al. Sertoli cell tight junction dynamics: their regulation during spermatogenesis. Biol Reprod 2003; 68(4):1087–1097.
98. Schuppe H-C, Meinhardt A. Immune privilege and inflammation of the testis. Chem Immunol Allergy 2005; 88:1–14.
99. Lui WY, Mruk D, Lee WM, et al. Sertoli cell tight junction dynamics: their regulation during spermatogenesis. Biol Reprod 2003; 68:1087–1097.
100. Huang HF, Yang CS, Meyerhofer M, et al. Disruption of sustentacular (Sertoli) cell tight junctions and regression of spermatogenesis of vitamin A-deficient rats. Acta Anat (Basel) 1988; 133:10–15.
101. Chubb C. Reversibility of damage to testicular vasculature resulting from exposure to toxic agents. In: Scialli AR, Clegg ED, eds. Reversibility in Testicular Toxicity Assessment. Boca Raton: CRC Press, 1992:127–158.
102. Al-Agha O, Axiotis C. An in-depth look at Leydig cell tumor of the testis. Arch Pathol Lab Med 2007; 131(2):311–317.
103. Ewing LL, Keeney DS. Leydig cells: structure and function. In: Desjardins C, Ewing LL, eds. Cell and Molecular Biology of the Testis. New York: Oxford University Press, 1993:137–167.
104. Mooradian AS, Morley JE, Korenman SG. Biological actions of androgens. Endocr Rev 1987; 8:1–27.
105. Clark BJ, Wells J, King SR, et al. The purification, cloning and expression of a novel luteinizing hormone-induce mitochondrial protein MA-10 mouse Leydig tumor cells. Characterization of the steroidogenic acute regulatory protein StAR. J Biol Chem 1994; 269(28):314–328.
106. Teerds K, Rijntjes E. Dynamics of Leydig cell regeneration after EDS. In: Payne AH, Hardy MP, eds. The Leydig Cell in Health and Disease. Totowa, New Jersey: Humana Press, 2007:91–116.
107. King SR, Rommerts FFG, Ford SL, et al. Ethane dimethyl sulfonate and NNN'N'-tetrakis-(2-pyridylmethyl) ethylenediamine inhibits steroidogenic acute regulatory (StAR) protein expression in MA-10 Leydig cells and rat Sertoli cells. Endocr Res 1998; 24:469–478.
108. Peterson RE, Theobald HM, Kimmel GL. Developmental and reproductive toxicity of dioxins and related compounds: cross-species comparisons. Crit Rev Toxicol 1993; 23(3): 283–335.
109. Mandal PK, McDaniel LR, Prough RA, et al. 7,12-Dimethylbenz[a]anthracene inhibition of steroid production in MA-10 mouse Leydig tumor cells is not directly linked to induction of CYP1B1. Toxicol Appl Pharmacol 2001; 175:200–208.
110. Akingbemi BT, Youker RT, Sottas CM, et al. Modulation of rat Leydig cell steroidogenic function by di(2-ethylhexyl)phthalate. Biol Reprod 2001; 65(4):1252–1259.
111. Jang MH, Shin HS, Kim KH, et al. Alcohol induces apoptosis in TM3 mouse Leydig cells via bax-dependent caspase-3 activation. Eur J Pharmacol 2002; 449(1–2):9–45.
112. Braude MC, Ludford JP. Marijuana effects on the endocrine and reproductive systems. NIDA Research Monograph Series, DHHS, Rockville, MD 20857, 1984.

113. Witorsch RJ, Hubbard JR, Kalimi M. Reproductive toxic effects of alcohol, tobacco, and substances of abuse. In: Witorsch RJ, ed. Reproductive Toxicology 2nd ed. New York: Raven Press, 1995:283–318.

114. Artifeksov SB, Artyukhin AA. Blood flow in the testes and Epididymides of male rats with experimental blockade of testicular veins. Bull Exp Biol Med 2007; 143(6):682–685.

115. Bergh A, Damber J-E. Vascular controls in testicular physiology. In: deKretser DM, ed. Molecular Biology of the Male Reproductive System. New York: Academic Press Inc. 1993:439–468.

116. Holash JA, Harik SI, Perry G, et al. Barrier properties of testis micro vessels. Proc Natl Acad Sci U S A 1993; 90:11069–11073.

117. Collin O, Bergh A. Leydig cells secrete factors which increase vascular permeability and endothelial cell proliferation. Int J Androl 1996; 19:221–228.

118. Rudolfsson SH, Wikström P, Jonsson A, et al. Hormonal regulation and functional role of vascular endothelial growth factor A in the rat testis. Biol Reprod 2004; 70(2):340–347.

119. Collin O, Bergh A, Damber J-E, et al. Control of testicular vasomotion by testosterone and tubular factors. J. Reprod Fertil 1993; 97:115–121.

120. Sharpe RM. Intratesticular factors controlling testicular function. Biol Reprod 1994; 30:29–49.

121. Pozor MA. Evaluation of testicular vasculature in stallions. Clin Tech Equine Pract 2007; 6(4):271–277.

122. Hew K-K, Heath GL, Jiwa AH, et al. Cadmium in vivo causes disruption of tight junction-associated microfilaments in rat Sertoli cells. Biol Reprod 1993; 49:840–849.

123. Hew K-K, Ericson WA, Jiwa AH, et al. A single low cadmium dose causes disruption of spermation in the rat. Toxicol Appl Pharmacol 1993; 121:15–21.

124. Prozialeck WC, Edwards JR, Nebert DW, et al. Review: the vascular system as a target of metal toxicity. Toxicol Sci 2008; 102(2):207–218.

125. Jones R. To store or mature spermatozoa? The primary role of the epididymis. Int J Androl 1999; 22(2):57–67.

126. Chudnovsky A, Niederberger CS. Copious pre-ejaculation. small glands major head-aches. J Androl 2007; 28(3):374–375.

127. Chughtai B, Sawas A, O'Malley RL, et al. A neglected gland: a review of Cowper's gland. Int J Androl 2005; 28(2):74–77.

128. Liao S, Kokontis J, Sai T, et al. Androgen receptors: structures, mutations, antibodies and cellular dynamics. J Steroid Biochem 1989; 34:1–6.

129. King RJB. Effects of steroid hormones and related compounds on gene transcription. Clin Endocinol 1992; 36:1–14.

130. Leger JG, LeGuellec R, Tenniswood MPR. Treatment with antiandrogens induces an androgen-repressed gene in the rat ventral prostate. Prostate 1988; 13:131–142.

131. Wilson MJ, Whitaker JN, Sinha AA. Immunocytochemical localization of cathepsin D in rat ventral prostate; evidence for castration-induced expression of cathepsin D in basal cells. Anat Rec 1991; 229:321–233.

132. Rao MV. Effects of methylmercury on mouse epididymis and spermatozoa. Biomed Biochim Acta 1989; 8:577–582.

133. Baratt CLR, Davies AG, Bansal MR, et al. The effects of lead on the male reproductive system. Andrologia 1989; 21:161–166.

134. Amann RP, Berndtson WE. Assessment of procedures for screening agents for effects on male reproduction: effects of dibromochloropropane (DBCP) on the rat. Fund Appl Toxicol 1986; 7:244–255.

135. Linder RE, Strader LF, Barbee RR, et al. Reproductive toxicity of a single dose of 1,3-dinitrobenzene in two ages of young male rats. Fund Appl Toxicol 1990; 14:284–298.

136. Klinefelter GR, Laskey JW, Robert NL, et al. Multiple effects of ethane dimethane-sulphonate on the epididymis of adult rats. Toxicol Appl Pharmacol 1990; 105:271–287.

137. Dodd DE, Fowler EH, Snellings WM, et al. Reproduction and fertility evaluations in CD rats following nitrobenzene inhalation. Fundam Appl Toxicol 1987; 8:493–505.

138. Chapin RE, George JD, Lamb JC IV. Reproductive toxicity of tricresyl phosphate in a continuous breeding protocol in Swiss (CDD-1) mice. Fund Appl Toxicol 1989; 10:344–354.

139. Gray LE Jr., Ostby J, Linder R, et al. Carbendazim-induced alterations of reproductive development and function in the rat and hamster. Fund Appl Toxicol 1990; 15:281–297.

140. Zanefield LJD, Waller P. Non-hormonal mediation of male reproductive tract damage: data from contraceptive drug research. In: Burger EJ, Tardiff RG, Scialli AR, et al., eds. Sperm Measures and Reproductive Success, Progress in Clinical and Biological Research. Vol. 302. Liss, New York: Alan R, 1989:129–149.
141. Fail PA, George JD, Seely JC, et al. Reproductive toxicity of boric acid in Swiss (CD-1) mice: assessment using the continuous breeding protocol. Fund Appl Toxicol 1991; 17:225–239.
142. Robaire B, Hales BF. Post-testicular mechanisms of male-mediated developmental toxicity. In: Olshan AF, Mattison DR, eds. Male Mediated Developmental Toxicity. New York: Plenum Press, 1994:93–104.
143. Gellert RJ, Wallace CA, Weismeier EM, et al. Topical exposure of neonates to hexachlorophene: long standing effects on mating behavior and prostatic development in rats. Toxicol Appl Pharmacol 1978; 43:339–349.
144. Chatterjee S, Roend K, Banerji TK. Morphological changes in some endocrine organs in rats following chronic lithium treatment. Anat Anz Jena 1990; 170:31–37.
145. Seethalakshmi L, Menon M, Malhotra RK, et al. Effects of cyclosporine A on male reproduction in rats. J Urol 1987; 138:991–995.
146. Boekelheide K, Darney SK, Daston GP, et al. NTP-CERHR expert panel report on the reproductive and developmental toxicity of 2-bromopropane. Reproductive Toxicology 2004; 18(2):189–217.
147. Urata K, Narahara H, Tanaka Y, et al. Effect of endotoxin-induced reactive oxygen species on sperm motility. Fertil Steril 2001; 76:163–166.
148. Aitken RJ, Krausz C. Oxidative stress, DNA damage and the Y chromosome. Reproduction 2001; 122:497–506.
149. Hinton BT, Palladino MA, Rudolph D, et al. The epididymis as protector of maturing spermatozoa. Reprod Fertil Dev 1995; 7:731–745.
150. Chen H, Cheung MP, Chow PH, et al. Protection of sperm DNA against oxidative stress in vivo by accessory sex gland secretions in male hamsters. Reproduction 2002; 124:491–499.
151. Amicarelli F, Ilio CD, Masciocco L, et al. Aging and detoxifying enzyme responses to hypoxic or hypertoxic treatment. Mech Aging Dev 1997; 97:215–226.
152. Zubkova, EV, Robaire B. Effect of glutathione depletion on antioxidant enzymes in the epididymis, seminal vesicles, and liver and on spermatozoa motility in the aging brown Norway rat. Biol Reprod 2004; 71(3):1002–1008.

15 Female reproductive toxicity

Ralph L. Cooper and Jerome M. Goldman

INTRODUCTION

The reproductive system of the female mammal is comprised of the brain-pituitary-ovarian axis and associated reproductive organs including the oviducts, uterus, cervix, vagina, and mammary glands. Perturbations of normal development and/or adult regulation of the complex processes controlling successful reproduction are major clinical problems. In this regard, there is growing concern that environmental chemicals may be contributing to reproductive failure and/or disease. Studies of wildlife populations offer compelling evidence that environmental contaminants can affect reproduction in a number of species (1). In addition, laboratory studies, as well as a limited number of occupational exposure studies, suggest that some naturally occurring (e.g., phytoestrogens) and man-made (xenobiotics) chemicals can interact with key molecular regulatory processes governing successful reproduction. Such data have led to speculation that impaired fertility or reproductive cancers are the result of chemical exposure to the developing or adult female.

Identifying the effects of environmental chemicals on reproductive function in the female requires both an understanding of the basic biology of the female and recognition that an assessment of toxicity must include an evaluation of the many susceptible periods present in the female. For example, exposure during sensitive periods of gestation can modify peripheral reproductive tract formation (2), or permanently alter central (brain and pituitary) control of ovarian development and function in adulthood (3). Likewise, in the adult female, brief exposure to toxicants during certain phases of the ovarian cycle can have serious adverse effects on the production and viability of the offspring (4). The course of pregnancy also involves numerous stages, all of which are susceptible to disruption to environmental chemicals (5). Finally, reproductive aging is a process that can be influenced by the culmination of environmental insults that occur during any or all of the preceding life stages.

In this chapter, we briefly discuss the test guidelines assessing potential reproductive and developmental toxicity required for the registration of food use pesticides and pharmaceuticals. This background is intended to familiarize the reader with many of the current issues and approaches germane to toxicity assessment in the female. Studies are presented that demonstrate the vulnerability of the developing animal to test chemicals, emphasizing both alterations in peripheral reproductive tract development and the central (hypothalamic) control of reproductive function. Finally, we discuss how environmental toxicants are able to affect the adult female.

REPRODUCTIVE TEST GUIDELINES

The process of reproductive and developmental toxicity testing for regulatory purposes is governed by a framework of guideline studies depicted in Table 1 (6–12). These whole animal bioassays are a critical part of human health risk assessment. In this regard, all of them reflect the combined efforts of research and regulatory experts to design protocols that are now available and in use. Furthermore, these guideline studies have been developed and issued by international organizations. For example, the Organisation for Economic Cooperation and Development (OECD) has developed internationally agreed on test guidelines for the reproductive and developmental toxicity testing of pesticides and industrial chemicals.

TABLE 1 Guideline Studies Required for Chemical Registration

OECD guideline number	US EPA guideline number	Other	Description
OECD 414 (7)	EPA 870-3700 (9)		Prenatal developmental toxicity study
OECD 415 (13)			One-generation reproduction toxicity study
OECD 416 (8)	EPA 870-3800 (10)		Two-generation reproduction toxicity study
OECD 421 (14)	EPA 870-3550 (15)		Reproduction/developmental toxicity screening test
OECD 422 (16)	EPA 870-3650 (17)		Combined repeated dose toxicity study with the reproduction/developmental toxicity screening test
OECD 426 (18)	EPA 870-6300 (19)		Developmental neurotoxicity study
		US FDA (11,12)	Multigeneration reproductive toxicity; developmental toxicity
		ICH (6)	Multigeneration reproductive toxicity

The purpose of the animal test protocols is to identify the reproductive and developmental effects of chemicals, including drugs, food additives, pesticides, and other agents to which humans are environmentally or occupationally exposed. These methods range from short-term exposure paradigms that examine specific developmental stages at risk to various long-term exposure scenarios that assess a chemical's potential effects over a broader portion of the animal's reproductive life span.

The use of such studies by regulatory agencies was formalized and introduced into the risk assessment process approximately 50 years ago in an effort to set acceptable exposure levels or tolerance levels. A basic premise is that these studies are predictive of potential adverse human health impacts, and that they provide the necessary information to identify potentially sensitive target organ systems, maternal toxicity, embryonic and fetal lethality, specific types of malformations, and altered birth weight and growth retardation. Using single generation or multigenerational designs, these assays also provide information on reproductive effects and pre- and postnatal functions by singling out critical endpoints that serve to determine a no-observed-adverse effect level (NOAEL) for the most sensitive measures, thus providing the basis for quantitative assessments. Some general considerations of the current reproductive and developmental toxicity testing merit discussion in view of the evaluation of female reproductive toxicity.

In assessing the potential reproductive and developmental toxicity of a compound, all steps in the reproductive process should be evaluated because, as noted earlier, a number of sensitive periods exist that may render either the developing or adult organism more susceptible to insult. In this regard, it is no accident that the guideline tests are designed to examine exposure during specific stages, such as preconception, embryonic and fetal, newborn, lactational (preweaning), and perijuvenile periods, as well as adulthood. Although the existing guidelines summarized in Table 1 do encompass all early life stages, there are still some major limitations. For example, the multigenerational protocols expose the F1 offspring to the test chemical during the juvenile period. However, a more detailed examination of this life stage is sometimes desirable, especially for pharmaceutical testing. Recognizing that the juvenile animal may have fundamental differences from adults with regard to susceptibility or sensitivity, the U.S. Food and Drug Administration (US FDA) has published Guidance on Nonclinical Safety Evaluation of Pediatric Drug Products (20). Similarly, the U.S. Environmental Protection Agency (US EPA) Endocrine Disruptors Screening and Testing Program has developed a male and female pubertal assay designed to evaluate more thoroughly the impact of environmental chemicals at this stage of reproductive development (21,22).

In addition to the above noted concerns for the evaluation of the perijuvenile animal, the current testing requirements, as designed, do not provide a detailed assessment of the effect of perinatal or continued exposure on reproductive aging in the male and female (i.e., altered reproductive senescence, altered reproductive organ diseases). This is because the oldest animal examined in the studies depicted in Table 1 is usually less than four months of age. At this point, the true impact of this limitation remains to be determined. Whereas several reproductive tissues are examined in the aged animals employed in the two-year carcinogenicity study (23), these protocols may miss effects on the aging process itself (i.e., changes that may contribute to the development of diseased tissue). At the same time, as we learn more from the basic literature about how environmental chemicals may impact a decline in reproductive activity with age, it is likely that mechanistic information obtained from the current protocols could trigger more focused examinations of the process of reproductive aging. In this regard, it is interesting to note that the test guidelines governing the evaluation of environmental agents differ from those evaluating reproductive/developmental testing for medicinal products in that "latent" effects (i.e., those adverse outcomes that may occur in adulthood as a consequence of exposure that is restricted to the gestational and/or lactational period) are examined only for medicinal products.

Another aspect of reproductive and developmental bioassays that has received considerable attention involves the inclusion of measurements of absorption, distribution, metabolism, and excretion (ADME). These parameters are an important source of information on the mechanism of action, as well as the risk, of a given compound. In the context of developmental studies, knowledge of whether a compound crosses the placenta and reaches the embryo/fetus, or the chemical is transferred in the milk to the offspring during lactation, is of crucial importance for the planning and interpretation of the data. In addition, for studies that employ dietary exposures, an understanding of how the internal dose may change as the dam's food intake is increased through gestation, and lactation is also critical to the interpretation of any adverse outcomes and can provide justification for modifying the dose during these periods. Such ADME measurements are proving to be a key component of multigenerational designs such as that proposed by the ILSI Health and Environmental Sciences Institute's (HESI) Agricultural Chemical Safety Assessment (ACSA) extended one-generation study (24).

It must also be recognized that the same agent can adversely affect different endpoints, depending on species, genetic background, sex, age, physiological state, route, reproductive stage, and duration of exposure, in addition to the dose absorbed and amount that reaches to target organ. For this reason, the developmental protocols can require testing in more than one species, although the rationale for the selection of the optimal species is not always clear.

Finally, the relative importance of any reproductive/developmental effect must be assessed within the context of the full toxicity profile of the agent. Reproductive effects occurring at dose levels greater than those that induce clear toxicity in non-reproductive parameters must be interpreted appropriately.

THE DEVELOPING FEMALE
Reproductive Tract

The primary function of the female reproductive tract is to produce viable gametes (oocytes) for fertilization to provide the hormonal support for the implantation and to assure the appropriate environment for intrauterine fetal development. This process begins in the fetus with the migration of the germ cells to the genital ridge and the formation of primary follicles. The formation of the female reproductive tract from the Müllerian (paramesonephric) ducts into the oviducts, uterine horns, cervix, and the anterior vagina is discussed in the chapter by Goldman and Cooper (25). During development, agenesis of the female reproductive tract has been estimated to occur in one in 4000 to 20,000 women per year (26). In addition, some diseases and malformations present in the adult appear to be the result of abnormal gene expression in the embryo. For example, Müllerian agenesis and the presence of a transverse or longitudinal vaginal septum (an abnormal partition of the vagina) are of fetal origin and prevent normal reproduction (27). Others have argued that disruption of embryonic gene expression is observed in some types of cancers in several organs (28–30). These late onset anomalies present a challenge to our understanding of the molecular mechanisms that regulate female reproductive tract development and the extent to which any disorders may result from environmental toxicants and thus remain largely unknown (31).

Follicular Toxicity

The follicle is the functional unit of the ovary that protects and nourishes the oocyte from the time the follicle is formed until of ovulation or natural atresia. During fetal development, primordial germ cells (oogonia) proliferate and become surrounded by granulosa cells forming primordial follicles. At this point, oogonia develop into oocytes and enter prophase of the first meiotic division, after which they become arrested until shortly before ovulation. Many pharmaceuticals and some environmental agents have been shown to be toxic to follicular formation or development. As the primordial follicles are the stock from which all growing follicles and ovulated oocytes are derived, a xenobiotic that interferes with their normal development will have obvious consequences on the reproductive ability of the female.

To mature, follicles need to survive recruitment and selection, during which they develop gonadotropin and steroid receptors to ensure their appropriate response to the cyclic fluctuation in hormones in adulthood. Only a very small fraction of primordial follicles that enter the growing phase will successfully undergo full maturation and ovulate. The remainder will become atretic at some stage (32,33). The factors control the size of the initial pool, follicular growth and, ultimately, the selection of those follicles destined to ovulate remains an important area of research (34–36).

Importantly, the primordial follicles present at birth represent the entire cohort of germ cells that the female will ever possess, and they serve as the reserve for her entire reproductive lifespan (37).[a] As such, the number of primordial follicles sets the limits on the fertility of the female. An early or premature depletion of primordial follicles can cause frank infertility or early premature reproductive senescence. For example, in mice with germ cell deficiencies, such as seen in genetic disorders in which deletions of the *atm*, *Sl*, or *W* genes have been observed, the females are sterile with poorly developed ovaries that are incapable of forming follicles (39,40). In humans, 45,XO Turner's syndrome is associated with a high rate of oocyte attrition during fetal life, leading to ovarian dysgenesis and sterility (41). In mice, as in humans, X chromosome monosomy induces a severe reduction in the number of oocytes. One observation from these studies and others using γ-irradiation to deplete the stock of primordial follicle in the developing ovary of the mouse demonstrates that these mice are initially fertile. In fact, Magre and colleagues (42–44) have shown that even though γ-irradiation exposure during fetal life (GD15) led to an almost total depletion of primordial follicles, the primary and secondary follicles assume the role of providing the follicular reserve,

---

[a]Although the a substantial number of studies support the concept that the female is born with all the follicles she will ever have, this concept has been challenged by Tilly and his colleagues (38). Whether the female is born with all the follicles she will ever have or the adult ovary is capable of some oogenesis does not modify the observations discussed in this section.

permitting the normal progression through puberty and a transient normal fertility. This observation emphasizes the need to include an evaluation of follicle number when examining toxicity and not relying exclusively on the more apical measures.

A number of chemicals have been shown to be ovotoxic because they extensively destroy primordial and primary follicles (see Ref. 38 for review). These include chemotherapeutic agents (e.g., cyclophosphamide), polycyclic aromatic hydrocarbons, (e.g., 3-methylcholanthrene and 9:10-dimethyl-1:2-benzanthracene), occupational chemicals 1,3-butadiene (BD), and related compounds such as 4-vinylcyclohexene (VCH) (45). In addition, a variety of other chemicals, fungal toxins, and antibiotics are capable of producing significant primordial follicle loss in mice. These compounds include methyl and ethyl methanesulfonate, busulfan, urethane, procarbazine HCl, 4-nitro-quinoline-1-oxide, dibromochloropropane, urethane, N-ethyl-N-nitrosourea, and bleomycin. The toxicity of these chemicals can be demonstrated at all life stages, but some have been shown to be ovotoxic (i.e., benzo[a]pyrene (46) as a result of gestational exposure (gestation days 7–16 to the female mouse offspring).

Hormones and Follicular Development

The role that estrogens and androgens play in follicular development has been studied extensively, as has the adult consequences of inappropriate steroid hormone signaling in the fetal ovary. In brief, although androgens may play important roles in the adult ovary by supporting ovarian follicle and oocyte maturation (47–49), testosterone receptor stimulation per se (and presumably androgen-dependent gene expression) is apparently not required during fetal life for normal ovarian development (50). This conclusion is supported by the observation that ovarian function shows little impairment following gestational exposure to AR receptor antagonist (e.g., flutamide) (51) and in the androgen receptor knockout (ARKO) mouse. In contrast, since testosterone is aromatized to estrogen in the ovary, a deficiency in female androgen biosynthesis induced by a congenital defect or experimentally induced knockouts of key steroidogenic enzymes [e.g., mitochondrial steroidogenic acute regulatory protein (StAR) (52) or 17α-hydroxylase/17,20 lyase (P450cyp17) (47,53)] produce obvious defects in ovarian development, deficiencies that persist into postnatal life.

As testosterone provides the substrate for ovarian estrogen, it is predictable that impaired testosterone biosynthesis (and thus decreased estradiol) will disrupt folliculogenesis in the developing animal. Many studies have demonstrated an important role for fetal estrogen in the normal development and function of the ovary. Estrogen is essential for granulosa cell development and follicle formation. Reducing serum estrogen in pregnant baboons with the aromatase inhibitor Letrozole during mid and late gestation significantly reduced the number of primordial follicles present in fetal ovaries at late gestation (54), and drastically decreased the numbers of oocyte-granulosa cell microvilli (55) that are essential for oocyte nutrition. Maintaining estrogen levels in the Letrozole-exposed pregnant mothers by a cotreatment with estradiol prevented this late gestation disruption of follicular development (54,55).

The use of knockout animals has also provided evidence for the role of estrogens in the developing ovary and whether these fetal ovarian deficiencies translate into abnormal adult ovarian function. Estrogen receptors α (ERα) and β (Erβ) are distributed differently in the ovary. ERα expression is found in theca cells and ovarian stroma, while the ERβ is located in granulosa cells of growing follicles (56,57). The phenotypes that emerge from the two different ER knockouts are dramatically different. The αERKO female exhibits chronic anovulation, cystic and hemorrhagic follicles, absent corpora lutea, interstitial/stromal hyperplasia, and elevated plasma estradiol and testosterone levels in the presence of an excess of luteinizing hormone (LH) (58). Importantly, these effects can be reversed by treatment with gonadotropin-releasing hormone (GnRH) analogues that "normalize" LH serum concentrations. Thus, with the disruption of estrogenic stimulation, many of the ovarian abnormalities appear secondary to the loss of ERα-mediated negative feedback regulation of LH at the hypothalamus-pituitary level (59,60). In contrast, the βER knockout mouse has somewhat reduced fertility, but it does ovulate, although the numbers of corpora lutea are reduced (61). The authors of these studies conclude that the outcomes reflect impairments in estrogen action within the ovary related to follicle maturation and differentiation (62,63). In aromatase knockout (ArKO) mice devoid of estrogen synthesis, the numbers of primordial follicles are substantially reduced, apparently due to their lack of formation from fetal germ cell nests and precocious activation of follicle growth from the primordial pool (64,65). In the complete absence of estrogen, ArKO females are anovulatory, folliculogenesis is arrested at the antral stage, and the ovaries manifest hemorrhagic cysts that develop as a result of increased gonadotropin secretion (66).

The absence of estrogenic stimulation during development results in the appearance of structures resembling testicular seminiferous tubules in the ovaries of ERα/ERβ double knockout (αβERKO) mice, as well as the presence of genes that are normally only expressed in the testes (e.g., sulfated glycoprotein-2 and *Sox9*, and an additional gene involved in androgen biosynthesis, *17β-Hsd-3*) (59,67). There is even more dramatic masculinization of the ovaries of ArKO mice. After puberty, the ArKO ovaries possess Sertoli and Leydig cells, express *Sox9*, and overexpress genes coding for androgen biosynthesis (59). Since both αβERKO and ArKO mice possess female and not male reproductive tracts with follicles that contain oocytes and no male-like germ cells, estrogen appears to play a key role in maintaining an adult ovarian somatic cell phenotype (59). In humans, a congenital aromatase deficiency leads to hyperandrogenic, multicystic ovarian phenotypes. LH and FSH levels are elevated but can be normalized by exogenous estrogen treatment (47).

SEXUAL DIFFERENTIATION OF THE BRAIN
Gender Specific Nuclei

There are sexually dimorphic differences in many brain structures. These differences appear at the cellular level as disparities in dendritic spine density (68), astrocyte complexity (69), or neurite branching (70). At the structural level, these differences are obvious in the size of the nuclei. For example, the sexually dimorphic nucleus of the preoptic area (SDN-POA) is larger in the male, as is the bed nucleus of the stria terminalis (3). In contrast, the anteroventral periventricular (AVPV) nucleus is larger in the female (71). Extrahypothalamic structures have been reported to be gender unique, including the amygdala, corpus callosum, and cerebellum (72). The fetal origins of these differences are primarily the result of the brain being masculinized or defeminized by the surge of testosterone from the fetal testis. Testosterone is aromatized to estradiol, as this enzyme is located in significant concentrations in the hypothalamic nuclei, and it is primarily this steroid that induces many of the permanent effects on the developing neuroarchitecture. This sensitive period of brain organization in the rat begins on embryonic day (ED) 18 and extends to postnatal day (PN) 10. The sensitive period is operationally defined by the onset of testicular androgen synthesis in males at ED 18 and the loss of sensitivity to exogenous androgens in females by around PN10 (see for review Refs. 73–75). Once differentiated, these structures serve to maintain the phenotypical hormonal and behavioral actions in the male and female.

The biological processes responsible for the creation of the sexually dimorphic nuclei include the creation of new neurons (neurogenesis), the movement of neurons or their processes from one region to another, and apoptosis. Estradiol can alter any of these processes and subsequently affect several sexually distinct structures throughout the brain. The structure most commonly studied is in the hypothalamic region. For example, the preoptic region of rats has two sexually dimorphic nuclei. As mentioned earlier, the sexually SDN-POA has more neurons and is larger in males than in females, whereas the AVPV has a higher cell density and is larger in females than in males. Sex differences in apoptosis during development contribute significantly to the differences in cell number in these nuclei in the two sexes (71). An inverse correlation exists between sex differences in apoptotic cell number and the number of living cells in the mature period. The SDN-POA of postnatal male rats exhibits a higher genetic expression of anti-apoptotic Bcl-2 and lower expression of pro-apoptotic Bax compared to that in females. Potentially, apoptotic cell death via caspase-3 activation occurs more frequently in the SDN-POA of females. The patterns of expression of Bcl-2 and Bax in the SDN-POA of postnatal female rats treated with estrogen are altered to that typically seen in the male. In the AVPV of postnatal rats, apoptotic regulation also differs between the sexes, although Bcl-2 expression is increased and Bax expression and caspase-3 activity are decreased in females (71).

In addition to regulating the programmed death of the neurons innervating these different regions, estradiol also alters the postsynaptic morphology by modifying the number of dendritic spines in the different sexually dimorphic regions. Furthermore, these cytoarchitectural changes result from different molecular events induced by estradiol in the developing male, events that are clearly different in the various nuclei. For example, McCarthy and her colleagues (3) have shown that the arcuate nucleus in the male has half as many dendritic spines as in the female. This difference is entirely dependent on elevated estradiol in the male brain during the perinatal period. At the same time, the astrocytes in the male arcuate vary in the complexity of their cytoarchitecture, with the male astroglia having more frequently branched processes than the female (76). This difference is regulated in the male by testosterone aromatized to estradiol binding to the nuclear estrogen receptor in the arcuate neurons, stimulating the synthesis of γ-aminobutyric acid (GABA) by increasing the activity of glutamate decarboxylase (GAD). In turn, GABA acts on the neighboring astrocytes to induce stellation (increased processes). It is speculated that the increased complexity of the astrocytes in males suppresses the formation of postsynaptic dendritic spine synapses creating fewer such processes than are present in the female (3).

In contrast to the molecular steps that lead to sex differences in the number of postsynaptic dendrites in the arcuate, the molecular steps contributing to the differences in the preoptic area do not involve GABA. The preoptic area (POA) is the critical brain region controlling expression of male sexual behavior and exhibits some of the most robust sex differences in the brain. In addition to the SDN, male POA neurons have about twice as many dendritic spines as females, and a greater number of spines can be induced in females by treatment with estradiol during the perinatal sensitive period. Dendritic spines are the primary site for excitatory synapses. Astrocytes are also more complex in the male POA, with longer and more frequently branching processes (3). Both of these morphological sex differences are the result of estradiol action in the neonatal brain. The molecular steps responsible for the male-female differences in spine number involve the induction of cyclooxygenase 2 (COX-2), a pivotal enzyme in the production of prostanoids and specifically linked to an increased synthesis and release of prostaglandin E2 (PGE2). Receptors for PGE2 are G-protein linked and can be found on astrocytes. Activation of these receptors can induce glutamate release from astrocytes in a calcium-dependent manner, and glutamate, in turn, induces the formation of dendritic spines in the downstream neurons (3).

Environmental Chemicals and Sexual Differentiation of the Brain

The above discussion focuses on only two of the many different molecular pathways associated with sexual differentiation in brain structures. The relevance of this is important to our understanding of how the developing brain may be susceptible to environmental compounds. It is clear that exogenous testosterone, estrogen, or synthetic steroids will masculinize the female brain, and it is not difficult to understand that environmental estrogens may also influence the molecular events responsible for sexual differentiation. For example, the phytoestrogens (77) and estrogenic toxicants such as bisphenol A (BPA), octylphenol, and nonylphenol (78–80) have been reported to influence different brain structures when administered during the sensitive period of development, if the dose delivered to the tissue is great enough. Moreover, treatments with environmental compounds that reduce the availability of testosterone to the brain during the sensitive period of pregnancy also pose a risk to appropriate phenotypic brain development (81).

Environmental estrogens have also been shown to influence the development of neuronal pathways involved in the regulation of GnRH release. It is known that neonatal injections of estradiol benzoate to male and female rats result in a dose-dependent decrease in hypothalamic KiSS-1 mRNA levels at the prepubertal stage, which is linked to a lowering of serum LH concentrations in both sexes. Navarro and his colleagues (79) have shown that subcutaneous exposure of prepubertal male and female rats to BPA at 100 and 500 µg/rat on PND 1 to 5, or to an estrogen positive control (estradiol benzoate), resulted in a dose-dependent decrease in the message (KiSS-1 mRNA) for the hypothalamic neuropeptide kisspeptin, an effect linked to a lowering of serum LH concentrations. These data indicate that the KiSS-1 system is sensitive to alterations in the sex steroid milieu during critical periods of brain sex differentiation and suggest that lowering of the endogenous kisspeptin tone induced by early exposures to xenoestrogens might be mechanistically relevant for disruption of gonadotropin secretion and puberty onset later in life.

Of equal importance is the finding that estrogenic stimulation of the cells in the sexual dimorphic nuclei induces a number of molecular changes critical for the complete remodeling of the cells. In the above two examples of how estrogen changes the arcuate and preoptic regions in the male rat, one could speculate that compounds that interfere with GABA synthesis, COX2 activity, PGE2 synthesis, or receptor binding or binding to the glutamate receptor would interfere with the normal phenotypic developmental changes. In fact, this has been demonstrated by studies of McCarthy and associates (3) who found that treating the developing male with indomethacin (blocking COX2), GAD inhibitors, and NMDA receptor blockers would block the appearance of the normal changes after estrogen stimulation.

In some instances, the effects of an endocrine-disrupting compound are straightforward and easily interpretable as mimicking estrogen action. A systematic comparison of the alkylphenol 4-*tert*-octylphenol, commonly used in paints, pesticides, herbicides, detergents, and plastics, with the synthetic estrogen Diethylstilbestrol (DES) found effects of comparable magnitude in preventing the ability to exhibit an LH surge and induce ovulation in rats (78). A similar parallel effect was observed between neonatal DES and two naturally occurring xenoestrogens: (*i*) genistein, an isoflavone found in many grains and (*ii*) zearalenone, a mycotoxin produced by the cereal pathogen *Fusarium graminearum*. Both compounds increased the volume of the SDN when administered to neonatal females and altered the responsiveness of the pituitary to GnRH in adult males and females (77). When genistein was examined for its effects on the AVPV, it did not mimic the effect of estradiol in females and exerted a demasculinizing effect on males, counter to what would be expected of an estrogen-mimicking compound. If another parameter, colocalization of ER with tyrosine hydroxylase, was used as the end point, genistein, as well as the plasticizer BPA, both behaved as estrogen mimetics (80,82). Interest in BPA is particularly high given its pervasiveness in the environment and the conflicting evidence regarding its disease-promoting potential. Administration of BPA in the drinking water to pregnant and lactating rats prevented the development of a sex difference in corticotropin-releasing hormone (CRH)-expressing neurons in one region of rat brain (83). This is an unusual and potentially important effect of BPA on the developing brain, but is hard to interpret in the context of endocrine disruption, since the hormonal basis of the sex difference in CRH expression has not been established. A similar caveat can be applied to prenatal BPA effects on "depression-like behavior" in which a sex difference was eliminated or prevented from forming (84).

THE ADULT FEMALE
The Ovarian Cycle
In the adult female, the healthy ovary produces viable gametes within each ovarian cycle. This process begins shortly after puberty and continues until reproductive senescence, unless interrupted by pregnancy, illness, or chemical insult. The female rat, like the female human, is a spontaneous ovulator. After puberty, rats display rhythmic four- to five-day estrous cycles. These cycles can be divided into three separate segments, a period of diestrus (lasting two to three days), proestrus (one day), and estrus (one day). Each segment is readily identified by observing the changes in vaginal cytology that occur in response to the fluctuating levels of ovarian steroids in the blood (85–87). The period of vaginal diestrus is associated with follicular growth, a process dependent on the tonic stimulation by the pituitary gonadotropins, primarily follicle-stimulating hormone (FSH). Follicular growth is accompanied by the secretion of estradiol. Serum levels of this ovarian steroid begin to increase late in diestrus and reach peak values around noon on vaginal proestrus. Circulating levels of ovarian progesterone also rise during this time. Follicular growth culminates on the day of vaginal proestrus, at which time the mature preovulatory follicle becomes the target site for the dramatic increase in pituitary LH secretion. During the estrous cycle, serum LH remains low except for a brief period late in the afternoon on the day of vaginal proestrus. At this time, serum concentration of LH rapidly increases approximately 20-to 40-fold. Ovulation occurs approximately 10 to 12 hours after this surge. Importantly, the female is sexually receptive during the intervening hours on the evening of vaginal proestrus. Thus, the synergistic feedback effects of ovarian estrogen and progesterone on the CNS serve to synchronize the physiological events responsible for ovulation with the behavior of the female, maximizing the chances that fertilization will occur.

After rupture of the follicle and discharge of the ovum, the follicle cells enlarge and LH stimulates the granulosa cells to luteinize to form the corpus luteum (CL), within which steroid hormone production favors progesterone over estrogen. In the rat, maintenance of the CL is under the influence of twice-daily surges of prolactin released from the anterior pituitary. In the rat, these prolactin surges are initiated as a result of the vaginal-cervical stimulation that occurs during mating. Sexual activity is restricted to the dark portion of the photoperiod on the night of vaginal proestrus and early hours of vaginal estrus. If mating does not occur, twice-daily diurnal surges do not ensue and the CL undergo atresia. It should be noted that the endocrine support of the CL shows considerable species variation. In some species, including humans, LH or human chorionic gonadotropin maintains the CL.

Examining Reproductive Toxicity in the Adult Rat
Relevant to this review, the Japanese National Institute of Health Sciences, in collaboration with the Japanese Pharmaceutical Manufacture Association, completed a series of studies to evaluate the duration of exposure necessary to identify adverse effects of the ovary, as well as other measures of the female reproductive tract (88). As this series of studies examined a wide range of female reproductive toxicants and a number of key endpoints (many of which are typically included in multigenerational studies), it is instructive to review the results. In the initial evaluation of the female, rats were dosed for

TABLE 2   Chemicals Tested in the Japanese NIEHS/JPMA Ovarian Toxicity Study

Hormone analogues
- Medroxyprogesterone (progestin)
- Mifipristone (progesterone receptor antagonist)
- Tamoxifen (mixed estrogen receptor antagonist/agonist)

Primary follicle toxicants
- 4-Vinylcyclohexane (group 2B carcinogen)
- Busulfan (antineoplastic alkylating agent)
- Cysplatin (antineoplastic)
- Cylcophosphamide (antineoplastic alkylating agent)

Induce metabolism imbalance
- Anastrozole (antineoplastic aromatase inhibitor)
- 2-(Diethylhexyl) adipate
- 2-(Diethylhexyl) phthalate
- Eethylglycol monomethyl ether
- Indomethacin (COX 1 and 2 inhibitor, NASID)
- Compound X (α/γ-PPAR duel agonist)

Endocrine distruptors
- Atrazine (unknown mechanism)
- Bromocriptine (dopamine agonist)
- Chlorpromazine (D2, 5-HT2A, H1 and α1)
- Sulpiride (D2 antagonist)

TABLE 3   Endpoints Examined in Japanese NIEHS/JPMA Studies of Ovarian Toxicity

Two- and four-week study
    Body weight
    Estrous cycle
    Ovary [H & E for histopathological examination and PNAS for primordial and primary follicle counts
      Muskhelishvili et al. (89)]
    Uterus
    Vagina
    Mammary gland
Fertility study
    Body weight
    Estrous cycle
    Number of animals mated
    Number of animals copulated (plug)
    Number of animals pregnant
    Number of animals
    Macroscopic findings
    Number of CL
    Number of implants
    Preimplantation loss
    Number of live embryos
    Number of dead embryos
    Postimplantation loss

two or four weeks to determine which time period would be appropriate for identifying the effect of the test chemicals (see Table 2, top). In the second phase of the study, females were mated and fertility examined (see Table 2, bottom). The chemicals tested in this collaborative study were selected because they were all reported to have adverse effects on female reproduction. These chemicals chosen for study are depicted in Table 3. In the female fertility study, dosing was initiated for two weeks prior to mating and throughout the mating period (until sperm positive or a maximum of 2 weeks). The females were killed on gestation day 7, and the uterine contents examined on gestation day 14.

These results demonstrated a number of key factors concerning the evaluation of the female reproductive system. First, a careful examination of the vaginal smear can readily detect adverse effects on the ovarian cycle. In addition, the ovarian histology matched the cycling pattern very well. For example, the persistent estrous vaginal smear noted in the mifepristone-treated animals matched the presence of persistent large follicles and a reduction in the number of CL. However, these data also demonstrated that significant effects on follicular dynamics can be present in the cycling female. Most interesting is the effect of indomethacin, a compound that did not interfere with vaginal cycling but blocked ovulation, as shown by the presence of retained ova in the luteinized follicle. Also, compounds such as busulfan and cyclophosphamide had minimal effects on cycling during the two- or four-week dosing period but did deplete the store of small follicles. These observations are similar to those discussed above in which perinatal exposure to γ-irradiation severely depleted the primordial follicles with minimal impact on the acquisition of vaginal opening and subsequent cycles. The observations from the fertility studies are also informative, as in many cases the number of offspring was not reduced following treatment although all the chemicals tested did affect ovarian histopathology. Lastly, the purpose of this series of studies was to determine whether short-term (2–4 weeks) repeat dose studies were sufficient to detect ovarian toxicity. Toward this end, the investigators reported that with the two-week study, the toxicity of 15 of the 17 chemicals tested was identified. All were identified in the four-week study.

ENVIRONMENTAL TOXICANTS ON HYPOTHALAMIC REGULATION OF PITUITARY HORMONE SECRETION

Understanding the extent to which the brain and pituitary serve as a target site for reproductive toxicants remains a challenge. There is clear evidence supporting the fact that hypothalamic neurotransmitters and neuropeptides are involved in the regulation of gonadotropin and prolactin secretion in females of many species. Thus, chemicals that influence normal function of these neuronal systems (gene expression, synthesis, receptor binding, membrane function, etc.) stand as candidates for disrupting female reproductive function.

Neuroendocrine Control of Ovulation

The sequence of neuroendocrine events surrounding the ovulatory surge of LH includes both positive and negative feedback effects of the ovarian steroids on the brain and pituitary. As noted, estradiol levels in the blood begin to rise on the afternoon of the second day of diestrus and reach their highest levels around noon on the day of proestrus. The serum progesterone concentration begins to rise on the morning of proestrus, and peak values are achieved later in the evening. The sequential pattern of rising estrogen followed by increasing progesterone levels has a synergistic effect on the brain and pituitary and serves to control the amount and timing of LH release.

The immediate stimulus for the surge of LH from the anterior pituitary is the pulsatile liberation of GnRH from the brain into the hypothalamic-hypophyseal portal system, the capillary network that bathes the pituitary. GnRH is synthesized in neuronal cell bodies located in specific brain regions (i.e., rostral hypothalamus). It is transported along axons projecting to the medial basal hypothalamus (median eminence area), where the terminals of the neurons come into contact with vessels of the hypothalamic-hypophyseal portal system. In turn, the release of GnRH into the portal system is under neurotransmitter and neuropeptide regulation.

The involvement of CNS in the regulation of the ovulatory surge of LH was demonstrated in the pioneering work of Everett (90). In these studies, it was demonstrated that compounds that can disrupt hypothalamic neuronal activity can also block pituitary hormonal secretion and ovulation (e.g., sodium pentobarbital, atropine, dibenamine) in the female rat (91). Importantly, Everett also demonstrated that such treatments were effective only if administered at a certain time of the estrous cycle, specifically during the early afternoon of vaginal proestrus (92). Subsequently, this has led to the identification of a "critical" or "sensitive" period for the neural generation of the LH surge in the rat, which occurs roughly between 1400 and 1600 hours on vaginal proestrus (i.e., immediately before the initiation of the LH surge). Furthermore, if the surge is blocked by drugs administered during the critical period on proestrus, this endocrine event will occur at approximately the same time on the following day, resulting in a corresponding 24-hour delay in ovulation (90,91). That is, the CNS mechanisms involved in the control of this circadian event become functional again 24 hours later (and are still influenced by the light-dark cycle). The significance of a 24- or 48-hour delay in ovulation in toxicology studies will be discussed further below.

Since the 1970s there have been a number of studies detailing the regulation of the LH surge by GnRH as well as the neurochemical modulation of GnRH release. As mentioned earlier, this decapeptide is released into the portal system in a pulsatile fashion. In turn the amount of LH released by the gonadotropins into systemic circulation is regulated by the frequency and amplitude of the GnRH pulses (92). The higher the frequency, the greater the concentration of LH observed in the female's systemic circulation.

The role of the various neurotransmitters and neuropeptides involved in the regulation of GnRH secretion has been elucidated by a number of pharmacological and biochemical studies. For example, norepinephrine (NE) (and epinephrine) has been shown to exert a direct stimulatory action on GnRH release (i.e., increased frequency) (93). Thus, agents that disrupt the synthesis of NE [e.g., tyrosine hydroxylase inhibitors such as α-methyl-paratyrosine and dopamine-β-hydroxylase inhibitors such as fusaric acid or diethyldithiocarbamate (DDC)] (94) or agents that interfere with α-noradrenergic receptor stimulation (e.g., phenoxybenzamine and phentolamine) (95–98) will disrupt the pattern of GnRH secretion and consequently impair the LH surge.

In addition to the involvement of the noradrenergic neurons, several other neurotransmitters and neuropeptides (i.e., kisspeptin, neuropeptide Y, endorphin, enkephalins) have been shown to play a significant role in the regulation of GnRH. An in-depth discussion of these is beyond the scope of this chapter and the reader is referred to several excellent reviews (94,99–101).

Disruption of CNS Control of Ovulation

Given the number of neurotransmitters and neuropeptides that are known to regulate the ovulatory surge of LH, it would be anticipated that environmental compounds targeting different hypothalamic mechanisms can interfere with the generation of the surge, delay ovulation, and result in an adverse effect on fertility.

The formamidine pesticides, chlordimeform (CDF) and amitraz (AMI), are acaricides reported to block NE binding to the α-2 receptor (102,103). It is, therefore, not surprising that a single exposure to CDF, administered during the critical presurge window, inhibited the ovulatory surge of LH (104). The site of action was most likely central in nature since systemically administered CDF does reach the brain (102,105) and has been found to suppress the stimulated release of GnRH in vitro without having a parallel direct effect on pituitary LH secretion (106). Importantly, exposure to CDF or AMI outside the sensitive period on vaginal proestrus was without effect on ovulation.

As mentioned previously, Everett and Sawyer (91) demonstrated that if the LH surge is blocked by drugs administered during the critical period on proestrus, this endocrine event will occur at approximately the same time on the following day, resulting in a corresponding 24-hour delay in ovulation with no reduction in the number of ova shed. Similar results are observed after acute exposure to formamidine pesticides. A single administration of AMI induced a one-day delay in the afternoon LH surge, thus causing a corresponding one-day delay in ovulation (107). Again, ovulation did occur and no effects were seen on the number of ova shed. The CDF-induced delay in the LH surge did not prevent the females from mating on the evening of the delayed surge (i.e., 24-hour later than normal) (107,108), thus allowing a substantial percentage of the CDF-delayed females to became pregnant. However, when evaluated on gestation day 20, there was a reduction in litter size of the treated dams.

Similar alterations in reproductive function occur following exposure to the dithiocarbamates including DDC (109), thiram (110), sodium dimethyldithiocarbamate (DMDC, primary metabolite of thiram) (111), and the structurally related sodium N-methyldithiocarbamate (metam sodium) (112). The dithiocarbamates are a group of fungicides known to inhibit dopamine-β-hydroxylase activity and thus decrease hypothalamic NE synthesis (112,113) by blocking its conversion from DA (114–116). This decrease, in turn, has a parallel effect with dose on GnRH neuronal activation and the LH surge (117). As with the NE-receptor blocking compounds, thiram induced a 24-hour delay in ovulation with no difference in the number of oocytes released and, when mated, there was no difference in the number of implantation sites. Importantly, however, this toxicant-induced delay in ovulation did reduce litter size (118). Also, the embryos in the delayed females had an overall delayed development noted as early as gestational day 11 (118). Thiram-treated animals also exhibited an increased incidence of polyspermic zygotes, which can lead to polyploidy, abnormal development, and early embryonic death (119). Thus, a very brief exposure to toxicants, such as the dithiocarbamates, affect the CNS noradrenergic mechanisms involved in pituitary LH release to delay ovulation. Although a normal complement of ova is released 24 hours later, the viability of the delayed oocyte is altered. This leads to a decrease in embryo survival and reduction in the number of live fetuses present late in gestation.

The CDF- or thiram-induced delays in the LH surge lead to intrafollicular oocyte changes and altered survivability of the embryo or fetus (119) that are similar to those reported by Butcher and coworkers (120). These investigators reported that a two-day delay of the LH surge with sodium pentobarbital led to decreased fertility and an increase in the number of abnormal fetuses (121). Ultrastructural studies of the released oocytes revealed the lack of a continuous layer of cortical granules lining the plasma membrane (122). An abnormal pattern of cortical granule exudate was also observed by confocal microscopy in thiram-delayed polyspermic zygotes (119), which may explain the oocyte's ineffective block against polyspermy (123). These observations in the rat model may be applicable to humans. Altered distribution of cortical granule exudate has been reported in human in vitro fertilization (124), and there is indirect evidence that increased intrafollicular aging of human eggs may be correlated to an increased rate of polyspermy after in vitro fertilization (125).

The above discussion demonstrates that environmental compounds affecting NE neurotransmission can adversely impact reproductive performance. The extent to which these compounds may impact other key neurotransmitter or neuropeptide systems remains to be fully determined. However, there are some indications that a toxicant-induced disruption of other neuronal pathways would induce similar adverse outcomes. As mentioned, centrally acting barbiturates, such as sodium pentobarbital, are routinely used to block the LH surge in animal models. Recently, this effect of pentobarbital has been attributed to a disruption of the GABA receptor leading to an opening of the $GABA_A$ receptor channels (126). GABA receptors have been identified on GnRH neurons (127), and GABA secretion has been found to suppress their electrical excitability (128). The GABA agonists Baclofen and muscimol have both been reported to block the LH surge in ovariectomized, steroid-primed rats (129). Interestingly, these authors reported that both agonists decreased hypothalamic NE concentrations as well as turnover rate over the course of several hours, suggesting that GABA was acting to modulate noradrenergic mechanisms involved in GnRH secretion.

Octamethylcyclotetrasiloxane, also called D4, is found in a variety of consumer products, including personal care products. Inhalation is considered the most relevant route of human exposure by pharmacokinetic modeling (130). A two-generation study of rats exposed to D4 by whole-body vapor inhalation revealed significant reductions in litter size at 500 or 700 ppm for both the F0 and F1 generations (131). The F1 generation also exhibited prolonged estrous cycles (increase in the number of diestrous days), and decreased mating and fertility. Similar exposures to D4 also attenuate the LH surge (132) and reduce or prevent ovulation (133,134). D4 is reported to be weakly estrogenic (134) and is in addition considered to be a potential dopamine D2 receptor agonist based on its ability to decrease prolactin levels in vivo (135). However, the mechanism leading to D4 attenuation of the LH surge is unknown.

Atrazine, a chlorotriazine herbicide introduced in the 1950s, also alters female reproductive function through a hypothalamic mechanism in the rodent model. Short-term atrazine oral exposure (21 days) induced repetitive pseudopregnancies (character-ized by prolonged periods of diestrus, several corpora lutea present in the ovaries, and elevated serum progesterone) at doses higher than 150 mg/kg/day (136). Six-month oral exposure to high doses led to a pattern of persistent estrus (137). This compound attenuates the estrogen-induced surges of LH and prolactin in ovariectomized female rats (50–300 mg/kg) (138). When given over the course of the estrous cycle, atrazine also inhibits the proestrous LH surge at much lower doses (6.25 mg/kg) (87), indicating that the intact animal may be more sensitive to the effect of chlorotriazines. This suggests that changes in the CNS regulation of pituitary function are an important component of the mechanism of action of these compounds. Indeed, increases in median eminence GnRH content are observed following atrazine exposure in intact females (91), with pituitary gonadotrophs still responsive to GnRH secretion (138), suggesting a hypothalamic site of action. However, the mechanism of atrazine's toxicity remains to be elucidated.

Molinate (S-ethyl hexahydro-1H-azepine-1-carbothioate), a selective preemergence thiocarbamate herbicide, attenuated the LH surge in the estrogen-primed ovariectomized female rat, as well as delayed ovulation for 24 hours and induced irregular estrous cyclicity in intact animals (139). A reduction of GnRH pulse frequency was also observed in long-term ovariectomized rats (139). LH release following a bolus dose of GnRH was comparable to control rats, suggesting that the effect of molinate is centrally mediated. The mechanism of action, however, is still unknown. Unlike the dithiocarbamates or chloro-triazines, in which hypothalamic NE concentrations were decreased following exposure, NE concentrations in the hypothalamus were significantly increased, and DA levels unchanged, three hours after oral gavage exposure to molinate.

The CNS is also involved in the regulation of prolactin and compounds that influence the normal secretion of prolactin may adversely impact reproductive function in the female. For example, in the rat, prolactin secretion is important for the initiation of CL formation and establishment of pregnancy. Compounds that interfere with this process [i.e., the dopamine agonsist bromocriptine (140) or chlorotriazines (141)] will block implantation. Other studies have shown that the dam's prolactin may enter the offspring via the milk and influence normal reproductive development in the offspring (142).

The steroid hormones also have a significant role in the regulation of anterior pituitary function. Thus, it is not surprising that environmental estrogens may modify this normal feedback influence on the hypothalamus and pituitary. Both the brain and pituitary are rich in steroid receptors. Xenobiotics with estrogenic, progestin, or androgenic activity (agonist or antagonist) could also influence the hypothalamic-pituitary control of gonadal function by interfering with the action of the endogenous steroids on their receptors. With the exception of some well-conducted developmental studies, there have been relatively few reports examining the effect of suspected xenosteroids at the level of the CNS and pituitary. It has been reported that exposure to either methoxychlor or chlordecone alters estrogen and progesterone receptor numbers in these tissues (143). Hong et al. (144) reported that the changes in hypothalamic and pituitary neuropeptide concentrations observed following chlordecone treatment were similar to those induced by estrogen, suggesting that the estrogenic action of this peptide is the mechanism responsible for the central effects. Likewise, Uphouse (145) has argued that since chlordecone can decrease serum LH, increase serum prolactin, and induce vaginal cornification in the ovariectomized female, it is reasonable to assume that this compound disrupts reproductive function because it is a weak estrogen. It remains unclear whether the dose necessary to elicit an estrogenic response is greater than that needed to alter CNS activity, including neurotransmitter or neurochemical mechanisms (146–149). Moreover, the rapid onset of various alterations, such as a reduction in sexual behavior (150), make it unlikely that steroidal mechanisms underlie the observed behavioral changes.

ACKNOWLEDGMENTS

The authors would like to express their appreciation to Drs Rochelle Tyl, John Rogers, and Tammy Stoker for their valuable comments on a previous draft of the manuscript.

REFERENCES

1. Cooper RL, Kavlock RJ. Endocrine disruptors and reproductive development: a weight-of-evidence overview. J Endocrinol 1997; 152:159–166.
2. Karlsson S. Histopathology and histomorphometry of the urogenital tract in 15-month old male and female rats treated neonatally with SERMs and estrogens. Exp Toxicol Pathol 2006; 58:1–12.
3. McCarthy MM. Estradiol and the developing brain. Physiol Rev 2008; 88:91–134.
4. Stoker TE, Goldman JM, Cooper RL. Delayed ovulation and pregnancy outcome: effect of environmental toxicants on the neuroendocrine control of the ovary. Environ Toxicol Pharmacol 2001; 9:117–129.
5. Narotsky MG, Best DS, Guidici DL, et al. Strain comparisons of atrazine-induced pregnancy loss in the rat. Reprod Toxicol 2001; 15:61–69.
6. ICH. International Conference on Harmonisation; Guideline on detection of toxicity to reproduction for medicinal products. Fed Regist 1994; 59:48746.
7. OECD. Test Guideline 414. OECD Guideline for Testing of Chemicals. Prenatal Developmental Toxicity Study. Paris, France: Organisation for Economic Co-operation and Development, 2001. Available at: http://caliban.sourceoecd.org/vl=8257865/cl=11/nw=1/rpsv/ij/oecdjournals/1607310x/v1n4/s15/p1
8. OECD. Test Guideline 416. OECD Guideline for Testing of Chemicals. Two-generation Reproduction Toxicity Study. Paris, France:Organisation for Economic Co-operation and Development, 2001. Available at: http://caliban.sourceoecd.org/vl=8257865/cl=11/nw=1/rpsv/ij/oecdjournals/1607310x/v1n4/s17/p1
9. US EPA. Health Effects Test Guidelines, OPPTS 870.3700, Prenatal Developmental Toxicity Study, EPA 712-C-98-207. Office of Prevention, Pesticides, and Toxic Substances. Washington, DC: US Environmental Protection Agency, 1998. Available at: http://www.epa.gov/opptsfrs/publications/OPPTS_Harmonized/870_Health_Effects_Test_Guidelines/Series/
10. US EPA. Health Effects Test Guidelines, OPPTS 870.3800, Reproduction and Fertility Effects, EPA 712-C-98-208. Office of Prevention, Pesticides, and Toxic Substances. Washington, DC: US Environmental Protection Agency, 1998. Available at: http://www.epa.gov/opptsfrs/publications/OPPTS_Harmonized/870_Health_Effects_Test_Guidelines/Series/
11. US FDA. Toxicological Principles for the Safety of Food Ingredients, Redbook 2000, IV. C.9.a. Guidelines for Reproduction Studies. Washington, DC: Center for Food Safety and Applied Nutrition, US Food and Drug Administration, 2000:80–117. Available at: http://www.cfsan.fda.gov/~redbook/red-toca.html
12. US FDA. Toxicological Principles for the Safety of Food Ingredients, Redbook 2B000, IV.C.9.b. Guidelines for Developmental Toxicity Studies. Washington, DC: Center for Food Safety and Applied Nutrition, US Food and Drug Administration, 2000:123–134. Available at: http://www.cfsan.fda.gov/~redbook/red-toca.html
13. OECD 1983 Test Guideline 415. OECD Guideline for Testing of Chemicals. One-Generation Reproduction Toxicity Study. Paris, France: Organisation for Economic Co-operation and Development. Available at: http://puck.sourceoecd.org/vl=2487060/cl=11/nw=1/rpsv/ij/oecdjournals/1607310x/v1n4/s15/p1.

14. OECD 1995 Test Guideline 421. OECD Guideline for Testing of Chemicals. Reproduction/Developmental Toxicity Screening Test. Paris, France: Organisation for Economic Co-operation and Development. Available at http://puck.sourceoecd.org/vl=2487060/cl=11/nw=1/rpsv/ij/oecdjournals/1607310x/v1n4/s21/p1.
15. US EPA 2000a Health Effects Test Guidelines, OPPTS 870.3550, Reproduction/Developmental Toxicity Screening Test, EPA 712-C-00-367. Washington, DC. Office of Prevention, Pesticides, and Toxic Substances, US Environmental Protection Agency, Washington, DC. Available at http://www.epa.gov/ocspp/pubs/frs/publications/Test_Guidelines/series870.htm
16. OECD 1996 Test Guideline 422. OECD Guideline for Testing of Chemicals. Combined Repeated Dose Toxicity Study with the Reproduction/Developmental Toxicity Screening Test. Paris, France: Organisation for Economic Co-operation and Development. Available at http://puck.sourceoecd.org/vl=2487060/cl=11/nw=1/rpsv/ij/oecdjournals/1607310x/v1n4/s22/p1.
17. US EPA 2000b Health Effects Test Guidelines, OPPTS 870.3650, Combined Repeated Dose Toxicity Study with the Reproduction/Developmental Toxicity Screening Test. EPA 712-C-00-368. Office of Prevention, Pesticides, and Toxic Substances, US Environmental Protection Agency, Washington, DC. Available at http://www.epa.gov/ocspp/pubs/frs/publications/Test_Guidelines/series870.htm
18. OECD 2007 Test Guideline 426. OECD Guideline for Testing of Chemicals. Developmental Neurotoxicity Study. Paris, France: Organisation for Economic Co-operation and Development. Available at http://puck.sourceoecd.org/vl=2487060/cl=11/nw=1/rpsv/ij/oecdjournals/1607310x/v1n4/s26/p1.
19. US EPA 1998e Health Effects Test Guidelines, OPPTS 870.6300, Developmental Neurotoxicity Study, EPA 712-C-98-239. Office of Prevention, Pesticides, and Toxic Substances. Washington, DC: US Environmental Protection Agency. Available at http://www.epa.gov/ocspp/pubs/frs/publications/Test_Guidelines/series870.htm
20. US FDA. Guidance for Industry. Nonclinical Safety Evaluation of Pediatric Drug Products. Washington, DC: Center for Drug Evaluation and Research, US Food and Drug Administration, 2006. Available at: http://www.fda.gov/cder/guidance/5671fnl.htm
21. US EPA. Review of the Endocrine Disruptor Screening Program (EDSP) Proposed Tier-1 Screening Battery. Washington DC: US Environmental Protection Agency, 2008. Available at: http://www.epa.gov/scipoly/sap/meetings /2008/032508_mtg.htm
22. US EPA. Endocrine Disruptor Screening Program (EDSP). Washington, DC: US Environmental Protection Agency, 2008. Available at: httBp://www.epa.gov/endo/pubs/edspoverview/background.htm
23. US EPA. Health Effects Test Guidelines, OPPTS 870.4200, Carcinogenicity, EPA 712-C-98-208. Washington, DC: Office of Prevention, Pesticides, and Toxic Substances, US Environmental Protection Agency, 1998. Available at: http://www.epa.gov/opptsfrs/publications/OPPTS_Harmonized/870_Health_Effects_Test_Guidelines/Series/
24. Cooper RL, Lamb JC, Barlow SM. A tiered approach to life stages testing for agricultural chemical safety assessment. Crit Rev Toxicol 2006; 36:69–98.
25. Goldman JM, Cooper RL. Normal development of the female reproductive system. In: Kapp RW, Tyl RW, eds. Reproductive Toxicology. 3rd ed.
26. Kim HH, Laufer MR. Developmental abnormalities of the female reproductive tract. Curr Opin Obstet Gynecol 1994; 6:518–525.
27. Gidwani G, Falcone T. Congenital Malformations of the Female Genital Tract: Diagnosis and Management. Philadelphia: Lippincott, Williams and Wilkins, 1999.
28. Peifer M, Polakis P. Wnt signaling in oncogenesis and embryogenesis—a look outside the nucleus. Science 2000; 287:1606–1609.
29. Chi N, Epstein JA. Getting your Pax straight: Pax proteins in development and disease. Trends Genet 2002; 18:41–47.
30. Ruiz i Altaba A, Sanchez P, Dahmane N. The hedgehog-Gli pathway in cancer. Tumors, embryos and stem cells. Nat. Rev. Cancer 2002; 2:361–372.
31. Kobayashi A, Behringer RR. Developmental genetics of the female reproductive tract in mammals. Nat Rev Genet 2003; 4:969–980.
32. Hsueh AJ, Billig H, Tsafriri A. Ovarian follicle atresia: a hormonally controlled apoptotic process. Endocr Rev 1994; 15:707–724.

33. McNatty KP, Fidler AE, Juengel JL, et al. Growth and paracrine factors regulating follicular formation and cellular function. Mol Cell Endocrinol 2000; 163:11–20.
34. Campbell BK, Scaramuzzi RJ, Webb R. Control of antral follicle development and selection in sheep and cattle. J Reprod Fertil Suppl 1995; 49:335–350.
35. Webb R, Campbell BK, Garverick HA, et al. Molecular mechanisms regulating follicular recruitment and selection. J Reprod Fertil Suppl 1999; 54:33–48.
36. Richards JS. Perspective: the ovarian follicle—a perspective in 2001. Endocrinology 2001; 142:2184–2193.
37. Hirshfield AN. Development of folllicles in the mammalian ovary. Int Rev Cytol 1991; 124:43–101.
38. Johnson J, Canning J, Kaneko T. Germline stem cells and follicular renewal in the postnatal mammalian ovary. Nature 2004; 428:145–150.
39. Besmer P, Manova K, Duttlinger R. The kit-ligand (steel factor) and its receptor c-kit/ W: pleiotropic roles in gametogenesis and melanogenesis. Dev Suppl 1993; 125–137.
40. Xu Y, Ashley T, Brainerd EE. Targeted disruption of ATM leads to growth retardation, chromosomal fragmentation during meiosis, immune defects, and thymic lymphoma. Genes Dev 1996; 10:2411–2422.
41. Swapp GH, Johnston AW, Watt JL. A fertile woman with non-mosaic Turner's syndrome: case report and review of the literature. Br J Obstet Gynaecol 1989; 96:876–880.
42. Guigon CJ, Mazaud S, Forest MG, et al. Role of the first wave of growing follicles in rat ovarian maturation. Ann Endocrinol 2003; 64:85.
43. Mazaud S, Guigon CJ, Lozach A, et al. Establishment of reproductive function and transient fertility of female rats lacking primordial follicle stock after fetal gamma irradiation. Endocrinology 2002; 143:4774–4787.
44. Guigon CJ, Mazaud S, Forest MG. Unaltered development of the initial follicular waves and normal pubertal onset in female rats after neonatal deletion of the follicular reserve. Endocrinology 2003; 144:3651–3662.
45. Hoyer PB. Damage to ovarian development and function. Cell Tissue Res 2005; 322:99–106.
46. MacKenzie KM, Angevine DM. Infertility in mice exposed in utero to benzo (a) pyrene. Biol Reprod 1981; 24:183–191.
47. Palter SF, Tavares AB, Hourvitz A, et al. Are estrogens of import to primate/human ovarian folliculogenesis? Endocr Rev 2001; 22:389–424.
48. Vendola KA, Zhou J, Adesanya OO, et al. Androgens stimulate early stages of follicular growth in the primateovary. J Clin Invest 1998; 101:2622–2629.
49. Vendola K, Zhou J, Wang J, et al. Androgens promote oocyte insulin-like growth factor I expression and initiation of follicle development in the primate ovary. Biol Reprod 1999; 61:353–357.
50. Abbott DH, Padmanabhan V, Dumesic DA. Contributions of androgen and estrogen to fetal programming of ovarian dysfunction. Reprod Biol Endocrinol 2006; 10:4–17.
51. Weil S, Vendola K, Zhou J, et al. Androgen and follicle-stimulating hormone interactions in primate ovarian follicle development. J Clin Endocrinol Metab 1999; 84:2951–2956.
52. Hasegawa T, Zhao L, Caron KM, et al. Developmental roles of the steroidogenic acute regulatory protein (StAR) as revealed by StAR knockout mice. Mol Endocrinol 2000; 14:1462–1471.
53. Miller WL. Disorders of androgen synthesis—from cholesterol to dehydroepiandrosterone. Med Princ Pract 2005; 14(suppl 1): 58–68.
54. Zachos NC, Billiar RB, Albrecht ED, et al. Developmental regulation of baboon fetal ovarian maturation by estrogen. Biol Reprod 2002; 67: 1148–1156.
55. Zachos NC, Billiar RB, Albrecht ED, et al. Regulation of oocyte microvilli development in the baboon fetal ovary by estrogen. Endocrinology 2004; 145:959–966.
56. Sar M, Welsch F. Differential expression of estrogen receptor beta and estrogen receptor-alpha in the rat ovary. Endocrinology 1999; 140:963–971.
57. Sharma SC, Clemens JW, Pisarska MD, et al. Expression and function of estrogen receptor subtypes in granulosa cells: regulation by estradiol and forskolin. Endocrinology 1999; 140:4320–4334.
58. Couse JF, Yates MM, Sanford R, et al. Formation of cystic ovarian follicles associated with elevated luteinizing hormone requires estrogen receptor-beta. Endocrinology 2004; 145:4693–4702.

59. Couse JF, Korach KS. Contrasting phenotypes in reproductive tissues of female estrogen receptor null mice. Ann N Y Acad Sci 2001; 948:1–8.

60. Couse JF, Bunch DO, Lindzey J, et al. Prevention of the polycystic ovarian phenotype and characterization of ovulatory capacity in the estrogen receptor-alpha knockout mouse. Endocrinology 1999; 140:5855–5865.

61. Krege JH, Hodgin JB, Couse JF, et al. Generation and reproductive phenotypes of mice lacking estrogen receptor beta. Proc Natl Acad Sci U S A 1998; 95:15677–15682.

62. Couse JF, Yates MM, Deroo BJ, et al. Estrogen receptor beta is critical to granulosa cell differentiation and the ovulatory response to gonadotropins. Endocrinology 2005; 146:3247–3262.

63. Emmen JM, Couse JF, Elmore SA, et al. In vitro growth and ovulation of follicles from ovaries of estrogen receptor (ER){alpha} and ER{beta} null mice indicate a role for ER {beta} in follicular maturation. Endocrinology 2005; 146: 2817–2826.

64. Britt KL, Saunders PK, McPherson SJ, et al. Estrogen actions on follicle formation and early follicle development. Biol Reprod 2004; 71:1712–1723.

65. Britt KL, Findlay JK. Estrogen actions in the ovary revisited. J Endocrinol 2002; 175: 269–276.

66. Couse JF, Yates MM, Walker VR, et al. Characterization of the hypothalamic-pituitary-gonadal axis in estrogen receptor (ER) Null mice reveals hypergonadism and endocrine sex reversal in females lacking ER-alpha but not ER-beta. Mol Endocrinol 2003; 17:1039–1053.

67. Britt KL, Stanton PG, Misso M. The effects of estrogen on the expression of genes underlying the differentiation of somatic cells in the murine gonad. Endocrinology 2004; 145:3950–3960.

68. Matsumoto A, Arai Y. Male-female differences in synaptic organization of the ventrome-dial nucleus of the hypothalamus in the rat. Neuroendocrinology 1986; 42:232–236.

69. Amateau SK, McCarthy, MM. Sexual differentiation of astrocyte morphology in the developing rat preoptic area. J Neuroendocrinol 2002; 14:904–910.

70. Kawashima S, Takagi, K. Role of sex steroids on the survival, neuritic outgrowth of neurons, and dopamine neurons in cultured preoptic area and hypothalamus. Horm Behav 1994; 28:305–312.

71. Tsukahara, S. Sex differences and the roles of sex steroids in apoptosis of sexually dimorphic nuclei of the preoptic area in postnatal rats. J Neuroendocrinol 2009; 21:370–376.

72. Dean SL, McCarthy MM. Steroids, sex and the cerebellar cortex: implications for human disease. Cerebellum 2008; 7:38–47.

73. Simerly RB. Wired for reproduction: organization and development of sexually dimorphic circuits in the mammalian forebrain. Annu Rev Neurosci 2002; 25:507–536.

74. De Vries GJ, Rissman EF, Simerly RB, et al. A model system for tudy of sex chromosomes effects on sexually dimorphic neural and behavioral traits. Neuroscience 2002; 22:9005–9014.

75. McCarthy MM, Albrecht ED. Steroid regulation of sexual behavior. Trends Endocrinol Metab 1996; 7:324–327.

76. Mong JA, Nunez JL, McCarthy MM. GABA mediates steroid induced astrocyte differentiation in the neonatal rat hypothalamus. J Neuroendocrinol 2002; 14:1–16.

77. Faber KA, Hughes CL Jr. The effect of neonatal exposure to diethylstilbestrol, genistein, zearalenone on pituitary responsiveness and sexually dimorphic nucleus volume in the castrated adult rat. Biol Reprod 1991; 45:649–653.

78. Willoughby KN, Sarkar AJ, Boyadjieva NI, et al. Neonatally administered tert-octylphenol affects onset of puberty and reproductive development in female rats. Endocrine 2005; 26:161–168.

79. Navarro VM, Sánchez-Garrido MA, Castellano JM, et al. Persistent impairment of hypothalamic KiSS-1 system after exposures to estrogenic compounds at critical periods of brain sex differentiation. Endocrinology 2009; 150:2359–2367.

80. Patisaul HB, Fortino AE, Polston EK. Neonatal genistein or bisphenol-A exposure alters sexual differentiation of the AVPV. Neurotoxicol Teratol 2006; 28:111–118.

81. Piffer RC, Garcia PC, Gerardin DC, et al. Semen parameters, fertility and testosterone levels in male rats exposed prenatally to betamethasone. Reprod Fertil Dev 2009; 21:634–639.

82. Patisaul HB, Todd KL, Mickens JA. Impact of neonatal exposure to the ERalpha agonist PPT, bisphenol-A or phytoestrogens on hypothalamic kisspeptin fiber density in male and female rats. Neurotoxicology 2009; 30:350–357.

83. Funabashi T, Kawaguchi M, Furuta M, et al. Exposure to bisphenol A during gestation and lactation causes loss of sex difference in corticotropin-releasing hormone-immunoreactive neurons in the bed nucleus of the stria terminalis of rats. Psychoneuroendocrinology 2004; 29:475–485.
84. Fujimoto T, Kubo K, Aou S. Prenatal exposure to bisphenol A impairs sexual differentiation of exploratory behavior and increases depression-like behavior in rats. Brain Res 2006; 1068:49–55.
85. Butcher RL, Collins WE, Fugo W. Plasma concentration of LH, FSH, prolactin, progesterone and estradiol-17beta throughout the 4-day estrous cycle of the rat. Endocrinology 1974; 94:1704–1708.
86. Cooper RL, Goldman JM, Rehnberg GL. Neuroendocrine control of reproductive function in the aging female rodent. J Am Geriatr Soc 1986; 34:735–751.
87. Cooper RL, Laws SC, Das PC, et al. Atrazine and reproductive function: mode and mechanism of action studies. Birth Defects Res B Dev Reprod Toxicol 2007; 80:98–112.
88. Kumazawa T, Nakajima A, Ishiguro T, et al. Collaborative work on evaluation of ovarian toxicity. (15) Two- or four-week repeated-dose studies and fertility study of bromocriptine in female rats. J Toxicol Sci 2009; 34(suppl 1):157–165.
89. Muskhelishvili L, Wingard SK, and Latendresse JR. Proliferating cell nuclear antigen—A marker for ovarian follicle counts. Toxicologic Pathology 2005; 33:365–368.
90. Everett JW. Neurobiology of Reproduction of the Female Rat. New York: Springer-Verlag, 1989.
91. Everett JW, Sawyer CH. A 24-hour periodicity in the "LH-release apparatus" of female rats, disclosed by barbiturate sedation. Endocrinology 1950; 47:198–218.
92. McCann SM, Lumpkin MD, Mizunuma H, et al. Recent studies on the role of brain peptides in control of anterior pituitary hormone secretion. Peptides 1984; 5(suppl 1):3–7.
93. Ching M, Krieg RJ Jr. Norepinephrine stimulates LH-RH secretion into the hypophysial portal blood of the rat. Peptides 1986; 7:705–708.
94. Kalra SP, Kalra PS. Neural regulation of luteinizing hormone secretion in the rat. Endocr Rev 1983; 4:311–351.
95. Ratner, A, Solomon S. Effect of phenoxybenzamine on luteinizing hormone release in the female rat. Proc Soc Exp Biol Med 1971; 138:995–998.
96. Sawyer, CH. First Geoffrey Harris Memorial lecture: some recent developments in brain-pituitary-ovarian physiology. Neuroendocrinology 1975; 17:97–124.
97. Plant TM, Nakai Y, Belchetz, P, et al. The sites of action of estradiol and phentolamine in the inhibition of the pulsatile, circhoral discharges of LH in the rhesus monkey (Macaca mulatta). Endocrinology 1978; 102:1015–1018.
98. Weick RF. Acute effects of adrenergic receptor blocking drugs and neuroleptic agents on pulsatile discharges of luteinizing hormone in the ovariectomized rat. Neuroendocrinology 1978; 26:108–117.
99. Herbison, AE. Noradrenergic regulation of cyclic GnRH secretion. Rev Reprod 1997; 2:1–6.
100. Kalra SP, Sahu A, Kalra PS, et al. Hypothalamic neuropeptide Y: a circuit in the regulation of gonadotropin secretion and feeding behavior. Ann N Y Acad Sci 1990; 611:273–283.
101. Tena-Sempere M. GPR54 and kisspeptin in reproduction. Hum Reprod Update 2006; 12:631–639.
102. Costa LG, Olibet G, Murphy SD. Alpha 2-adrenoceptors as a target for formamidine pesticides: in vitro and in vivo studies in mice. Toxicol Appl Pharmacol 1988; 93:319–328.
103. Altobelli D, Martire M, Maurizi S, et al. Interaction of , formamidine pesticides with the presynaptic alpha(2)-adrenoceptor regulating. Toxicol Appl Pharmacol 2001; 172:179–185.
104. Goldman JM, Cooper RL, Edwards TL, et al. Suppression of the luteinizing hormone surge by chlordimeform in ovariectomized, steroid-primed female rats. Pharmacol Toxicol 1991; 68:131–136.
105. Johnson TL, Knowles CO. Influence of formamidines on biogenic amine levels in rat brain and plasma. Gen Pharmacol 1983; 14:591–596.
106. Goldman JM, Cooper RL, Laws, SC, et al. Chlordimeform-induced alterations in endocrine regulation within the male rat reproductive system. Toxicol Appl Pharmacol 1990; 104:25–35.
107. Goldman JM, Cooper RL. Assessment of toxicant-induced alterations in the luteinizing hormone control of ovulation in the rat. In: Heindel JJ, Chapin RE, eds. Female Reproductive Toxicology. San Diego, CA: Academic Press, 1993:79–91.

108. Cooper RL, Barrett MA, Goldman, JM, et al. Pregnancy alterations following xenobiotic-induced delays in ovulation in the female rat. Fundam Appl Toxicol 1994; 22:474–480.
109. Kalra SP, McCann SM. Effects of drugs modifying catecholamine synthesis on plasma LH and ovulation in the rat. Neuroendocrinology 1974; 15:79–91.
110. Stoker TE, Goldman JM, Cooper, RL. The dithiocarbamate fungicide thiram disrupts the hormonal control of ovulation in the female rat. Reprod Toxicol 1993; 7:211–218.
111. Goldman, JM, Parrish MB, Cooper, RL, et al. Blockade of ovulation in the rat by systemic and ovarian intrabursal administration of the fungicide sodium dimethyldithiocarbamate. Reprod Toxicol 1997; 11:185–190.
112. Goldman JM, Stoker TE, Cooper RL, et al. Blockade of ovulation in the rat by the fungicide sodium N-methyldithiocarbamate: relationship between effects on the luteinizing hormone surge and alterations in hypothalamic catecholamines. Neurotoxicol Teratol 1994; 16:257–268.
113. Stoker TE, Cooper RL, Goldman JM, et al. The effect of thiram on the CNS control of luteinizing hormone in the male and female rat. The Toxicologist 1995; 15:294.
114. Lippmann W, Lloyd K. Dopamine-β-hydroxylase inhibition by dimethyldithiocarbamate and related compounds. Biochem Pharmacol 1969; 18:2507–2516.
115. Lippmann W, Lloyd K. Effects of tetramethylthiuram disulfide and structurally-related compounds on the dopamine-β-hydroxylase activity in the rat and hamster. Arch Int Pharmacodyn Ther 1971; 189:348–357.
116. Prewlocka B, Sarnek J, Szmigielski A, et al. The effect of some dithiocarbamic acids on dopamine-beta-hydroxylase and catecholamines level in rat's brain. Pol J Pharmacol Pharm 1975;27:555–559.
117. Goldman JM, Murr AS, Buckalew AR, et al. Suppression of the steroid-primed luteinizing hormone surge in the female rat by sodium dimethyldithiocarbamate: relationship to hypothalamic catecholamines and GnRH neuronal activation. Toxicol Sci 2008; 104:107–112.
118. Stoker TE, Cooper RL, Goldman JM, et al. Characterization of pregnancy outcome following thiram-induced ovulatory delay in the female rat. Neurotoxicol Teratol 1996; 18:277–282.
119. Stoker TE, Jeffay SC, Zucker RM, et al. Abnormal fertilization is responsible for reduced fecundity following thiram-induced ovulatory delay in the rat. Biol Reprod 2003; 68:2142–2149.
120. Butcher RL, Fugo NW. Overripeness and the mammalian ova. II. Delayed ovulation and chromosome anomalies. Fertil Steril 1967; 18:297–302.
121. Fugo NW, Butcher RL. Overripeness and the mammalian ova. I. Overripeness and early embryonic development. Fertil Steril 1966; 17:804–814.
122. Peluso JJ, Butcher RL. The effect of follicular aging on the ultrastructure of the rat oocyte. Fertil Steril 1974; 25:494–502.
123. Dandekar, P, Talbot P. Perivitelline space of mammalian oocytes: extracellular matrix of unfertilized oocytes and formation of a cortical granule envelope following fertilization. Mol Reprod Dev 1992; 31:135–143.
124. Santhanathan AH, Trounson AO. Ultrastructure of cortical granule release and zona interaction in monospermic and polyspermic human ova fertilized in vitro. Gamete Res 1982; 6:225–234.
125. Ben-Rafael Z, Meloni F, Strauss JF III, et al. Relationships between polypronuclear fertilization and follicular fluid hormones in gonadotropin-treated women. Fertil Steril 1987; 47:284–288.
126. Mercado J, Czajkowski C. Gamma-aminobutyric acid (GABA) and pentobarbital induce different conformational rearrangements in the GABA A receptor alpha1 and beta2 pre-M1 regions. J Biol Chem 2008; 283:15250–15257.
127. Spergel DJ, Kruth U, Hanley DF, et al. GABA- and glutamate-activated channels in green fluorescent protein-tagged gonadotropin-releasing hormone neurons in transgenic mice. J Neurosci 1999; 19:2037–2050.
128. Han SK, Todman MG, Herbison AE. Endogenous GABA release inhibits the firing of adult gonadotropin-releasing hormone neurons. Endocrinology 2004; 145:495–499.
129. Adler BA, Crowley WR. Evidence for gamma-aminobutyric acid modulation of ovarian hormonal effects on luteinizing hormone secretion and hypothalamic catecholamine activity in the female rat. Endocrinology 1986; 118:91–97.

130. Andersen ME, Sarangapani R, Reitz RH, et al. Physiological modeling reveals novel pharmacokinetic behavior for inhaled octamethylcyclotetrasiloxane in rats. Toxicol Sci 2001; 60:214–231.
131. Siddiqui WH, Stump DG, Plotzke KP, et al. A two-generation reproductive toxicity study of octamethylcyclotetrasiloxane (D4) in rats exposed by whole-body vapor inhalation. Reprod Toxicol 2007; 23: 202–215.
132. Quinn AL, Dalu A, Meeker LS, et al. Effects of octamethylcyclotetra-siloxane (D4) on the luteinizing hormone (LH) surge and levels of various reproductive hormones in female Sprague-Dawley rats. Reprod Toxicol 2007; 23:532–540.
133. Meeks RG, Stump DG, Siddiqui WH, et al. An inhalation reproductive toxicity study of octamethylcyclotetrasiloxane (D4) in female rats using multiple and single day exposure regimens. Reprod Toxicol 2007; 23:192–201.
134. Quinn AL, Regan JM, Tobin JM, et al. In vitro and in vivo evaluation of the estrogenic, androgenic, and progestagenic potential of two cyclic siloxanes. Toxicol Sci 2007; 96:145–153.
135. Jean PA, McCracken KA, Arthurton JA, et al. Investigation of octamethylcyclotetrasiloxane (D_4) and decamethylcyclopentasiloxane (D_5) as dopamine D2-receptor agonists. Proceedings 44th Annual Meeting of Society of Toxicology, 2005. Toxicologist 84(S-1):370.
136. Cooper RL, Stoker TE, Goldman JM, et al. Effect of atrazine on ovarian function in the rat. Reprod Toxicol 1996; 10:257–264.
137. Eldridge JC, Wetzel LT, Tyrey L. Estrous cycle patterns of Sprague-Dawley rats during acute and chronic atrazine administration. Reprod Toxicol 1999; 13:491–499.
138. Cooper RL, Stoker TE, Tyrey, L, et al. Atrazine disrupts the hypothalamic control of pituitary-ovarian function. Toxicol Sci 2000; 53:297–307.
139. Stoker TE, Perreault SD, Bremser K, et al. Acute exposure to molinate alters neuroendocrine control of ovulation in the rat. Toxicol Sci 2005; 84:38–48.
140. Cummings AM, Perreault SD, Harris, ST. Use of bromoergocryptine in the validation of protocols for the assessment of mechanisms of early pregnancy loss in the rat. Fundam Appl Toxicol 1991; 17:563–574.
141. Cummings AM, Rhodes BE, Cooper RL. Effect of atrazine on implantation and early pregnancy in four strains of rats. Toxicol Sci 2000; 58:135–143.
142. Stoker TE, Robinette CL, Cooper RL. Maternal exposure to atrazine during lactation suppresses suckling-induced prolactin release and results in prostatitis in the adult offspring. Toxicol Sci 1999; 52:68–79.
143. Laws SC, Carey SA, Hart, DW, et al. Lindane does not alter the estrogen receptor or the estrogen-dependent induction of progesterone receptors in sexually immature or ovariectomized adult rats. Toxicology 1994; 92:127–142.
144. Hong JS, Hudson PM, Yoshikawa, K, et al. Effects of chlordecone administration on brain and pituitary peptide systems. Neurotoxicology 1985; 6:167–182.
145. Uphouse L. Effects of chlordecone on neuroendocrine function of female rats. Neurotoxicology 1985; 6:191–210.
146. Gandolfi O, Cheney DL, Hong, JS, et al. On the neurotoxicity of chlordecone: a role for gamma-aminobutyric acid and serotonin. Brain Res 1984; 303:117–123.
147. Chen PH, Tilson HA, Marbury GD, et al. Effect of chlordecone (Kepone) on the rat brain concentration of 3-methoxy-4-hydroxyphenylglycol: evidence for a possible involvement of the norepinephrine system in chlordecone-induced tremor. Toxicol Appl Pharmacol 1985; 77:158–164.
148. Williams J, Uphouse L. Vaginal cyclicity, sexual receptivity, and eating behavior of the female rat following treatment with chlordecone. Reprod Toxicol 1991; 5:65–71.
149. Hoskins B, Ho IK. Chlordecone-induced alterations in content and subcellular distribution of calcium in mouse brain. J Toxicol Environ Health 1982; 9:535–544.
150. Brown HE, Salamanca S, Stewart G, et al. Chlordecone (Kepone) on the night of proestrus inhibits female sexual behavior in CDF-344 rats. Toxicol Appl Pharmacol 1991; 110:97–106.

Disclaimer: The material described in this chapter has been reviewed by the National health and Environmental Effects Research Laboratory, U.S. Environmental Protection Agency and approved for publication. Approval does not signify that the contents necessarily reflect the views and policies of the Agency, nor does mention of trade names or commercial product is constitute endorsement or recommendation for use.

16   Toxicity of the pregnant female reproductive system

Mallikarjuna Basavarajappa, Jackye Peretz, Tessie Paulose, Rupesh Gupta, Ayelet Ziv-Gal, and Jodi A. Flaws

INTRODUCTION

Pregnancy is an elaborate physiological process, which involves many events such as fertilization, implantation, placentation, and development of the embryo and fetus. Each of these events must proceed normally for pregnancy and positive birth outcomes. Unfortunately, environmental chemicals often disrupt the processes required for pregnancy and lead to adverse pregnancy outcomes. The purpose of this chapter is to provide a brief overview of each of these events during pregnancy and to describe how toxicants alter the events, leading to adverse pregnancy and birth outcomes.

CURRENT STATUS

Fertilization

Fertilization is a complex process in which male and female gametes unite to produce a genetically distinct individual. In humans, the entire process of fertilization takes place in the fallopian tube (oviduct) (Fig. 1). The oviduct is composed of distinct parts, namely the infundibulum, ampulla, and isthmus. The infundibulum terminates with the ostium surrounded by finger-like structures called fimbriae. For fertilization to occur, an egg must be released from the follicle during ovulation. This egg then must be moved into the fallopian tube by the cilia present on fimbriae. Once the egg is in the fallopian tube, it must interact with sperm to form a zygote.

The movement of the egg into the fallopian tube is a complex process. When the egg is released from the ovary, it is surrounded by cumulus cells and becomes known as an oocyte-cumulus complex (OCC) (2,3). The OCC sticks to the cilia of the fimbriae and passes smoothly over the surface of the infundibulum. The OCC, however, is too large to move through the oviduct when it reaches the ostium. Thus, it must be compacted to move easily into the lumen of the oviduct (3). The oocyte then moves from the lumen of the infundibulum to the lumen of the ampulla, where it is fertilized. It takes roughly two hours for the oocyte to reach the ampulla from the time of ovulation.

For fertilization to occur, sperm must also reach the ampulla of the oviduct. The sperm that enter the female reproductive tract undergo a process known as capacitation, which results in a pool of activated sperm (4). The capacitated sperm gain the ability to acrosome react in response to signals from the zona pellucida (ZP) of the egg (5). The acrosome-intact sperm penetrate the cumulus matrix of the OCC. This process is aided by hyaluronidase activity present on the plasma membrane of the sperm (6). Specifically, β1,4-galactosyl transferase on sperm binds to serine/threonine-linked oligosaccharide chains (i.e., O-linked oligosaccharides) of ZP3 glycoprotein residing in the ZP (7). The binding of sperm to the ZP induces the acrosome reaction and releases Ca^{2+}, ultimately leading to acrosome exocytosis (8,9). Once the sperm pass through the ZP, sperm plasma membrane proteins interact with the plasma membrane of the oocyte, leading to fusion of sperm and egg and thereby resulting in formation of a zygote.

Little is known about which chemicals alter fertilization. However, several studies have focused on the effects of alcohol and tobacco smoking on fertilization because

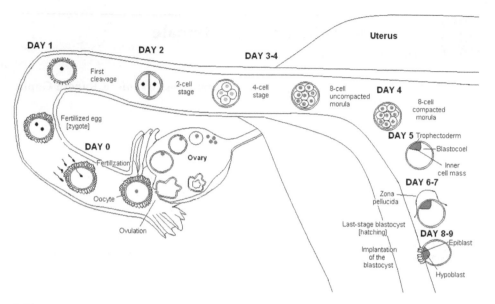

FIGURE 1 Once released from the folliclo, the egg enters the infundibulum of the fallopian tube with the help of cilia of the infundibulum, and then it travels to the ampulla where it is fertilized by a sperm. The zygote then migrates through the fallopian tube and into the uterus where it becomes a blastocyst and begins implantation into the uterine endometrial wall. *Source*: From Ref. 1.

alcohol and cigarettes are broadly used on a daily basis. The effects of alcohol on fertilization were tested by Cebral et al. who chronically exposed young female mice to moderate ethanol levels (10%) starting at the onset of sexual maturity (10). After 30 days of exposure, they examined the effects of moderate ethanol exposure on in vitro fertilization (IVF) and embryo development. Their findings indicate that moderate ethanol levels cause reduced fertilization rates and arrested embryo development. They also found that moderate ethanol exposure causes embryonic morphological abnormalities and embryonic loss.

Another toxicant that alters the fertilization is tobacco smoke. Cigarette smoke contains more than 4000 chemicals including dangerous toxins, carcinogens, metals, nicotine, and polycyclic aromatic hydrocarbons (PAHs). The effects of chemicals in cigarette smoke were summarized by Cooper and Moley who performed a literature review analysis on the effects of tobacco smoking by either or both parents on fertilization (11). Some of the fertilization-oriented impacts of smoking that they described are alterations of cilia and smooth muscle in the oviduct, which result in changes in embryo transport that could preclude implantation itself. Consistent with these findings, one study in which female mice were chronically exposed prior to conception to 7,12-dimethylbenz[a]anthracene (DMBA), a common PAH, showed a 35% loss in litters (11).

Several studies have reported that various anesthetics used during different reproductive procedures have effects on fertilization. Many general anesthetics such as nitrous oxide, isoflurane, enflurane, thiopental, fentanyl, curare derivatives, and succinyl choline used during oocyte retrieval for IVF have been associated with a decreased rate of fertilization (12). In addition, enflurane–nitrous oxide given to patients during gamete intrafallopian tube transfer reduced IVF rates compared with enflurane alone or sufentanil–nitrous oxide (12).

Embryogenesis

After fertilization, a zygote is formed and the cells in the zygote undergo mitosis to form a multicell structure. When the zygote reaches the four- to eight-cell stage, it becomes capable of expressing its own genes (13–15). The embryonic cells then compact and eventually form a 16-cell morula. The morula contains a group of inner cells covered by a group of external cells. The external cells form the trophoblast (trophectoderm) cells, which eventually form the embryonic portion of the placenta known as the chorion. The inner cells, with the supplementation of the outer cells, form the inner cell mass (ICM). The ICM gives rise to the embryo proper and its associated yolk sac, allantois, and amnion during the transition from a 16-cell stage to 32-cell stage embryo (16,17).

After the 16-cell stage, the morula develops a secretion-filled cavity called the blastocoel. As the blastocoel expands, the ICM moves to one side of the ring of trophoblast cells, resulting in a structure called a blastocyst. The ICM then differentiates into two layers: the lower hypoblast and upper epiblast. Together, the layers form a structure called the bilaminar germ disc. The hypoblast cells eventually separate from the ICM and line the blastocoel cavity, forming the extraembryonic endoderm. The extraembryonic endoderm then becomes the yolk sac. Part of the epiblast cell layer stratifies into the embryonic epiblast and the amniotic epiblast (amnionic ectoderm). The amniotic epiblast surrounds the amnionic cavity and is eventually filled with a secretion called amniotic fluid. The amniotic fluid serves as shock absorber for the developing embryo, provides nourishment, protects from heat loss, and allows easier movement of the fetus. The embryonic epiblast contains all the cells required to generate the actual embryo (18).

Few studies have examined the effects of chemicals on embryonic development. In one study, isoflurane at a concentration of 0.75% significantly blocked development of the mouse embryo to the blastocyst stage (19). Further, sera collected from patients treated with isoflurane–nitrous oxide during IVF procedures significantly impaired mouse embryo development compared with sera collected from patients given nitrous oxide and narcotics (20). Other studies show that alpha momorcharin (αMMC), a protein used for induction of abortion, anti-HIV, and inhibition of tumor growth, and extracted from bitter gourd, affects mouse embryonic development (21). These activities may be due to the ribosome inhibiting activity of the protein. αMMC affects compaction of blastomeres, impairing the morula stage; thus, leading to inhibition of early pregnancy (21). Other studies indicate that pentoxifylline, a chemical used to enhance sperm function during in vitro cultures, is embryotoxic in mice (22). When zygotes are incubated with pentoxifylline, mitosis of the zygote cells is inhibited, leading to a decreased number of cells in embryos reaching the blastocyst stage (22). Drugs such as acetaminophen at high doses depress embryonic development (23). Nitric oxide (NO) is another compound that is toxic to embryos. NO is involved in many physiological functions in the body when produced within physiological limits. At high concentrations, however, NO inhibits mouse embryo development in vitro (24). Metals have also been shown to affect embryonic development. In one study in mice, zinc was toxic at the early developmental stages of the embryo, copper was toxic to all the developmental stages of the embryo, and together the metals were strongly embryotoxic (25). Some vitamins can be toxic to embryos at high concentrations. Nicotinamide is a B-group vitamin routinely used in complex cultures. High levels of nicotinamide inhibit growth and reduce the viability of the mouse embryo (26). Another vitamin, retinoic acid, has been shown to affect the formation of the two-layer ICM, thus affecting mouse embryonic development (27). Shigella dysenteriae toxin is also known to inhibit embryonic development in mice (28). It blocks development of the embryo to the blastocyst stage by significantly decreasing adenosine triphosphate content and protein synthesis in the embryo (28). Chronic moderate alcohol treatment inhibits embryonic development in mice by blocking development of the compacted morula, and thus causing abnormal embryonic morphology and embryo loss (29). Several studies also show that pesticides affect embryonic development. At high doses, the pesticide lindane causes loss of two-cell embryos and decreases mitosis, leading to a decreased number of blastomere cells in the morula (30). The pesticide methoxychlor suppresses mouse embryonic development to the blastocyst stage, decreases embryo cell numbers, and causes abnormal blastocyst formation (31). The agricultural herbicide paraquat affects development of preimplantation embryos, but without affecting glutathione levels, which are required for protection of cells in blastocyst (32).

Implantation

When the embryo reaches the blastocyst stage, implantation occurs in the uterus. During implantation, contact is established between the blastocyst and the uterus. The process of implantation is initiated when embryo development is synchronized with the onset of uterine receptivity.

Implantation can be classified into three stages: apposition, attachment, and penetration. During apposition, trophoblast cells of the blastocyst closely attach to the uterine luminal epithelium. This process is followed by closer contact between the blastocyst and uterine layers and is termed adhesion/attachment. Attachment occurs on days 4 and 5 after fertilization in rodents and on day 8 after fertilization in humans (33–35). After attachment, the embryo penetrates the uterine epithelium and basal lamina of the uterus and enters the stroma, where it establishes a vascular relationship with the mother. In humans, implantation is intrusive because the trophoblast cells penetrate through the luminal epithelium, and then reach and extend through the basal lamina (36,37). In rodents, implantation occurs via displacement because the basal lamina underlying the luminal epithelium is dispersed to accommodate the expanding trophoblasts through the epithelium (36,37).

Various hormones are required for establishing uterine receptivity during implantation. In rodents, both estradiol (E_2) and progesterone (P_4) act in a coordinated manner to establish a window of uterine receptivity (38). The E_2 stimulates the luminal and the glandular epithelium, whereas both E_2 and P_4 are required for stromal cell proliferation during the process of implantation (33,39). In baboons and humans, chorionic gonadotrophin (CG) released by the blastocyst, in conjunction with E_2 and P_4, acts on luminal, glandular, and stromal cells to prepare the uterus for implantation (40).

Many endocrine-disrupting chemicals might interrupt implantation, leading to loss of the embryo. Some studies in animal models report that use of indomethacin, a prostaglandin inhibitor, blocks blastocyst implantation (41). Therefore, clinicians advise that pregnant women do not use analgesics or anti-inflammatory drugs such as prostaglandin synthetase inhibitors. Acute intoxication of ethanol has been shown to affect the implantation of day 4 embryos in mice (42). Pentoxifylline, a chemical used to enhance sperm function in vitro, impairs implantation of cultured mouse blastocysts (22). The protein αMMC inhibits implantation in mice when injected intraperitoneally (21). Higher doses of the organochlorine pesticide methoxychlor (800 μg/g body weight) significantly inhibit the implantation of embryos in mice, whereas lower doses of methoxychlor (400 μg/g body weight) initiate implantation in the delayed implanting mice (43). Finally, polychlorinated biphenyls (PCBs) delay implantation of the mouse embryo, resulting in delayed growth of the conceptus (44).

Placentation

The placenta develops during pregnancy in sharks and all mammals, except marsupials. It connects the mother to the conceptus and eventually allows for substance exchange between the mother and fetus. It also acts as an endocrine organ once developed, producing the necessary steroids to maintain pregnancy. Further, it protects the fetus from exposure to endogenous or xenobiotic compounds.

There are two types of placenta: a yolk sac placenta and a chorioallantoic placenta. In humans, the yolk sac placenta functions as a nutrient source for the fetus until the chorioallantoic placenta is developed during the fourth week of gestation (45). The yolk sac is also beneficial for hematopoesis, protein secretion, and primordial germ cell development. Around week 8 in humans, the yolk sac recedes into the umbilical cord. In rodents, the yolk sac placenta nourishes the fetus until the 12th day of gestation when the chorioallantoic placenta takes over (45). Unlike in humans, the rodent yolk sac placenta never recedes. Instead, it remains the principal nutrient supplying organ and allows for immunoglobulin uptake from the mother and for the protein synthesis and degradation that are required for embryonic development (46).

The chorioallantoic placenta morphologically varies among species. For example, in the sheep, the placenta has cotyledons, the vascular houses of the placenta, covering the outside, whereas in the nonhuman primate, the cotyledons are enclosed within two placental discs. In humans, rodents, guinea pigs, and rabbits, the cotyledons are encased within one placental disc. In dogs, the placenta is cylindrical in shape, not discus. Furthermore, the placenta can be hemochorial (bathed in mother's blood) as in humans, or nonhemochorial (not bathed in mother's blood) as in rodents (46).

The human hemochorial chorioallantoic placenta is formed from the trophoblast cells, blastocyst cells, and endometrial cells (Fig. 2). There are two main types of trophoblast cells: the cytotrophoblasts and the syncytiotrophoblasts. The cytotrophoblasts are small stem cells, which serve to support the syncytiotrophoblasts. The cytotrophoblasts decrease in number as gestation continues. Syncytiotrophoblasts are cytotrophoblasts that continually fuse together to make multinucleated, lateral membrane-deficient cells (48). The syncytiotrophoblasts line the outside of the finger-like villi of the apical brush border membrane that is bathed in maternal blood from the blood sinuses in the uterine endometrium. The syncytiotrophoblasts also lie adjacent to fetal capillary cells on the interior of the villi, which are physically separate from interacting with maternal blood. Syncytiotrophoblasts are the main functional cells of the placenta, acting as the interface between the maternal and fetal circulatory systems.

Although previous studies suggest that the number of cell layers between the maternal and fetal circulation influences nutrient exchange and overall placental function, recent studies indicate that it is not the number of cell layers that affects nutrient exchange and placental function. Instead, it is thought that the size of the placenta cells affects nutrient exchange and placental function. The early placenta is very thick, far less permeable, and smaller in surface area than the more developed placenta, but as the membranes of the placenta continue to grow and thin out, individual cells begin to swell and increase their surface area. This increases permeability and diffusion across the membranes, leading to increases in all placental conductances between the mother and developing fetus (49).

The placenta is also a metabolic organ that synthesizes hormones. The syncytiotrophoblast cells produce E_2 and P_4. They also produce hCG and human placental lactogen (hPL), with hCG being produced in early gestation and hPL being produced later in pregnancy (46). P_4 is made by syncytiotrophoblast cells that contain cytochrome P450 side-chain cleavage (CYP450scc) in their mitochondrial membranes. This enzyme

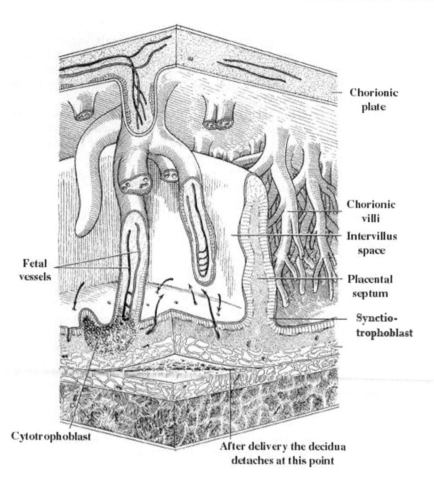

Chorionic
plate

Chorionic
villi

Intervillus
space

Placental
septum

Synctio-
trophoblast

Fetal
vessels

Cytotrophoblast

After delivery the decidua
detaches at this point

FIGURE 2 The structure of the human hemochorial chorioallantoic placenta. Trophoblasts, blastocyst cells, and endometrial cells form the placenta. The chorionic villi protrusions within the placenta are bathed in maternal blood. Vital nutrients and gases are then transferred into the villi where fetal blood vessels are present to bring them to the fetus. *Source*: From Ref. 47.

converts cholesterol into pregnenolone. The pregnenolone is then converted to P_4 by 3-beta-hydroxysteroid dehydrogenase isoform 1 (HSD3β1). P_4 is required for maintaining pregnancy by promoting relaxation of the myometrium and by curbing the labor-inducing effects of E_2, prostaglandins, and oxytocin (50). The placenta cannot synthesize E_2 without the use of androgen substrates from the fetal and maternal adrenal glands. This is because the enzyme responsible for converting pregnenolone and P_4 to dehydroepiandrosterone (DHEA) and androstenedione, CYP45017α (CYP17), is not present in the placenta (50).

While the placenta previously has been regarded as a barrier between the mother and fetus, many compounds can cross the placenta effectively. Molecules go from maternal blood, through the cytoplasm of syncytiotrophoblast cells into the embryonic villous system, then to the fetal blood through veins in the umbilical cord (51). After passing through the placenta, most chemicals enter the fetal circulation by way of the umbilical vein. Most of the blood then enters the liver of the conceptus, but a portion is directed toward the inferior vena cava and then to the heart, and therefore, the rest of the body (46).

Almost any maternally administered substance/compound has the ability to cross the placenta and affect the fetus unless it is metabolized or detoxified before placental transfer. CYP450 enzymes, as well as other metabolic enzymes, are expressed in the placenta early in gestation to protect the fetus, especially during organogenesis, from adverse substrate exposure (52). Phase 1 metabolic enzymes catalyze oxidation, reduction, and hydrolysis reactions, whereas phase 2 enzymes catalyze molecular conjugation with endogenous moieties after phase 1 reactions (46). Phase 2 enzymes are minimally active in the placenta, while phase 1 enzymes metabolize many foreign chemicals and drugs either through bioactivation or detoxification.

The metabolic activity in the placenta helps limit the accessibility of some xenobiotics and adverse substrates to the fetus. Some compounds, however, like those found in cigarette smoke, are converted to more toxic metabolites by placental cells. These toxic metabolites can pass from the mother through the placental barrier to the fetus (46). Nicotine induces expression of CYP1A1 and CYP2E1, which bioactivate metabolically inert compounds into toxic metabolites, such as benzo(a)pyrene metabolites and reactive oxygen species (ROS) (53,54). CYP1A1 is involved in biotransformation of toxicants related to oxidative stress and CYP2E1 bioactivates compounds important for creating ROS. ROS increase oxidative stress and have adverse fetal effects such as intrauterine growth retardation. CYP2E1 can also metabolize compounds that are not natural substrates in a process called futile cycling (53).

The transfer of toxic metabolites can be controlled somewhat by binding of substrates to maternal and fetal serum proteins. The binding influences rate of transfer because only free chemicals can cross the placenta. If protein binding is reversible, the substrate will cross the placenta, but at a relatively slow rate (46).

The permeability of chemicals in the placenta depends on morphological factors such as placental thickness or surface area, as well as chemical properties such as ionization constant (pK_a), lipid solubility, protein binding, and molecular weight (46). Permeability is mostly controlled by blood flow. It is thought to be rapid for nonionized, low molecular weight, lipid soluble compounds such as oxygen and carbon monoxide. Highly ionized compounds at a normal blood pH can cross the placenta easily as well (46).

While some substances can diffuse through the placental membranes (49), others like carbon dioxide or cholesterol must rely on facilitated diffusion, active transport, pinocytosis, and filtration methods to cross into fetal circulation (55). The main way the placenta protects the fetus from teratogenic compounds is through efflux transporters. The syncytiotrophoblast cells express many efflux transporters both on the apical brush border membrane and on the fetal basal membrane (56). Such transporters are the ATP-binding cassette (ABCs) drug efflux proteins, multidrug-resistance proteins (MRPs), and breast cancer–resistance proteins (BRCPs) (56). P-glycoprotein (PgP) is a member of the ABC family of efflux transporters. It is an ATP-dependent pump that solely exports xenobiotics from the fetal circulation back into the maternal blood flow, thereby reducing teratogenic exposure to the fetus. Fetal to maternal transport of PgP substrates normally exceeds that of maternal to fetal transport, but when PgP inhibitors are administered, transport is negated (48). PgP transporters are the only receptors with such specific action in the placenta. They are regulated by *mdr1* gene expression, stimulated by P_4, and localized in the syncytiotrophoblast cell layers, but not within fetal capillaries. The substrates size for PgP ranges from 200 to 1900 Da, and such substrates are usually organic, uncharged, weakly basic compounds, though some acidic compounds can bind to the transporter as well. As gestation continues, PgP expression decreases and the fetus becomes more susceptible to xenobiotic exposure (56).

Several studies reviewed by Young et al. suggest that PgPs are necessary for limiting xenobiotic exposure and removal. PgP knockout murine models have increased xenobiotic accumulation in the fetuses (48). Additionally, PgP blockers increase xeno-biotic accumulation in PgP null mice when compared with the PgP heterozygote and intact models (48).

MRPs have similar substrates to PgP, but MRPs transport cyclic nucleotides, while PgPs prefer organic cations (48). Studies indicate that there are five isoforms of MRPs, but it is unclear where all five isoforms are located in the placenta. MRP1, however, has been found mostly in the fetal capillaries and blood vessel endothelia, not in the syncytiotrophoblast cells (48).

BCRP is localized in the syncytiotrophoblast surface layer of placental chorionic villi and highly expressed in the placenta. Placental BCRP may contribute to placental protection of the fetus by transporting cytotoxic drugs and xenobiotics out of the fetus in a similar fashion to PgP (56).

While the majority of substances reach the fetus by crossing the placenta, some can pass through the amnion and into the amniotic fluid, increasing fetal exposure to adverse compounds (46). The fetus can swallow portions of the amniotic fluid during gestation, intaking up to 500 mL daily at term. Cocaine, methadone, nicotine, and their subsequent metabolites have all been found in the amniotic fluid of fetal lambs (57). While it has been suggested fetal swallowing is not the most important route for adverse chemical exposure, incessant deglutination of the amniotic fluid may cause the fetus to experience prolonged toxic exposure, as toxic chemical concentrations in the amniotic fluid are significantly correlated with both maternal and fetal serum concentrations of the xenobiotics (57).

Xenobiotic compounds, including PAHs, dithiocarbamates, metals, and endogenous opioid peptides, that bypass the placental barrier adversely affect fetal development through a variety of mechanisms of action. The major mechanisms of action include the ability of teratogenic compounds to decrease the transfer of necessary growth factors and nutrients across the placenta, decrease blood flow in the placenta, inhibit steroidogenesis in the placenta, or bioactivate inert compounds that will affect one or more of the previous areas (58). These are not exclusive events either. For example, steroidogenesis in the placenta will be inhibited if placental blood flow is decreased because the sympathetic stress response has been activated, causing vaso-constriction in the placenta, a lack of O_2 transferred to the fetus, and therefore reduced activity of CYP450scc, a necessary enzyme in steroidogenesis (58).

Some chemicals such as morphine and other opioids adversely affect placental functions by decreasing growth factor transfer across the placenta (59). The placenta is a major source of epidermal growth factors (EGFs) and insulin-like growth factors (IGF) that reach the fetus (59). Failure to transport these growth factors is detrimental to development of the fetus (59). EGFs are positively correlated with placental weight at term and EGF receptor mRNA is decreased in intrauterine growth–restricted babies (59). IGF1 levels are lower in small for gestational age babies and low–birth weight babies compared with normal weight babies. IGF1 stimulates uptake of nonmetabolized amino acid and hPL production (59), factors that are important for normal fetal growth and development.

Some chemicals also interfere with transfer of required nutrients across the placenta. Ethanol consumption inhibits vitamin B6 materno-fetal transfer, and is thought to be associated with low–birth weight babies (60). Cadmium inhibits zinc transfer across the placenta (55). This is of concern because zinc is essential for normal growth and development, and for secretion of necessary growth factors, human prolactin and hCG (55). Further, it is needed to protect against the toxic effects of cadmium (55). As the placenta protects the fetus by accumulating cadmium, more zinc is required in the placenta, decreasing the amount available for transfer to the fetus. Cocaine has been shown to prevent the influx of Ca^{2+} into cells, and to subsequently block acetylcholine (Ach) efflux (61). Cocaine used during pregnancy also down-regulates expression of the Ach receptor. Morphine activates opiate kappa receptors in cells that inhibit Ach release (62). This is a concern because Ach causes vasodilation and maintains blood flow between the maternal-placental-fetal units. Thus, any inhibition of Ach release from cells adversely affects blood flow and substance transport.

Possibly the most pronounced mechanism by which chemicals affect the transfer of substrates in the placenta is by decreasing blood flow through the villous system. Some chemicals cause vasoconstriction, which leads to restricted blood flow followed by a decrease in fetal uptake of amino acids, oxygen species, and nutrients. This has been shown to result in low–birth weight babies, intrauterine growth retardation, placental abruptions, and other malformations. Cocaine causes vasoconstriction by blocking the reuptake of norephinephrine/epinephrine, triggering the sympathetic nervous system. Nicotine from smoking releases norephinephrine/epinephrine from the adrenal medulla in the mother and releases norepinephrine from sympathetic nerve terminals. This activates the sympathetic nervous system, causing vasoconstriction in the placenta (59). The sympathetic stress response also can be activated indirectly by blocking the cellular release of Ach or uptake of Ca^{2+}. Ca^{2+} is required to release Ach, just as Ach is required for norepinephrine/epinephrine uptake (59).

Vasoconstriction also indirectly causes a decrease in hormone biosynthesis, mainly P_4 biosynthesis. Oxygen is an essential substrate for CYP450scc (50), and it is depleted during vasoconstriction. Without functioning CYP450scc, cholesterol cannot be converted to pregnenolone and then further converted into P_4. Since placenta-produced P_4 is necessary to maintain pregnancy, inhibition of P_4 leads to an increase in spontaneous abortion (50). Since P_4 also mediates PgP expression, a decrease in P_4 causes a decrease in PgP and xenobiotic efflux, thereby leading to an increase in fetal exposure to toxicants and teratogenic effects (56). Ethanol use is thought to have an inhibitory effect on P_4 secretion and to be associated with an increase in spontaneous abortion (60).

Exposure to several gases such as carbon monoxide (CO), hydrogen cyanide (HCN), and nitrogen oxides (NO_2, NO, N_2O) can deplete O_2, causing hypoxia and leading to placental problems. Cigarette smoking causes hypoxia by increasing carboxyhemoglobin (COHb) in the blood of smokers. The COHb cannot carry on respiratory function. Cigarette smoke also contains HCN and NO_2 species, which bind to hemoglobin with more affinity than O_2. Thus, there is a decrease in O_2 in the blood, leading to hypoxia. Hypoxia inhibits facilitated oxygen transfer systems and carbonic anhydrase in the placenta, increasing CO_2 levels in the fetus (59,61).

Some conceptuses are genetically predisposed to compensate for adverse effects of xenobiotics. For example, in maternal smokers, some conceptuses are genetically equipped to physiologically adapt to the teratogenic effects of smoking. Such adaptations include an increase in placenta size to sustain a small fetus, an increase in amino acid carriers in the membranes of syncytiotrophoblasts to take up more amino acids, and a decrease in membrane fluidity of the blood vessels so they are less responsive to vasoconstriction. Mostly, if the fetus can compensate for teratogenic effects and maintain normal amino acid, nutrient, and growth factor levels, they can grow and develop as if there were no adverse toxic exposure (56,59).

Parturition
Once the fetus grows and fully develops, parturition can occur. Parturition is the process by which the fetus is expelled from the uterus into the extrauterine environment

(63). In humans, there are four different stages of parturition known as phase 0, phase 1, phase 2, and phase 3 (64,65). During phase 0, also known as pregnancy or gestation, the myometrium is unresponsive to uterotonins such as prostaglandins and oxytocin. The myometrium is also relatively quiescent with low-frequency, low-amplitude, and long-duration contractions occurring throughout gestation. The quiescent nature of the myometrium is maintained by many different inhibitors such as P_4, prostacyclin, relaxin, parathyroid hormone–related protein, and NO (64,66).

Phase 1 parturition is also known as near-term activation or activation of parturition. During this phase, many changes take place in the cervix and myometrium. The cervix, which is hard and unyielding during pregnancy, softens during the activation of parturition. The cervix has an extensive extracellular matrix composed of collagen, elastin, proteoglycans, smooth muscle, fibroblast epithelial cells, and blood vessels (67–70). These cervical components undergo extensive changes just before labor, resulting in changes in the interactions between various components and infiltration of inflammatory cells.

Many factors such as relaxin and metalloproteinases (MMPs) are involved in the remodeling of the cervix (71). Further, infiltrating inflammatory cells help in cervix remodeling by increasing NO synthase activity (70). In turn, this leads to the production of prostaglandins, factors that degrade collagen and produce cytokines. The cytokines act on fibroblasts and smooth muscle cells to release proteases, which cause proteolysis of cervical cells. The inflammatory cells that invade the cervix at term also produce cytokines. In addition, the cytokines interleukin-β (IL-β) and IL-8 increase the activity of MMPs. Further, IL-β downregulates the tissue inhibitors of MMP expression, while IL-8 acts as a chemotactic factor involved in inflammatory cell invasion of the cervix (68,72–76). Thus, the inflammation, proteolysis, and inhibition of MMPs by cytokines lead to cervical opening.

As labor approaches, the myometrium becomes highly excitable and responsive to uterotonins, generating high-frequency contractions that rapidly spread to the entire uterus. Further, a marked increase in the concentration of gap junctions in the myometrium leads to high-amplitude synchronous contractions (77).

In humans and nonhuman primates, E_2 and P_4 are the major steroid hormones required for regulatory contractions. P_4 from the placenta is involved in inhibition of myometrial contractions, and E_2 from the placenta induces myometrial contractions (78). As term approaches, the concentration of P_4 plateaus and the concentration of E_2 increases, leading to a high plasma E_2-to-P_4 ratio (79). In rodents, P_4 withdrawal due to luteolysis leads to onset of labor and delivery. In humans, estrogen (estriol) stimulates production of prostaglandins, contractile-associated proteins, oxytocin, and increases expression of estrogen receptor α. This stimulates myometrial contractions ultimately leading to parturition (80,81).

Phase 2 of parturition is known as stimulation of parturition. It is characterized by strong myometrial contractions, which are stimulated by uterotonins such as prostaglandins and oxytocin. In baboons, prostaglandin E_2 (PGE_2) contracts the upper part of myometrium known as fundus, whereas prostaglandin $F_{2\alpha}$ ($PGF_{2\alpha}$) contracts both the upper and lower segments, inducing strong contractions for easier expulsion of fetus. The upper segment expresses high levels of prostaglandin receptor 3 (EP3) and low levels of EP2. The EP3 stimulates contractions in fundus. The lower part of uterus expresses high levels of EP2, but PGE_2 receptor and EP4 are expressed equally in both the parts of myometrium (82,83). In rodents, both PGE_2 and $PGF_{2\alpha}$ stimulate the upper part of myometrium, while PGE_2 inhibits the contractility of the lower part. In addition, expression of PGE_2 receptor is decreased and EP4 is increased in the lower segment, allowing the lower segment to relax and upper segment to contract at the same time for easier expulsion of the fetus (84,85). In humans, oxytocin from the posterior pituitary stimulates prostaglandin production and prostaglandin H synthase-2, and acts by increasing intracellular Ca^{2+} concentrations. Along with placental induction of corticotrophin-releasing hormone (CRH) during preterm and term labor, the CRH-binding protein concentrations fall, resulting in a net increase in concentration of CRH. The CRH stimulates prostaglandin H synthase-2 expression and androgens from the fetal adrenals. The androgens can be metabolized to estrogens that ultimately act on the myometrium (86–88). Thus CRH, oxytocin, and PGs act synergistically to produce strong myometrial contractions or relaxation for easier and successful delivery.

During phase 3 of parturition, the uterus involutes. During involution, many retrograde changes occur to return the uterus to a cyclic or nonpregnant size. Involution involves dissolution of the fetal and maternal layers of the placenta, expulsion of the placenta, and reepithelialization of the uterus (89).

Chemical exposures that affect parturition have been associated with early onset of parturition or delivery. Early onset of parturition leads to spontaneous abortion, preterm birth, and low birth weight. The sections below provide more detail on spontaneous abortion, preterm birth, and low birth weight. Further, chemical exposures that affect the pregnancy stages described above may result in birth defects. Thus, the section below also provides information on birth defects associated with chemical exposures.

Spontaneous Abortion

A spontaneous abortion is defined as an interruption of gestation before the fetus becomes viable on its own. In humans, it usually occurs before the 20th week of age or at a weight lower than 500 g. Alternatively, spontaneous loss of fetus before 20 weeks of pregnancy is also known as miscarriage (90). Early spontaneous abortions are defined as those that occur before week 12 of gestation, with late spontaneous abortions being those that occur between 12 to 20 weeks of pregnancy, and at 500 g or less (91). Numerous factors have been reported to cause spontaneous abortion, but the majority of spontaneous abortions are thought to be due to chromosomal abnormalities, anatomical uterine defects, immunotoxicity, endogenous hormonal imbalance, and environmental toxicants.

Most spontaneous abortions are thought to be due to chromosomal abnormalities (92–94). The factors that cause chromosomal abnormalities are not completely understood, but are likely to involve environmental exposure. Occupational or residential exposure to pesticides in men has been associated with an increased risk of sperm aneuploidy, which could result in spontaneous abortions if aneuploid sperm fertilize an egg (95,96). Padungtod et al. found that the crude proportion of all aneuploidies in workers exposed to organophosphates was higher than in the control group (95). Recio et al. found significant associations between organophosphate concentrations and increased frequency of sperm aneuploidies (96). Regarding possible mechanisms for these aneuploidies, it has been postulated that chemicals induce cross-links (DNA:DNA and/or DNA:protein), increasing the probability of chromosomal nondisjunction, possibly through disturbances in recombination or kinetochore and microtubule perturbations during cell division (97,98).

Spontaneous abortions may be due to structural defects in the uterus. Such uterine anomalies include congenital malformations (unicornuate uterus, uterus didelphys, bicornuate uterus, and septate uterus) and acquired uterine defects (Asherman's syndrome). Uterine defects may also be secondary to leiomyomata (fibroids) and cervical incompetence (99). Interestingly, several studies indicate that chemical exposures may be associated with uterine defects. Kaufman et al. reported that women who were exposed to diethylstilbesterol (DES) in utero had uterine abnormalities that were detected on hysterosalpingography (100). The most common DES abnormality was a hypoplastic T-shaped cavity. These uterine abnormalities are thought to be responsible for the increased rates of ectopic pregnancy and spontaneous abortion in DES-exposed patients (100).

Spontaneous abortion also may be associated with chemically induced changes in endogenous hormone levels. Given the important roles that E_2 and P_4 play in pregnancy, it is plausible that chemicals in the environment that affect these hormones might induce spontaneous abortions. Exogenous chemicals such as PCBs, bisphenol A, pesticides [specifically 1,1,1-trichloror-2,2-bis(p-chlorophenyl)ethane (DDT), and DDT metabolites, methoxychlor, lindane, endosulfan, toxaphene, and dieldrin] (101,102), cigarette smoke (103), and lead (104,105) have been shown to exhibit estrogenic and anti-estrogenic activity. The estrogenic and anti-estrogenic activity of these chemicals may be associated with an increased risk of spontaneous abortion. In a small case-control study in India, families in which the mothers or fathers worked in grape gardens where organochlorine pesticides were used experienced a nearly sixfold increase in the risk of spontaneous abortions compared with the control nonexposed group (106).

Further, spontaneous abortion may be associated with exposure to environmental contaminants such as chlorine by-products present in drinking water (107). Thus, numerous studies have assessed spontaneous abortion rates in relation to drinking water sources. When chlorine is added to drinking water, it reacts with residual organic matter to form chlorination disinfection by-products, including trihalomethanes, halo-acetic acids, and 3-chloro-4-(dichloromethyl)-5-hydroxy-2(5H)-furanone. A prospective study in California found a significant association between spontaneous abortion in women who were exposed to high levels of total trihalomethanes and to one of its constituents, bromodichloromethane (107).

Spontaneous abortion also may be associated with exposure to environmental contaminants such as dioxins. Dioxins are produced through the incomplete combustion of a variety of natural and industrial processes, and are ubiquitous in the environment. Once in the body, dioxins tend to accumulate in adipose tissue, where they remain for several years. In the late 1990s, dioxins were measured in air, soil, drinking water, and human milk and blood in Chapaevsk, Russia. In one study, drinking water levels of dioxins were the highest ever found in Chapaevsk, and the mean frequency of spontaneous abortions was found to be significantly higher in Chapaevsk relative to other towns (108).

PCB exposure has also been associated with spontaneous abortion. PCBs are a group of synthetic compounds that make up a complex mixture of up to 209 different individual chlorinated congeners (109). PCBs were used in electrical, heat transfer, and hydraulic equipment from the 1930s through to the mid-1970s because of their inflammability and insulating properties. Nonhuman primates exposed to PCBs via concrete sealant in their cages or via ingestion of 2.5 to 5 ppm PCBs in their diet had an increased incidence of spontaneous abortions and stillbirths compared with controls (110,111). The mechanism of action of PCBs is still under investigation, but studies suggest that PCBs increase the frequency of contractions of pregnant rat uteri, suggesting a possible mechanism for decreased gestational age and increased spontaneous abortion in women and animals exposed to PCBs (112).

Preterm Birth

Preterm birth in humans is defined as the birth of a baby prior to 37 weeks of gestation. More than 400,000 babies are born prematurely each year in the United States (113). Moreover, within the last two decades, there has been a steady increase in the occurrence of preterm birth in the United States (Fig. 3). Various factors can play a role in

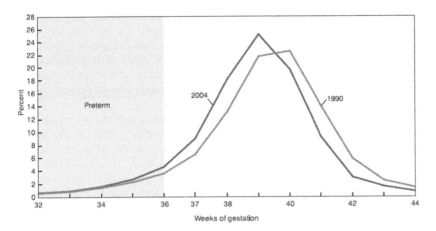

FIGURE 3 Percent distribution of births by gestational age (32–44 weeks): United States, 1990 and 2004. *Source*: From Ref. 114.

increasing the risks of a preterm birth such as pyschosocial environment, socioeconomic status, race, and maternal infections. In recent years, however, another potential risk factor that has raised a significant amount of concern is the ever-increasing levels of potential toxicants in the environment. The developing fetus is susceptible to preterm birth as a consequence of either parent being exposed to toxicants in the environment prior to conception as well as during gestation (113).

Exposure to environmental toxicants can occur either at the workplace or through air, water, and food that is contaminated with toxicants. Several studies have demonstrated that occupational exposures of either parent to toxicants can increase the risk of preterm birth (113,115–119). For example, one study showed that maternal exposure to lead was strongly associated with preterm birth (115). Paternal employment in the glass, ceramic, clay and stone, textile, and mining industries also was associated with an increase in the incidence of preterm birth. Moreover, paternal exposure to X ray and polyvinyl alcohol was associated with a 1.5-fold increase in preterm birth (115,116). In a study done in Norway, paternal exposure to radiofrequency electromagnetic fields increased the risk of preterm birth (117). Further, women working in certain industries such as metal, electrical, janitorial, food services, and textile were also found to be at a considerable risk of delivering prematurely (113). One study also showed that women working with solvents in laboratories faced a higher risk of preterm birth than women not working with solvents in laboratories (118). Similarly, another study showed that dental assistants exposed to ethylene oxide are at risk of delivering preterm babies (119).

Apart from occupational hazards, rising evidence links ambient air pollution to preterm birth. The compositions of pollutants that are found in air vary by geographical location as well as the sources that give rise to it. Some of the most common pollutants found in air are carbon monoxide, particulate matter, ozone, nitrogen dioxide, sulfur dioxide, and polyaromatic hydrocarbons (120). Numerous studies have reported that exposure to these air pollutants during early and late pregnancies is associated with preterm birth as well as other adverse reproductive and developmental outcomes (120–123). Although most of these studies were done in highly polluted areas such as the Los Angeles basin, studies done in less polluted areas such as Brisbane, Australia, showed an association between maternal exposure to low levels of ambient air pollution and preterm birth (124). Studies done in China and the Czech Republic showed that exposure to particulate matter and sulfur dioxide produced by burning coal for industrial purposes or for heating and cooking in homes increased the risk of preterm birth (125,126). One recent study done in Shanghai, China, showed that particulate matter, carbon monoxide, sulfur dioxide, and ozone were associated with an increase in preterm birth when the mother was exposed only eight weeks prior to preterm birth, but not if the mother was exposed just before birth (127).

Researchers have attempted to explain the mechanism of action that could link the exposure of air pollutants to preterm birth. One hypothesis is that all the afore-mentioned air pollutants are capable of inducing the expression of inflammatory cytokines and chemokines, leading to the production of ROS, which can in turn cause oxidative stress (120). Since the normal process of parturition involves the production of cytokines, women exposed to the air pollutants may trigger an early inflammatory response, causing them to deliver preterm (120).

The organochlorine pesticides DDT, its metabolite DDE [1,1-dichloro-2,2-bisbis(p-chlorophenyl)ethylene], as well as hexachlorobenzene (HCB) have been well studied in humans and animals and are associated with preterm birth (128–134). DDT was used for its potent antimalarial properties, but was banned in many countries including the United States after it was established as a reproductive toxicant. However, DDT is still used in countries where malaria is prevalent. Early studies with DDT and DDE demonstrated that rabbits and sea lions exposed to these chemicals have a higher rate of premature birth compared with controls (128,129). Similarly, in a small popu-lation of humans, higher levels of DDE in the serum were found in women who gave birth prematurely (130–132). Recently, researchers have conducted studies on a larger population of women in the United States and other countries exposed to DDT, and found the same result (133,134). Longnecker et al. analyzed DDE levels in frozen serum samples of mothers who delivered between 1959 and 1966, a time when DDT was being used in the United States (134). They observed that DDE concentrations were several fold higher than the current U.S. concentrations and that the adjusted odds ratio for preterm birth increased steadily with increasing concentrations of DDE in the maternal serum (134). Another study done in Spain found that premature newborns had higher concentrations of HCB, DDE, and PCB in cord serum compared with normal term babies (135). Conversely, other researchers have found no association between maternal or paternal exposure to chlorinated hydrocarbons and preterm birth (115). Similarly, in a study conducted in Mexico, there was no evidence of preterm birth in mothers who had high serum levels of HCB or DDE, but there was a nonsignificant increase in preterm birth in mothers with high levels of β-hexachlorocyclohexane (β-HCH) (133).

The lack of good animal models to study preterm birth makes it complicated to elucidate the mechanism of action of these toxicants in inducing premature labor. While it is not easy to study premature birth in rodents because monitoring the exact time of conception is difficult, Loch-Caruso et al. have approached this in an interesting manner. They performed in vitro studies by suspending uterine strips in muscle baths containing various chemicals and recorded the changes in force and frequency of spontaneous oscillatory contractions in response to the test chemical. They reported that PCB mixtures, lindane, β-HCH, DDT and its isomers, all exert direct actions on uterine contractions. Moreover, an estrogenic PCB, 4-hydroxy-2',4',6'-trichlorobiphenyl, increased uterine contractions induced by oxytocin in an estrogen receptor–dependent manner, showing that environmental chemicals may play a role in inducing preterm birth via the estrogen receptor (136).

Some other sources of toxicants that could increase the risk of preterm birth are through food and water contaminants. Some of the contaminants commonly found in drinking water are disinfection by-products and arsenic. A few studies report an increase in the risk of preterm birth in women exposed to drinking water that was contaminated with arsenic and chemicals that are by-products of disinfection (137–139). Certain kinds of food such as seafood contaminated with toxicants may be unknowingly eaten by pregnant women, causing reproductive toxicity. Although fish consumption during pregnancy has been shown to have several benefits, high mercury levels in predatory fish, sport-caught fish, and canned fish have raised concerns about the ability of fish to cause adverse reproductive outcomes. In the Pregnancy Outcomes and Community Health (POUCH) study, Holzman et al. measured the amount of total mercury in women at mid-pregnancy and examined whether there was an association of preterm birth with mercury levels (140). Women enrolled in this study resided in communities surrounding the Great Lakes, and maternal hair mercury levels were used as a measure of exposure. The study showed that there is a direct association between high hair mercury levels and preterm birth. Although the biological explanation of this needs more research, it has been shown that methylmercury produces oxidative stress at the cellular level, and that oxidative stress can induce premature labor (140,141).

Low Birth Weight

Often premature birth or pregnancy-related exposures are associated with low birth weight. Low birth weight is considered as weight less than 2500 g for a full-term baby. Low birth weight has been associated with greater risks for fetal and neonatal mortality, delayed cognitive development, adverse developments in childhood, and an increased risk of some diseases during adulthood (142). Various studies suggest that early exposure to toxicants is associated with low birth weight. One major group of toxicants of concern is PCBs because studies indicate that increased maternal serum concentrations of PCBs are associated with low birth weight (143). More specifically, a significant reduction in birth weight of approximately 500 g was found in offspring of mothers with serum PCB levels of ≥25 μg/L (143).

In addition, other chemical compounds that can be found in environmental sources (e.g., water, soil, air) may affect birth weight. Such compounds include PAHs, which are present in the air and have been strongly associated with low birth weight (144). Commonly consumed chemicals such as tobacco cigarette and caffeine have been linked to low birth weight. For example, Ward et al. found that compared to no tobacco exposure during pregnancy, domestic environmental and maternal smoking significantly lowered infants' adjusted mean birth weight by 36 g (95% CI, 5–67 g) and 146 g (122–171 g), respectively (145). Bracken et al. found that mean birth weight was reduced by 28 g for every 100 mg/day of caffeine consumed and added that caffeine consumption of ≥600 mg/day (~six 10-ounce cups of coffee) is of significant clinical concern (146). Another factor that increases the risk for low birth weight, but is less well known is the use of methadone maintenance therapy as the standard of care for opiate-addicted pregnant women (147).

Birth Defects

Toxicant exposure during pregnancy may affect different tissues or organs in the developing fetus, leading to specific birth defects (Fig. 4). The thalidomide tragedy in 1960s increased the awareness of the potentially devastating effects of toxicants in pregnant women. Thalidomide was widely used in many countries as a hypnotic and sedative in early pregnant women to treat morning sickness, anxiety, and insomnia. The children born to those women taking thalidomide had severe limb deformities, which ranged from shortened limbs to complete absence of arms and legs. The drug also caused abnormalities of the heart, genitals, kidneys, arms and legs, and digestive tract

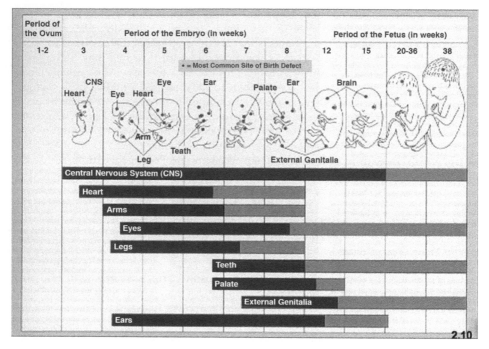

FIGURE 1 Timeframe for human development in utero including ovum, embryonic, and fetal periods of growth. *Source*: From Ref. 148.

including lips and mouth (149,150). This led to the establishment of the testing of other toxicants and drugs in pregnant animals and women. Since then many environmental chemicals such as cigarette smoke, alcohol, pesticides, and heavy metals have been tested in pregnant women or animals to determine whether they are associated with congenital malformations in the offspring.

Cigarette smoking is one of the most common exposures among the pregnant women. The chemicals in cigarette smoke have been associated with toxicity in different organs, including the urinary tract, kidney, gastrointestinal tract, and limbs (151,152). Cigarette smoking has also been associated with craniosynostosis (153). Craniosynostosis is a fetal malformation, characterized by premature closure of one or more of the cranial sutures. Smoking is also associated with testes abnormalities such as cryptorchidism in the offspring. In newborn babies, the testes spontaneously descend within six months of age. Failure of the testes to descend results in an abnormality known as cryptorchidism (154). Cryptorchidism is of concern because it is linked to low sperm count and testicular cancer. Several studies, but not all, have shown a positive association between maternal smoking and cryptorchidism in offspring (155–157).

Maternal smoking also may be associated with oral cleft defects. Oral cleft defects are birth defects affecting the formation of the mouth and/or lip during the fetal development. The incidence of oral cleft defects in the United States is 1 in every 800 infants, making it one of the most common birth defects. Although some studies suggest no direct association between maternal smoking and cleft lip or palate (158), some studies show that maternal smoking is associated with an increased risk of cleft lip or palate when accompanied by other deformities associated with smoking (158,159).

Many studies in humans have shown that maternal smoking and maternal exposure to second-hand smoke increase the risk of neural tube defects (NTD) in the offspring. The relationship between maternal smoking and the NTDs is dose dependent, with heavier smokers (half pack per day) showing higher odds of NTDs than lighter smokers (less than half pack per day). The associations between second-hand smoke exposures and NTDs and of smoking less than half a pack per day and NTDs are similar (160–163).

Smoking may also be associated with congenital heart defects. Although the etiology of many congenital heart defects is unknown, one study reported a significantly increased risk for heart defects in infants born to smoking mothers compared with nonsmoking mothers (164). However, the study did not find any association between maternal smoking and specific heart defects because of the low statistical power of the study (164). One Swedish study found an association between maternal smoking and some types of congenital heart defects such as truncus abnormalities, atrial septal defects, and persistent ductus arteriosus. The study did not find any significant association between maternal smoking and ventricular septal defects (165).

Maternal smoking during pregnancy also has been shown to affect limb development in the offspring. One study showed that maternal smoking of 10 or more cigarettes per day is associated with terminally transverse limb deficiencies (absence of the distal structures of a limb) (166). A few other studies have also found a significant association between maternal smoking during pregnancy and limb reduction malformations (167,168).

Many pesticides have been associated with several reproductive abnormalities in newborn babies. One study investigated the possible association of 27 organochlorine pesticides with the cryptorchidism in male children and found no single chemical was significantly associated with the cryptorchidism. However, additional analysis showed that the levels of eight pesticides in breast milk were higher in women with cryptorchid sons than in women without cryptorchid sons, suggesting that maternal exposure of chemicals may affect the testicular descent in newborns (169). In another study, DDT exposure was shown to reduce anogenital distance, and to increase hypospadias, cryptorchidism, and epididymal agenesis in male rats and rabbits in a dose-dependent manner (170,171). Similarly, Brucker-Davis et al. found an association between congenital cryptorchidism and fetal exposure to PCBs and DDE in humans (172). A study in Spain assessed the relationship between occupational exposure to pesticides as a result of agricultural work and the occurrence of congenital malformations (173). The data suggested a possible risk of congenital malformations with paternal pesticide exposure, namely exposure to aliphatic hydrocarbons, pyridils, inorganic compounds, and glufosinate. However, the study did not find an increased risk of birth defects with paternal exposure to organophosphates, organochlorines, carbamates, chloroalkylthio fungicides, and organosulfurs (173).

Other occupational exposures have been associated with birth defects. In one study, mothers working in the leather industry had an increased risk of having a child with congenital defects (oral cleft, cardiac defects of closure, nervous system, epispadia or hypospadia, and multiple anomalies) compared with mothers not working in leather industry (174). Similarly, maternal occupational exposure to organic solvents has been associated with an increased risk for oral clefts, digestive defects, and multiple anomalies (175).

Environmental contaminants such as mercury, HCB, and PCBs have been associated with birth defects. One study found an association between maternal fish consumption from the Bay of Augusta and risk of malformed children (176). The study determined that the fish and water were contaminated with toxic effluents (mercury, HCB, and PCBs) from nearby industries (176).

Many studies have shown that PCB and dioxin exposures in pregnant women are associated with neurodevelopment of infants, but the data are inconsistent. For example, several studies have shown a significant negative association between PCB and dioxin exposure in pregnant mothers and motor development in the newborn babies (177–179). One study in the United States has shown a positive association between prenatal exposure of PCBs and dioxins and motor development in newborn babies (180). Another study in Japan found that several isomers of dioxins significantly affect the motor development and mental development in six-month-old infants (181).

Some pharmaceutical agents have been associated with birth defects. The daughters of DES-exposed mothers have an increased risk of infertility, ectopic pregnancy, spontaneous abortion, and premature labor (182–184). Low-dose DES exposure causes several abnormalities in the reproductive tract. Specifically, low-dose DES causes vaginal adenocarcinoma, epithelial stromal hyperplasia, cystic endometrial hyperplasia, and cystic ovaries in rodents (185). High-dose DES exposure also causes reproductive abnormalities, but the abnormalities tend to be excessive keratinization of vagina, hypospadias (urethra opens into the vagina rather than the vulva), and epidermoid tumors (185). Interestingly, recent studies suggest that DES has transgenerational effects. For example, when F1 female mice born to DES-treated mothers were mated to normal untreated males, the resulting F2 male pups developed rete testes tumors and the F2 female pups developed adenocarcinoma (186).

FUTURE DIRECTIONS

Pregnancy is an intricate process, which involves many stages: fertilization, embryogenesis, implantation, placentation, and parturition. Each stage has been studied with respect to its physiology and there have been recent advancements in understanding the stages of pregnancy at the biochemical and molecular levels. Little is known, however, about which chemicals are associated with adverse pregnancy stages, particularly in humans. Little is also known about the mechanisms by which known toxicants cause toxicity of the pregnant female reproductive system. Thus, future studies should identify chemicals that cause pregnancy toxicity, and studies should investigate the mechanisms of action using animal models and humans.

Finally, spontaneous abortion, preterm birth, low birth weight, and birth defects continue to be public health issues and yet little is known about their etiology and the role that environmental chemicals play in causing these adverse pregnancy outcomes. Thus, there is a need for (*i*) improved animal and cell models to study the etiologies of these adverse pregnancy outcomes, (*ii*) genomic-based assays to classify the molecular stages of preterm birth, low birth weight, and birth defects, (*iii*) ways to identify the sources of exposure associated with increased incidence of adverse pregnancy outcomes, and (*iv*) studies to elucidate biological pathways involved in normal and adverse birth outcomes.

CONCLUSIONS

Exposure of humans and wildlife to environmental toxicants is of growing concern because of the known detrimental effects of some environment toxicants on the reproductive system. Developmental and chronic lifetime exposure to certain chemicals can compromise the fertility of mammals as well as the health of their offspring. Exposure to chemicals during pregnancy may lead to spontaneous abortion, preterm birth, low birth weight, and birth defects. It is important to be aware of potential toxicants so that exposure during pregnancy can be avoided or minimized to prevent adverse pregnancy outcomes. It is also important to understand the mechanism of action of toxicants so that treatments for toxicant-induced adverse pregnancy outcomes can be developed.

REFERENCES

1. Mader S. Biology. 9th ed. McGraw-Hill Companies, 2006.
2. Lam X, Gieseke C, Knoll M, et al. Assay and importance of adhesive interaction between hamster (mesocricetus auratus) oocyte-cumulus complexes and the oviductal epithelium. Biol Reprod 2000; 62(3):579–588.
3. Talbot P, Geiske C, Knoll M. Oocyte pickup by the mammalian oviduct. Mol Biol Cell 1999; 10(1):5–8.
4. Cohen-Dayag A, Tur-Kaspa I, Dor J, et al. Sperm capacitation in humans is transient and correlates with chemotactic responsiveness to follicular factors. Proc Natl Acad Sci U S A 1995; 92(24):11039–11043.
5. Ward CR, Storey BT. Determination of the time course of capacitation in mouse spermatozoa using a chlortetracycline fluorescence assay. Dev Biol 1984; 104(2):287–296.
6. Lin Y, Mahan K, Lathrop WF, et al. A hyaluronidase activity of the sperm plasma membrane protein PH-20 enables sperm to penetrate the cumulus cell layer surrounding the egg. J Cell Biol 1994; 125(5):1157–1163.
7. Florman HM, Wassarman PM. O-linked oligosaccharides of mouse egg ZP3 account for its sperm receptor activity. Cell 1985; 41(1):313–324.
8. Saling PM, Sowinski J, Storey BT. An ultrastructural study of epididymal mouse spermatozoa binding to zonae pellucidae in vitro: sequential relationship to the acrosome reaction. J Exp Zool 1979; 209(2):229–238.
9. Saling PM, Storey BT. Mouse gamete interactions during fertilization in vitro. Chlortetracycline as a fluorescent probe for the mouse sperm acrosome reaction. J Cell Biol 1979; 83(3):544–555.
10. Cebral E, Lasserre A, Rettori V, et al. Deleterious effects of chronic moderate alcohol intake by female mice on preimplantation embryo growth in vitro. Alcohol Alcohol 1999; 34(4):551–558.
11. Cooper AR, Moley KH. Maternal tobacco use and its preimplantation effects on fertility: more reasons to stop smoking. Semin Reprod Med 2008; 26(2):204–212.
12. Hood A, Brown J, Serafina P, et al. The effect of anesthesia on GIFT outcome. In: Witorsch RJ, ed. Reproductive Toxicology. 2nd ed. New York: Raven Press, 1995:175–193.
13. Piko L, Clegg KB. Quantitative changes in total RNA, total poly (A), and ribosomes in early mouse embryos. Dev Biol 1982; 89(2):362–378.
14. Braude P, Bolton V, Moore S. Human gene expression first occurs between the four- and eight-cell stages of preimplantation development. Nature 1988; 332(6163):459–461.
15. Prather RS. Nuclear transplantation in mammals and amphibians: nuclear equivalence, species specificity? In: Schatten H and Schatten G, eds. The Molecular Biology of Fertilization. Orlando: Academic Press, 1989:323–340.
16. Pedersen RA, Wu K, BaLakier H. Origin of the inner cell mass in mouse embryos: Cell lineage analysis by microinjection. Dev Biol 1986; 117(2):581–595.
17. Fleming TP. A quantitative analysis of cell allocation to trophectoderm and inner cell mass in the mouse blastocyst. Dev Biol 1987; 119(2):520–531.
18. Gilbert SF. The early development of vertebrates: fish, birds, and mammals. In: Gilbert SF, ed. Developmental Biology. 8th ed. Sunderland: Sinauer Associates, 2006:325–369.
19. Chetkowski RJ, Nass TE. Isofluorane inhibits early mouse embryo development in vitro. Fertil Steril 1988; 49(1):171–173.
20. Matt DW, Steingold KA, Dastvan CM, et al. Effects of sera from patients given various anesthetics on preimplantation mouse embryo development in vitro. J In Vitro Fertil Embryo Transf 1991; 8(4):191–197.
21. Tam PP, Law LK, Yeung HW. Effects of alpha-momorcharin on preimplantation development in the mouse. J Reprod Fertil 1984; 71(1):33–38.
22. Tournaye H, Van der Linden M, Van den Abbeel E, et al. Effect of pentoxifylline on implantation and post-implantation development of mouse embryos in vitro. Hum Reprod 1993; 8(11):1948–1954.

23. Laub DN, Elmagbari NO, Elmagbari NM, et al. Effects of acetaminophen on preimplantation embryo glutathione concentration and development in vivo and in vitro. Toxicol Sci 2000; 56(1):150–155.
24. Barroso RP, Osuamkpe C, Nagamani M, et al. Nitric oxide inhibits development of embryos and implantation in mice. Mol Hum Reprod 1998; 4(5):503–507.
25. Vidal F, Hidalgo J. Effect of zinc and copper on preimplantation mouse embryo development in vitro and metallothionein levels. Zygote 1993; 1(3):225–229.
26. Tsai FC, Gardner DK. Nicotinamide, a component of complex culture media, inhibits mouse embryo development in vitro and reduces subsequent developmental potential after transfer. Fertil Steril 1994; 61(2):376–382.
27. Huang FJ, Shen CC, Chang SY, et al. Retinoic acid decreases the viability of mouse blastocysts in vitro. Hum Reprod 2003; 18(1):130–136.
28. Olsen WM, Storeng R. Effect of Shigella toxin on preimplantation mouse embryos in vitro. Teratology 1986; 33(2):243–246.
29. Cebral E, Lasserre A, Rettori V, et al. Alterations in preimplantation in vivo development after preconceptional chronic moderate alcohol consumption in female mice. Alcohol Alcohol 2000; 35(4):336–343.
30. Scascitelli M, Pacchierotti F. Effects of lindane on oocyte maturation and preimplantation embryonic development in the mouse. Reprod Toxicol 2003; 17(3):299–303.
31. Amstislavksy SY, Kizilova EA, Eroschenko VP. Preimplantation mouse embryo development as a target of the pesticide methoxychlor. Reprod Toxicol 2003; 17(1):79–86.
32. Hausburg MA, DeKrey GK, Salmen JJ, et al. Effects of paraquat on development of preimplantation embryos in vivo and in vitro. Reprod Toxicol 2005; 20(2):239–246.
33. Psychoyos A. Endocrine control of egg implantation. In: Greep RO, Astwood EG, & Geiger SR, eds. Handbook of Physiology. Washington, D.C.: American Physiological Society, 1973:187–215.
34. Das SK, Wang XN, Paria BC, et al. Heparin-binding EGF-like growth factor gene is induced in the mouse uterus temporally by the blastocyst solely at the site of its apposition: a possible ligand for interaction with blastocyst EGF receptor in implantation. Development 1994; 120(5):1071–1083.
35. Enders AC. Schlafke S. Implantation in nonhuman primates and in the human. Comp Prim Biol 1986; 3:453–459.
36. Schlafke S, Enders AC. Cellular basis of interaction between trophoblast and uterus at implantation. Biol Reprod 1975; 12(1):41–65.
37. Sudhansu KD, Hyunjung K. Implantation. 2006; Third (4):147–188.
38. Huet YM, Andrews GK, Dey SK. Modulation of c-myc protein in the mouse uterus during pregnancy and by steroid hormones. Prog Clin Biol Res 1989; 294:401–412.
39. Yoshinaga K. Uterine receptivity for blastocyst implantation. Ann N Y Acad Sci 1988; 541:424–431.
40. Fazleabas AT, Donnelly KM, Srinivasan S, et al. Modulation of the baboon (Papio anubis) uterine endometrium by chorionic gonadotrophin during the period of uterine receptivity. Proc Natl Acad Sci U S A 1999; 96(5):2543–2548.
41. Kennedy TG, Squires PM, Yee M. Mediators involved in decidualization. In: Yoshinaga K, ed. Blastocyst Implantation. Boston: Adarns Publishing Group, 1989:135–143.
42. Checiu M, Sandor S. The effect of ethanol upon early development in mice and rats. IV. The effect of acute ethanol intoxication of day 4 of pregnancy upon implantation and early postimplantation development in mice. Morphol Embryol (Bucur) 1982; 28(2):127–133.
43. Hall DL, Payne LA, Putnam JM, et al. Effect of methoxychlor on implantation and embryo development in the mouse. Reprod Toxicol 1997; 11(5):703–708.
44. Torok P. Delayed implantation and early developmental defects in the mouse caused by PCB: 2,2'-dichlorobiphenyl. Arch Toxicol 1978; 40(4):249–254.
45. Woollett LA. Maternal cholesterol in fetal development: transport of cholesterol from the maternal to the fetal circulation. Am J Clin Nutr 2005; 82(6):1155–1161.
46. Slikker W Jr., Miller RK. Placental metabolism and transfer: role in developmental toxicology. In: Kimmel CA, Buelke-Sam J, eds. Developmental Toxicology. New York: Raven Press, 1994:245–283.
47. Junqueira LC, Carneiro J, Kelley RO. The female reproductive system. In: Basic Histology. Norwalk, CA: Appleton and Lange, 1980:478.

48. Young AM, Allen CE, Audus KL. Efflux transporters of the human placenta. Adv Drug Deliv Rev 2003; 55(1):125–132.
49. Korach KS, Quarmby VE. Morphological, Physiological, and Biochemical Aspects of Female Reproduction. 1985; 1(4):47–68.
50. Tuckey RC. Progesterone synthesis by the human placenta. Placenta 2005; 26(4):273–281.
51. Gedeon C, Koren G. Designing pregnancy centered medications: drugs which do not cross the human placenta. Placenta 2006; 27(8):861–868.
52. Pavek P, Dvorak Z. Xenobiotic-induced transcriptional regulation of xenobiotic metabolizing enzymes of the cytochrome P450 superfamily in human extrahepatic tissues. Curr Drug Metab 2008; 9(2):129–143.
53. Wang T, Chen M, Yan YE, et al. Growth retardation of fetal rats exposed to nicotine in utero: possible involvement of CYP1A1, CYP2E1, and P-glycoprotein. Environ Toxicol 2009; 24(1):33–42.
54. Gonzalez FJ. Role of cytochromes P450 in chemical toxicity and oxidative stress: studies with CYP2E1. Mutat Res 2005; 569(1–2):101–110.
55. Myllynen P, Pasanen M, Pelkonen O. Human placenta: a human organ for developmental toxicology research and biomonitoring. Placenta 2005; 26(5):361–371.
56. Ceckova-Novotna M, Pavek P, Staud F. P-glycoprotein in the placenta: expression, localization, regulation and function. Reprod Toxicol 2006; 22(3):400–410.
57. Bradman A, Barr DB, Claus Henn BG, et al. Measurement of pesticides and other toxicants in amniotic fluid as a potential biomarker of prenatal exposure: a validation study. Environ Health Perspect 2003; 111(14):1779–1782.
58. Harbison RD, Borgert CJ, Teaf MC. Placental metabolism of xenobiotics. In: Sastry BVR, ed. Placental Toxicology. Boca Raton: CRC Press, 1995;1(10):213–238.
59. Sastry BVR. Opiod addiction and placental function. 1995; 1(4):83–106.
60. Henderson GI, Schenker S. Alcohol, placental function, and fetal growth. 1995; 1(2):27–44.
61. Pastrakuljic A, Derewlany LO, Koren G. Maternal cocaine use and cigarette smoking in pregnancy in relation to amino acid transport and fetal growth. Placenta 1999; 20 (7):499–512.
62. Sastry BVR. Placental Toxicology—Tobacco-smoke, abused drugs, multiple chemical interactions, and placental function. Reprod Fertil Dev 1991; 3(4):355–372.
63. Challis JRG, Lye S. Parturition. In: Knobil E, Neill JD, eds. The Physiology of Reproduction. New York: Raven Press, 1994:985–1031.
64. Challis JRG, Matthews SG, Gibb W, et al. Endocrine and paracrine regulation of birth at term and preterm. Endocr Rev 2000; 21(5):514–550.
65. Challis JRG. Characteristics of parturition. In: Creasly R, Resnik R, eds. Maternal-Fetal Medicine: Principles and Practice. Philadelphia: WB Saunders, 1998:484–497.
66. Al Matubsi HY, Eis ALW, Brodt-Eppley J, et al. Expression and localization of the contractile prostaglandin f receptor in pregnant rat myometrium in late gestation, labor, and postpartum. Biol Reprod 2001; 65(4):1029–1037.
67. Leppert PC, Keller S, Cerreta J, et al. Conclusive evidence for the presence of elastin in human and monkey cervix. Am J Obstet Gynecol 1982; 142(2):179–182.
68. Ludmir J, Sehdev HM. Anatomy and physiology of the uterine cervix. Clin Obstet Gynecol 2000; 43(3):433–439.
69. Huszar GB. The physiology of the uterine cervix in reproduction. Introduction. Semin Perinatol 1991, 15(2):95–96.
70. Leppert PC, Woessner JF. The extracellular matrix of the uterus cervix and fetal membranes: synthesis, degradation and hormonal regulation. In: Phyllis C, Leppert PC, Woessner F, eds. The Extracellular Matrix of the Uterus Cervix and Fetal Membranes. Ithaca, NY: Perinatology Press, 1991:68–76.
71. Goldsmith L, Palejwala S, Weiss G. The role of relaxin in preterm labor rupture of the fetal membranes independent of infection. In: Charleston S, ed. Preterm Birth: Etiology, Mechanisms and Prevention. Birmingham: University of Alabama Press, 1997.
72. Chwalisz K, Garfield RE. Regulation of the uterus and cervix during pregnancy and labor. Role of progesterone and nitric oxide. Ann N Y Acad Sci 1997; 828:238–253.
73. Osman I, Young A, Ledingham MA, et al. Leukocyte density and pro-inflammatory cytokine expression in human fetal membranes, decidua, cervix and myometrium before and during labour at term. Mol Hum Reprod 2003; 9(1):41–45.

74. Yellon SM, Mackler AM, Kirby MA. The role of leukocyte traffic and activation in parturition. J Soc Gynecol Investig 2003; 10(6):323–338.
75. Kelly RW. Inflammatory mediators and cervical ripening. J Reprod Immunol 2002; 57 (1–2):217–224.
76. Watari M, Watari H, DiSanto ME, et al. Pro-inflammatory cytokines induce expression of matrix-metabolizing enzymes in human cervical smooth muscle cells. Am J Pathol 1999; 154(6):1755–1762.
77. Sakai N, Tabb T, Garfield RE. Modulation of cell-to-cell coupling between myometrial cells of the human uterus during pregnancy. Am J Obstet Gynecol 1992; 167(2):472–480.
78. Cunningham FG, MacDonald PC, Grant NF. The placental hormones. In: Williams Obstetrics. 18th ed. Norwalk, CT: Appleton and Lange, 1989:67–85.
79. Tulchinsky D, Hobel CJ, Yeager E, et al. Plasma estrone, estradiol, estriol, progesterone, and 17-hydroxyprogesterone in human pregnancy. I. Normal pregnancy. Am J Obstet Gynecol 1972; 112(8):1095–1100.
80. Liggins GC, Forster CS, Ggrieves SA, et al. Control of parturition in man. Biol Reprod 1977; 16(1):39–56.
81. Mesiano S, Chan EC, Fitter JT, et al. Progesterone withdrawal and estrogen activation in human parturition are coordinated by progesterone receptor A expression in the myometrium. J Clin Endocrinol Metab 2002; 87(6):2924–2930.
82. Smith GC, Baguma-Nibasheka M, Wu WX, et al. Regional variations in contractile responses to prostaglandins and prostanoid receptor messenger ribonucleic acid in pregnant baboon uterus. Am J Obstet Gynecol 1998; 176(6 pt 1):1545–1552.
83. Smith GC, Wu WX, Nathanielsz PW. Effects of gestational age and labor on the expression of prostanoid receptor genes in pregnant baboon cervix. Prostaglandins Other Lipid Mediat 2001; 63(4):153–163.
84. Wikland M, Lindblom B, Iqvist N. Myometrial response to prostaglandins during labor. Gynecol Obstet Invest 1984; 17(3):131–138.
85. Wiqvist N, Bryman I, Lindblom B, et al. The role of prostaglandins for the coordination of myometrial forces during labour. Acta Physiol Hung 1985, 65(3):313–322.
86. McLean M, Smith R. Corticotrophin-releasing hormone and human parturition. Reproduction 2001; 121(4):493–501.
87. Smith R, Mesiano S, McGrath S. Hormone trajectories leading to human birth. Regul Pept 2002; 108(2–3):159–164.
88. Smith R, Mesiano S, Chan EC, et al. Corticotropin-releasing hormone directly and preferentially stimulates dehydroepiandrosterone sulfate secretion by human fetal adrenal cortical cells. J Clin Endocrinol Metab 1998; 83(8):2916–2920.
89. Gray CA, Stewart MD, Johnson GA, et al. Postpartum uterine involution in sheep: histoarchitecture and changes in endometrial gene expression. Reproduction 2003; 125 (2):185–198.
90. U.S. National Library of Medicine and National Institutes of Health. Available at: http://www.nlm.nih.gov/medlineplus/ency/article/001488.htm.
91. Källén B. Epidemiology of Human Reproduction. Boca Raton, FL: CRC Press, 1988: 11–17.
92. Eiben B, Bartels I, Bahr-Porsch S, et al. Cytogenetic analysis of 750 spontaneous abortions with the direct-preparation method of chorionic villi and its implications for studying genetic causes of pregnancy wastage. Am J Hum Genet 1990; 47:656–663.
93. Ohno M, Maeda T, Matsunobu A. A cytogenetic study of spontaneous abortions with direct analysis of chorionic villi. Obstet Gynecol 1991; 77:394–398.
94. Zhou CR. Cytogenetic studies of spontaneous abortions in humans. Zhonghua Fu Chan Ke Za Zhi 1990; 25(2):89–91, 124.
95. Padungtod C, Hassold TJ, Millie E, et al. Sperm aneuploidy among Chinese pesticide factory workers: scoring by the FISH method. Am J Ind Med 1999; 36:230–238.
96. Recio R, Robbins WA, Borja-Aburto V, et al. Organophosphorous pesticide exposure increases the frequency of sperm sex null aneuploidy. Environ Health Perspect 2001; 109(12):1237–1240.
97. Baumgartner A, Van Hummelen P, Lowe XR, et al. Numerical and structural chromosomal abnormalities detected in human sperm with a combination of multicolor FISH assays. Environ Mol Mutagen 1999; 33(1):49–58.

98. Preston RJ. Aneuploidy in germ cells: disruption of chromosome mover components. Environ Mol Mutagen 1996; 28(3):176–181.
99. Garcia-Enguidanos A, Calle ME, Valero J, et al. Risk factors in miscarriage: a review. Eur J Obstet Gynecol Reprod Biol 2002; 102(2):111–119.
100. Kaufman RH, Adam E, Binder GL, et al. Upper genital tract changes and pregnancy outcome in offspring exposed in utero to diethylstilbestrol. Am J Obstet Gynecol 1980; 137(3):299–308.
101. National Research Council. Hormonally Active Agents in the Environment. Washington, D.C.: National Academy of Sciences, 1999.
102. Soto AM, Chung KL, Sonnenschein C. The pesticides endosulfan, toxaphene, and dieldrin have estrogenic effects on human estrogen-sensitive cells. Environ Health Perspect 1994; 102(4):380–383.
103. Key TJ, Pike MC, Brown JB, et al. Cigarette smoking and urinary oestrogen excretion in premenopausal and post-menopausal women. Br J Cancer 1996; 74(8):1313–1316.
104. Ronis MJ, Badger TM, Shema SJ, et al. Endocrine mechanisms underlying the growth effects of developmental lead exposure in the rat. J Toxicol Environ Health A 1998; 54(2):101–120.
105. Ronis MJ, Gandy J, Badger T. Endocrine mechanisms underlying reproductive toxicity in the developing rat chronically exposed to dietary lead. J Toxicol Environ Health A 1998; 54(2):77–99.
106. Rita P, Reddy PP, Reddy SV. Monitoring of workers occupationally exposed to pesticides in grape gardens of Andhra Pradesh. Environ Res 1987; 44(1):1–5.
107. Waller K, Swan SH, DeLorenze G, et al. Trihalomethanes in drinking water and spontaneous abortion. Epidemiology 1998; 9(2):134–140.
108. Revich B, Aksel E, Ushakova T, et al. Dioxin exposure and public health in Chapaevsk, Russia. Chemosphere 2001; 43(4–7):951–966.
109. McKinney JD, Waller CL. Molecular determinants of hormone mimicry: halogenated aromatic hydrocarbon environmental agents. J Toxicol Environ Health B 1998; 1(1):27–58.
110. McConnell EE, Hass JR, Altman NH, et al. A spontaneous outbreak of polychlorinated biphenyl (PCB) toxicity in rhesus monkeys (*Macaca mulatta*): toxicopathology clinical observations. Lab Anim Sci 1979; 29(5):666–673.
111. Barsotti DA, Marlar RJ, Allen JR. Reproductive dysfunction in rhesus monkeys exposed to low levels of polychlorinated biphenyls (Aroclor 1248). Food Cosmet Toxicol 1976; 14(2):99–103.
112. Bae J, Peters-Golden M, Loch-Caruso R. Stimulation of pregnant rat uterine contraction by the polychlorinated biphenyl (PCB) mixture aroclor 1242 may be mediated by arachidonic acid release through activation of phospholipase A2 enzymes. J Pharmacol Exp Ther 1999; 289(2):1112–1120.
113. Mattison DR, Wilson S, Coussens C, et al. The Role of Environmental Hazards in Premature Birth (Workshop Summary). Washington, D.C.: The National Academies Press, 2003.
114. Martin JA, Hamilton BE, Sutton PD, et al. Births: Final data for 2004. National Vital Statistics Reports. Vol. 55, No. 1. Hyattaville, MD: National Center for Health Statistics, 2006.
115. Savitz DA, Whelan EA, Kleckner RC. Effect of parents' occupational exposures on risk of stillbirth, preterm delivery, and small-for-gestational-age infants. Am J Epidemiol 1989; 129(6):1201–1218.
116. Savitz DA, Brett KM, Baird NJ, et al. Male and female employment in the textile industry in relation to miscarriage and preterm delivery. Am J Ind Med 1996; 30(3):307–316.
117. Mjøen G, Sætre DO, Lie RT, et al. Paternal occupational exposure to radiofrequency electromagnetic fields and risk of adverse pregnancy outcome. Eur J Epidemiol 2006; 21:529–535.
118. Wennborg H, Bonde JP, Stenbeck M, et al. Adverse reproduction outcomes among employees working in biomedical research laboratories. Scand J Work Environ Health 2002; 28(1):5–11.

119. Rowland AS, Baird DD, Shore DL, et al. Ethylene oxide exposure may increase the risk of spontaneous abortion, preterm birth and postterm birth. Epidemiology 1996; 7(4):363–368.
120. Ritz B, Wilhelm M. Ambient air pollution and adverse birth outcomes: methodologic issues in an emerging field. Basic Clin Pharmacol Toxicol 2008; 102:182–190.
121. Sram RJ, Binkova B, Dejmek J, et al. Ambient air pollution and pregnancy outcomes: a review of the literature. Environ Health Perspect 2005; 113(4):375–382.
122. Ritz B, Wilhelm M, Hoggatt KJ, et al. Ambient air pollution and preterm birth in the environment and pregnancy outcomes study at the University of California, Los Angeles. Am J Epidemiol 2007; 166:1045–1052.
123. Ritz B, Yu F, Chapa G, et al. Effect of air pollution on preterm birth among children born in Southern California between 1989 and 1993. Epidemiology 2000; 11(5):502–511.
124. Hansen C, Neller A, Williams G, et al. Maternal exposures to low levels of ambient air pollution and preterm birth in Brisbane, Australia. BJOG 2006; 113(8):935–941.
125. Xu X, Ding H, Wang X. Acute effects of total suspended particles and sulfur dioxides on preterm delivery: a community-based chort study. Arch Environ Health 1995; 50(6):407–415.
126. Bobak M. Outdoor air pollution, low birth weight and prematurity. Environ Health Perspect 2000; 108(2):173–176.
127. Jiang LL, Zhang YH, Song GX, et al. A time series analysis of outdoor air pollution and preterm birth in Shanghai, China. Biomed Environ Sci. 2007; 20(5):426–431.
128. Hart MM, Adamson RH, Fabro S. Prematurity and intrauterine growth retardation induced by DDT in the rabbit. Arch Int Pharmacodyn Ther 1971; 192(2):286–290.
129. DeLong R, Gilmartin WG, Simpson JG. Premature births in California sea lions: association with high organochlorine pollutant residue levels. Science 1973; 181(105):1168–1170.
130. Saxena MC, Siddiqui MK, Seth TD, et al. Organochlorine pesticides in specimens from women undergoing spontaneous abortion, premature of full-term delivery. J Anal Toxicol 1981; 5(1):6–9.
131. Procianoy RS, Schvartsman S. Blood pesticide concentration in mothers and their newborn infants. Relation to prematurity. Acta Paediatr Scand 1981; 70(6):925–928.
132. Berkowitz GS, Lapinski RH, Wolff MS. The role of DDE and polychlorinated biphenyl levels in preterm birth. Arch Environ Contam Toxicol 1996; 30(1):139–141.
133. Torres-Arreola L, Berkowitz G, Torres-Sánchez L, et al. Preterm birth in relation to maternal organochlorine serum levels. Ann Epidemiol 2003; 13(3):158–162.
134. Longnecker MP, Klebanoff MA, Zhou H, et al. Association between maternal serum concentration of the DDT metabolite DDE and preterm and small-for-gestational-age babies at birth. Lancet 2001; 358(9276):110–114.
135. Ribas-Fitó N, Sala M, Cardo E, et al. Association of hexachlorobenzene and other organochlorine compounds with anthropometric measures at birth. Paediatr Res 2002; 52(2):163–167.
136. Loch-Caruso R. A mechanistic-based approach for assessing chemical hazards to parturition. J Womens Health 1999; 8(2):235–248.
137. Yang CY, Chang CC, Tsai SS, et al. Arsenic in drinking water and adverse pregnancy outcome in an arseniasis-endemic area in northeastern Taiwan. Environ Res 2003; 91 (1):29–34.
138. Lewis C, Suffet IH, Hoggatt K, et al. Estimated effects of disinfection by-products on preterm birth in a population served by a single water utility. Environ Health Perspect 2007; 115(2):290–295.
139. Bove F, Shim Y, Zeitz P. Drinking water contaminants and adverse pregnancy outcomes. Environ Health Perspect 2002; 110(suppl 1):61–74.
140. Holzman C, Bullen B, Fisher R, et al. Pregnancy outcomes and community health: the POUCH study of preterm delivery. Paediatr Perinat Epidemiol 2001; 15(suppl 2): 136–158.
141. Xue F, Holzman C, Rahbar MH, et al. Maternal fish consumption, mercury levels and risk of preterm delivery. Environ Health Perspect 2007; 115(1):42–47.
142. Suzuki K, Tanaka T, Kondo N, et al. Is maternal smoking during early pregnancy a risk factor for all low birth weight infants? J Epidemiol 2008; 18(3):89–96.

143. Karmaus W, Zhu X. Maternal concentration of polychlorinated biphenyls and dichlorodiphenyl dichlorethylene and birth weight in Michigan fish eaters: a cohort study. Environ Health 2004; 3(1):1.

144. Choi H, Jedrychowski W, Spengler J, et al. International studies of prenatal exposure to polycyclic aromatic hydrocarbons and fetal growth. Environ Health Perspect 2006; 114 (11):1744–1750.

145. Ward C, Lewis S, Coleman T. Prevalence of maternal smoking and environmental tobacco smoke exposure during pregnancy and impact on birth weight: retrospective study using Millennium Cohort. BMC Public Health 2007; 7:81.

146. Bracken MB, Triche EW, Belanger K, et al. Association of maternal caffeine consumption with decrements in fetal growth. Am J Epidemiol 2003; 157(5):456–466.

147. Pritham UA, Troese M, Stetson A. Methadone and buprenorphine treatment during pregnancy: what are the effects on infants? Nurs Womens Health 2007; 11(6):558–567.

148. Moore KL, Persaud TVN. Before We Are Born: Essentials of Embryology Birth Defects. Philadelphia: WB Saunders Company, 1998.

149. Franks ME, Macpherson GR, Figg WD. Thalidomide. Lancet 2004; 363(9423):1802–1811.

150. Perri AJ III, Hsu S. A review of thalidomide's history and current dermatological applications. Dermatol Online J 2003; 9(3):5.

151. Li DK, Mueller BA, Hickok DE, et al. Maternal smoking during pregnancy and the risk of congenital urinary tract anomalies. Am J Public Health 1996; 86(2):249–253.

152. Kallen K. Maternal smoking and urinary organ malformations. Int J Epidemiol 1997; 26 (3):571–574.

153. Kallen K. Maternal smoking and craniosynostosis. Teratology 1999; 60(3):146–150.

154. Mayr JM, Lawrenz K, Berghold A. Undescended testicles: an epidemiological review. Acta Paediatr 1999; 88(10):1089–1093.

155. Jensen MS, Toft G, Thulstrup AM, et al. Cryptorchidism according to maternal gestational smoking. Epidemiology 2007; 18(2):220–225.

156. Akre O, Lipworth L, Cnattingius S, et al. Risk factor patterns for cryptorchidism and hypospadias. Epidemiology 1999; 10(4):364–369.

157. Biggs ML, Baer A, Critchlow CW. Maternal, delivery, and perinatal characteristics associated with cryptorchidism: a population-based case-control study among births in Washington State. Epidemiology 2002; 13(2):197–204.

158. Kallen K. Maternal smoking and orofacial clefts. Cleft Palate Craniofac J 1997; 34(1):11–16.

159. Lieff S, Olshan AF, Werler M, et al. Maternal cigarette smoking during pregnancy and risk of oral clefts in newborns. Am J Epidemiol 1999; 150(7):683–694.

160. Kelsey JL, Dwyer T, Holford TR, et al. Maternal smoking and congenital malformations: an epidemiological study. J Epidemiol Community Health 1978; 32(2):102–107.

161. Evans DR, Newcombe RG, Campbell H. Maternal smoking habits and congenital malformations: a population study. Br Med J 1979; 2(6183):171–173.

162. Seidman DS, Ever-Hadani P, Gale R. Effect of maternal smoking and age on congenital anomalies. Obstet Gynecol 1990; 76(6):1046–1050.

163. Suarez L, Felkner M, Brender JD, et al. Maternal exposures to cigarette smoke, alcohol, and street drugs and neural tube defect occurrence in offspring. Matern Child Health J 2008; 12(3):394–401.

164. Alberman ED, Goldstein H. Possible teratogenic effect of cigarette smoking. Nature 1971; 231:529–530.

165. Kallen K. Maternal smoking and congenital heart defects. Eur J Epidemiol 1999; 15 (8):731–737.

166. Czeizel AE, Kodaj I, Lenz W. Smoking during pregnancy and congenital limb deficiency. BMJ 1994; 308(6942):1473–1476.

167. Aro T. Maternal diseases, alcohol consumption and smoking during pregnancy associated with reduction limb defects. Early Hum Dev 1983; 9(1):49–57.

168. Kallen K. Maternal smoking during pregnancy and limb reduction malformations in Sweden. Am J Public Health 1997; 87(1):29–32.

169. Damgaard IN, Skakkebaek NE, Toppari J, et al. Persistent pesticides in human breast milk and cryptorchidism. Environ Health Perspect 2006; 114(7):1133–1138.

170. Gray LE, Ostby J, Furr J, et al. Effects of environmental antiandrogens on reproductive development in experimental animals. Hum Reprod Update 2001; 7(3):248–264.

171. Gray LE, Ostby J, Furr J, et al. Toxicant-induced hypospadias in the male rat. Adv Exp Med Biol 2004; 545:217–241.
172. Brucker-Davis F, Ducot B, Wagner-Mahler K, et al. Environmental pollutants in maternal milk and cryptorchidism. Gynecol Obstet Fertil 2008; 36(9):840–847.
173. Garcia AM, Fletcher T, Benavides FG, et al. Parental agricultural work and selected congenital malformations. Am J Epidemiol 1999; 149(1):64–74.
174. Garcia AM, Fletcher T. Maternal occupation in the leather industry and selected congenital malformations. Occup Environ Med 1998; 55(4):284–286.
175. Cordier S, Goujard J. Occupational exposure to chemical substances and congenital anomalies: state of the art. Rev Epidemiol Sante Publique 1994; 42(2):144–159.
176. Madeddu A, Sciacca S. Biological tracking on presence of Hg, PCB, and HCG in milk and hair of a women resident in a region with high incidence of children born with malformation (Augusta). Ann Ig 2008; 20(3 suppl 1):59–64.
177. Rogan WJ, Gladen BC. PCBs, DDE, and child development at 18 and 24 months. Ann Epidemiol 1991; 1(5):407–413.
178. Vreugdenhil HJ, Lanting CI, Mulder PG, et al. Effects of prenatal PCB and dioxin background exposure on cognitive and motor abilities in Dutch children at school age. J Pediatr 2002; 140(1):48–56.
179. Vreugdenhil HJ, Mulder PG, Emmen HH, et al. Effects of perinatal exposure to PCBs on neuropsychological functions in the Rotterdam cohort at 9 years of age. Neuropsychology 2004; 18(1):185–193.
180. Daniels JL, Longnecker MP, Klebanoff MA, et al. Prenatal exposure to low-level polychlorinated biphenyls in relation to mental and motor development at 8 months. Am J Epidemiol 2003; 157(6):485–492.
181. Nakajima S, Saijo Y, Kato S, et al. Effects of prenatal exposure to polychlorinated biphenyls and dioxins on mental and motor development in Japanese children at 6 months of age. Environ Health Perspect 2006; 114(5):773–778.
182. Bibbo M, Gill WB, Azizi F, et al. Follow-up study of male and female offspring of DES-exposed mothers. Obstet Gynecol 1977; 49(1).1–0.
183. Cousins L, Karp W, Lacey C, et al. Reproductive outcome of women exposed to diethylstilbestrol in utero. Obstet Gynecol 1980; 56(1):70–76.
184. Sandberg EC, Riffle NL, Higdon JV, et al. Pregnancy outcome in women exposed to diethylstilbestrol in utero. Am J Obstet Gynecol 1981; 140(2):194–205.
185. McLachlan JA, Newbold RR, Bullock BC. Long-term effects on the female mouse genital tract associated with prenatal exposure to diethylstilbestrol. Cancer Res 1980; 40(11):3988–3999.
186. Newbold RR, Padilla-Banks E, Jefferson WN. Adverse effects of the model environmental estrogen diethylstilbestrol are transmitted to subsequent generations. Endocrinology 2006; 147(6):11–17.

17 Epigenetic reproductive toxicants

Retha Newbold and Harriet Karimi Kinyamu

INTRODUCTION

Epigenetics is the study of gene regulation that is independent of DNA sequence. The term, reintroduced from classical embryology by Conrad Waddington in the 1950s, is what we now refer to as "developmental biology," in that an organism's phenotype arises from its genotype through programmed "epigenetic" events. For example, cells from reproductive tract tissues and hepatocytes from the same individual have identical genomes at the level of the nucleotide sequence, yet these cells have large differences in gene expression as evidenced by their differentiated cell types and function; epigenetic mechanisms are responsible for providing stable (or semi-stable) regulation of gene expression that is separate from nucleotide sequence. In fact, differentiation processes described in developmental biology largely depend on epigenetic mechanisms to orchestrate the formation of different tissues and organs, which come from a single fertilized egg at conception and grow to a fully formed individual at the end of gestation who is ready to be born, because all cells, except just a few, have the same nucleotide sequence. However, the modern definition of epigenetics used today is more narrow and specific (1) and refers to stable and heritable changes in gene expression that do not involve a change in DNA sequence or create mutations. These changes encompass an array of molecular modifications to both DNA and chromatin; the most extensively investigated are DNA methylation and histone modifications.

DNA methylation is the covalent addition of methyl groups to a cytosine residue in a CpG site (i.e., where a cytosine lies next to a guanine in the DNA sequence). CpG sites are usually clustered in high frequency near gene promoters, and these regions are referred to as CpG islands. The methylation states of CpG islands, in turn, may affect gene activity and expression. Another epigenetic mechanism responsible for modulation of gene expression is modification of histones, which affects chromatin structure. Concurrent, multiple modifications of various histones create a complex pattern, often referred to as the histone code. These modifications permit transitions between chromatin states and alterations in transcriptional activity. DNA methylation usually works together with histone modifications to activate or silence genes by influencing chromatin structure and stability, and therefore its accessibility by transcriptional factors (2). Both DNA methylation and histone modification are heritable from one cell to another. As described by Callinan and Feinberg (2), potentially hundreds of methylated cytosines in multiple genes and dozens of posttranslational chromatin modifications can arise.

Epigenetic alterations are believed to mainly occur during prenatal life or just shortly after birth. These developmental periods are characterized by high phenotypic plasticity. Some studies show that epigenetic alterations can also occur during early childhood and adolescence (3,4) but still influence gene expression differentially throughout life. Further, behavioral studies with adult rats, which underwent reversal of stable epigenetic programming associated with behavioral responses, point out that epigenetic mechanisms can also be reprogrammed or deprogrammed even in adults (5).

Such behavioral studies also help document the potential for environmental impact on the epigenome and further our understanding of the role of epigenetics in development. For example, studies demonstrate that epigenetic changes can be retained during mitosis in somatic cells since the generation-to-generation acquisition of maternal nurturing behaviors of rats is passed on to their offspring directly from the mother during the first week of postnatal life (6); for example, rat pups that received extended nurturing (licking and grooming) from their moms had reduced stress responses later in life. So, the behavioral consequences of a nurturing rat mom were calmer offspring that grew up to be less disturbed in stressful situations as compared to rat pups who were less nurtured by their moms (6). These effects in the pups, due to maternal behavior, were shown to be epigenetic involving DNA methylation and histone modifications of nerve growth factor inducible protein A binding motif in the promoter region of the brain-specific glucocorticoid receptor (GR) (7). These epigenetic changes were correlated with alterations in GR expression and behavioral responses to stress. Furthermore, histone deacetylase (HDAC) inhibitors were able to reverse the stress effects in adults, suggesting a dynamic interaction between histone modification and phenotypic or epigenetic changes.

Further, recent behavioral studies suggest that epigenetic mechanisms are ultimately a significant determinant factor in evolution since they can also affect mate preference in wildlife populations (8). Taken together, these data in experimental rat models and wildlife populations point out the complexity of the role of epigenetic modifications on the overall behavioral phenotype of an individual animal and its potential long-range consequences. These studies and others in animals provide a rich area of research linking epigenetic events during development to behavioral outcomes later in life and form a framework to study similar epigenetic events in the acquisition of human behaviors.

Even though epigenetic processes are clearly involved in genomic imprinting or programming, the exact signals for the imprint are not well understood, but it is likely that hormones may provide one of these signals. Thus, exposure to environmental factors including chemicals with hormone like activity could easily disrupt epigenetic programming occurring during development.

Epigenetics is a rapidly expanding research area, and it is becoming increasingly apparent that the original definition of epigenetics is not really different from the current modern definition in that development is largely epigenetic in origin, since different cell types maintain their distinction during cell division, while their DNA sequence remain essentially the same. Most importantly, we now know that epigenetic changes provide a potential mechanism by which environmental factors can interact with the genome and have long-lasting effects on gene expression (9) and increase susceptibility to disease/dysfunction throughout life.

Just as our ideas about epigenetic mechanisms are evolving, our understanding about their role in toxicology research continues to expand. In the past, traditional toxicology focused on the interaction of genes and the environment in susceptibility to disease by searching for DNA mutations in the coding and promoter regions of target genes. Hence, numerous oncogenes and tumor suppressor genes were subsequently described that played a role in identifying environmental factors associated with carcinogenesis. Such research efforts highlighted the importance of genotype in human disease. However, it has become clear that a full understanding of environmental interactions with the genome requires that epigenetic mechanisms also be considered. Epigenetics provides an additional layer of complexity in toxicology studies since, in addition to genetic variation, epigenetic variation might mediate the relationship between genotype and environmental factors, both internal and external. Further, we also now recognize that reproductive toxicants need to be expanded to include those that cause nonneoplastic disease, subfertility/infertility, as well as cancer. Numerous epidemiological and experimental animal studies provide ample evidence that prenatal and early postnatal environmental factors/stressors can generate developmental abnormalities or functional changes via alterations in epigenetic programming and thereby influence the risk of developing various illnesses such as reproductive disease/dysfunction as well as cardio-vascular disease, diabetes, obesity, and cancers later in life.

DEVELOPMENTAL ORIGINS OF ADULT DISEASE/DYSFUNCTION

In the early 1990s, Barker and colleagues postulated that during fetal life, organs undergo developmental programming that predetermines subsequent physiological and metabolic adaptations during adult life. This research mainly focused on in utero nutritional requirements and its role in susceptibility to disease later in life (for review, see Refs. 10,11). In brief, epidemiology studies described that "low birth weight" babies resulting from poor nutrition of their mothers had latent appearance of disease in adult life that included increased susceptibility to noncommunicable diseases, coronary heart disease, obesity/overweight, type 2 diabetes, osteoporosis, and metabolic dysfunction (12). Subsequently, chronic stress during development was also associated with similar latent responses; for example, experimental studies using macque monkeys demonstrated that early life stress resulted in obesity and increased incidences of metabolic diseases later in life (13). Maternal smoking, another fetal stressor, was also linked to the development of obesity and disease later in life (14). These studies represent some examples in the literature that have lead to a substantial research effort focusing on perinatal influences and subsequent chronic disease (15). Although these diseases may result from alterations in epigenetic programming during development, for the most part, the exact molecular alterations have not been identified.

However, other seminal work (16–18) shows that dietary modifications can have a profound effects on DNA methylation and genomic imprinting. Dietary methyl supplementation during pregnancy with folic acid, vitamin B_{12}, and other agents influenced the heritable phenotype of agouti mice offspring by altering their coat color distribution by increasing CpG methylation of the region upstream of the agouti gene (16,17). Further research showed that the early postnatal diet induced epigenetic regulation of the imprinted gene *Igf2* (19), which has been associated with several human cancers in murine animal models (20).

Interestingly, supporting evidence for the "developmental basis of adult disease" concept occurred quite independently from the field of nutrition and behavior; the concept developed a strong foundation in the field developmental toxicology where it has been long recognized that chemical exposure during sensitive periods of development could lead to long-term adverse effects. In fact, between 2% and 5% of all live births experienced major developmental abnormalities, and up to 40% of these defects were estimated to result from maternal exposures to harmful environmental agents that impacted the intrauterine environment (21,22). Although a spectrum of adverse effects can occur that range from fetal death or frank structural malformations to functional defects not readily apparent, the last may be the most common and may result in increased susceptibility to disease/dysfunction later in life. Further, these functional defects are probably the most difficult to detect because of the length of time, which may be years, between exposure and detection of the abnormality. However, there are numerous examples in experimental animals and wildlife populations that document that perinatal exposure to environmental chemicals, especially to those with hormone-like or endocrine disrupting activities (EDCs), can alter the developing organism and cause long-term effects including infertility/subfertility, retained testes, altered puberty, premature menopause, and increased cancer rates (for review, see Refs. 23–25). Taken together, nutritional studies describing an association of restricted fetal growth with the subsequent development of obesity and metabolic diseases, and experimental toxicology studies showing a correlation of perinatal exposure to EDCs with multiple long-term adverse effects, provide an attractive framework to understand delayed effects of environmental toxicant exposures.

The "developmental origins of adult disease and dysfunction" concept now incorporates ideas that are common to nutritional, behavioral, and environmental exposures. These have been discussed in detail (26) but briefly, they are as follows:

- Time-specific (vulnerable window) and tissue-specific effects may occur with nutritional, behavioral, and environmental chemical exposures.
- The initiating developmental insult (nutritional, behavioral, or environmental chemical) may act alone or in combination with other environmental agents/stressors.
- Long-term effects may be manifested in various ways such as the occurrence of a disease/dysfunction that otherwise would not have happened; an increase in risk for a disease/dysfunction that would normally be of low prevalence; an earlier onset of a disease/dysfunction that normally would have occurred; or an exacerbation of the disease/dysfunction.
- The disease/dysfunction may have a variable latency period depending on numerous factors such as the environmental stressor, time of exposure, and tissue/organ affected.
- Altered nutrition, behavioral, and/or environmental chemical exposures can lead to aberrant developmental programming that permanently alters tissue, organ or system potential. Altered or compromised function (regardless of the stressor-nutritional or chemical) can result from epigenetic changes such as altered gene expression due to effects on imprinting and the underlying methylation-related protein-DNA relationships associated with chromatin remodeling. The end result is increased susceptibility to disease/dysfunction later in life.
- Developmental nutrition, behavioral, or environmental chemical exposures can have transgenerational effects.
- Extrapolation of risk from nutritional, behavioral, and chemical exposures may be difficult because effects may not follow a monotonic dose-response relationship. Nutritional effects that result in low birth weight are different from those that result in high birth weight. Similarly, low-dose effects of environmental chemicals may not be the same as the effects that occur at higher doses.
- Environmental chemical and/or nutritional and/or behavioral effects may have entirely different manifestations in the embryo, fetus, or perinatal organism, compared to the adult.
- Exposure of one individual to an environmental stressor (environmental chemical or nutritional or combinations) may have little effect, whereas another individual will develop overt disease or dysfunctions due to differences in genetic background including genetic polymorphisms.
- The toxicant, behavioral, or nutritional-induced responses may result from altered gene expression or altered protein regulation associated with altered cell production and differentiation that are involved in the interactions between cell types and the establishment of cell lineages. These may lead to altered morphological and/or functional characteristics of tissues or organ systems. These changes may be due to altered epigenetic programming and produce changes that are irreversible.

These are key concepts in the "developmental origins of the adult disease and dysfunction" paradigm that need to be considered. The list of diseases/dysfunctions that are supported by animal data indicating a role for environmental agents during development is extensive, and it includes reproductive disorders and disease such as uterine fibroids, endometriosis, early reproductive senescence, altered fertility in the male and female, and reproductive cancers. These are summarized in Table 1. To date, there are more reported reproductive diseases/dysfunctions associated with environmental chemical exposure than with altered nutrition.

TABLE 1 Various Reported Reproductive Diseases/Dysfunctions Associated with Environmental Epigenetic Toxicants

		Disease/dysfunction
Female	Ovary	Aneuploidy
		Polycystic ovarian syndrome
		Multioocyte follicles
	Oviduct	Malformation
	Uterus/cervix	Malformation
		Endometriosis
		Fibroids(leiomyomas)
		Pregnancy loss
		Implantation disorders and miscarriage
		Altered fucundancy
		Fetal growth restriction
		Preneoplastic or neoplastic lesions
	Vagina	Adenosis
		Neoplasia
	Hypothalamic/pituitary/ovarian axis	Altered timing of puberty
		Altered menstrual/estrous cyclicity
		Premature reproductive senescence
Male	Testes	Testicular dysgenesis syndrome
		Decreased sperm numbers
		Decreased sperm fertilizability
		Subfertility/infertility
		Neoplasia
	Prostate	BPH
		PIN, preneoplasia, neoplasia
	Seminal vesicle	Abnormal gene expression
		Preneoplasia /neoplasia
Female and male		Metabolic disorders

The main focus of this review on epigenetic toxicants is on experimental animal studies, although human studies are discussed where data are available. The topics discussed are not inclusive but are provided as examples of reproductive toxicants to show "proof of principle" that environmental chemicals can alter the epigenome and have long-lasting effects. These studies were chosen to highlight the state of science in the area and to suggest future research directions. Thus, to further examine and to provide additional support for the concept, a well-known example of perinatal chemical exposure that influences developmental plasticity and results in altered programming is described.

DES: Reproductive Tract Diseases/Dysfunction in Females
The classic and prototypical example of "the developmental basis of adult disease/dysfunction" is perinatal DES exposure and its long-lasting effects on reproductive tract tissues in both humans and experimental animals. DES was prescribed to pregnant women with high-risk pregnancies in the late 1940s to 1970s, but later was found to cause a rare but significant increase in vaginal cancer in their female offspring. Subsequently, numerous benign reproductive abnormalities and dysfunctions in both the male and female offspring were shown (27,28). Since then, epidemiology studies showed that prenatal DES treatment was associated with increased risk of miscarriage, ectopic pregnancy, infertility, premature birth, uterine fibroids, premature menopause (28),

and increased breast cancer with age (29,30). Over the years, researchers have conducted extensive studies on developmental reprogramming, also referred to as "developmental estrogenization" or "estrogen imprinting" (31) using experimental animal models of DES exposure that replicated many of the abnormalities reported in DES-exposed humans. In some cases, DES animal models were successfully used to predict effects that were later shown in DES-exposed humans (e.g., early reproductive senescence, paraovarian cysts, oviductal malformations and uterine fibroids in females, and retained testes and lesions in the prostate and seminal vesicles in males) (32,33). The mechanisms responsible for many of the DES adverse effects involve epigenetic mechanisms (34–36).

DES is no longer prescribed for use during pregnancy, but it is a useful prototype estrogenic chemical that acts during development to increase disease incidence later in life. For example, treatment of outbred CD-1 mice with low doses of DES (0.1 µg/kg) only during neonatal life resulted in altered uterine response and gene expression at puberty; these alterations were irreversible (37). Further, these developmentally DES-exposed mice developed uterine carcinoma in adulthood (35).

Most importantly, increased susceptibility for tumors was also seen in second-generation (F2) DES mice (38,39); these were the offspring of mice that were exposed during fetal or early neonatal life. These F2 mice did not receive DES during development themselves; only their germ cells could have received DES exposure. Similar findings of second-generation DES effects have also been reported by other laboratories in rodent models (40–42) and in DES-exposed humans (43–45). The ability of DES to induce a transgenerational effect suggests that an epigenetic change occurred during development, and that it was transmitted through the germ line. Consistent with this hypothesis, it was shown that the increased uterine carcinomas observed in DES-treated mice were accompanied by altered methylation of specific estrogen-responsive genes such as lactoferrin (*ltf*) (34–36). Other genes also showing permanent alterations included transforming growth factor beta-inducible (*Tgfb1*), cyclin D1 (*Ccnd1*), and secreted frizzled-related protein 4 (*Sprr2f*) (34–36) among others in the uterus. (Hox and Wnt genes were reported to be permanently repressed after DES treatment (46,47) but these were related to structural malformations rather than susceptibility to tumors later in life). Neonatal DES treatment was also shown to result in increased and persistent phosphorylation of EGFR, erbB2 and ERα along with an elevation of c-fos expression, effects capable of activating the receptor-mediated pathways in a ligand-independent manner, thus providing plausible epigenetic mechanisms for the adverse effects of neonatal DES exposure (34); it is likely that these effects are related to epigenetic changes in estrogen responsive genes.

Recently, using the neonatal DES mouse model just described (33), DES or the phytoestrogen genistein were shown to cause persistent hypomethylation in the promoter of nucleosome-binding protein 1 (*Nsbp1*), the gene for nucleosome-core-particle–binding protein important for its central role in chromatin remodeling; hypomethylation resulted in the gene being persistently overexpressed throughout life (36). However, if neonatal DES- or genistein-treated mice were ovariectomized before puberty, the *Nsbp1* promoter CpG island remained minimally to moderately methylated, and gene expression was subdued (36). Thus, the permanent reprogramming of uterine *Nsbp1* expression by neonatal DES or genistein appeared to be moderated by an epigenetic mechanism that, in turn, interacted with ovarian hormones in adulthood. This demonstrates that, in this particular case, epigenetic reprogramming involved a two-step process where estrogens during development first altered gene expression and then ovarian estrogens at puberty influenced it again. Interestingly, a "second hit model" has long been used when referring to genetic changes and the development of cancer (48), but now it appears that a parallel two-step model may also apply to epigenetic alterations as well (36,49–51). All together, these novel findings with DES provide important evidence that epigenetic reprogramming by early life exposure to estrogens may involve multiple and complex epigenetic pathways that combine together to initiate disease later in life.

Uterine Fibroids
Neonatal treatment of mice with DES was also associated with uterine leiomyomas (fibroids) later in life as reported following prenatal treatment (32,52). Together, these prenatal and neonatal animal studies provided the background for the recently reported findings in DES-exposed women showing a significant increased risk of prenatal DES being associated with the development of fibroids in adulthood (53). The etiology of fibroids is currently unclear, but additional investigation into prenatal exposures may provide a link and perhaps suggest altered epigenetic programming during development.

Results with other experimental animal models have also shown a link between developmental exposure to DES and uterine fibroids. Eker rats carrying a defect in a tumor suppressor gene develop abnormalities including uterine fibroids (54). Early life exposures of the Eker rat to DES increased this susceptibility for uterine fibroids because the defective tumor suppressor gene resulted in reprogramming of the myometrium leading to an increased expression of estrogen-responsive genes (55). Thus, later in life, DES-exposed Eker rats developed high incidence uterine fibroids. This experimental animal model provides still another example of gene-environment interactions in the "developmental basis of adult disease/dysfunction" paradigm.

Ovarian Differentiation
DES and other environmental estrogens including genistein also affect ovarian follicle development in outbred mice if given during critical windows of differentiation (56); the result is an increase in polyovular follicles that is associated with increased problems with fertility (57) and that have reduced fertilization potential (25). Developmental exposure of Sprague–Dawley rats to DES also induces morphological and functional abnormalities in the adult ovary including altered gene expression of steroidogenic enzymes (58). Again, it is likely that epigenetic mechanisms are responsible.

Thus, DES serves as "proof of principle" that environmental chemicals can impact the epigenome and cause a multitude of diseases/dysfunctions when exposure occurs during development. The mechanisms involved in DES-induced effects in animals have been shown to be associated with altered gene expression due to altered epigenetic marks. Although similar epigenetic events occur in humans during development, more studies are necessary to show these epigenetic mechanisms are also involved in misprogramming for disease later in life in the DES-exposed human population.

Methoxychlor and Vinclozolin: Decreased Sperm Count and Fertility
Other compelling data showing an impact of environmental chemicals on the epigenome resulting in long-term reproductive toxicity come from studies with methoxychlor and vinclozolin. Prenatal exposure to either of these two chemicals was reported to cause adverse effects in adult male mice that manifested as increased testicular apoptosis of developing germ cells leading to infertility starting at 100 days of age in up to 20% of the exposed mice (59,60). Unlike transgenerational effects of DES that occurred through the maternal germ line (39), developmental exposure to methoxychlor or vinclozolin caused adverse effects to be transmitted through the male germ line and further, these effects were seen for at least four generations (60). The transgenerational effects were noted only when prenatal exposure correlated with critical developmental processes such as testicular cord formation, sex determination, and the time of germ cell

remethylation. If exposure occurred later during testis development, the transgenerational effects could not be seen; however, different adverse effects were still shown in the adult. Although these data with methoxychlor and vinclozolin remain to be replicated in other labs, the specific timing of exposure appears to be essential in altering epigenetic programming and it speaks to the idea of critical windows of exposure and/or various windows of susceptibility depending on the particular differentiating tissues. Although many other investigators have shown significant reproductive effects of developmental exposure to environmental chemicals, this is a unique example, along with DES, where the effects were demonstrated to pass along to subsequent generations. Interestingly, prenatal dioxin exposure has been shown to stunt mammary gland development and increase tumor incidence later in life; further, these effects were also shown in second-generation pups (61).

Unfortunately, often times, experimental animal studies are not designed to look for transgenerational reproductive effects, thus the dosing may be too early or too late during development to study altered epigenetic programming. Timing of exposure and evaluation are crucial factors that have to be considered.

DES, Dioxin, Estradiol, and Bisphenol A: Increased Prostate Cancer Susceptibility

It has been known for over 70 years that estrogen is involved in both the normal process of prostate development and adult prostate disease. However, the major role of estrogens in benign prostatic hyperplasia (BPH) and prostatic cancer was thought to be direct and involving DNA-altering events; thus, most studies focused on measuring levels of estrogen in the tissue or serum, or determining estrogen markers at the time prostatic disease was identified. This approach failed to detect consistent differences and, in some cases, the findings appeared to be contradictory between human and animal studies, and sometimes among animal studies. Thus, an alternative hypothesis was proposed that estrogens could be a predisposing factor for prostatic disease in which the critical time for estrogen action is in early life when estrogens are most influential on prostate development (62). In this scenario, the critical time for estrogen exposure would precede (sometimes by many years) the clinical appearance of the disease; thus, the differences in estrogen concentrations and/or markers may not be detectable at the time of tumor/disease manifestation. Conceptually, estrogens were proposed to cause defects in the differentiation of prostatic cells, either genetic and/or epigenetic, that would then lead to altered growth and neoplastic transformation later in life. This hypothesis was consistent with prostatic disease and reproductive tract abnormalities identified in males following developmental exposure to DES in both animal models and humans (reviewed in Ref. 63). In light of the current interest of endocrine disrupting chemicals, we are again reminded that excessive estrogen exposure during development may contribute to the high incidence of prostate cancer later in life.

In utero and lactational exposure to low doses of dioxin at levels close to background body burdens in humans also alters prostate development; the alterations seem to be lobe specific with the ventral prostate being the most sensitive followed by the anterior and then the dorsolateral lobe (64). Dioxin exposure caused decreased formation of prostatic buds and agenesis of ventral buds in mice, essentially preventing the ventral prostate from developing. This effect was found to be related to activation of the aryl hydrocarbon (Ah) receptor and was susceptible to regulation by estrogen levels in the developing fetus (65). Thus, androgen-dependent prostate growth that occurs normally with age was prevented by in utero and lactational dioxin exposure. This is another example of permanent altered gene expression by in utero exposure to dioxin. Interestingly, there are also data implicating increasing WNT5 signaling in the TCDD inhibition of ventral prostate bud formation (66). The mechanisms for these permanent alterations in gene expression may be altered tissue imprinting.

Other studies also suggested altered prostate development following prenatal exposure to various chemicals. In utero exposure to the environmental estrogens BPA, methoxychlor (an estrogenic pesticide), or DES at low levels produced an increase in number and size of dorsolateral prostate ducts in the male fetus and an overall increase in prostate volume due to increased proliferation of basal epithelial cells resulting in increased prostate weight in adulthood (67,68). Further, there was a permanent increase in prostatic androgen receptors in adult male mice due to fetal exposure to low doses of DES, BPA, estradiol, or estradiol (69). Again epigenetic factors may play a role in the altered programming of developing prostate cells.

The most compelling evidence for altered epigenetic programming in prostatic disease comes from the collaborations of the Ho and Prins laboratories (70). As noted (51), since the prostate is highly dependent on steroid hormones during development, it is likely that alterations in these levels by exogenous environmental chemicals can result in aberrant prostate growth. These investigators showed that brief exposure of rats to natural or synthetic estrogens permanently alters the prostate resulting in reduced growth, differentiation defects, aberrant gene expression, and perturbations in cell signaling mechanisms (70). They further proposed that excessive estrogen exposure during the critical period of development was a predisposing factor for prostate disease later in life. Thus, to determine if in utero or neonatal exposures are associated with prostate cancer-like effects later in life, effects that could be extrapolated to human prostate hyperplasia and cancers, they developed a "two hit model" in which rats were transiently exposed to low doses of BPA during development and then further exposed to a "second hit" of estrogen in the form of estradiol on day 90. Animals were followed until 28 weeks of age and then examined for hyperplasia, inflammation, and prostatic intraepithelial neoplasia (PIN), the presumed precursor of prostate cancer. Neonatal exposure to low-dose BPA plus estradiol produced a significant increase in PIN lesions characterized by severe cellular atypia later in life. Most importantly, these histological changes were correlated with permanent alterations in DNA methylation patterns of several cell-signaling genes; for example, methylation of the gene coding for the enzyme phosphodiesterase 4 was permanently altered in prostatic tissue. Normally, this gene is expressed at very low levels in adults, but after early exposure to estrogenic chemicals, the animals continue to have high levels of expression of the gene throughout life. Interestingly, additional studies using cell cultures showed that high levels of this gene were identified in prostate cancer cell lines. Thus, altogether these studies indicate that exposure to low-dose estrogens during development can affect the prostate epigenome and promote prostate disease with aging.

Phthalates: Testicular Dysgenesis Syndrome

Developmental exposure to certain phthalate esters (dibutyl phthalate and diethylhexyl phthalate) causes a syndrome of effects in the offspring including malformations of the epididymis, vas deferens, seminal vesicles, prostate, hypospadias, retention of the nipples/areolae, cryptorchidism, and reduced anogenital distance (71). These abnormalities are thought to be due to the ability of these phthalates to reduce testosterone production by way of altering gene expression of a number of genes involved in testosterone biosynthesis in testicular Leydig cells. The decrease in testosterone also causes functional alterations later in life including decreased sperm production and testicular tumors. The phthalates syndrome in the rodent model has many similarities to the testicular dysgenesis syndrome observed in humans although there is little evidence for involvement of phthalates in the human syndrome.

EPIGENETICS AND TRANSMISSION OF DEFECTS
TO SUBSEQUENT GENERATIONS

Probably one of the most important emerging concepts is that epigenetic disease has the potential to be passed to subsequent generations. There are numerous studies in animal models that show the transmission of susceptibility to tumors from one generation to the next following exposure during development to estrogenic chemicals; many have already been discussed. DES involved transmission through the female germ line, and effects were seen in both the granddaughters and the grandsons (39,41,42). Further, multigenerational effects in humans have also been reported in DES-exposed grand-daughters and grandsons (43–45,72). This phenomenon is not estrogen specific since exposure to methoxychlor and vinclozolin also caused germ cell alterations and reduced fertility across multiple generations (59,60). Likely, other chemicals can have similar transgenerational effects. Thus, the implications for environmental exposures to impact future "unexposed" populations cannot be underestimated.

The studies above suggest that environmental toxicants can alter the epigenome to affect gene expression that can impact on later generations resulting in dysfunction/disease later in life. The caveat of most of the current studies is the lack of definitive epigenetic mechanisms that lead to these effects. Only a few studies, however, have attempted to address this issue.

For example, recent studies connected exposure to compounds that mimic estrogen and androgen to phenotypic changes by monitoring DNA methylation of the germ line loci (59,60,73,74). In support for a role of DNA methylation in the distribution of color among the offspring, maternal dietary supplementation, with either methyl donors like folic acid or the phytoestrogen genistein, negated the DNA hypomethylating effect of BPA to preserve agouti color (74). In trying to associate epigenetic mechanisms to uterine adenocarcinoma after neonatal exposure of CD-1 mice to DES or genistein, Tang et al. (36) have used methylation-sensitive restriction fingerprinting and demonstrated that nucleosomal binding protein 1 gene is hypomethylated, after treatment leading to persistent over expression this gene throughout life. And, as described earlier, studies on prostatic disease have shown permanent alterations in DNA methylation patterns of several cell signaling genes following developmental exposure to environmental estrogens (70).

Another example specifically addressing epigenetic mechanisms comes from behavioral studies that have demonstrated that maternal care alters epigenetic programming of GR gene expression in the hippocampus (7). These studies have been previously described, but in brief, using sodium bisulfite mapping and chromatin immunoprecipitation (ChIP), the authors demonstrated that postnatal maternal licking/grooming behavior inhibited DNA methylation and enhanced acetylated histone H3 occupancy on the GR promoter within the rat hippocampus. HDAC inhibitors were able to reverse GR expression at the hypothalamus-pituitary axis, further supporting a dynamic interplay between histone modification and phenotypic changes or epigenetic alteration (7). Such reports provide evidence that environmental factors can cause epigenetic alterations including DNA hyper- and hypomethylation and posttranslational modification of histones that provide another layer of epigenetic regulation.

In addition to chromatin modification, there is overwhelming recognition that noncoding RNAs are a part of epigenetic regulation. A good example is the Xist RNA that silences the X chromosome (reviewed in Ref. 1). This could be especially important in sex determination and expression of imprinted genes. For example, embryonic exposure to the endocrine disruptor vinclozolin at the time of gonadal sex determination was previously found to promote transgenerational disease states implying epigenetic alterations in the male germ line are transmitted to subsequent generations

(60). Analysis of the transgenerational epigenetic effects on the male germ line (i.e., sperm) identified 25 candidate DNA sequences with altered methylation patterns in the vinclozolin generation sperm that are mapped to specific genes and noncoding DNA regions (60). The authors suggest this reprogramming of the male germ line involves the induction of new imprinted genes/DNA sequences that acquire an apparent permanent DNA methylation pattern that is passed at least through the paternal allele.

Other studies show evidence of nongenetic transgenerational inheritance with conspicuous marked sexual dimorphism, mainly influenced by imprinted genes, suggesting that epigenetic marks including noncoding RNA/DNA may contribute to environmental and nutritional factors or endocrine disruptors to sex-specific phenotypes (reviewed in Ref. 75).

The importance of chromatin modifications and noncoding RNA/DNA is only just beginning to be realized, and the need for a more complete understanding of these mechanisms involved in environmental alteration of the epigenome is critical for our understanding of human disease.

FUTURE DIRECTIONS—WHAT ARE THE KEY RESEARCH QUESTIONS?

Many basic research questions will benefit from a focus on epigenetics. The evidence is overwhelming that environmental factors may alter the epigenome, but only a few studies have tried to address the mechanisms involved (see the preceding text). The most relevant question now is how to address those mechanisms. There is a recent explosion of techniques that can measure epigenetic events in genome wide scale. The combination of ChIP and DNA microarrays (ChIP-chip) and more recently various sequencing-based protocols that analyze ChIP samples such as the combination of ChIP with massive parallel sequencing (ChIP-Seq) allows toxicology researchers to survey the whole genome for epigenetic marks and to unveil new aspects of biology after various treatment with environmental toxicants (reviewed in Ref. 76). These experiments could first be done with some of the known epigenetic marks known to influence various diseases. For example, posttranslation modifications of histone H4: Loss of acetylation of histone H4 lysine 16 and trimethylation of lysine 20 are common hallmarks of human cancer (77). Use of antibodies against these marks to perform ChIP using tissue samples from challenged animals at various stages of life could give an idea where these marks are within the genome and the characteristics of these loci in absence and presence of environmental toxicants. These techniques can also be modified to study DNA methylation at the whole genome level. For example, immunoprecipitation of methylated DNA (mCIP) can be monitored by immunuprecipitating DNA with an antibody against methylated cytosines that pulls down methylated regions and combining this with direct sequencing can be used to study DNA methylation on a genomic scale (76). A combination of new technology and newly expanding understanding of chromatin marks will undoubtedly expand our knowledge on the effects of environmental toxicants on the epigenome and impact on disease states.

CONCLUSIONS

Clearly, there are many other things that remain to be done to link epigenetically altered genes and the ensuing dysfunction or disease. This is true for cancers of the uterus, breast, testes, prostate, etc. and diseases that result from imprinting. Is it possible that there are secondary triggers for repressed epigenetic memories aside from hormones? Can an "epigenetic fingerprint" be developed that consists of multiple genes that may be similar or unique for different chemicals, end organs, or second hits? If so, these could be used in the future for early detection of altered disease susceptibility as a result of known or unknown early-life exposures. Epigenetic fingerprinting could also be used to screen and identify environmental factors/chemicals that cause specific alterations and thereby allow the development of sensitive risks assessment biomarkers. Ultimately, novel interventions and novel disease therapies focused on the epigenome, generally or targeted to specific pathways, could be developed. Numerous possibilities provide exciting areas of new research in the field of epigenetics.

SUMMARY

Epigenetics is defined as the study of heritable changes in gene function that occur without a change in the DNA sequence. Environmental chemicals especially those with endocrine-disrupting activity can alter epigenetic programming that normally occurs during embryogenesis and development, and result in dysfunction/disease later in life. The epigenome is very vulnerable to dysregulation by environmental factors during critical windows of differentiation because of the high degree of developmental plasticity and the high rate of DNA synthesis that is occurring. Most importantly, elaborate epigenetic programming such as DNA methylation patterning required for normal tissue development is established during this period. Alterations in epigenetic events including changes in DNA methylation, posttranslational modifications of histones involved in chromatin structure, and noncoding RNA/DNA may be permanent and transmitted to subsequent generations causing adverse consequences in "unexposed" populations. Epigenetic reprogramming by early life exposure to endocrine disrupting chemicals is likely complex and may involve multiple epigenetic pathways that combine together to initiate disease/dysfunction later in life. Examples of reproductive epigenetic toxicants including diethylstilbestrol (DES), methoxychlor, vinclozolin, dioxin, estradiol, and bisphenol A are discussed. Finally, the significance of epigenetic changes in chemical toxicity studies is summarized and future directions for research are outlined.

ACKNOWLEDGMENTS

The author thanks Dr. Wendy Jefferson for her comments and critical review of this manuscript. This research was supported by the Intramural Research Program of the NIH, National Institute of Environmental Health Sciences.

REFERENCES

1. Berger SL, Kouzarides T, Shiekhattar R, et al. An operational definition of epigenetics. Genes Dev 2009; 23(7):781–783.
2. Callinan PA, Feinberg AP. The emerging science of epigenomics. Hum Mol Genet 2006; 15(Spec No 1):R95–R101.
3. Kurukulaaratchy RJ, Matthews S, Arshad SH. Does environment mediate earlier onset of the persistent childhood asthma phenotype? Pediatrics 2004; 113(2):345–350.
4. Mandhane PJ, Greene JM, Cowan JO, et al. Sex differences in factors associated with childhood- and adolescent-onset wheeze. Am J Respir Crit Care Med 2005; 172(1):45–54.
5. Weaver IC, Champagne FA, Brown SE, et al. Reversal of maternal programming of stress responses in adult offspring through methyl supplementation: altering epigenetic marking later in life. J Neurosci 2005; 25(47):11045–11054.
6. Szyf M, Weaver I, Meaney M. Maternal care, the epigenome and phenotypic differences in behavior. Reprod Toxicol 2007; 24(1):9–19.
7. Weaver IC, Cervoni N, Champagne FA, et al. Epigenetic programming by maternal behavior. Nat Neurosci 2004; 7(8):847–854.
8. Crews D, Gore AC, Hsu TS, et al. Transgenerational epigenetic imprints on mate preference. Proc Natl Acad Sci USA 2007; 104(14):5942–5946.
9. Wade PA, Archer TK. Epigenetics: environmental instructions for the genome. Environ Health Perspect 2006; 114(3):A140–A141.

10. Gluckman PD, Hanson MA, Beedle AS. Early life events and their consequences for later disease: a life history and evolutionary perspective. Am J Hum Biol 2007; 19(1):1–19.
11. Hanson MA, Gluckman PD. Developmental origins of health and disease: new insights. Basic Clin Pharmacol Toxicol 2008; 102(2):90–93.
12. Barker DJ, Eriksson JG, Forsen T, et al. Fetal origins of adult disease: strength of effects and biological basis. Int J Epidemiol 2002; 31(6):1235–1239.
13. Kaufman D, Banerji MA, Shorman I, et al. Early-life stress and the development of obesity and insulin resistance in juvenile bonnet macaques. Diabetes 2007; 56(5):1382–1386.
14. Levin ED. Fetal nicotinic overload, blunted sympathetic responsivity, and obesity. Birth Defects Res A Clin Mol Teratol 2005; 73(7):481–484.
15. Gluckman PD, Hanson MA, Pinal C. The developmental origins of adult disease. Matern Child Nutr 2005; 1(3):130–141.
16. Cooney CA, Dave AA, Wolff GL. Maternal methyl supplements in mice affect epigenetic variation and DNA methylation of offspring. J Nutr 2002; 132(8 suppl): 2393S–2400S.
17. Waterland RA, Jirtle RL. Transposable elements: targets for early nutritional effects on epigenetic gene regulation. Mol Cell Biol 2003; 23(15):5293–5300.
18. Waterland RA, Jirtle RL. Early nutrition, epigenetic changes at transposons and imprinted genes, and enhanced susceptibility to adult chronic diseases. Nutrition 2004; 20(1):63–68.
19. Waterland RA, Lin JR, Smith CA, et al. Post-weaning diet affects genomic imprinting at the insulin-like growth factor 2 (Igf2) locus. Hum Mol Genet 2006; 15(5):705–716.
20. Feinberg AP, Tycko B. The history of cancer epigenetics. Nat Rev Cancer 2004; 4(2):143–153.
21. Heindel JJ, Levin E. Developmental origins and environmental influences—Introduction. NIEHS symposium. Birth Defects Res A Clin Mol Teratol 2005; 73(7):469.
22. Heindel JJ, McAllister KA, Worth L Jr., et al. Environmental epigenomics, imprinting and disease susceptibility. Epigenetics 2006; 1(1):1–6.
23. Colborn T, vom Saal FS, Soto AM. Developmental effects of endocrine-disrupting chemicals in wildlife and humans. Environ Health Perspect 1993; 101(5):378–384.
24. Colborn T, Dumanoski D, Myers JP. Our Stolen Future. New York: Penguin Group, 1996:1–306.
25. Crain DA, Janssen SJ, Edwards TM, et al. Female reproductive disorders: the roles of endocrine-disrupting compounds and developmental timing. Fertil Steril 2008; 90(4): 911–940.
26. Newbold RR, Heindel JJ. Developmental origins of health and disease: the importance of environmental exposures. In: Newnham JP, Ross MG, eds. Early Life Origins of Human Health and Disease. Basel: Karger, 2009.
27. NIH. DES Research Update: NIH Publication No. 00-4722. Bethesda, MD, 1999.
28. Palmlund I. Exposure to a xenoestrogen before birth: the diethylstilbestrol experience. J Psychosom Obstet Gynaecol 1996; 17(2):71–84.
29. Palmer JR, Hatch EE, Rosenberg CL, et al. Risk of breast cancer in women exposed to diethylstilbestrol in utero: preliminary results (United States). Cancer Causes Control 2002; 13(8):753–758.
30. Palmer JR, Wise LA, Hatch EE, et al. Prenatal diethylstilbestrol exposure and risk of breast cancer. Cancer Epidemiol Biomarkers Prev 2006; 15(8):1509–1514.
31. McLachlan JA, Newbold RR, Burow ME, et al. From malformations to molecular mechanisms in the male: three decades of research on endocrine disrupters. APMIS 2001; 109(4):263–272.
32. Newbold R. Cellular and molecular effects of developmental exposure to diethylstilbestrol: implications for other environmental estrogens. Environ Health Perspect 1995;103(suppl 7):83–87.
33. Newbold RR. Lessons learned from perinatal exposure to diethylstilbestrol. Toxicol Appl Pharmacol 2004; 199(2):142–150.
34. Li S, Hansman R, Newbold R, et al. Neonatal diethylstilbestrol exposure induces persistent elevation of c-fos expression and hypomethylation in its exon-4 in mouse uterus. Mol Carcinog 2003; 38(2):78–84.
35. Newbold RR, Jefferson WN, Grissom SF, et al. Developmental exposure to diethylstilbestrol alters uterine gene expression that may be associated with uterine neoplasia later in life. Mol Carcinog 2007; 46(9):783–796.

36. Tang WY, Newbold R, Mardilovich K, et al. Persistent hypomethylation in the promoter of nucleosomal binding protein 1 (Nsbp1) correlates with overexpression of Nsbp1 in mouse uteri neonatally exposed to diethylstilbestrol or genistein. Endocrinology 2008; 149(12):5922–5931.
37. Newbold RR, Jefferson WN, Padilla-Banks E, et al. Developmental exposure to diethylstilbestrol (DES) alters uterine response to estrogens in prepubescent mice: low versus high dose effects. Reprod Toxicol 2004; 18(3):399–406.
38. Newbold RR, Hanson RB, Jefferson WN, et al. Increased tumors but uncompromised fertility in the female descendants of mice exposed developmentally to diethylstilbestrol. Carcinogenesis 1998; 19(9):1655–1663.
39. Newbold RR, Padilla-Banks E, Jefferson WN. Adverse effects of the model environmental estrogen diethylstilbestrol are transmitted to subsequent generations. Endocrinology 2006; 147(6 suppl):S11–S17.
40. Tomatis L, Narod S, Yamasaki H. Transgeneration transmission of carcinogenic risk. Carcinogenesis 1992; 13(2):145–151.
41. Turusov VS, Trukhanova LS, Parfenov Yu D, et al. Occurrence of tumours in the descendants of CBA male mice prenatally treated with diethylstilbestrol. Int J Cancer 1992; 50(1):131–135.
42. Walker BE, Haven MI. Intensity of multigenerational carcinogenesis from diethylstilbestrol in mice. Carcinogenesis 1997; 18(4):791–793.
43. Blatt J, Van Le L, Weiner T, et al. Ovarian carcinoma in an adolescent with transgenerational exposure to diethylstilbestrol. J Pediatr Hematol Oncol 2003; 25(8):635–636.
44. Titus-Ernstoff L, Troisi R, Hatch EE, et al. Offspring of women exposed in utero to diethylstilbestrol (DES): a preliminary report of benign and malignant pathology in the third generation. Epidemiology 2008; 19(2):251–257.
45. Titus-Ernstoff L, Troisi R, Hatch EE, et al. Menstrual and reproductive characteristics of women whose mothers were exposed in utero to diethylstilbestrol (DES). Int J Epidemiol 2006; 35(4):862–868.
46. Ma L, Benson GV, Lim H, et al. Abdominal B (AbdB) Hoxa genes regulation in adult uterus by estrogen and progesterone and repression in mullerian duct by the synthetic estrogen diethylstilbestrol (DES). Dev Biol 1998; 197(2):141–154.
47. Smith CC, Taylor HS. Xenoestrogen exposure imprints expression of genes (Hoxa10) required for normal uterine development. FASEB J 2007; 21(1):239–246.
48. Knudson AG. Two genetic hits (more or less) to cancer. Nat Rev Cancer 2001; 1(2):157–162.
49. Ruden DM, Xiao L, Garfinkel MD, et al. Hsp90 and environmental impacts on epigenetic states: a model for the trans-generational effects of diethylstibesterol on uterine development and cancer. Hum Mol Genet 2005; 14(Spec No 1):R149–R155.
50. Prins GS. Estrogen imprinting: when your epigenetic memories come back to haunt you. Endocrinology 2008; 149(12):5919–5921.
51. Prins GS, Birch L, Tang WY, et al. Developmental estrogen exposures predispose to prostate carcinogenesis with aging. Reprod Toxicol 2007; 23(3):374–382.
52. Newbold RR, Moore AB, Dixon D. Characterization of uterine leiomyomas in CD-1 mice following developmental exposure to diethylstilbestrol (DES). Toxicol Pathol 2002; 30(5):611–616.
53. Baird DD, Newbold R. Prenatal diethylstilbestrol (DES) exposure is associated with uterine leiomyoma development. Reprod Toxicol 2005; 20(1):81–84.
54. Everitt JI, Wolf DC, Howe SR, et al. Rodent model of reproductive tract leiomyomata. Clinical and pathological features. Am J Pathol 1995; 146(6):1556–1567.
55. Cook JD, Davis BJ, Cai SL, et al. Interaction between genetic susceptibility and early-life environmental exposure determines tumor-suppressor-gene penetrance. Proc Natl Acad Sci USA 2005; 102(24):8644–8649.
56. Jefferson W, Newbold R, Padilla-Banks E, et al. Neonatal genistein treatment alters ovarian differentiation in the mouse: inhibition of oocyte nest breakdown and increased oocyte survival. Biol Reprod 2006; 74(1):161–168.
57. Jefferson WN, Padilla-Banks E, Newbold RR. Disruption of the developing female reproductive system by phytoestrogens: genistein as an example. Mol Nutr Food Res 2007; 51(7):832–844.
58. Nagai A, Ikeda Y, Aso T, et al. Exposure of neonatal ratsto diethylstilbestrol affects the expression of genes involved in ovarian differentiation. J Med Dent Sci 2003; 50(1):35–40.

59. Anway MD, Cupp AS, Uzumcu M, et al. Epigenetic transgenerational actions of endocrine disruptors and male fertility. Science 2005; 308(5727):1466–1469.
60. Anway MD, Leathers C, Skinner MK. Endocrine disruptor vinclozolin induced epigenetic transgenerational adult-onset disease. Endocrinology 2006; 147(12):5515–5523.
61. Fenton SE, Hamm JT, Birnbaum LS, et al. Persistent abnormalities in the rat mammary gland following gestational and lactational exposure to 2,3,7,8-tetrachlorodibenzo-p-dioxin (TCDD). Toxicol Sci 2002; 67(1):63–74.
62. Santti R, Pylkkanen L, Newbold R, et al. Developmental oestrogenization and prostatic neoplasia. Int J Androl 1990; 13(2):77–80.
63. Newbold R. Influence of estrogenic agents on mammalian male reproductive tract differentiation. In: Korach K, ed. Reproductive and Developmental Toxicology. New York: Marcel Dekker, Inc, 1998:531–552.
64. Lin TM, Rasmussen NT, Moore RW, et al. Region-specific inhibition of prostatic epithelial bud formation in the urogenital sinus of C57BL/6 mice exposed in utero to 2,3,7,8-tetrachlorodibenzo-p-dioxin. Toxicol Sci 2003; 76(1):171–181.
65. Timms BG, Peterson RE, vom Saal FS. 2,3,7,8-tetrachlorodibenzo-p-dioxin interacts with endogenous estradiol to disrupt prostate gland morphogenesis in male rat fetuses. Toxicol Sci 2002; 67(2):264–274.
66. Allgeier SH, Vezina CM, Lin TM, et al. Estrogen signaling is not required for prostate bud patterning or for its disruption by 2,3,7,8-tetrachlorodibenzo-p-dioxin. Toxicol Appl Pharmacol 2009; 239(1):80–86.
67. Ramos JG, Varayoud J, Sonnenschein C, et al. Prenatal exposure to low doses of bisphenol A alters the periductal stroma and glandular cell function in the rat ventral prostate. Biol Reprod 2001; 65(4):1271–1277.
68. Timms BG, Howdeshell KL, Barton L, et al. Estrogenic chemicals in plastic and oral contraceptives disrupt development of the fetal mouse prostate and urethra. Proc Natl Acad Sci U S A 2005; 102(19):7014–7019.
69. vom Saal FS, Timms BG, Montano MM, et al. Prostate enlargement in mice due to fetal exposure to low doses of estradiol or diethylstilbestrol and opposite effects at high doses. Proc Natl Acad Sci USA 1997; 94(5):2056–2061.
70. Ho SM, Tang WY, Belmonte de Frausto J, et al. Developmental exposure to estradiol and bisphenol A increases susceptibility to prostate carcinogenesis and epigenetically regulates phosphodiesterase type 4 variant 4. Cancer Res 2006; 66(11):5624–5632.
71. Foster PM. Disruption of reproductive development in male rat offspring following in utero exposure to phthalate esters. Int J Androl 2006; 29(1):140–147; discussion 181–145.
72. Klip H, Verloop J, van Gool JD, et al. Hypospadias in sons of women exposed to diethylstilbestrol in utero: a cohort study. Lancet 2002; 359(9312):1102–1107.
73. Dolinoy DC, Weidman JR, Waterland RA, et al. Maternal genistein alters coat color and protects Avy mouse offspring from obesity by modifying the fetal epigenome. Environ Health Perspect 2006; 114(4):567–572.
74. Dolinoy DC, Huang D, Jirtle RL. Maternal nutrient supplementation counteracts bisphenol A-induced DNA hypomethylation in early development. Proc Natl Acad Sci USA 2007; 104(32):13056–13061.
75. Vige A, Gallou-Kabani C, Junien C. Sexual dimorphism in non-Mendelian inheritance. Pediatr Res 2008; 63(4):340–347.
76. Schones DE, Zhao K. Genome-wide approaches to studying chromatin modifications. Nat Rev Genet 2008; 9(3):179–191.
77. Fraga MF, Ballestar E, Villar-Garea A, et al. Loss of acetylation at Lys16 and trimethylation at Lys20 of histone H4 is a common hallmark of human cancer. Nat Genet 2005; 37(4):391–400.

Competing Interest: The authors declare they have no competing financial interest.

18 Cumulative effects of in utero administration to mixtures of reproductive toxicants in the male rat: a systems biology framework

L. Earl Gray, Jr., Cynthia V. Rider, Vickie S. Wilson,
Kembra L. Howdeshell, Andrew K. Hotchkiss, Chad Blystone,
Johnathan R. Furr, Christy R. Lambright, and Paul Foster

INTRODUCTION

Although risk assessments are typically conducted on a chemical-by-chemical basis, regulatory agencies are beginning to consider cumulative risk of chemicals that act via a common mechanism of toxicity. It is well known that humans (1–8), fish (9–12), and wildlife (13) are continuously exposed to multiple contaminants. The chemicals found in some aquatic systems not only include pesticides (14,15) and industrial chemicals (13) but also include pharmaceuticals and hormones (16,17).

The effects of mixtures of chemicals like the phthalates are a concern since humans are exposed to multiple phthalates at one time (3,18). To address this issue, in 2006 the U.S. Environmental Protection Agency (EPA) requested that the National Academy of Sciences (NAS) establish a panel to provide the agency with recommendations on how to address the cumulative effects of the phthalates. This review will present the conclusions of the NAS panel (Box, from p. 116 of the NAS report) and then discuss the data from our laboratory on the reproductive effects of mixtures. Our working hypothesis is that the chemicals included in a cumulative risk assessment should include all the chemicals that disrupt a common system or tissue during development rather than only those that act via a common mechanism of toxicity.

The results of our studies support this hypothesis and are in agreement with the conclusions of a recent National Academy of Science Report (2008):

> The report concludes that a cumulative risk assessment should be conducted for phthalates and identifies other chemicals that also affect development of the male reproductive system, and therefore should be considered for inclusion in this risk assessment. Phthalates reduce concentrations of testosterone, an important androgen (or male sex hormone) that contributes to the development of male sex organs. This androgen deficiency causes the phthalate syndrome in laboratory animals if it occurs during time periods that are critical for male reproductive development. A number of other chemicals (often referred to as antiandrogens) can cause similar effects, although by different mechanisms or biological pathways, and it is difficult to differentiate between the effects of phthalates and the effects of these other chemicals. Therefore, focusing solely on phthalates to the exclusion of other antiandrogens in assessing risk would be artificial and could seriously underestimate cumulative risk.

National Academy of Sciences Committee on Health Risks of Phthalates Conclusions (p. 116) from the Report "Phthalates and Cumulative Risk Assessment: The Task Ahead"

CONCLUSIONS

"A major challenge to conducting cumulative risk assessment is choosing an approach to predict mixture effects. However, evidence from the recent peer-reviewed scientific literature shows not only that phthalates produce mixture effects but that the effects are often predicted well by using the dose-addition concept. That is also true for other classes of antiandrogens and for combinations of phthalates with such antiandrogens. Although a variety of molecular mechanisms are at play, dose addition provided equal or better approximations of mixture effects compared with independent action (when such comparisons were performed). In no example in the literature did independent action produce a mixture-effect prediction that proved to be correct and differed substantially from that produced with dose addition. The evidence that supports adoption of a physiologic approach is strong. Experimental evidence demonstrates that toxic effects of phthalates and other antiandrogens are similar despite differences in the molecular details of the mechanisms, including metabolism, distribution, and elimination.

The criteria recommended by EPA (2000) for guiding decisions between dose addition and independent action appear too narrow when applied to phthalates and other antiandrogens, particularly those requiring similarity in uptake, metabolism, distribution, and elimination and congruent dose-response curves for application of dose addition. The requirements are not met by combinations of phthalates with other antiandrogens, but the dose-addition principle applies. The case for using dose addition as an approximation for mixture risk assessment of phthalates and other antiandrogens is strong.

When risks posed by low-level exposures need to be evaluated, there are substantial differences between the single-chemical approach and cumulative risk assessment. There is good evidence that combinations of phthalates and of other antiandrogens produce combined effects at doses that when administered alone do not have significant effects. In some cases, those doses are similar to those used as PODs to estimate tolerable human exposure. The results highlight the problem that may arise when PODs for individual chemicals are used as the basis of human-health risk assessment in situations in which exposure to other chemicals with similar effects also occurs. The results emphasize the necessity of conducting cumulative risk assessment of phthalates and other antiandrogens to assess risks posed by exposure to mixtures of these compounds. Assessments based solely on the effects of single phthalates and other antiandrogens may lead to considerable underestimation of risks to the developing fetus.

In this chapter, the committee has provided recommendations on various aspects of conducting cumulative risk assessment. The recommendations were designed specifically to deal with phthalates and antiandrogens. However, the conceptual framework that the committee has used is generic and lends itself to dealing with other groups of chemicals, provided that the relevant toxicologic data are available."

Note: PODs—points of departure

As a result of growing concerns about mixtures, the field of "mixtures toxicology" is emerging as an area of increasing scientific and regulatory focus in the United States and abroad. For example, in 1996 the EPA began considering the cumulative risk of chemicals that act via a common mechanism of toxicity as mandated in the Food Quality Protection Act (FQPA). The EPA's Offices of Water and Research and Development and the EPA Superfund, Solid Waste, and Air Programs also have ongoing programs in this area. In this regard, the research from our laboratory, described herein, is intended to contribute to the development of a guidance framework for assessing cumulative risks to reproduction and development from exposures during pregnancy.

Developmental Reproductive Toxicants

Estrogens
  Methoxychlor
  Ethinyl Estradiol
  Bisphenol A

AR Antagonists

Compete with natural hormones T and DHT for AR, prevent AR-DNA binding in vitro, inhibit AR-dependent gene expression in vivo, and may induce malformations in male reproductive tract and delay puberty in male rat

- **Vinclozolin**
- **Procymidone**
- **Linuron**
- **p,p′ DDE and other o,p′- and p,p′ DDT metabolites**
- **Prochloraz**

Inhibitors of fetal androgen synthesis

Prevent the synthesis of natural hormones T and DHT and induce malformations in male reproductive tract and delay puberty in male rat

- **DEHP**
- **DPP**
- **BBP**
- **DBP, DiBP**
- **DiHP**
- **DINP**
- **Linuron**
- **Prochloraz**

Fetal Germ Cell Toxicants
  Busulfan
  Diazo dyes

Steroidogenesis inhibitors
  Prochloraz
  Linuron
  Ketoconazole
  Fenarimol

Androgens
  Testosterone
  Trenbolone

Dioxins and PCBs
  Dioxin
  PCB I 69 congener

FIGURE 1 Chemicals that we have studied for their ability to produce postnatal effects after in utero exposure in rats. *Abbreviations*: AR, androgen receptor; DHT, 5α-dihydrotestosterone; DEHP, di-*n*-ethyl hexyl phthalate; DPP, di-*n*-pentyl phthalate; BBP, benzyl butyl; DBP, di-*n*-butyl phthalate; DiBP, di-iso-butyl phthalate; DiHP, di-iso-heptyl phthalate; DDE, dichlorodiphenyldichloroethylene; DDT, dichlorodiphenyltrichloroethane; PCB, Polychlorinated biphenyls.

Our research has included mixtures of pesticides, phthalates, and 2,3,7,8 tetra-chlorodibenzo dioxin (TCDD), chemicals that disrupt sexual differentiation by acting as androgen receptor (AR) antagonists, inhibitors of fetal testosterone synthesis, or as an aryl hydrocarbon receptor (AhR) agonist (Figs. 1 and 2).

We initially conducted a series of binary mixture studies exposing pregnant dams during fetal sexual differentiation with pairs of chemicals (19–23). In the binary studies, rats were dosed with chemicals singly or in pairs at dosage levels equivalent to about one half of the ED50 for hypospadias or epididymal agenesis (2 × 2 factorial designs). We predicted that by itself each chemical in a pair would cause a minimal rate of malformations, whereas when two chemicals were mixed together they would produce predictable, dose-additive effects on common target tissues. We found that the binary combinations produced cumulative, dose-additive effects on the androgen-dependent tissues that were a common component of each chemical's phenotype. The binary mixtures are as follows:

- Vinclozolin plus procymidone (common mechanism: both AR antagonists)
- Di-*n*-butyl phthalate (DBP) plus di-*n*-ethyl hexyl phthalate (DEHP) (common mechanism: both inhibit fetal androgen production)
- Benzyl butyl (BBP) plus DEHP (common mechanism: both inhibit fetal androgen production)
- DBP plus procymidone (different mechanisms)
- Linuron plus BBP (different mechanisms)
- DBP plus TCDD (different mechanisms)

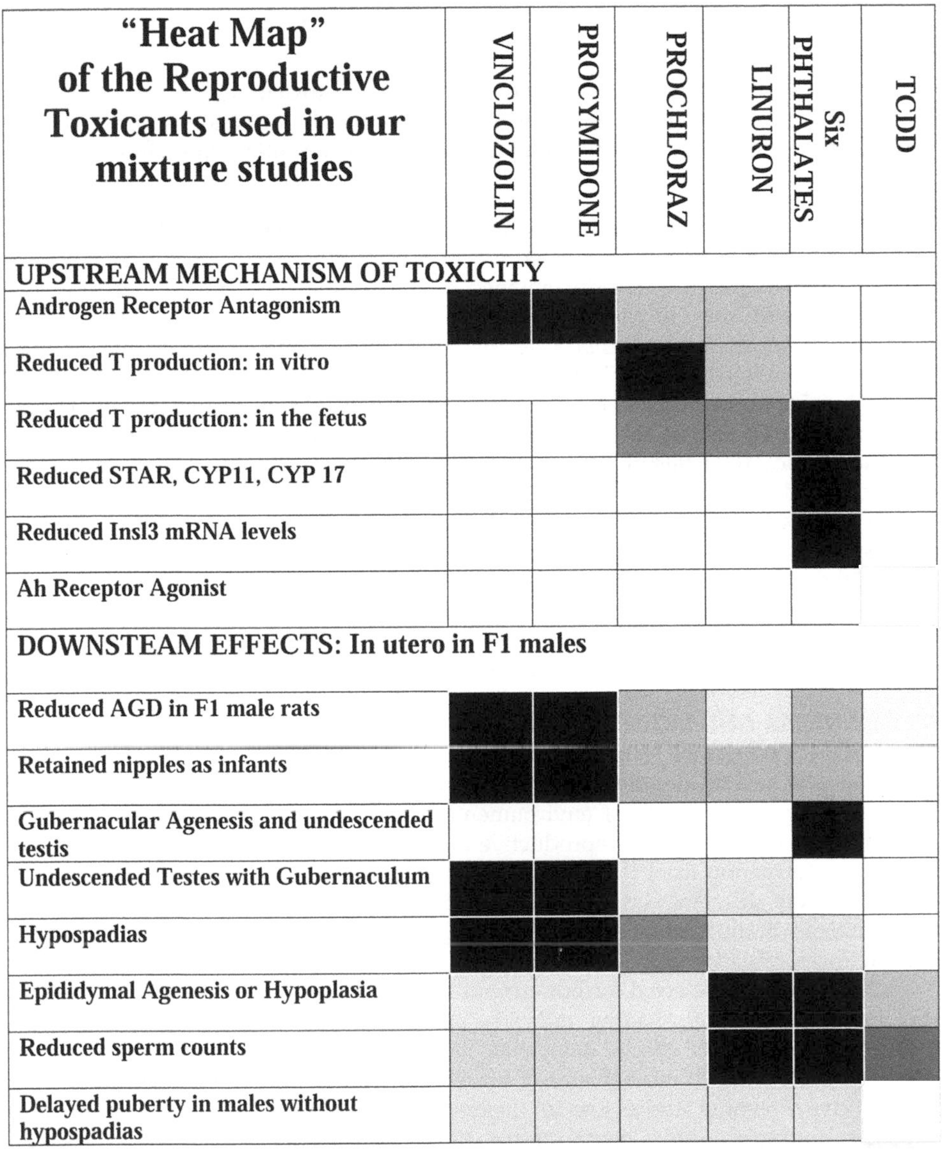

"Heat Map" of the Reproductive Toxicants used in our mixture studies	VINCLOZOLIN	PROCYMIDONE	PROCHLORAZ	LINURON	Six PHTHALATES	TCDD
UPSTREAM MECHANISM OF TOXICITY						
Androgen Receptor Antagonism						
Reduced T production: in vitro						
Reduced T production: in the fetus						
Reduced STAR, CYP11, CYP 17						
Reduced Insl3 mRNA levels						
Ah Receptor Agonist						
DOWNSTEAM EFFECTS: In utero in F1 males						
Reduced AGD in F1 male rats						
Retained nipples as infants						
Gubernacular Agenesis and undescended testis						
Undescended Testes with Gubernaculum						
Hypospadias						
Epididymal Agenesis or Hypoplasia						
Reduced sperm counts						
Delayed puberty in males without hypospadias						

FIGURE 2 "Heat map" displaying the intensity of effects of each chemical in vitro, in short-term in vivo screens, and on F1 male offspring after exposure in utero during sexual differentiation. The darker shading indicates stronger effects whereas the white areas indicate the absence of effects. Reviewing the columns associated with each chemical describes the (*i*) known mechanisms of toxicity as determined from in vitro and short-term in vivo screening studies and (*ii*) the overall phenotype in the male offspring after exposure during fetal life. Comparing the different chemicals by rows (the endpoints) allows one to compare the relative potencies displayed by the toxicants to one another. For each endpoint, we predict that all the chemicals with shading (*black* to *light gray*) will interact jointly when combined in a mixture, but those with white squares will have no effect when included in the mixture. *Abbreviation*: TCDD, 2,3,7,8 tetrachlorodibenzo dioxin.

In addition to the binary studies, we also have conducted three mixture studies with larger numbers of chemicals in the mixtures.

- Five phthalate esters with a common mechanism of toxicity (24)
- Seven chemicals which disrupt androgen signaling via multiple mechanisms (25)
- Ten chemicals which disrupt androgen signaling via multiple mechanisms

One study combined five phthalates together in the mixture to determine if they produced dose-additive effects on fetal testosterone production (24) and fetal testis gene expression. These phthalates all shared a common mechanism of toxicity. The remaining two complex mixture studies combined chemicals that elicit antiandrogenic effects at two different sites in the androgen signaling pathway (i.e., AR antagonist or inhibition of androgen synthesis). One study combined seven chemicals together and the second combined 10 chemicals in the mixture.

This chapter (*i*) describes the mechanisms and modes of toxicity in vitro and in vivo of the individual chemicals that we selected to study in our research program, (*ii*) describes the mathematical modeling procedures that we use to describe the effects of the individual chemicals and derive predictions for how the mixtures will behave, (*iii*) presents the results of our mixture studies on chemicals that act via common mechanisms of toxicity (see sect. "Mixtures Section I"), chemicals that act via disparate mechanisms of toxicity but disrupt the same signaling pathway (see sect. "Mixtures Section II"), and chemicals that disrupt common developing tissues via alteration of diverse signaling pathways (see sect. "Mixtures Section III"), and (*iv*) describes a proposed framework for cumulative risk assessment based on disruption of a common developing system.

MECHANISMS AND MODES OF ACTION OF THE INDIVIDUAL CHEMICALS USED IN THE CURRENT MIXTURE STUDIES

Over the past several decades, our laboratory has studied the postnatal effects of in utero exposure to a variety of environmental chemicals with diverse endocrine and nonendocrine mechanisms of reproductive toxicity (Fig. 1). We are now using the dose response information from these studies to design mixture studies to determine how chemicals from different classes with similar and diverse mechanisms and modes of action interact in the mixture. The objective of our research is to define a framework for conducting cumulative risk assessments.

The chemicals selected for our current mixture studies are shown in Figure 2. In this figure, the darker shading indicates stronger effects whereas the white areas indicate the absence of effects. Reviewing the columns associated with each chemical describes the (*i*) known mechanisms of toxicity as determined from in vitro and short-term in vivo screening studies and (*ii*) the overall phenotype in the male offspring after exposure during fetal life. Comparing the different chemicals by rows (the endpoints) allows one to compare the relative potencies displayed by the toxicants to one another. For each endpoint, we predict that all the chemicals with shading (black to light gray) will interact jointly when combined in a mixture, but those with white squares will have no effect when included in the mixture. For example, of the chemicals in this chart, only the phthalates reduce insulin-like peptide hormone-3 (insl3) and cause gubernacular agenesis. Therefore, combining a phthalate with any of the other chemicals in the chart will not enhance the phthalate-induced reduction in insl3 mRNA levels in the fetal testis or increase the incidence of gubernacular agenesis.

Dicarboximide Fungicides: Vinclozolin and Procymidone

Of the dicarboximide fungicides, vinclozolin (26), iprodione (27), and procymidone (28–31) act as AR antagonists in vitro and/or in vivo. These pesticides, or their metabolites, competitively inhibit the binding of androgens to AR, which leads to an inhibition of androgen-dependent gene expression in vitro and in vivo (32).

Vinclozolin, iprodione, and procymidone also antagonize the action of androgens in vivo as AR antagonists (27,33–39). For example, in a Hershberger assay using castrated immature testosterone-treated male rats, vinclozolin and procymidone (0, 25, 50, and 100 mg/kg/day) antagonize testosterone stimulated growth of the androgen-dependent tissues (ventral prostate, seminal vesicles, and levator ani-bulbocavernosus muscles) (19,29).

Peripubertal administration of antiandrogens can alter the onset of pubertal landmarks and reduce androgen-dependent organ weights in the young intact male rat (40,41).

Administration of vinclozolin or procymidone during sexual differentiation demasculinizes and feminizes the male rat offspring causing a variety of malformations (30,42); however, even at high dosage levels (200 mg/kg/day), epididymal hypoplasia was rare and gubernacular agenesis was not displayed.

At lower doses, vinclozolin administration reduces neonatal anogenital distance (AGD) (3.125 mg/kg/day) and increases the incidence of retained nipples/areolae in infant male rats (6.25 mg/kg/day) (43). In adult life, ventral prostate weight is permanently reduced (6.25 mg/kg/day) and male offspring display permanent female-like nipples (50 mg/kg/day). Treatment at 50 and 100 mg/kg/day induces hypospadias and other reproductive tract malformations (43,44).

Similarly, perinatal procymidone treatment causes shortened AGD in F1 male pups, and they display retained nipples, hypospadias, cleft phallus, a vaginal pouch, reduced sex accessory gland size, and undescended testes (30).

These two dicarboximide pesticides not only induce reproductive tract malformations and permanent reductions in androgen-dependent organ weights, but they also program the differentiating prostatic and vesicular tissues abnormally such that F1 male offspring develop high rates of inflammation in these tissues later in life. In utero, procymidone treatment induced fibrosis, cellular infiltration, and epithelial hyperplasia in the dorsolateral and ventral prostatic and seminal vesicular tissues in the offspring when examined as adults (30). Similar histopathological lesions were seen in males exposed to vinclozolin in utero (45) with 100% of the male rats displaying prostatitis after puberty.

Linuron (Herbicide)

While some toxicants disrupt sexual differentiation act predominantly via one mechanism of toxicity (i.e., AR antagonists or inhibitors of testosterone synthesis), some pesticides including linuron and prochloraz act via dual mechanisms of toxicity. These pesticides display AR antagonist activity and inhibit testosterone synthesis with varying potencies.

The herbicide linuron is an AR antagonist in vitro (46–50). In contrast to some AR antagonists, neither short-term linuron (46–51) nor long-term (52) administration induced elevated serum luteinizing hormone (LH) levels.

Linuron administration to the dam or in vitro also inhibits fetal male rat testosterone synthesis during sexual differentiation, demonstrating that linuron is antiandrogenic via dual mechanisms of action (Fig. 1) inhibiting androgen synthesis and as an AR antagonist (23,53).

When administered in utero, linuron exposure causes malformations in male rat offspring. More than half of the males exposed to 100-mg linuron/kg/day [gestation days (GD) 14–18] display epididymal and testicular abnormalities (43), with effects having been seen at dosage levels as low as 12.5 mg/kg/day (exposed from GD 10–22) (49). The testicular effect seen in the adult F1 males results from epididymal lesions rather than a direct effect of linuron on testis morphology. When male rat fetuses or offspring were necropsied on GD 17, 19, and 21, and on postnatal days (PND) 7 and 14, epididymal malformations were not observed in fetuses from linuron-treated dams but were seen in linuron-exposed male offspring on PND 7 and 14 (48). The testicular lesions are only seen in adults and not younger animals. These lesions appear to develop as a consequence of pressure-atrophy induced by fluid accumulation in the postpubertal testis in animals with epididymal lesions. Testicular lesions were not observed at any time point during fetal or infant life.

In contrast to the effects of vinclozolin and procymidone, malformed external genitalia and undescended testes were rarely displayed by linuron-exposed males. The syndrome of effects induced by linuron is atypical of an AR antagonist and more closely resembles the phthalate syndrome (23).

Prochloraz (Fungicide)

Prochloraz is a fungicide that also disrupts reproductive development and function by several modes of action (54–56). Prochloraz inhibits the steroidogenic enzymes 17, 20 lyase and aromatase and it also is an AR antagonist (57,58). In a study in which rat dams were dosed from GD 14 to 18, Wilson et al. (53) found that prochloraz reduced fetal testis testosterone levels and increased progesterone production 10-fold on GD 18 without affecting Leydig cell insl3 mRNA levels.

In a transgenerational study, prochloraz treatment from GD 14 to 18 at doses of 62.5, 125, 250, and 500 mg/kg/day delayed parturition and altered reproductive development in the male offspring in a dose-related manner (54). Treated males displayed reduced AGD and female-like areolas (33%, 71%, and 100% in 62.5, 125, and 250 mg/kg groups, respectively), and males in the 250 mg/kg treatment group also displayed hypospadias. However, the epididymides and gubernacular ligaments were relatively unaffected by prochloraz exposure.

In conclusion, the profile of effects in the male rat offspring induced by prenatal prochloraz exposure appears to more closely resemble that of an AR antagonist, like vinclozolin, rather than an inhibitor of fetal testosterone synthesis, like a phthalate or linuron.

Phthalates

The phthalates represent a class of high production volume chemicals that alter reproductive development. This class of chemicals does not appear to act via estrogen or androgen nuclear receptors. While a few studies suggest that some of the phthalates are estrogenic (59), DBP injections do not induce a uterotropic response or estrogen-dependent sex behavior (lordosis) in the ovariectomized adult female rats (60), and oral DBP treatment fails to accelerate vaginal opening or induce constant estrus in intact female rats (52). The phthalate diesters and their monoester metabolites also do not compete significantly with androgens for binding to AR at environmentally relevant concentrations (61). In vivo, the phthalate diesters fail to display consistent AR antagonist activity. DBP and BBP produce negative results in a Hershberger assay, whereas DEHP causes equivocal reductions in androgen-induced tissue growth even at 1000 mg/kg/day (Gray, personal communication; (62)).

In utero, some phthalate esters alter the development of the male rat in an antiandrogenic manner. Prenatal exposure to DBP, BBP, di-iso-nonyl phthalate (DINP), and DEHP treatment cause a syndrome of effects, including underdevelopment and agenesis of the epididymis and other androgen-dependent tissues and testicular abnormalities characterized as the "phthalate syndrome" (63,64). Prenatal exposure to DBP from day 10 to 22 of gestation produces effects nearly identical to those seen with DEHP, with effects occurring at dosage levels of 50 to 100 mg/kg/day (64–66). Among the antiandrogenic endocrine disrupting chemicals (EDCs), the phthalates are unique in their ability to induce agenesis of the gubernacular cords, a tissue whose development is dependent on the peptide hormone insulin-like peptide-3, and critical for testis descent.

Concerns about some phthalates in toys led to passage of the 2008 U.S. Consumer Protection Agency Modernization Act (Public Law No: 110-314, section 108), which prohibits the sale of certain products containing phthalates [DEHP, DBP, BBP, DINP, di-iso-octyl phthalate (DIOP), and di-*n*-octyl phthalate (DNOP)]. Additionally, the law established a Chronic Hazard Advisory Panel charged with examining "the potential health effects of each of these phthalates both in isolation and in combination with other phthalates" and "to consider the cumulative effect of total exposure to phthalates, both from children's products and from other sources, such as personal care products."

2,3,7,8 Tetrachlorodibenzo Dioxin

TCDD binds with high affinity to the cytosolic AhR. This orphan receptor acts as a transcription factor in a manner similar to the steroid hormone receptors. Although the normal role of this transcription factor in development is unknown, gestational TCDD treatment induces malformations of the external genitalia and subfertility in female rat (67–69) and hamster (70) offspring and alters reproductive development of the male rat (71–76) and hamster (71).

Several studies have suggested that gestational TCDD administration does *not* reduce adult (67,77) or fetal androgen levels in the male rat or mouse fetus (78–80), even though several of the effects in the male offspring include suppressed development of some androgen-dependent tissues. In mice, it is known that TCDD inhibits the androgen-dependent processes by which the urogenital sinus of fetal mice forms prostatic epithelial buds. This inhibition is mediated by AhR in urogenital sinus mesenchyme and causes prostate lobes to develop abnormally. Furthermore, TCDD appears to act directly on the urogenital sinus to cause AhR-dependent inhibition of prostatic epithelial bud formation. TCDD inhibits prostatic budding primarily via direct effects on the urogenital sinus rather than indirectly through effects on other organs (81). Furthermore, TCDD exposure sufficient to inhibit prostatic budding on GD 13 had no effect on testicular testosterone content on GD 16 or 18 and did not inhibit the conversion of testosterone to 5α-dihydrotestosterone (DHT) by the urogenital sinus (78). Additional mechanistic studies proposed that TCDD acted directly on AhR, ARNT, and AhR-induced transcripts in the periprostatic mesenchyme (82,83). This tissue intimately contacts urogenital sinus epithelium where buds are specified, and they proposed that activation of AhR signaling disrupted dorsoventral patterning of the urogenital sinus, reprogramming the areas where prostatic buds are specified and prostate lobes are formed.

Administration of TCDD on the 15th day of gestation (GD 15) at doses lower than or equal to 1 µg/kg both demasculinizes and feminizes male rat reproductive morphology and behavior. In our first series of studies, Long Evans hooded rats were dosed by gavage with 1-µg TCDD/kg on GD 8 (a period of major organogenesis), or GD 15, and Syrian hamsters, a species relatively insensitive to the lethal effects of TCDD, were dosed on GD 11 (a period equivalent to GD 15 in the rat), with TCDD at 2 µg/kg. In Long Evans and Holtzman F1 male rats exposed on GD 15 or hamsters dosed on GD 11, puberty (preputial separation) was delayed by about three days, ejaculated sperm counts were reduced by at least 58% and epididymal sperm storage was reduced by

30%. Testicular sperm production was less affected. The sex accessory glands were also reduced in size in Long Evans offspring treated on GD 15 in spite of the fact that serum testosterone levels, testosterone production by the testis in vitro, and AR levels were not reduced. Some reproductive measures, like anogenital distance and male sex behavior, were altered by TCDD treatment in rat but not hamster offspring.

In Long Evans hooded rats exposed to 0, 0.05, 0.20, or 0.80-µg TCDD/kg maternal bodyweight on GD 15, growth and viability of the pups were reduced at 0.80-µg TCDD/kg, while the onset of eye-opening was accelerated in all dosage groups, including 0.05 µg/kg. Preputial separation, a landmark of puberty in rodents, was significantly delayed in the 0.20- and 0.80-µg TCDD/kg dosage groups. At 49 days of age, male progeny (at 0.8 µg/kg) displayed reductions in ventral prostate and seminal vesicle weights and epididymal sperm counts. The reductions in sex accessory gland size were transient, while the effect on the epididymal sperm numbers was permanent. For example, at 63 days, epididymal sperm numbers were reduced by 10% at 0.20-µg TCDD/kg and by 25% at 0.80-µg TCDD/kg, and reduction in cauda epididymal sperm numbers was seen in middle-aged male offspring in these dosage groups. Ejaculated sperm numbers were reduced to a greater degree by TCDD treatment than were cauda epididymal or testicular sperm numbers, being reduced by 45% in the 0.8 and by 25% in the 0.05- and 0.2-µg TCDD/kg dosage groups.

In the F1 female offspring, transplacental, but not lactational, TCDD exposure (1 µg/kg on GD 15) induces cleft phallus, vaginal thread formation, and reduces ovarian weight. TCDD treatment on GD 15 at 0, 0.05, 0.20, or 0.80 µg/kg delays vaginal opening and induces malformations of the external genitalia in the female progeny (69,71,72).

MIXTURE MODELING

Over the past several decades, research in several laboratories including our own has described the effects of individual chemicals on the reproductive development of male rats. However, recent concerns have shifted focus from individual chemical effects to mixtures effects (16,84,85). Individual chemical data can be used in mathematical models to make predictions about the potential effects of mixtures on male reproductive tract development. Predicted mixture responses can then be compared with the observed effects of the mixtures to determine the type of joint action (dose addition, response addition, synergy, antagonism) exhibited by the mixture.

Mixture toxicity models require dose-response data from individual chemical exposures. For our mixture studies, the data were compiled from studies conducted in our laboratory over the past 20 years. While the historical data included studies conducted by different researchers with several rat strains and slightly different dosing schedules, these studies included exposure to the chemical during the critical in utero window of male rat reproductive tract differentiation.

Using GraphPad Prizim 5.0 software, we transformed the data to fit a 0% to 100% scale. For continuous endpoints (AGD and organ weights), we converted the data to percent change from the control value. For malformation data, we presented the data as percent incidence. We then graphed the data on a log-linear scale and fit the data with a logistic equation (see example in equation 1):

$$R = \frac{1}{1 + (ED50/D)^{\rho}} \tag{1}$$

where R is the response, D is the chemical dose, ρ is the power or Hill slope of the curve, and ED50 is the exposure dose eliciting a 50% response. The parameters (Hill slope and ED50) generated from the logistic fit to the individual chemical data were used in models to make predictions of the mixture responses.

Predicted Responses from Models Vs. Observed Responses

We modeled the data and generated predictions of the mixture effects with three types of models (described in more detail in Ref. 25):

- Dose addition
- Response addition
- Integrated addition

The dose addition model is commonly applied to mixtures of chemicals that have the same mechanism of action (86,87), whereas the response addition model has generally been used with mixtures containing chemicals with different mechanisms of action (88). The integrated addition model is typically used to predict the effects of mixtures containing chemicals with similar and dissimilar mechanisms of toxicity (25,89,90).

In the dose addition (DA) model, mixture components can be thought of as dilutions of one another. Once the potencies of the individual chemicals are accounted for, their doses can simply be added together to determine the total dose of the mixture. From this total mixture dose, we can then calculate the predicted response.

The dose addition equation that we used to calculate predicted responses of mixtures is as follows:

$$R = \frac{1}{1 + \left\{ 1 \Big/ \left[\sum_{i=1}^{n} (D_i/\mathrm{ED}50_i) \right]^{\rho'} \right\}} \tag{2}$$

where R is the response to the mixture, D_i is the concentration of chemical i in the mixture, $\mathrm{ED}50_i$ is the concentration of chemical i that causes a 50% response, and ρ' is the average power (Hill slope) associated with the chemicals.

The response addition (RA) model, also referred to as independent action model, was first introduced by Bliss (91) and has been used to describe mixtures of chemicals with different mechanisms of action. The equation for response addition is based on probability theory and is expressed as

$$R = 1 - \prod_{i=1}^{n} (1 - R_i) \tag{3}$$

The integrated addition (IA) model, introduced relatively recently by several different groups (89,90,92), combines the dose and response addition models. In this approach, chemicals with the same mechanism of action are grouped and the total dose associated with each group is calculated using dose addition. The groups are then combined using response addition.

The integrated addition model is expressed mathematically as

$$R = 1 - \prod_{i=1}^{n} \left(1 - \frac{1}{1 + \left\{ 1 \Big/ \left[\sum_{i=1}^{n} (D_i/\mathrm{ED}50_i) \right]^{\rho'} \right\}} \right) \tag{4}$$

MIXTURES SECTION
Identifying a Framework for Cumulative Risk Assessment

The following discussion will present the data that supports the conclusions of the NAS report and our hypothesis that the framework for cumulative risk assessments should be more broadly based on disruption of common systems or target tissues.

Mixtures Section I. The mixture included pairs of chemicals that acted via common mechanisms of toxicity

- AR antagonists (vinclozolin plus procymidone)
- Phthalate esters (DBP plus BBP)
- Phthalate esters (DEHP plus DBP)

Mixtures Section II. Chemicals that disrupted a common pathway via disparate mechanisms of toxicity. Disruption of the androgen signaling pathway

- A phthalate ester plus an AR antagonist (DBP plus procymidone)
- A pesticide plus a phthalate (linuron plus BBP)

Mixtures Section III. Chemicals that disrupted a common tissue via different signaling pathways and diverse mechanisms of toxicity. Disruption of the androgen and AhR signaling pathways in the fetal male rat reproductive tract

- 2,3,7,8 Tetrachlorodibenzo dioxin plus a phthalate (TCDD plus DBP).

Several different strategies have been proposed for including chemicals in a cumulative risk assessment ranging from one limited to chemicals that clearly display a common cellular and molecular mechanism of action, to a much broader strategy based on disruption of a common target tissue (Figs. 2 and 3). The former approach is typically used by the EPA, Office of Chemical Safety and Pollution Prevention in assessing the cumulative risk of food use pesticides, while the latter, broader strategy, was recommended by a recent NAS panel report.

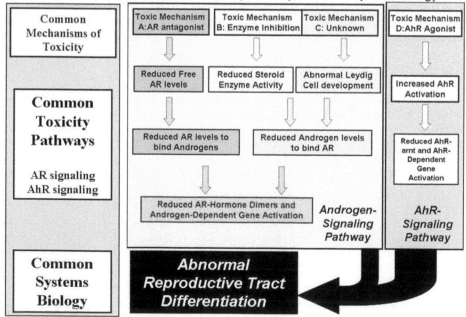

FIGURE 3 Mechanisms and modes of action of the chemicals included in our current studies. *Abbreviations*: AR, androgen receptor; AhR, aryl hydrocarbon receptor.

Three different strategies for conducting cumulative risk assessments will be presented here including grouping chemicals in a cumulative risk assessment based on (*i*) common mechanisms of toxicity, (*ii*) disruption of common signaling pathways (a toxicity pathway approach), or (*iii*) a broader approach, based on disruption of common reproductive target tissues in the fetus (a systems approach including multiple signaling pathways) (Fig. 3).

If cumulative risk assessments were conducted on the 11 chemicals discussed herein using the framework based on common mechanisms of toxicity, then they would be clustered into four different assessments, the assumption being that chemicals with different mechanisms of toxicity interact independently and the outcome can be predicted by response addition modeling, whereas dose addition models would over-predict the effects.

Using this framework, results in the following assessments:

Cumulative risk assessments	Assessment 1	Assessment 2	Assessment 3	Assessment 4
Mechanism of toxicity	AR antagonist	Altered Leydig cell development and testosterone synthesis	Dual mechanisms of action	AhR agonist
	Vinclozolin Procymidone	Six phthalates	Linuron Prochloraz	TCDD

Grouping the chemicals for cumulative risk assessments based on disruption of a common signaling pathway rather than common mechanisms of toxicity assumes that chemicals that disrupt androgen signaling and AhR signaling will act jointly, and the effects can be predicted by dose addition modeling, but these chemicals would act independently from those that disrupt the same tissue via the AhR pathway.

Using this framework, results in the following assessments:

Cumulative risk assessments	Assessment 1	Assessment 2
Toxicity pathways	Androgen signaling Vinclozolin Procymidone Linuron Prochloraz Six phthalates	AhR signaling TCDD

Grouping the chemicals for cumulative risk assessments based on disruption of a common system via multiple pathways and mechanisms of toxicity assumes that chemicals that disrupt androgen signaling and AhR signaling will act jointly, and the effects can be predicted by dose addition modeling. Using this framework, results in a single cumulative risk assessment:

Cumulative risk assessments	Assessment 1
Common target tissue	Epididymal differentiation Vinclozolin Procymidone Linuron Prochloraz Six phthalates TCDD

Mixtures Section I

I.a Mixtures of AR Antagonists: Vinclozolin and Procymidone (Fungicides)

Disruption of reproductive differentiation by common mechanisms of toxicity. In the late 1990s, the agency began an examination of whether some or all members of the dicarboximide class of fungicides, which includes vinclozolin, iprodione, and procymidone, shared a common mechanism of toxicity. At this time, the scientific information on the mechanisms of toxicity of this class of fungicides was incomplete. For this reason, the EPA (http://www.epa.gov/opp00001/reregistration/vinclozolin/) concluded in 2000 that

> The Agency does not currently have a fully developed understanding of whether vinclozolin shares a common mechanism of toxicity with iprodione and procymidone because the androgen system is highly complex. As a result, the Agency has not determined if it would be appropriate to include them in a cumulative risk assessment. Therefore, for the purposes of this assessment, the Agency has assumed that vinclozolin does not share a common mechanism of toxicity with iprodione and procymidone.

In addition, in a risk assessment on procymidone in 2005 the Agency (http://www.epa.gov/oppsrrd1/REDs/procymidone_tred.pdf) concluded that "EPA has not made a common mechanism of toxicity finding and therefore, has not assumed that procymidone has a common mechanism of toxicity with other substances for the purposes of this tolerance action."With the encouragement of the program office, we initiated studies at EPA's National Health and Environmental Effects Laboratory to better elucidate the mechanism of toxicity for these antiandrogenic fungicides as well as mixture studies on how they interact. Since then several studies from our laboratory and other laboratories have been completed that address these uncertainties. These studies demonstrate that vinclozolin and procymidone share a common mechanism of toxicity and interact in a cumulative manner (19).

Vinclozolin plus procymidone mixtures In a Hershberger assay using castrated immature testosterone-treated male rats, oral administration of vinclozolin or procymidone for 7 to 10 days in combination inhibited testosterone-induced growth of androgen-dependent tissues (ventral prostate, seminal vesicles, and levator ani-bulbocavernosus muscles) in a dose-additive fashion (19,29). In our AR antagonist binary study (19), in utero exposure to vinclozolin alone resulted in 10% incidence of hypospadias and 0% incidence of vaginal pouch development in male rats, while procymidone exposure resulted in 0% incidence of either malformation. The combination exposure, however, resulted in 96% incidence of hypospadias and 54% incidence of vaginal pouch in treated animals (Fig. 4A).

I.b Disruption of the Androgen Signaling Pathway by a Common Mechanism of Toxicity

Alteration of fetal Leydig cell differentiation and hormone production by phthalates. The effects of mixtures of phthalates are a concern since humans are exposed to multiple phthalates at one time (3,18). To address this issue, in 2006 the EPA requested that the NAS establish a panel to provide the agency with recommendations on how to address the cumulative effects of the phthalates. The data from several of the phthalate mixture studies presented herein were given to the NAS panel for their reanalysis and evaluation (http://www8.nationalacademies.org/cp/projectview.aspx?key=48860).

Mixtures of phthalates: three studies We conducted three studies with mixtures of phthalate esters. The first study combined two phthalate esters with a common active metabolite (DBP and BBP). The second combined two phthalate esters with different active metabolites (DEHP and DBP) (21). In the third phthalate mixture study, we

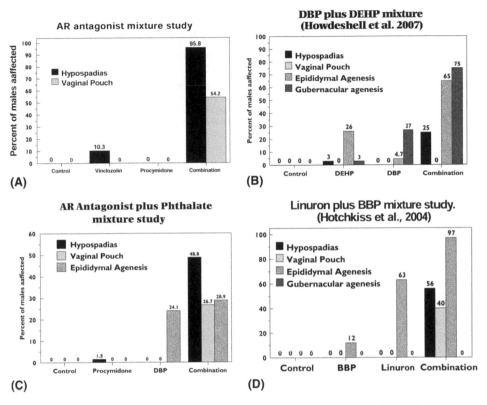

FIGURE 4 Dose-additive effects of mixtures from pairs of chemicals that disrupt the androgen signaling pathway. (**A**) Vinclozolin and procymidone, two AR antagonists; (**B**) DBP and DEHP, two phthalates that alter fetal Leydig cell development and hormone (testosterone and insl3) production; (**C**) DBP plus procymidone, toxicants with different mechanisms of toxicity; (**D**) Linuron plus BBP, toxicants with different mechanisms of toxicity. *Abbreviations*: AR, androgen receptor; DBP, di-*n*-butyl phthalate; DEHP, di-*n*-ethyl hexyl phthalate; BBP, benzyl butyl. *Source*: Modified from Ref. 22.

combined five phthalates and examined the effects of the mixture on fetal testosterone synthesis (93). The results from all three of these studies were consistent with predictions based on a dose addition model for inducing abnormal development in utero in male rats.

In both binary phthalate mixture studies, exposure to the individual chemicals resulted in no malformations or low incidences of malformations, and the combination exposures typically resulted in 50% or greater incidences of malformations (Fig. 4B) (21).

In the third mixture study, we assessed the cumulative effects on fetal testosterone production following in utero exposure to a mixture of five phthalates: DBP, di-iso-butyl phthalate (DiBP), BBP, DEHP, and di-*n*-pentyl phthalate (DPP) (22). First, we characterized the dose response effects of six individual phthalates [BBP, DBP, DEHP, diethyl phthalate (DEP), DiBP, and DPP] on GD 18 testicular testosterone production following exposure of Sprague Dawley rats on GD 8 to 18. BBP, DBP, DEHP, and DiBP were equipotent (ED50 of 440 ± 16 mg/kg/day), DPP was about threefold more potent (ED50 = 130 mg/kg/day), and DEP had no effect on fetal testosterone production.

We hypothesized that coadministration of these five antiandrogenic phthalates would reduce testosterone production in a dose-additive fashion since they act via a

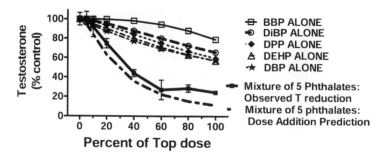

FIGURE 5 Dose-additive effect of the mixture of five phthalates. The observed effects are shown along with the dose addition model prediction and the dose response of each of the five individual phthalates based on each phthalates concentration in each dilution of the mixture. The mixture is a fixed ratio design with dilutions of the top dose (100% dose). *Abbreviations*: BBP, benzyl butyl; DiBP, di-iso-butyl phthalate; DPP, di-*n*-pentyl phthalate; DEHP, di-*n*-ethyl hexyl phthalate; DBP, di-*n*-butyl phthalate.

common mode of toxicity. In the five-phthalate mixture study, dams were dosed at 100%, 80%, 60%, 40%, 20%, 10%, 5%, or 0% of the mixture. The top dose contained 1300 mg of total phthalates/kg/day, including BBP, DBP, DEHP, DiBP (300 mg/kg/day per chemical) and DPP (100 mg DPP/kg/day). This mixture ratio was selected such that each phthalate would contribute equally to the reduction in testosterone. As hypothesized, testosterone production was reduced in a dose-additive manner (Fig. 5).

From these mixture studies, we conclude that chemicals which target the androgen signaling pathway via the same mechanism of action are dose additive when present in a mixture. These results indicate that these chemicals would be good candidates for cumulative risk assessment and supports the use of the dose addition model for determining the effects of these mixtures.

Mixtures Section II
Disruption of the Androgen Signaling Pathway by Disparate Mechanisms of Toxicity
The pesticides prochloraz and linuron disrupt androgen signaling by acting as AR antagonists and by inhibiting fetal testosterone synthesis. Environmental chemicals can disrupt the androgen signaling pathway in the male rat fetus during sexual differentiation and alter sexual differentiation via several diverse endocrine mechanisms of toxicity (Fig. 2). Knowledge of the mechanisms and modes of toxicity of EDCs allows us to make some predictions about how individual tissues will be affected when antiandrogens are combined. The classes of EDCs known to interfere with the androgen signaling pathway include dicarboximide fungicides [e.g., vinclozolin (26)], organochlorine-based insecticides [e.g., *p,p'* DDT and *p,p'* DDE (94)], conazole fungicides [e.g., prochloraz (31,54)], plasticizers (e.g., phthalates), polybrominated diphenyl ethers [PBDEs (95,96)], and urea-based herbicides (e.g., linuron) (46,49). We hypothesized that chemicals with different mechanisms of toxicity, which target the same signaling pathway, would exhibit cumulative effects that conform to a model of dose addition, not response addition. To test this hypothesis, we first assessed the joint effects of binary mixtures of chemicals with different mechanisms of action.

Mixture studies with pesticides and phthalates with diverse modes of toxicity We conducted two binary mixture studies to determine how procymidone and linuron interacted in mixtures with phthalates. The first binary mixture consisted of a fetal testosterone inhibitor (BBP) and an antiandrogen with multiple mechanisms of action (linuron) (20). The second binary mixture consisted of DBP and the AR antagonist procymidone. In these studies, pregnant rats were dosed on GD 14 to 18 with either the individual compounds or the binary mixture at a dose level equivalent to approximately one half of the ED50 value for malformations. Following this, we conducted a study with a mixture of seven chemicals that included vinclozolin, procymidone, linuron, prochloraz, and three phthalates DBP, DEHP, and BBP (25), and we have initiated a similar complex mixture study with ten chemicals.

II.a Benzyl Butyl Phthalate Plus Linuron
In the BBP and linuron study, in utero exposure to BBP alone elicited a 0% incidence of hypospadias and vaginal pouch formation and 12% incidence of epididymal agenesis in male rats (Fig. 4D). In utero exposure to linuron alone resulted in 0% incidence of hypospadias and vaginal pouch development and 63% incidence of epididymal agenesis. However, exposure to the combination resulted in cumulative effects with males displaying 56%, 40%, and 97% incidence of hypospadias, vaginal pouch, and epididymal agenesis, respectively (20).

II.b Di-n-Butyl Phthalate Plus Procymidone
In the procymidone plus DBP study, procymidone or DBP alone induced low incidences of hypospadias (1.5% and 0%, respectively) and vaginal pouch (0% and 0%, respectively), whereas the males treated with the combination of procymidone and DBP displayed 49% and 27% incidences of hypospadias and vaginal pouch, respectively, indicating that the interaction was at least dose additive. We are currently conducting an expanded binary study including multiple doses of a fixed ratio mixture of DBP and procymidone. Initial results demonstrate that responses to the binary mixture conform to a model of dose addition, not response addition (AK Hotchkiss, J Furr, and LE Gray, Jr., unpublished data; data not shown).

II.c Seven Chemical and Ten Chemical Complex Mixture Studies, Including Vinclozolin, Procymidone, Prochloraz, Linuron, and Three Phthalates or Six Phthalates
To further test our hypothesis, we designed two complex mixture studies, one with seven antiandrogenic chemicals and the other with ten chemicals. The chemicals in these complex mixtures alter the androgen signaling pathway via diverse mechanisms of action, including AR antagonists (vinclozolin and procymidone), mixed mechanism chemicals that bind to the AR, and decrease testosterone production (linuron and prochloraz) and testosterone synthesis inhibiting phthalates [BBP, DBP, and DEHP in the seven chemical study (25) and BBP, DBP, DiBP, DEHP, DPP, and di-iso-heptyl phthalate (DiHP) in the ten chemical study].

According to the current mixtures paradigm, these mixtures should conform to a model of integrated addition. However, we found that models of integrated addition or response addition consistently underestimated the effects of these mixtures. The malformations induced by the mixture of seven chemicals are shown in Figure 6. The ten chemical study results were entirely consistent with those seen in the seven chemical study. The dose addition models, provided estimates of mixture responses that closely approximated the observed responses. For example, in the study with seven chemicals, hypospadias was observed in all of the high-dose animals, dose addition and toxic equivalency models predicted 70% affected whereas integrated and response addition models predicted only a 0% affected. Clearly, integrated addition and response addition models grossly under predict the effects of such mixtures indicating that a framework based narrowly on common mechanisms of toxicity underestimates the risk of exposure to complex mixtures such as those used here.

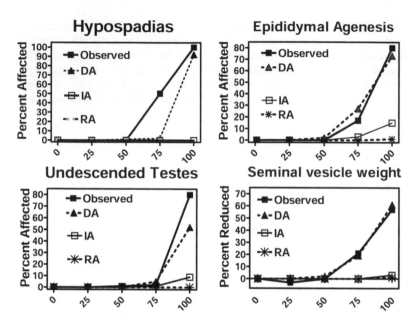

FIGURE 6 Dose-additive effects of the complex mixture study with seven "antiandrogens." The observed effects are generally well predicted by dose addition models, whereas the integrated and response addition models grossly underpredict the adverse effects of the complex mixture. Abbreviations: DA, dose addition; IA, integrated addition; RA, response addition.

Since all the chemicals used in these mixtures disrupt the androgen signaling pathway, albeit via different mechanisms of toxicity, it is clear that a cumulative risk assessment framework based on mechanisms of toxicity would not adequately predict the effects of these two complex mixtures of chemicals with different mechanisms of toxicity.

Mixtures Section III
Chemicals That Disrupted a Common Tissue Via Different Signaling Pathways and Diverse Mechanisms of Toxicity
Disruption of the androgen and AhR signaling pathways in the fetal male rat reproductive tract (97).

- 2,3,7,8 Tetrachlorodibenzo dioxin plus a phthalate (TCDD plus DBP).

We recently conducted a study to determine if TCDD, which alters epididymal differentiation and reduces sperm counts would interact with DBP in an independent or joint manner. The purpose of the study was to determine if a mixture of two chemicals that disrupted the same developing system in the fetal male rat but via different signaling pathways would produce dose-additive effects. If these two chemicals acted jointly, then we expected to see dose-additive effects on epididymal sperm numbers, epididymal morphology, and epididymal malformations since these are targets common to both chemicals. We selected at relatively high dose level of TCDD (2.0 µg/kg) because a study by Wilker et al. had not only reported large effects on total number of sperm numbers in the whole epididymis (decreased ∼30%) with a larger percent decrease in the cauda epididymis (∼50%), but they also observed epididymal agenesis in a number of F1 male rat offspring (98).

Doses of each chemical were selected based on individual chemical effects on epididymal development, so that each chemical would contribute to the mixture effect on endpoints associated specifically with epididymal function. Timed pregnant Sprague Dawley rats in the mixture group were dosed with 500 mg DBP/kg on GD 14 to 18 and with 2 µg TCDD/kg on GD 15 by oral gavage. In the study, mixture responses for testis weight, epididymal sperm counts and vas deferens, testis and epididymal malformations exceeded the values predicted by response addition modeling. We plan to build on this study by examining a more complete mixture dosing regime to further elucidate the target-dependent differences in mixture responses observed in this study.

CONCLUSIONS

Our results indicate that compounds that act by disparate mechanisms of toxicity display cumulative, dose-additive effects when present in combination and suggest that a modification of the approach for cumulative risk assessments from one based on "common mechanism of toxicity" to one that includes the cumulative assessment of chemicals that disrupt development of the same reproductive tissues during sexual differentiation. We propose that the primary focus should be on the biological system or the target tissue rather than the mechanism of toxicity or even a single signaling pathway and should include all chemicals that target that system during the same critical developmental period.

REFERENCES

1. Landrigan PJ, Claudio L, Markowitz SB, et al. Pesticides and inner-city children: exposures, risks, and prevention. Environ Health Perspect 1999; 107(suppl 3):431–437.
2. Eskenazi B, Bradman A, Castorina R. Exposures of children to organophosphate pesticides and their potential adverse health effects. Environ Health Perspect 1999; 107(suppl 3):409–419.
3. Silva. Urinary levels of seven phthalate metabolites in the US population from the National Health and Nutrition Examination Survey (NHANES) 1999–2000 (vol. 112, pg 331, 2004). Environ Health Perspect 2004; 112(5):A270.
4. Silva MJ, Reidy JA, Herbert AR, et al. Detection of phthalate metabolites in human amniotic fluid. Bull Environ Contam Toxicol 2004; 72(6):1226–1231.
5. Wolff MS, Britton JA, Boguski L, et al. Environmental exposures and puberty in inner-city girls. Environ Res 2008; 107(3):393–400.
6. Wolff MS, Engel S, Berkowitz G, et al. Prenatal pesticide and PCB exposures and birth outcomes. Pediatric Res 2007; 61(2):243–250.
7. Wolff MS, Engel SM, Berkowitz GS, et al. Prenatal phenol and phthalate exposures and birth outcomes. Environ Health Perspect 2008; 116(8):1092–1097.
8. Calafat AM, Ye X, Wong LY, et al. Exposure of the U.S. population to bisphenol A and 4-tertiary-octylphenol: 2003–2004. Environ Health Perspect 2008; 116(1):39–44.
9. Ankley GT, Brooks BW, Huggett DB, et al. Repeating history: pharmaceuticals in the environment. Environ Sci Technol 2007; 41(24):8211–8217.
10. Jobling S, Nolan M, Tyler CR, et al. Widespread sexual disruption in wild fish. Environ Sci Technol 1998; 32(17):2498–2506.
11. Jobling S, Tyler CR. Introduction: the ecological relevance of chemically induced endocrine disruption in wildlife. Environ Health Perspect 2006; 114(suppl 1):7–8.
12. Jobling S, Williams R, Johnson A, et al. Predicted exposures to steroid estrogens in U.K. rivers correlate with widespread sexual disruption in wild fish populations. Environ Health Perspect 2006; 114(suppl 1):32–39.
13. Hall AJ, Thomas GO. Polychlorinated biphenyls, DDT, polybrominated diphenyl ethers, and organic pesticides in United Kingdom harbor seals (Phoca vitulina)—mixed exposures and thyroid homeostasis. Environ Toxicol Chem 2007; 26(5):851–861.
14. Hela DG, Lambropoulou DA, Konstantinou IK, et al. Environmental monitoring and ecological risk assessment for pesticide contamination and effects in Lake Pamvotis, northwestern Greece. Environ Toxicol Chem 2005; 24(6):1548–1556.

15. Jaspers VL, Covaci A, Voorspoels S, et al. Brominated flame retardants and organo-chlorine pollutants in aquatic and terrestrial predatory birds of Belgium: levels, patterns, tissue distribution and condition factors. Environ Pollut 2006; 139(2):340–352.

16. Kolpin DW, Furlong ET, Meyer MT, et al. Pharmaceuticals, hormones, and other organic wastewater contaminants in US streams, 1999–2000: a national reconnaissance. Environ Sci Technol 2002; 36(6):1202–1211.

17. Durhan EJ, Lambright CS, Makynen EA, et al. Identification of metabolites of trenbolone acetate in androgenic runoff from a beef feedlot. Environ Health Perspect 2006; 114(suppl 1): 65–68.

18. Wolff MS, Teitelbaum SL, Windham G, et al. Pilot study of urinary biomarkers of phyto-estrogens, phthalates, and phenols in girls. Environ Health Perspect 2007; 115(1):116–121.

19. Gray LE, Ostby J, Furr J, et al. Effects of environmental antiandrogens on reproductive development in experimental animals. Hum Reprod Update 2001; 7(3):248–264.

20. Hotchkiss A, Parks-Saldutti L, Ostby J, et al. A mixture of the "antiandrogens" linuron and betyl benzyl phthalate alters sexual differentiation of the male rat in a cumulative fashion. Biol Reprod 2004; 71:1852–1861.

21. Howdeshell KL, Furr J, Lambright CR, et al. Cumulative effects of dibutyl phthalate and diethylhexyl phthalate on male rat reproductive tract development: altered fetal steroid hormones and genes. Toxicol Sci 2007; 99(1):190–202.

22. Howdeshell KL, Rider CV, Wilson VS, et al. Mechanisms of action of phthalate esters, individually and in combination, to induce abnormal reproductive development in male laboratory rats. Environ Res 2008; 108(2):168–176.

23. Rider CV, Wilson VS, Howdeshell KL, et al. Cumulative effects of in utero administra-tion of mixtures of "antiandrogens" on male rat reproductive development. Toxicol Pathol 2009; 37(1):100–113.

24. Howdeshell K, Wilson VS, Furr J, et al. A mixture of five phthalate esters inhibits fetal testicular testosterone production in the Sprague Dawley rat in a cumulative, dose-additive manner. Toxicol Sci 2008; 105(1):153–165.

25. Rider CV, Furr J, Wilson VS, et al. A mixture of seven antiandrogens induces repro-ductive malformations in rats. Int J Androl 2008; 31:249–262.

26. Kelce WR, Monosson E, Gamcsik MP, et al. Environmental hormone disruptors: evidence that vinclozolin developmental toxicity is mediated by antiandrogenic metab-olites. Toxicol Appl Pharm 1994; 126:276–285.

27. Blystone CR, Lambright C S, Cardon MC, et al. Cumulative and antagonistic effects of a mixture of the antiandrogens vinclozolin and iprodione in the pubertal male rat. Toxicol Sci 2009; 111:179–188.

28. Hosokawa S, Murakami M, Ineyama M, et al. The affinity of procymidone to androgen receptor in rats and mice. J Toxicol Sci 1993; 18(2):83–93.

29. Nellemann C, Dalgaard M, Lam HR, et al. The combined effects of vinclozolin and procymidone do not deviate from expected additivity in vitro and in vivo. Toxicol Sci 2003; 71(2):251–262.

30. Ostby J, Kelce WR, Lambright C, et al. The fungicide procymidone alters sexual differentiation in the male rat by acting as an androgen-receptor antagonist in vivo and in vitro. Toxicol Ind Health 1999; 15(1–2):80–93.

31. Vinggaard AM, Joergensen EC, Larsen JC. Rapid and sensitive reporter gene assays for detection of antiandrogenic and estrogenic effects of environmental chemicals. Toxicol Appl Pharmacol 1999; 155(2):150–160.

32. Kelce WR, Wilson EM. Environmental antiandrogens: developmental effects, molecular mechanisms, and clinical implications. J Mol Med 1997; 75(3):198–207.

33. Ashby J, Odum J. Gene expression changes in the immature rat uterus: effects of uterotrophic and sub-uterotrophic doses of bisphenol A. Toxicol Sci 2004; 82(2):458–467.

34. Bullesbach EE, Schwabe C. LGR8 signal activation by the relaxin-like factor. J Biol Chem 2005; 280(15):14586–14590.

35. Kang IH, Kim HS, Shin JH, et al. Comparison of anti-androgenic activity of flutamide, vinclozolin, procymidone, linuron, and p, p′-DDE in rodent 10 day Hershberger assay. Toxicology 2004; 199(2–3):145–159.

36. Kennel PF, Pallen CT, Bars RG. Evaluation of the rodent Hershberger assay using three reference endocrine disrupters (androgen and antiandrogens). Reprod Toxicol 2004; 18(1): 63–73.

37. Owens W, Gray LE, Zeiger E, et al. The OECD program to validate the rat Hershberger bioassay to screen compounds for in vivo androgen and antiandrogen responses: phase 2 dose-response studies. Environ Health Perspect 2007; 115(5):671–678.

38. NRC. Toxicity Testing in the Twenty-first Century: A Vision and a Strategy. Washington D.C.: National Academy of Sciences, 2007.

39. Yamasaki K, Takeyoshi M, Sawaki M, et al. Immature rat uterotrophic assay of 18 chemicals and Hershberger assay of 30 chemicals. Toxicology 2003; 183(1–3):93–115.

40. Blystone CR, Lambright CS, Furr J, et al. Iprodione delays male rat pubertal development, reduces serum testosterone levels, and decreases ex vivo testicular testosterone production. Toxicol Lett 2007; 174(1–3):74–81.

41. Monosson E, Kelce WR, Lambright C, et al. Peripubertal exposure to the antiandrogenic fungicide, vinclozolin, delays puberty, inhibits the development of androgen-dependent tissues, and alters androgen receptor function in the male rat. Toxicol Ind Health 1999; 15(1–2):65–79.

42. Gray LE Jr., Ostby JS, Kelce WR. Developmental effects of an environmental antiandrogen: the fungicide vinclozolin alters sex differentiation of the male rat. Toxicol Appl Pharmacol 1994; 129(1):46–52.

43. Gray LE Jr., Ostby J, Monosson E, et al. Environmental antiandrogens: low doses of the fungicide vinclozolin alter sexual differentiation of the male rat. Toxicol Ind Health 1999; 15(1–2):48–64.

44. Hellwig J, van Ravenzwaay B, Mayer M, et al. Pre- and postnatal oral toxicity of vinclozolin in Wistar and Long-Evans rats. Regul Toxicol Pharmacol 2000; 32(1):42–50.

45. Cowin PA, Foster P, Pedersen J, et al. Early-onset endocrine disruptor-induced prostatitis in the rat. Environ Health Perspect 2008; 116(7):923–929.

46. Lambright C, Ostby J, Bobseine K, et al. Cellular and molecular mechanisms of action of linuron: an antiandrogenic herbicide that produces reproductive malformations in male rats. Toxicol Sci 2000; 56(2):389–399.

47. McIntyre BS, Barlow NJ, Foster PM. Male rats exposed to linuron in utero exhibit permanent changes in anogenital distance, nipple retention, and epididymal malformations that result in subsequent testicular atrophy. Toxicol Sci 2002; 65(1):62–70.

48. McIntyre BS, Barlow NJ, Sar M, et al. Effects of in utero linuron exposure on rat Wolffian duct development. Reprod Toxicol 2002; 16(2):131–139.

49. McIntyre BS, Barlow NJ, Wallace DG, et al. Effects of in utero exposure to linuron on androgen-dependent reproductive development in the male Crl:CD(SD)BR rat. Toxicol Appl Pharmacol 2000; 167(2):87–99.

50. Turner KJ, McIntyre BS, Phillips SL, et al. Altered gene expression during rat Wolffian duct development in response to in utero exposure to the antiandrogen linuron. Toxicol Sci 2003; 74(1):114–128.

51. O'Connor JC, Frame SR, Ladics GS. Evaluation of a 15-day screening assay using intact male rats for identifying antiandrogens. Toxicol Sci 2002; 69(1):92–108.

52. Gray LE, Wolf C, Lambright C, et al. Administration of potentially antiandrogenic pesticides (procymidone, linuron, iprodione, chlozolinate, p,p'-DDE, and ketoconazole) and toxic substances (dibutyl- and diethylhexyl phthalate, PCB 169, and ethane dimethane sulphonate) during sexual differentiation produces diverse profiles of reproductive malformations in the male rat. Toxicol Ind Health 1999; 15(1–2):94–118.

53. Wilson VS, Lambright C, Furr J, et al. Phthalate ester-induced gubernacular lesions are associated with reduced insl3 gene expression in the fetal rat testis. Toxicol Lett 2004; 146(3):207–215.

54. Noriega NC, Ostby J, Lambright C, et al. Late gestational exposure to the fungicide prochloraz delays the onset of parturition and causes reproductive malformations in male but not female rat offspring. Biol Reprod 2005; 72(6):1324–1335.

55. Vinggaard AM, Christiansen S, Laier P, et al. Perinatal exposure to the fungicide prochloraz feminizes the male rat offspring. Toxicol Sci 2005; 85(2):886–897.

56. Vinggaard AM, Hass U, Dalgaard M, et al. Prochloraz: an imidazole fungicide with multiple mechanisms of action. Int J Androl 2006; 29(1):186–191.

57. Blystone CR, Furr J, Lambright CS, et al. Prochloraz inhibits testosterone production at dosages below those that affect androgen-dependent organ weights or the onset of puberty in the male Sprague Dawley rat. Toxicol Sci 2007; 97(1):65–74.

58. Blystone CR, Lambright CS, Howdeshell KL, et al. Sensitivity of fetal rat testicular steroidogenesis to maternal prochloraz exposure and the underlying mechanism of inhibition. Toxicol Sci 2007; 97:65–74.

59. Harris CA, Henttu P, Parker MG, et al. The estrogenic activity of phthalate esters in vitro. Environ Health Perspect 1997; 105(8):802–811.

60. Gray LE Jr., Ostby J. Effects of pesticides and toxic substances on behavioral and morphological reproductive development: endocrine versus nonendocrine mechanisms. Toxicol Ind Health 1998; 14(1–2):159–184.

61. Parks LG, Ostby JS, Lambright CR, et al. The plasticizer diethylhexyl phthalate induces malformations by decreasing fetal testosterone synthesis during sexual differentiation in the male rat. Toxicol Sci 2000; 58(2):339–349.

62. Stroheker T, Cabaton N, Nourdin G, et al. Evaluation of anti-androgenic activity of di-(2-ethylhexyl)phthalate. Toxicology 2005; 208(1):115–121.

63. Foster PM, Mylchreest E, Gaido KW, et al. Effects of phthalate esters on the developing reproductive tract of male rats. Hum Reprod Update 2001; 7(3):231–235.

64. Gray LE Jr., Ostby J, Furr J, et al. Perinatal exposure to the phthalates DEHP, BBP, and DINP, but not DEP, DMP, or DOTP, alters sexual differentiation of the male rat. Toxicol Sci 2000; 58(2):350–365.

65. Mylchreest E, Sar M, Cattley RC, et al. Disruption of androgen-regulated male reproductive development by di(n-butyl) phthalate during late gestation in rats is different from flutamide. Toxicol Appl Pharmacol 1999; 156(2):81–95.

66. Mylchreest E, Foster PM. DBP exerts its antiandrogenic activity by indirectly interfering with androgen signaling pathways. Toxicol Appl Pharmacol 2000; 168(2):174–175.

67. Gray LE Jr., Ostby JS. In utero 2,3,7,8-tetrachlorodibenzo-p-dioxin (TCDD) alters reproductive morphology and function in female rat offspring. Toxicol Appl Pharmacol 1995; 133(2):285–294.

68. Gray LE, Wolf C, Mann P, et al. In utero exposure to low doses of 2,3,7,8-tetrachlorodibenzo-p-dioxin alters reproductive development of female Long Evans hooded rat offspring. Toxicol Appl Pharmacol 1997; 146(2):237–244.

69. Flaws JA, Sommer RJ, Silbergeld EK, et al. In utero and lactational exposure to 2,3,7,8-tetrachlorodibenzo-p-dioxin (TCDD) induces genital dysmorphogenesis in the female rat. Toxicol Appl Pharmacol 1997; 147(2):351–362.

70. Wolf CJ, Ostby JS, Gray LE Jr. Gestational exposure to 2,3,7,8-tetrachlorodibenzo-p-dioxin (TCDD) severely alters reproductive function of female hamster offspring. Toxicol Sci 1999; 51(2):259–264.

71. Gray LE Jr., Kelce WR, Monosson E, et al. Exposure to TCDD during development permanently alters reproductive function in male Long Evans rats and hamsters: reduced ejaculated and epididymal sperm numbers and sex accessory gland weights in offspring with normal androgenic status. Toxicol Appl Pharmacol 1995; 131(1):108–118.

72. Gray LE, Ostby JS, Kelce WR. A dose-response analysis of the reproductive effects of a single gestational dose of 2,3,7,8-tetrachlorodibenzo-p-dioxin in male Long Evans Hooded rat offspring. Toxicol Appl Pharmacol 1997; 146(1):11–20.

73. Mably TA, Bjerke DL, Moore RW, et al. In utero and lactational exposure of male rats to 2,3,7,8-tetrachlorodibenzo-p-dioxin. 3. Effects on spermatogenesis and reproductive capability. Toxicol Appl Pharmacol 1992; 114(1):118–126.

74. Mably TA, Moore RW, Goy RW, et al. In utero and lactational exposure of male rats to 2,3,7,8-tetrachlorodibenzo-p-dioxin. 2. Effects on sexual behavior and the regulation of luteinizing hormone secretion in adulthood. Toxicol Appl Pharmacol 1992; 114(1):108–117.

75. Mably TA, Moore RW, Peterson RE. In utero and lactational exposure of male rats to 2,3,7,8-tetrachlorodibenzo-p-dioxin. 1. Effects on androgenic status. Toxicol Appl Pharmacol 1992; 114(1):97–107.

76. Simanainen U, Haavisto T, Tuomisto JT, et al. Pattern of male reproductive system effects after in utero and lactational 2,3,7,8-tetrachlorodibenzo-p-dioxin (TCDD) exposure in three differentially TCDD-sensitive rat lines. Toxicol Sci 2004; 80(1):101–108.

77. Theobald HM, Roman BL, Lin TM, et al. 2,3,7,8-tetrachlorodibenzo-p-dioxin inhibits luminal cell differentiation and androgen responsiveness of the ventral prostate without inhibiting prostatic 5alpha-dihydrotestosterone formation or testicular androgen production in rat offspring. Toxicol Sci 2000; 58(2):324–338.

78. Ko K, Theobald HM, Moore RW, et al. Evidence that inhibited prostatic epithelial bud formation in 2,3,7,8-tetrachlorodibenzo-p-dioxin-exposed C57BL/6J fetal mice is not

due to interruption of androgen signaling in the urogenital sinus. Toxicol Sci 2004; 79(2): 360–369.

79. Haavisto TE, Myllymaki SA, Adamsson NA, et al. The effects of maternal exposure to 2,3,7,8-tetrachlorodibenzo-p-dioxin on testicular steroidogenesis in infantile male rats. Int J Androl 2006; 29(2):313–322.

80. Haavisto T, Nurmela K, Pohjanvirta R, et al. Prenatal testosterone and luteinizing hormone levels in male rats exposed during pregnancy to 2,3,7,8-tetrachlorodibenzo-p-dioxin and diethylstilbestrol. Mol Cell Endocrinol 2001; 178(1–2):169–179.

81. Lin TM, Rasmussen NT, Moore RW, et al. 2,3,7,8-tetrachlorodibenzo-p-dioxin inhibits prostatic epithelial bud formation by acting directly on the urogenital sinus. J Urol 2004; 172(1):365–368.

82. Vezina CM, Allgeier SH, Moore RW, et al. Dioxin causes ventral prostate agenesis by disrupting dorsoventral patterning in developing mouse prostate. Toxicol Sci 2008; 106(2): 488–496.

83. Vezina CM, Lin TM, Peterson RE. AHR signaling in prostate growth, morphogenesis, and disease. Biochem Pharmacol 2009; 77(4):566–576.

84. Squillace PJ, Scott JC, Moran MJ, et al. VOCs, pesticides, nitrate, and their mixtures in groundwater used for drinking water in the United States. Environ Sci Technol 2002; 36(9): 1923–1930.

85. CDC. National Report on Human Exposure to Environmental Chemicals, 2008. Available at: http://www.cdc.gov/exposurereport/biomonitoring_results.htm.

86. Silva E, Rajapakse N, Kortenkamp A. Something from "nothing"—eight weak estrogenic chemicals combined at concentrations below NOECs produce significant mixture effects. Environ Sci Technol 2002; 36(8):1751–1756.

87. Altenburger R, Backhaus T, Boedeker W, et al. Predictability of the toxicity of multiple chemical mixtures to Vibrio fischeri: mixtures composed of similarly acting chemicals. Environ Toxicol Chem 2000; 19(9):2341–2347.

88. Backhaus T, Altenburger R, Boedeker W, et al. Predictability of the toxicity of a multiple mixture of dissimilarly acting chemicals to Vibrio fischeri. Environ Toxicol Chem 2000; 19(9):2348–2356.

89. Rider CV, Leblanc GA. An integrated addition and interaction model for assessing toxicity of chemical mixtures. Toxicol Sci 2005; 87(1):520–528.

90. Altenburger R, Schmitt H, Schuurmann G. Algal toxicity of nitrobenzenes: combined effect analysis as a pharmacological probe for similar modes of interaction. Environ Toxicol Chem 2005; 24(2):324–333.

91. Bliss CI. The toxicity of poisons applied jointly. Ann Appl Biol 1939; 26:585–615.

92. Teuschler LK, Rice GE, Wilkes CR, et al. Feasibility study of cumulative risk assessment methods for drinking water disinfection by-product mixtures. J Toxicol Environ Health A 2004; 67(8–10):755–777.

93. Howdeshell KL, Wilson VS, Furr J, et al. A mixture of five phthalate esters inhibits fetal testicular testosterone production in the Sprague Dawley rat in a cumulative, dose-additive manner. Toxicol Sci 2008; 105(1):153–165.

94. Kelce WR, Monosson E, Gray LE Jr. An environmental antiandrogen. Recent Prog Horm Res 1995; 50:449–453.

95. Stoker TE, Cooper RL, Lambright CS, et al. In vivo and in vitro anti-androgenic effects of DE-71, a commercial polybrominated diphenyl ether (PBDE) mixture. Toxicol Appl Pharmacol 2005; 207(1):78–88.

96. Gray LE Jr., Wilson V, Noriega N, et al. Use of the laboratory rat as a model in endocrine disruptor screening and testing. ILAR J 2004; 45(4):425–437.

97. Rider CV, Furr JR, Wilson VS, et al. Cumulative effects of in utero administration of mixtures of reproductive toxicants that disrupt common target tissues via diverse mechanisms of toxicity. Int J Androl 2010; 33:443–462.

98. Wilker C, Johnson L, Safe S. Effects of developmental exposure to indole-3-carbinol or 2,3,7,8-tetrachlorodibenzo-p-dioxin on reproductive potential of male rat offspring. Toxicol Appl Pharmacol 1996; 141(1):68–75.

Disclaimer: The research described in this article has been reviewed by the National Health and Environmental Effects Research Laboratory, ORD, U. S. Environmental Protection Agency, and approved for publication. Approval does not signify that the contents necessarily reflect the views and policies of the Agency nor does the mention of trade names or commercial products constitute endorsement or recommendation for use.

19

Metals and metal compounds in reproductive and developmental toxicology

Mari S. Golub

INTRODUCTION

Most elements (88 of 112 known elements) are classified as metals in the periodic table. Metals and metal compounds that are commonly found in biological systems or commonly used in commerce have been studied for their reproductive toxicity, while little or no information is available for many other metals. Most of these studies are in the open literature, but metals have also been candidates for guideline toxicity testing. Risk assessments for metal toxicity are widely available. Documents from regulatory agencies that include review of the developmental and reproductive toxicity of metals are shown in Table 1. A number of other reviews of metal reproductive toxicity are available covering a literature that extends back more than 50 years (1–3), including newer reviews on topical issues (4).

The present review organizes metals by category and emphasizes metals with a vigorous contemporary research effort. Metalloids (arsenic, boron, silicon, antimony, tellurium, polonium, and germanium) are not reviewed here.

In addressing the topic of reproductive and developmental toxicology, this review is limited to pregnancy outcome measures as indices of developmental toxicology. The broader developmental toxicology literature looking at functional deficits arising throughout the life span as a consequence of exposure to metals from conception to puberty is not covered. Many metals are neurotoxicants and developmental neurotoxicants, and these topics are most appropriately reviewed together (5).

Contemporary issues in toxicology have provided a fresh perspective on the well-established area of metal toxicity. Some issues discussed here in connection with metal reproductive toxicity include endocrine disruption, nanotoxicology, and gene-environment interactions.

HEAVY METALS: LEAD, MERCURY, CADMIUM, AND TIN WITH A SMALL LITERATURE ON GOLD, SILVER, PLATINUM, AND TUNGSTEN

As a toxicant of long-standing societal concern, *lead* has generated a fairly large literature on reproductive toxicity in laboratory animals and human populations.

As reviewed recently by Sokol (6), lead effects on the male reproductive system are highly dependent on age of exposure and demonstrate both adaptation and reversibility, making integration of studies difficult. Lead accumulation in testes and effects on hormone levels, sperm parameters, and fertilization capacity have been demonstrated in laboratory animals. Sites of action in the brain and testes have been documented. In humans, associations of paternal lead exposure with spontaneous abortion and malformation have been reported, while associations with reduced fertility or fecundity have less empirical support. The literature on female reproductive toxicity is more limited; direct effects on ovaries and ovulation are not extensively reported in animal studies although some effects on follicle numbers have been shown in mice (7). A growing area of research concerns vulnerability of the developing female reproductive tract to in utero exposure to lead, as well as postnatal exposures through

TABLE 1 Toxicology Document Series Containing Information on the Reproductive Toxicity of Individual Metals

	Current volume	ATSDR[a]	IPCS[b]	HC[c]	OEHHA[d]	USGS[e]
Aluminum	x	x	x	x	PHG	
Antimony		x				
Arsenic	x	x	x	x	HID,REL	x
Barium		x	x			
Beryllium		x	x		DS	
Boron		x	x		DS	x
Cadmium	x	x	x	x	HID, PHG	x
Chromium	x	x	x	x	HID, PHG, DS	x
Cobalt	x	x				
Copper	x	x	x		PHG, REL, DS	x
Lead	x	x	x		PHG	x
Lithium						
Manganese		x	x		REL, DS	
Mercury	x	x	x		PHG, REL	x
Molybdenum	x					
Nickel	x	x	x	x	PHG, REL	x
Osmium						
Platinum	x		x			
Radium		x				
Selenium		x	x			x
Silver	x	x				x
Thallium	x	x	x		PHG	
Thorium		x				
Titanium		x	x			
Tin	x	x	x	x		x
Uranium	x	x			PHG	
Vanadium	x	x	x		REL	
Zinc	x	x	x			x

[a]Agency for Toxic Substances and Disease Control. Center for Disease Control. Toxicological Profiles. Available at: http://www.atsdr.cdc.gov/toxpro2.html.
[b]International Programme on Chemical Safety: World Health Organization. Environmental Health Criteria. Available at: http://www.who.int/pcs/pubs/pub_ehc.alph.htm.
[c]Health Canada: Existing Substances Division. Priority Substances List. Available at: http://www.hc-so.go.ca/hecs-sesc/exsd/psap.htm.
[d]Office of Environmental Health Hazard Assessment. Available at: http://www.oehha.ca.gov.
Health risk assessment documents from the OEHHA include those directed exclusively at reproductive toxicity [Hazard Identification Documents (HIDs) and Data Summaries (DS)] as well as more general toxicity reviews that include reproductive toxicity [Public Health Goals (PHG) for drinking water, Reference Exposure Levels (REL) for air contaminants].
[e]Patuxent Health Center: United States Geological Survey. Contaminants Hazard Review. Available at: http://www.pwrc.usgs.gov/contaminants.
Source: From Ref. 1.

puberty. Sensitive endpoints in these animal studies were altered steroidogenesis, delayed vaginal opening, and disrupted early estrous cycles. Studies in humans also reported associations between lead exposure and delayed puberty as reflected in age at menarche (8,9) and vaginal opening in rodents (10–12), bringing lead into focus as a contemporary endocrine disruptor. Recently, lead exposure has also been implicated in delayed puberty in boys (13). Because the effects of developmental lead exposure on growth can accompany effects on puberty timing (14,15), indirect effects on mechanisms of puberty initiation must be considered. The remarkable sensitivity of neurobehavioral indices to low level lead exposure may carry over to measures of reproductive tract maturation.

Like many metals, lead has not been the subject of a complete set of guideline reproductive toxicity studies. However, a thorough toxicity study with exposure of female rats from weaning through mating and pregnancy and including both fetal and postnatal assessments (10–12) did not suggest that malformation rate or fetotoxicity were sensitive endpoints, although growth parameters were affected. Delayed vaginal opening was also seen in this study in the breeders.

As is the case for lead, a major focus of the toxicity literature for *mercury* has been on postnatal neurobehavioral endpoints and brain development. These studies are mainly concerned with organic mercury (methyl mercury and, more recently, thimerosal) exposures. Rodent studies with methyl mercury administration during pregnancy report a spectrum of results from fetal viability and malformation at high doses to skeletal ossification and external hemorrhage at lower doses (16). In contrast, reproductive toxicity studies have more frequently focused on inorganic mercury (17). The provocation for these studies was occupational exposures to elemental mercury, particularly in dentistry. Studies in female dental assistants identified an association between high levels of exposure and lower fertility (18), but studies of female dentists did not find this association. Male reproductive toxicity has received less attention. While there are some studies in male dentists, occupationally exposed men, and male patients in infertility clinics, the data are very limited on the subject of mercury-induced male reproductive impairment in humans (17). In a series of gavage studies in mice, inorganic mercury effects on testes and sperm parameters were described in some detail (19).

Organotin compounds, or stannanes, form a large group of biologically active compounds, many of which have been identified as reproductive toxicants (Table 2). The stimulus for research on the reproductive toxicity of butyl tins has largely come from the discovery of imposex (male sex organs "imposed" on females) in marine snails exposed to tributyltin, used as an antifouling agent on marine vessels and pilings (20). Triphenyltin as well as tributyltin were effective in this regard, whereas dibutyltin and monophenyltin were not. However, dibutyl tin, as well as tributyl tin, has been shown to disrupt pregnancy in the rat. In hypothesis-testing research, Ema et al. (20) have

TABLE 2 Organotins Studied for Reproductive Toxicity

Methyl compounds
  Monomethyl tin (MMT)
  Dimethyl tin (DMT)
  Trimethyl tin (TMT)
Ethyl compounds
  Triethyltin (TET)
Butyl compounds
  Monobutyl tin (MBT)
  Dibutyl tin (DBT)
  Tributyl tin (TBT)
  Tetrabutyl tin (TeBT)
Hexl compounds
  Trihexyl tin (THT)
Octyl compounds
  Monooctyl
  Dioctyl
Phenyl compounds
  Monophenyl tin
  Diphenyl tin
  Triphenyl tin
Mixed compounds
  Phenyl butyl tin (fenbutatin oxide)

demonstrated implantation failure when di- or tributyltin was administered during the preimplantation period to rats; monobutyl tin was not effective. Because implantation failure was counteracted by progesterone, the authors suggested that effects on steroid hormone metabolism were responsible. Some information on reproductive toxicity in mammalian or aquatic species is available for many of the other organotins (20). Structural requirements for toxicity and common mechanisms have not yet been established. Phenyl tins have been the subject of guideline regulatory studies because of the use of triphenyltin as a pesticide (21). Multigeneration studies from the 1980s demonstrated reduced pregnancy rates, reduced litter sizes, and decreased live pups at birth in rats fed triphenyltin in diet at doses ≤12 mg/kg/day (21). Di- and triphenyltin have been shown to damage testes as well as ovaries in laboratory rodents. Recent mechanism work has focused on direct phenyltin activation of nuclear hormone receptors (22). Phenyl butyl tin (fenbutatin oxide) is another organotin pesticide for which guideline studies are available (23). Finally, trimethyl- and triethyl tin are well known neurotoxicants in adults, leading to research concerning developmental neurotoxicity (24,25). In recent studies, developmental neurobehavioral toxicity was not seen when monomethyl tin was given to rats in drinking water prior to mating and through weaning, although some brain histopathology changes were noted (26). When dibutyltin was examined for developmental neurotoxicity, histopathological changes were documented; behavioral evaluations were not conducted (27). Studies of inorganic tin were not located for this review and no human studies appear to be available.

Of the metals, *cadmium* has the most extensive reproductive toxicity literature. It is particularly well characterized in animal models as an ovarian toxicant, testicular toxicant, and placental toxicant as well as a teratogen in animal models (7,28,29). A number of human studies have assessed a possible role of cadmium in intrauterine growth retardation (17). Mechanism studies show that cadmium can interfere with cell viability, cell differentiation, cell function, and cell adhesion (29) in target organs. Cadmium is a mutagenic and genotoxic metal, so interactions with DNA need to be considered. Cadmium is thought to compete with zinc, calcium, and other essential elements in metallomolecules and also for binding to metallothionein. In an interesting series of experiments, Martin and associates (30–32) have documented a direct estrogenic action of cadmium in vitro and in vivo at low doses, earning cadmium premier status as a "metalloestrogen" and a potentially important endocrine disruptor. Cadmium is an effective teratogen when administered by the injection route to rodents and other test species. Because cadmium accumulation is much greater in placenta than in embryo, placental mediation has been suggested. Oral and inhalation exposures during pregnancy affect primarily fetal growth and viability. More recent literature has focused on possible prevention of these effects, especially by antioxidants (33) and essential trace elements (34,35), and studies of gene expression (36), and gene-toxicant interactions (37). Although not as widely recognized for neurotoxicity as other heavy metals, cadmium developmental neurobehavioral toxicity has been the subject of a number of studies over the years (28,38,39).

Cadmium is of interest as a component of cigarette smoke and environmental tobacco smoke, as a pigment, and as an environmental contaminant. Studies correlating cadmium exposure with pregnancy outcome in women have appeared in the literature (40–48), with effects on birth weight a common theme. Populations studied include women smokers, industrial workers, and women living in areas with high cadmium contamination of soil and water. Cadmium-induced zinc deficiency has often been studied as a possible mechanism.

Two of the three "precious" heavy metals (platinum, silver, and gold) are of concern because of new exposures due to their extensive use in nanoparticles, silver as an antibacterial, and platinum in fuel cells. Information on *platinum* reproductive toxicity has come from research on platinum-based chemotherapeutic agents (cisplatin, carboplatin, oxyplatin). Information is available from patient populations as well as studies of reproductive toxicity in animals. The studies in animals focused on testicular effects (sperm and testosterone measures) due to the observation of residual infertility in patients treated with platinum-based chemotherapy for testicular cancer (49). Using intraperitoneal dosing relevant to chemotherapy, cisplatin-treated rats showed a full spectrum of male reproductive toxicity including reduced testosterone, damage to Sertoli cells, lower numbers and motility of sperm, and lower testes weights (50–55). Upon mating increased pre- and post-implantation loss, malformation and growth retardation were seen in fetuses (52). A primary effect on steroidogenesis was indicated by the continued response of the testes to gonadotrophic hormones and suppression of the rate-limiting step in testosterone synthesis (P-450scc) (50,52). Continuing fertility and reversal of effects after discontinuation of treatment were characteristic of the syndrome. No studies with oral administration or with forms of platinum other than chemotherapeutic drugs were located.

Some data on pregnancy effects was also generated with platinum chemotherapeutic agents. In mice, embryolethality, growth retardation, and delayed ossification were the main components of the syndrome that occurred in a dose-dependent pattern when cisplatin was given on individual days of pregnancy by intraperitoneal injection (56). The authors cite studies in rats with similar results (embryotoxicity with minimal indication of malformation).

A guideline study of *silver* (as silver acetate) developmental toxicity was conducted in 2001 (Ref. 57 reviewed as NTIS abstract). The study found minimal developmental and maternal toxicity up to the highest dose tested (64.6 mg silver/kg/day). Significant dose-related trends were seen for late fetal deaths and fetal body weight. The mechanism of silver developmental toxicity has been referred to an interference with copper metabolism (58). As is the case for many other metals, silver accumulates in testes (59), but testicular toxicity studies were not found. Silver nanoparticles were tested in a screening assay using a mouse spermatogonial stem cell line and mitochondrial function, membrane integrity, and activation of apoptosis as endpoints. In this assay, silver nanoparticles were more toxic on all three endpoints than the corresponding silver carbonate concentrations (60). This study suggests that metals as nanoparticles cannot be expected to interact with biological systems in the same way as metal ions. The U.S. Food and Drug Administration (USFDA) has nominated silver nanoparticles for toxicity testing by the National Toxicology Program that would include in utero exposures (61).

Gold, the third of the "precious" metals, has received negligible attention in toxicology literature. No studies of reproductive toxicity were identified.

Tungsten, along with vanadium, has recently received attention for normalization of insulin production as a treatment for diabetes (62). Tungsten (as sodium tungstate in drinking water, 2 mg/mL) was found to reverse the reproductive impairments of streptozotocin-induced female and male diabetic rats, but had no adverse effect on normal rats (63,64). Tungsten administered to pregnant mice is transferred to the embryo/fetus (65) but adverse effects of this exposure have not been studied. One recent study screened for neurobehavioral effects of exposures beginning before mating and extending throughout pregnancy and lactation in rats and found effects on locomotor activity in the absence of effects on fertility and pregnancy outcome (66). The Agency for Toxic Substances and Disease Registry (ATSDR) (67) describes additional studies in the Russian language that could not be obtained for review here.

TRANSITION METALS: ESSENTIAL NUTRIENTS (CHROMIUM, COBALT, COPPER, IRON, MANGANESE, MOLYBDENUM, ZINC) AND BENEFICIAL NUTRIENTS (NICKEL, VANADIUM)

Transition metals can assume multiple valence states making them valuable for electron transfer in biological systems. In particular, large families of enzymes rely on iron, copper, and zinc for their reactions, and a smaller number on chromium and manganese, molybdenum, nickel, and vanadium. A subgroup of the transition metals have been classified by the Food and Nutrition Board of the National Academy of Sciences as essential nutrients based on identification of a deficiency syndrome when they are reduced or deleted from diets. Consumption of nutritional supplements is a unique source of exposure for these metals. In some cases, however, the valence state of the metal in supplements (Cr+3, V+4) is different from the valence state in which it has been studied as a toxicant (Cr+6, V+5). Several of the essential metal deficiency syndromes include teratogenesis (zinc, copper, manganese) and interference with reproduction (68). Many of the essential metals compete for common uptake, transport, and binding sites in biological systems. As a consequence, a common theme in the toxicology of these metals is the induction of deficiency of one trace metal by overload with another metal, resulting in induction of a deficiency syndrome (69). Another theme, common to most metals, is the study of male reproductive toxicity focusing on metal accumulation in testes and subsequent effects on hormone production, spermatogenesis, and fertility.

Iron has received little formal attention as a reproductive toxicant. Genetic syndromes that lead to iron accumulation include infertility and reproductive organ damage (70). A single animal experiment with iron overload during pregnancy in rabbits and rats found effects on fetal and neonatal viability (71). There is currently concern about associations between population-based iron supplementation during infancy and adverse effects on growth and immune defense. Animal studies with iron administration during the early postnatal period (72,73) have looked for potential toxic effects, but similar studies have not been done in pregnancy.

Chromium (+6) and *nickel* are both known human carcinogens and are both implicated in toxic exposures occurring in welding occupations. Consequently, human occupational and environmental exposures are monitored and a number of human studies of reproductive outcome and markers of reproductive toxicity have been produced (74). In some of these studies, differential effects of the metals can be inferred from exposure and biomonitoring data as well as comparison of workers using different types of welding processes. Animal studies contribute to the biological plausibility of associations seen in worker populations.

In the case of chromium (+6), a series of studies in rats and mice using the same drinking water concentrations (0, 250, 500, and 750 µg/mL) evaluated fertility, pregnancy outcome, and ovarian histology (follicles, corpora lutea) (75–82). The chromium was given either prior to mating, at different periods during gestation, or during lactation. These studies reported chromium effects on fetal viability, growth and skeletal ossification. With exposure prior to mating, estrous cycle lengths increased, and corpora lutea and conceptions decreased. In addition, effects on fetal viability and growth were seen. Development of the female reproductive tract was studied by giving chromium in drinking water during lactation. Only one chromium concentration (200 µg/mL) was used. Vaginal opening was markedly delayed, early estrous cycles were lengthened, and the number of ovarian follicles was reduced. Some studies of male reproductive toxicity in rodents with oral dosing are also available (83). Many of the endpoints assessed were affected by the chromium exposures: reproductive organ weights, testicular sperm counts, seminiferous tubule diameter, epididymal sperm counts, numbers of abnormal sperm, and mating behavior. However, in a single study with mating trials, fertility was not reduced (84).

Single or multigeneration studies of chromium (+6) with both parents exposed were not located. ATSDR has identified multigeneration studies of chromium via the oral route as an "unfilled priority data need" (85).

The chromium male reproductive toxicity literature is unique in containing a study in nonhuman primates (bonnet macaque monkeys) (82,86–89). This model is valuable in allowing successive evaluations of sperm parameters in individuals during the dosing period. Chromium was administered in drinking water. The authors reported a dose and time-dependent decrease in sperm counts. At the end of the study, testes and epididymis histopathology was indicative of spermatotoxicity. No mating trials were conducted.

For *nickel*, there is an older multigeneration study with drinking water administration (90) as well as a more recent single- and two-generation studies with gavage administration (Refs. 91,92 described in Ref. 93). The sensitive endpoints were offspring viability (postimplantation loss, dead pups at birth, smaller litter size) at the higher doses in these studies. The two-generation study (92) included estrous cycle and sperm parameters, neither of which showed effects of the nickel treatment. Studies in which only females were treated (94–96) also found offspring viability effects. These studies also used drinking water administration. Male-only exposures by drinking water for four weeks (95) reported that nickel exposure led to lower pregnancy rates (50%) at a drinking water concentration of 30 ppm and effects on testes (number of spermatogonia). However, these effects did not appear with a longer (six week) exposure.

Nickel is among the metals studied for fertility effects in worker populations (74). Correlations between sperm parameters (some indices of motility, morphology, and viability) and blood nickel were found in welders (97), but a single study of fertility with limited exposure assessment found no effects of nickel (98).

Molybdenum was studied in an early multigeneration studies in rodents (90,99). At the highest doses, about 7 mg/kg/day, few litters were produced and males demonstrated testicular degeneration. General toxicity including mortality was also present in these studies. In a later study of male reproductive toxicity (100), sodium molybdate administered orally for 60 days to rats led to accumulation in reproductive organs, lower weights of reproductive organs (testes, epididymis, seminal vesicles, and ventral prostate), decreased sperm numbers and motility, reduced testicular enzyme activity, and testicular histopathology at 30 and 50 mg/kg/day. The lowest dose in the study, 10 mg/kg/day, was an apparent no observed adverse effect level (NOAEL). Sensitive endpoints of female reproductive toxicity in rats were prolonged estrous cycles and, when administered during pregnancy, increased resorption, and delay in some aspects of embryogenesis (101). The molybdenum-based chelating agent ammonium tetrathiomolybdate has also been studied for its reproductive toxicity but the importance of the metal component of this compound is not clear. Molybdenum may be of interest in the future because of its use in nanomaterials (as molybdenum trioxide).

Vanadium is not widely surveyed in environmental media. Attention to its toxicity has come about in connection with its possible use in diabetes therapy. Reproductive toxicity studies were recently reviewed by Domingo (2005) (102). Two key rodent studies of vanadate (vanadium +5) are available. When both male and female rats were exposed to vanadate in a single-generation study design with three oral doses, the sensitive endpoint was decreased offspring growth during lactation, seen at all doses (103). Vanadate in drinking water of male mice for 64 days before mating led to lower pregnancy rates at the highest doses, and decreased epididymis but not testes weights, accompanied by lower sperm counts but no changes in motility at all three doses studied (104). Domingo (102) estimated the NOAEL for male reproductive effects at 7.5 mg V/kg/day as compared to an estimated human daily intake through diet and drinking water of 0.03 µg/kg/day. The lowest observed adverse effect level (LOAEL) for postnatal growth retardation was about 1 mg/kg/day. Vanadium is used extensively in production of stainless steel, which leads to occupational exposures via inhalation. Although cancer studies have been conducted with vanadium pentoxide via inhalation, no reproductive toxicity studies were located using this route of administration. Vanadium has been nominated by the National Institute of Environmental Health Sciences and the U.S. Environmental Protection Agency for toxicity testing including multigeneration studies by the National Toxicology Program (105).

A number of small animal studies failed to find effects of *zinc* excess on reproductive endpoints, although implantation was reduced in a study where zinc at 200 mg/kg/day was fed to rats beginning on gd 0 (106). Recently, more comprehensive multigeneration studies of zinc have been undertaken in rats (107). Zinc was administered as $ZnCl_2$ in drinking water. Reduced fertility, offspring growth, and viability were identified at the high dose (30 mg/kg/day), which can be compared to the "tolerable upper limits" established by the Food and Nutrition Board (about 0.6 mg/kg/day). Interpretation of this study, however, is complicated by relatively pronounced mortality and weight gain effects in the zinc-treated group. A similar study in mice has been reported in abstract but is not yet published (108). A single study of male reproductive toxicity was located (109). At an exposure of 4000 ppm in diet, reduced sperm motility and fertility were reported in rats.

The *cobalt* literature is limited to a fairly extensive study of male reproductive toxicity in male mice by Pedigo and colleagues (110–112). The cobalt chloride was administered in drinking water. The studies varied in length from 7 to 12 weeks of dosing and found reduced testes and epididymal weights, seminiferous tubule degeneration sperm parameters (number and motility), but increased serum testosterone. Reduced fertility was detected in these studies in terms of fertilized ova retrieved after mating, an effect that first worsened then recovered after cessation of exposure. The effective dose was 50 mg/kg/day and included minimal apparent general toxicity. In view of the performance of cobalt in screening for estrogenic activity (113) studies of female toxicity may be valuable.

GROUP 13 METALS: ALUMINUM, GALLIUM, INDIUM, AND THALLIUM

These elements have one valence state, +3, limiting their interaction with biological systems. They have in common the ability to bind to transferrin (which transports ferric iron) and be taken up by cells by transferrin-dependent mechanisms. The group 13 metalloid boron is a well-studied male reproductive toxicant, and there is also an emphasis on male reproduction in studies of other group 13 metals.

Reproductive toxicology of *gallium* has been studied as a variety of relevant gallium compounds: gallium arsenide, a semiconductor material; copper gallium diselenide, a photovoltaic material; and gallium nitrate, a chemotherapeutic agent. In considering these gallium-containing compounds together, it is important to note that each has unique chemical properties and interactions with biological systems. Gallium arsenide by intratracheal administration in male hamsters and rats caused reduced epididymal sperm and testicular sperm retention in the hamsters and rats (114,115). The rat study also demonstrated sperm head abnormalities suggesting a late-stage interference with spermatogenesis. Controls in the study indicated that the effects were not entirely due to arsenic, although an influence of arsenic on gallium toxicity could not be entirely ruled out. Gallium nitrate by subcutaneous injection in mice did not affect fertility, sperm parameters, or testes histopathology (116). Copper gallium diselenide administered to male rats by gavage did not alter fertility or sperm parameters (117).

As regards developmental toxicity, gallium arsenide by inhalation in a guideline developmental toxicity produced growth retardation and skeletal variations in rats, and growth retardation, skeletal variations, and also malformation in mice (118). Copper gallium diselenide by gavage in rats during organogenesis produced smaller litters but no effect on growth (117). Corpora lutea and implantations were lowered when dosing began prior to mating. Gallium sulfate by intravenous administration to pregnant hamsters also reduced embryo viability without inducing malformations (119).

Thallium was shown to induce premature sperm release when administered in drinking water at 10 ppm to rats, presumably mediated by Sertoli cells that showed histopathological changes (120). A follow-up study of testicular cell culture also demonstrated release of developing sperm in response to thallium (121). In a developmental toxicity study by intraperitoneal injection in rats, thallium sulphate produced growth retardation and developmental delay but no increase in structural malformations (122). Thallium exposures can occur from environmental contamination and use of thallium as a rat poison. Case reports of pregnancy outcome after thallium poisoning in pregnancy are available (123).

Indium has uses as a biological tracer as well as a component of liquid crystal displays (LCDs) and semiconductors. Indium radionuclides used in medicine have been studied for testicular toxicity in animal models (124–126), but nonradioactive indium has only been studied as indium arsenide by intratracheal installation, and no testicular toxicity was reported (115).

Indium is the most striking teratogen among the group 13 metals. In early studies, indium was found to produce a high incidence of malformations in hamsters after intravenous administration (119), particularly digit and tail abnormalities. By this route indium was highly toxic leading to no live pups at doses above 1 mg/kg/day and maternal death at 20 mg/kg/day. More recently, this teratogenic action was confirmed with oral exposure in rats and rabbits at doses over 100 mg/kg/day (127) although another group has failed to identify the teratogenic effect in rats by the oral route (128). In follow-up work, indium was found to reach the fetus and accumulate in bone where its presence was thought to cause cartilage ossification and mediate fetal skeletal malformations (129,130). Interestingly, teratogenic effects were not seen in mice in an intravenous (131) or oral (132) administration studies. Embryotoxicity and growth were sensitive endpoints in mice (131,132). Indium is transferred from maternal to fetal compartments during gestation, and embryo culture studies suggest that indium developmental *toxicity* is a direct effect on the conceptus (127,132–134).

Aluminum has been fairly well-defined male reproductive toxicity in labortatory rodents (102), although no information in humans is available. Smaller reproductive organs, lower sperm counts, and reduced fertility were seen after four-week oral exposure at doses ≥100 mg/kg/day in male mice (135). A longer-term exposure in drinking water impaired mating behavior in rats along with the effect on reproductive organ weights (136). A similar pattern of effects was seen in rabbits treated orally every other day for 16 weeks with 34 mg/kg/day, with the exception that sperm motility and morphology were also affected in the rabbits (137), an effect not seen in the mice. These investigators associated the testicular toxicity of aluminum with oxidative damage, using measure of oxidative damage and protection by antioxidants (137). However, applicability to oral exposure is uncertain given the low bioavailability of Al. Potential ovarian effects of aluminum and exposure levels affecting fertility in females have not been studied. A one-generation study (both males and females rats were treated) using doses <100 mg/kg/day did not find effects on fertility (138,139). Teratogenicity in terms of malformation has not been demonstrated, but effects on birth weight and survival have been documented (102). Aluminum is neurotoxic and developmental neurotoxicity of aluminum has also been studied (140). Aluminum exposures secondary to water treatment, food additives, pharmaceuticals and biomedical devices have been the major source of concern in humans. In one study (141), premature infants assigned randomly to receive parenteral fluids high in aluminum showed delayed neurobehavioral development. Human studies of fertility were not located.

Various group 13 metals are found together in semiconductor materials. For example, aluminium gallium indium phosphide (AlGaInP) is used in light emitting diodes (LEDs). The reproductive toxicity of this interesting metal compound have not been studied. Testicular toxicity would be of particular interest.

ACTINOIDS AND LANTHANOIDS: URANIUM

The reproductive and developmental toxicity of radiation is widely recognized although contemporary studies are minimal. Highly radioactive uranium (enriched in isotope U235) is not considered here. However, attention has been directed recently at depleted uranium toxicity based on chemical properties.

The small number of studies conducted on prior to 2000 have been reviewed (142,143). A series of basic, well-conducted studies by Domingo and colleagues used *uranium* as the uranyl ion (UO_2^{2+}) in mice. Male exposures led to reduced fertility without any apparent testicular or epididymal pathology. This study used drinking water exposure. However, when both males and females were exposed via gavage before mating, no fertility effects were observed. Administration during embryogenesis by the oral route led to effects on growth, skeletal development, and external anomalies; when given by injection fetal viability was also affected and maternal toxicity including maternal mortality occurred. When exposures continued during gestation and lactation, offspring viability was the most sensitive endpoint. The NOAEL for developmental and reproductive effects was identified at 5 mg/kg/day (102). Female reproductive toxicity, ovarian toxicity, and reproductive tract maturation were not included in these investigations.

After 2000, public health concern (144,145) and toxicology research have focused on depleted uranium. This form of uranium is a byproduct of uranium enrichment, contains a low percent of U235, and is not of concern for radioactive emissions. It is valued for its density and has a number of uses including munitions and other materials used in warfare. One practical concern regarding uranium is leaching from shrapnel in war veterans and civilians in war zones. An early animal study with implanted pellets of depleted uranium during pregnancy found no effects (142) as did a later single generation study (both parents implanted) and a male reproductive toxicity study in rats (146,147).

Another concern is environmental exposure to uranium in abandoned uranium mines in the southwest. Using drinking water exposures in mice, uranium has recently been examined for in utero estrogenic effects and shown to be uterotrophic, and to accelerate vaginal opening, effects that are counteracted by estrogen receptor blockers (148). This study included ovarian follicle counts of mice exposed to uranium in drinking water beginning at 28 days of age (initiation of puberty) at four doses. After 30-day exposure the mice showed fewer large primary follicles at the two lower doses and more secondary follicle at the next highest dose, with no change in follicle counts at the highest dose. The authors relate this finding to known estrogen effects in the uterus, but no comparison estrogen or antiestrogen was used in this experiment. In studies where exposure occurred in utero, a reduction of primary ovarian follicles was seen in female offspring (assessed at the end of gestation).

TABLE 3 Metalloestrogens from Screening Assays

Tributyltin, cadmium, antimony, lithium, barium, chromium	Choe et al. 2003 (149)
Aluminum	Darbre 2005 (150)
Copper, chromium, cobalt, lead, mercury, nickel, tin, vanadate	Martin et al. 2003 (113)
Arsenic, selenium (metalloids)	Stoica et al. 2000 (151,152)

ENDOCRINE DISRUPTION

Concern about endocrine disruption has led to government actions including establishment of screening tests for endocrine disruption. These screening tests are beginning to be used on metals and metal compounds (Table 3). A series of 20 metals and metal compounds were screened using a transcription assay and a proliferation assay in a transformed breast cancer cell line (MCF-7). Six metals showed substantial estrogenicity in both assays: tributyltin, cadmium, antimony, lithium, barium, chromium (descending order of potency). The metal salt used also made a difference; chromium chloride and lithium hydroxide were effective but not as potassium chromate or lithium chloride. Five metals showed little or no activity in either assay: arsenic, cobalt, copper, lead, and potassium chromate. Metals and metal compounds showing minimal or inconsistent effects across the two tests were mercury (dimethyl), lead (acetate), lithium chloride, magnesium, manganese (chloride and also methylcyclopentadienyl manganese tricarbonyl, MMT), tellurium, molybdenum, selenium, stannous chloride, silver titanium tungsten, and zinc. Using the same cell line, Martin et al. (113) performed a cell proliferation assay, an ER binding assay, ERα expression, and ER transcription for the following metals (all as inorganic salts): cobalt, copper, nickel, vanadate, mercury, lead, and tin. The metals were used at a single molar concentration except for the transcription assay. All the metals were active in the proliferation assay; copper, cobalt, lead, and vanadium were active in the ERα expression assay. Nickel was most potent in the transcription assay, with the other metals showing a similar potency relative to estradiol (E2) (0.2–0.3). Finally, similar experiments in the MCF-7 cells demonstrated stimulating effects of aluminum in a transcription assay, competition in an ER receptor–binding assay, but no effect on cell proliferation (150). In general, metals can be described as estrogenic due to their effects on cell proliferation and gene expression. There is evidence that these actions take place at the estrogen receptor. In vivo effects are beginning to be demonstrated, for example, uterotrophic effects of cadmium (30), mercuric chloride (153), and uranium (148).

LOOKING FORWARD

Metallurgy is one of the oldest human scientific enterprises. Contemporary materials science is constantly innovating new forms of metals for practical applications. Knowledge of metals as nutrients and development of metal-based pharmaceuticals and pesticides are active areas of the biomedical sciences. Each new metal application carries with it the possibility of toxic exposure. Consequently, the field of developmental and reproductive toxicity of metals is constantly reenergized by new demands and continues to grow.

REFERENCES

1. Golub M, ed. Metals, Fertility and Reproductive Toxicity. Boca Raton, FL: Taylor and Francis, 2005.
2. Clarkson T, Nordberg G, Sager P. Reproductive and developmental toxicity of metals. Scand J Work Environ Health 1985; 11(3):145–154.

3. Goyer R, Clarkson T. Toxic effects of metals. In: Klaasen C, ed. Casarett and Doull's Toxicology: The Basic Science of Poisons. 5th ed. New York: McGraw-Hill, 2001:811–868.
4. Dyer C. Heavy metals as endocrine-disrupting chemicals. In: Gore A, ed. Endocrine-Disrupting Chemicals: From Basic Research to Clinical Practice. Totawa, NJ: Humana Press, 2007:111–134.
5. Wright RO, Baccarelli A. Metals and neurotoxicology. J Nutr 2007; 137(12):2809–2813.
6. Sokol R. Lead exposure and its effects on the reproductive system. In: Golub M, ed. Metals, Fertility, and Reproductive Toxicity. Boca Raton FL: Taylor and Francis, 2005:117–154.
7. Hoyer P. Impact of metals on ovarian function. In: Golub M, ed. Metals, Fertility, and Reproductive Toxicity. Boca Raton, FL: Taylor and Francis, 2005:155–174.
8. Selevan SG, Rice DC, Hogan KA, et al. Blood lead concentration and delayed puberty in girls. N Engl J Med 2003; 348(16):1527–1536.
9. Wu T, Buck GM, Mendola P. Blood lead levels and sexual maturation in U.S. girls: the Third National Health and Nutrition Examination Survey, 1988–1994. Environ Health Perspect 2003; 111(5):737–741.
10. Fowler BA, Kimmel CA, Woods JS, et al. Chronic low-level lead toxicity in the rat. III. An integrated assessment of long-term toxicity with special reference to the kidney. Toxicol Appl Pharmacol 1980; 56(1):59–77.
11. Grant LD, Kimmel CA, West GL, et al. Chronic low-level lead toxicity in the rat. II. Effects on postnatal physical and behavioral development. Toxicol Appl Pharmacol 1980; 56(1):42–58.
12. Kimmel CA, Grant LD, Sloan CS, et al. Chronic low-level lead toxicity in the rat. I. Maternal toxicity and perinatal effects. Toxicol Appl Pharmacol 1980; 56(1):28–41.
13. Hauser R, Sergeyev O, Korrick S, et al. Association of blood lead levels with onset of puberty in Russian boys. Environ Health Perspect 2008; 116(7):976–980.
14. Ronis MJ, Badger TM, Shema SJ, et al. Effects on pubertal growth and reproduction in rats exposed to lead perinatally or continuously throughout development. J Toxicol Environ Health A 1998; 53(4):327–341.
15. Ronis MJ, Gandy J, Badger T. Endocrine mechanisms underlying reproductive toxicity in the developing rat chronically exposed to dietary lead. J Toxicol Environ Health A 1998; 54(2):77–99.
16. National Research Council. Toxicological Effects of Methyl Mercury. Washington, DC: National Academies Press, 2000.
17. Golub M. Reproductive toxicity of mercury, cadmium, and arsenic. In: Golub M, ed. Metals, Fertility and Reproductive Toxicity. Boca Raton, FL: Taylor and Francis, 2005:5–22.
18. Rowland A, Baird D, Weinberg C, et al. The effect of occupational exposure to mercury vapour on the fertility of female dental assistants. Occup Environ Med 1994; 51:28–34.
19. Rao MV, Sharma PS. Protective effect of vitamin E against mercuric chloride reproductive toxicity in male mice. Reprod Toxicol 2001; 15(6):705–712.
20. Ema M, Hirose A. Reproductive and developmental toxicity of organotin compounds. In: Golub M, ed. Metals, Fertility and Reproductive Toxicity. Boca Raton, FL: Taylor and Francis, 2005:23–64.
21. Golub M, Doherty J. Triphenyltin as a potential human endocrine disruptor. J Toxicol Environ Health 2004; 7(4):281–295.
22. Nakanishi T. Endocrine disruption induced by organotin compounds; organotins function as a powerful agonist for nuclear receptors rather than an aromatase inhibitor. J Toxicol Sci 2008; 33(3):269–276.
23. Morgan J. Evidence for developmental and reproductive toxiciy of fenbutatin oxide. Office of Environmental Health Hazard Assessment, California Environmental Protection Agency, 1999. Available at: http://www.oehha.ca.gov.
24. Paule MG, Reuhl K, Chen JJ, et al. Developmental toxicology of trimethyltin in the rat. Toxicol Appl Pharmacol 1986; 84(2):412–417.
25. Reuhl KR, Cranmer JM. Developmental neuropathology of organotin compounds. Neurotoxicology 1984; 5(2):187–204.
26. Moser VC, Barone S Jr., Phillips PM, et al. Evaluation of developmental neurotoxicity of organotins via drinking water in rats: monomethyltin. Neurotoxicology 2006; 27(3):409–420.

27. Jenkins SM, Ehman K, Barone S Jr. Structure-activity comparison of organotin species: dibutyltin is a developmental neurotoxicant in vitro and in vivo. Brain Res 2004; 151(1–2):1–12.
28. Office of Environmental Health Hazard Assessment. Evidence on the reproductive and developmental toxicity of cadmium: Office of Environmental Health Hazard Assessment, California Environmental Protection Agency, 1996. Available at: http://www.oehha.ca.gov.
29. Thompson J, Bannigan J. Cadmium: toxic effects on the reproductive system and the embryo. Reprod Toxicol 2008; 25(3):304–315.
30. Johnson MD, Kenney N, Stoica A, et al. Cadmium mimics the in vivo effects of estrogen in the uterus and mammary gland. Nat Med 2003; 9(8):1081–1084.
31. Stoica A, Katzenellenbogen BS, Martin MB. Activation of estrogen receptor-alpha by the heavy metal cadmium. Mol Endocrinol (Baltimore, MD) 2000; 14(4):545–553.
32. Garcia-Morales P, Saceda M, Kenney N, et al. Effect of cadmium on estrogen receptor levels and estrogen-induced responses in human breast cancer cells. J Biol Chem 1994; 269(24):16896–16901.
33. Wimmer U, Wang Y, Georgiev O, et al. Two major branches of anti-cadmium defense in the mouse: MTF-1/metallothioneins and glutathione. Nucleic Acids Res 2005; 33(18):5715–5727.
34. Kim DW, Kim KY, Choi BS, et al. Regulation of metal transporters by dietary iron, and the relationship between body iron levels and cadmium uptake. Arch Toxicol 2007; 81(5):327–334.
35. Ishitobi H, Watanabe C. Effects of low-dose perinatal cadmium exposure on tissue zinc and copper concentrations in neonatal mice and on the reproductive development of female offspring. Toxicol Lett 2005; 159(1):38–46.
36. Fernandez EL, Dencker L, Tallkvist J. Expression of ZnT-1 (Slc30a1) and MT-1 (Mt1) in the conceptus of cadmium treated mice. Reprod Toxicol 2007; 24(3–4):353–358.
37. Chen H, Boontheung P, Loo RR, et al. Proteomic analysis to characterize differential mouse strain sensitivity to cadmium induced forelimb teratogenesis. Birth Defects Res 2008; 82(4):187–199.
38. Salvatori F, Talassi CB, Salzgeber SA, et al. Embryotoxic and long-term effects of cadmium exposure during embryogenesis in rats. Neurotoxicol Teratol 2004; 26(5): 673–680.
39. Sun TJ, Miller ML, Hastings L. Effects of inhalation of cadmium on the rat olfactory system: behavior and morphology. Neurotoxicol Teratol 1996; 18(1):89–98.
40. Kuhnert BR, Kuhnert PM, Debanne S, et al. The relationship between cadmium, zinc, and birth weight in pregnant women who smoke. Am J Obstet Gynecol 1987; 157(5):1247–1251.
41. Ronco AM, Arguello G, Munoz L, et al. Metals content in placentas from moderate cigarette consumers: correlation with newborn birth weight. Biometals 2005; 18(3): 233–241.
42. Nishijo M, Nakagawa H, Honda R, et al. Effects of maternal exposure to cadmium on pregnancy outcome and breast milk. Occup Environ Med 2002; 59(6):394–396; discussion 7.
43. Frery N, Nessmann C, Girard F, et al. Environmental exposure to cadmium and human birthweight. Toxicology 1993; 79(2):109–118.
44. Bonithon-Kopp C, Huel G, Grasmick C, et al. Effects of pregnancy on the interindividual variations in blood levels of lead, cadmium and mercury. Biol Res Pregnancy Perinatol 1986; 7(1):37–42.
45. Huel G, Boudene C, Ibrahim MA. Cadmium and lead content of maternal and newborn hair: relationship to parity, birth weight, and hypertension. Arch Environ Health 1981; 36(5):221–227.
46. Berlin M, Blanks R, Catton M, et al. Birth weight of children and cadmium accumulation in placentas of female nickel-cadmium (long-life) battery workers. IARC Sci Publ 1992; (118):257–262.
47. Sikorski R, Radomanski T, Paszkowski T, et al. Smoking during pregnancy and the perinatal cadmium burden. J Perinat Med 1988; 16(3):225–231.
48. Loiacono NJ, Graziano JH, Kline JK, et al. Placental cadmium and birthweight in women living near a lead smelter. Arch Environ Health 1992; 47(4):250–255.

49. Drasaga R, Einhorn L, Williams S, et al. Fertility after chemotherapy for testicular cancer. J Clin Oncol 1983; 1:179–183.
50. Maines MD, Sluss PM, Iscan M. cis-platinum-mediated decrease in serum testosterone is associated with depression of luteinizing hormone receptors and cytochrome P-450scc in rat testis. Endocrinology 1990; 126(5):2398–2406.
51. Kinkead T, Flores C, Carboni AA, et al. Short term effects of cis-platinum on male reproduction, fertility and pregnancy outcome. J Urol 1992; 147(1):201–206.
52. Seethalakshmi L, Flores C, Kinkead T, et al. Effects of subchronic treatment with cis-platinum on testicular function, fertility, pregnancy outcome, and progeny. J Androl 1992; 13(1):65–74.
53. Pogach LM, Lee Y, Gould S, et al. Characterization of cis-platinum-induced Sertoli cell dysfunction in rodents. Toxicol Appl Pharmacol 1989; 98(2):350–361.
54. Pogach LM, Lee Y, Giglio W, et al. Zinc acetate pretreatment ameliorates cisplatin-induced Sertoli cell dysfunction in Sprague-Dawley rats. Cancer Chemother Pharmacol 1989; 24(3):177–180.
55. Azouri H, Bidart JM, Bohuon C. In vivo toxicity of cisplatin and carboplatin on the Leydig cell function and effect of the human choriogonadotropin. Biochem Pharmacol 1989; 38(4):567–571.
56. Kopf-Maier P, Erkenswick P, Merker HJ. Lack of severe malformations versus occurrence of marked embryotoxic effects after treatment of pregnant mice with cis-platinum. Toxicology 1985; 34(4):321–331.
57. Price CJ, George JD. Final study report on the developmental toxicity evaluation for silver acetate (CASNo. 563-63-3) administered by gavage to Sprague-Dawley rats on gestational days 6 through 19. NTIS Technical Reports 2002. Document Number NTIS/PB2002–109208.
58. Shavlovski MM, Chebotar NA, Konopistseva LA, et al. Embryotoxicity of silver ions is diminished by ceruloplasmin—further evidence for its role in the transport of copper. Biometals 1995; 8(2):122–128.
59. Ernst E, Rungby J, Baatrup E. Ultrastructural localization of silver in rat testis and organ distribution of radioactive silver in the rat. J Appl Toxicol 1991; 11(5):317–3121.
60. Braydich-Stolle L, Hussain S, Schlager JJ, et al. In vitro cytotoxicity of nanoparticles in mammalian germline stem cells. Toxicol Sci 2005; 88(2):412–419.
61. National Toxicology Program. Nominations to the Testing Program. Available at: http://ntp.niehs.nih.gov/go/29287. Accessed January 2009.
62. Domingo JL. Vanadium and tungsten derivatives as antidiabetic agents: a reviewe of their toxic effects. Biol Trace Elem Res 2002; 88(2):97–112.
63. Ballester J, Dominguez J, Munoz MC, et al. Tungstate treatment improves Leydig cell function in streptozotocin-diabetic rats. J Androl 2005; 26(6):706–715.
64. Ballester J, Munoz MC, Dominguez J, et al. Tungstate administration improves the sexual and reproductive function in female rats with streptozotocin-induced diabetes. Hum Reprod 2007; 22(8):2128–2135.
65. Wide M, Danielsson BR, Dencker L. Distribution of tungstate in pregnant mice and effects on embryonic cells in vitro. Environ Res 1986; 40(2):487–498.
66. McInturf SM, Bekkedal MY, Wilfong E, et al. Neurobehavioral effects of sodium tungstate exposure on rats and their progeny. Neurotoxicol Teratol 2008; 30(6):455–461.
67. Agency for Toxic Substances and Disease Registry. Toxicological Profile for Tungsten. Atlanta, GA: Department of Health and Human Services, 2005.
68. Keen C. Teratogenic effects of essential trace metals: deficiencies and excesses. In: Chang L, ed. Toxicology of Metals, 1996. Boca Raton, FL: CRC Lewis Publishers, 1996:977–1001.
69. Goyer RA. Toxic and essential metal interactions. Annu Rev Nutr 1997; 17:37–50.
70. Golub M. Intrauterine and reproductive toxicity of nutritionally essential metals. In: Golub M, ed. Metals, Fertility, and Reproductive Toxicity. Boca Raton, FL: Taylor and Francis, 2005:93–116.
71. Beliles RP, Palmer AK. The effect of massive transplacental iron loading. Toxicology 1975; 5(2):147–158.
72. Archer T, Schroder N, Fredriksson A. Neurobehavioural deficits following postnatal iron overload. II. Instrumental learning performance. Neurotox Res 2003; 5(1–2):77–94.
73. Fredriksson A, Schroder N, Archer T. Neurobehavioural deficits following postnatal iron overload: I spontaneous motor activity. Neurotox Res 2003; 5(1–2):53–76.

74. Robbins W. Epidemiological and occupational studies of metals in male reproductive toxicity. In: Golub M, ed. Metals, Fertility, and Reproductive Toxicity. Boca Raton, FL: Taylor and Francis, 2005:175–212.
75. Trivedi B, Saxena DK, Murthy RC, et al. Embryotoxicity and fetotoxicity of orally administered hexavalent chromium in mice. Reprod Toxicol 1989; 3(4):275–278.
76. Kanojia RK, Junaid M, Murthy RC. Embryo and fetotoxicity of hexavalent chromium: a long-term study. Toxicol Lett 1998; 95(3):165–172.
77. Kanojia RK, Junaid M, Murthy RC. Chromium induced teratogenicity in female rat. Toxicol Lett 1996; 89(3):207–213.
78. Murthy RC, Junaid M, Saxena DK. Ovarian dysfunction in mice following chromium (VI) exposure. Toxicol Lett 1996; 89(2):147–154.
79. Junaid M, Murthy RC, Saxena DK. Embryo- and fetotoxicity of chromium in pregestationally exposed mice. Bull Environ Contam Toxicol 1996; 57(2):327–334.
80. Junaid M, Murthy RC, Saxena DK. Embryotoxicity of orally administered chromium in mice: exposure during the period of organogenesis. Toxicol Lett 1996; 84(3):143–148.
81. Junaid M, Murthy RC, Saxena DK. Chromium fetotoxicity in mice during late pregnancy. Vet Hum Toxicol 1995; 37(4):320–323.
82. Banu SK, Samuel JB, Arosh JA, et al. Lactational exposure to hexavalent chromium delays puberty by impairing ovarian development, steroidogenesis and pituitary hormone synthesis in developing Wistar rats. Toxicol Appl Pharmacol 2008; 232(2):180–189.
83. Campbell M. Evidence on the developmental and reproductive toxicity of chromium (hexavalent compounds): Office of Environmental Health Hazard Assessment, California Environmental Protection Agency, 2008. Available at: http://www.oehha.ca.gov.
84. Bataineh H, al-Hamood MH, Elbetieha A, et al. Effect of long-term ingestion of chromium compounds on aggression, sex behavior and fertility in adult male rat. Drug Chem Toxicol 1997; 20(3):133–149.
85. Agency for Toxic Substances and Disease Registry. Available at: http://www.atsdr.cdc.gov/pdns/unfilled.html. Accessed January 10, 2009.
86. Aruldhas MM, Subramanian S, Sekhar P, et al. In vivo spermatotoxic effect of chromium as reflected in the epididymal epithelial principal cells, basal cells, and intraepithelial macrophages of a nonhuman primate (Macaca radiata Geoffroy). Fertil Steril 2006; 86(4 suppl):1097–1105.
87. Subramanian S, Rajendiran G, Sekhar P, et al. Reproductive toxicity of chromium in adult bonnet monkeys (Macaca radiata Geoffrey). Reversible oxidative stress in the semen. Toxicol Appl Pharmacol 2006; 215(3):237–249.
88. Aruldhas MM, Subramanian S, Sekar P, et al. Chronic chromium exposure-induced changes in testicular histoarchitecture are associated with oxidative stress: study in a non-human primate (Macaca radiata Geoffroy). Hum Reprod 2005; 20(10):2801–2813.
89. Aruldhas MM, Subramanian S, Sekhar P, et al. Microcanalization in the epididymis to overcome ductal obstruction caused by chronic exposure to chromium—a study in the mature bonnet monkey (Macaca radiata Geoffroy). Reproduction (Cambridge, England) 2004; 128(1):127–137.
90. Schroeder HA, Mitchener M. Toxic effects of trace elements on the reproduction of mice and rats. Arch Environ Health 1971; 23(2):102–106.
91. Springborn Laboratory. A one-generation reproduction range-finding study in rats with nickel sulfate heahydrate. Final Report. Springborn Laboratory, Inc. Study No. 3472.3. 2000.
92. Springborn Laboratory. An oral(gavage) two-generation reproduction toxicity study in Sprague-Dawley rats with nickkel sulfate hexahydrate. Final report. Springborn Laboratory, Inc. Study No. 3472.4. 2000. Available at: http://www.oehha.ca.gov.
93. Ting D. Public Health Goals for Chemicals in Drinking Water: Nickel: Office of Environmental Health Hazard Assessment, California Environmental Protection Agency, 2001.
94. Al-Hamood MH, Elbetieha A, Bataineh H. Sexual maturation and fertility of male and female mice exposed prenatally and postnatally to trivalent and hexavalent chromium compounds. Reprod Fertil Dev 1998; 10(2):179–183.

95. Kakela R, Kakela A, Hyvarinen H. Effects of nickel chloride on reproduction of the rat and possible antagonistic role of selenium. Comp Biochem Physiol C Pharmacol Toxicol Endocrinol 1999; 123(1):27–37.
96. Smith MK, George EL, Stober JA, et al. Perinatal toxicity associated with nickel chloride exposure. Environ Res 1993; 61(2):200–211.
97. Danadevi K, Rozati R, Reddy PP, et al. Semen quality of Indian welders occupationally exposed to nickel and chromium. Reprod Toxicol 2003; 17(4):451–456.
98. Figa-Talamanca I, Petrelli G. Reduction in male births among workers exposed to metal fumes. Int J Epidemiol 2000; 29(2):381.
99. Jeter MA, Davis GK. The effect of dietary molybdenum upon growth, hemoglobin, reproduction and lactation of rats. J Nutr 1954; 54(2):215–220.
100. Pandey R, Singh SP. Effects of molybdenum on fertility of male rats. Biometals 2002; 15(1):65–72.
101. Fungwe TV, Buddingh F, Yang MT, et al. Hepatic, placental, and fetal trace elements following molybdenum supplementation during gestation. Biol Trace Elem Res 1989; 22(2):189–199.
102. Domingo JL. Adverse effects of aluminum, uranium, and vanadium on reproduction and intrauterine development in mammals. In: Golub M, ed. Metals, Fertility, and Reproductive Toxicity. Boca Raton, FL: Taylot and Francis, 2005:65–92.
103. Domingo JL, Paternain JL, Llobet JM, et al. Effects of vanadium on reproduction, gestation, parturition and lactation in rats upon oral administration. Life Sci 1986; 39(9): 819–824.
104. Llobet JM, Colomina MT, Sirvent JJ, et al. Reproductive toxicity evaluation of vanadium in male mice. Toxicology 1993; 80(2–3):199–206.
105. National Toxicology Program. Available at: http://www.ntp.niehs.nih.gov/index.cfm?objectid=C9288A2C-F1F6-975E-7CB4DC2FAC392BB8. Accessed January 17, 2009.
106. Pal N, Pal B. Zinc feeding and conception in the rats. Int J Vitam Nutr Res (Internationale Zeitschrift fur Vitamin- und Ernahrungsforschung) 1987; 57(4):437–440.
107. Khan AT, Graham TC, Ogden L, et al. A two-generational reproductive toxicity study of zinc in rats. J Environ Sci Health Part 2007; 42(4):403–415.
108. Ogden L, Graham T, Mahboob M, et al. Effects of zinc chloride on reproductive parameters of cd-1 mice. Toxicologist 2002; 66(1-S):33.
109. Samanta K, Pal B. Zinc feeding and fertility of male rats. Int J Vitam Nutr Res (Internationale Zeitschrift fur Vitamin- und Ernahrungsforschung) 1986; 56(1):105–108.
110. Anderson MB, Pedigo NG, Katz RP, et al. Histopathology of testes from mice chronically treated with cobalt. Reprod Toxicol 1992; 6(1):41–50.
111. Pedigo NG, George WJ, Anderson MB. Effects of acute and chronic exposure to cobalt on male reproduction in mice. Reprod Toxicol 1988; 2(1):45–53.
112. Pedigo NG, Vernon MW. Embryonic losses after 10-week administration of cobalt to male mice. Reprod Toxicol 1993; 7(2):111–116.
113. Martin MB, Reiter R, Pham T, et al. Estrogen-like activity of metals in MCF-7 breast cancer cells. Endocrinology 2003; 144(6):2425–2436.
114. Omura M, Hirata M, Tanaka A, et al. Testicular toxicity evaluation of arsenic containing binary compound semiconductors, gallium arsenide and indium arsenide, in hamsters. Toxicol Lett 1996; 89(2):123–129.
115. Omura M, Tanaka A, Hirata M, et al. Testicular toxicity of gallium arsenide, indium arsenide, and arsenic oxide in rats by repetitive intratracheal instillation. Fundam Appl Toxicol 1996; 32(1):72–78.
116. Colomina MT, Llobet JM, Sirvent JJ, et al. Evaluation of the reproductive toxicity of gallium nitrate in mice. Food Chem Toxicol 1993; 31(11):847–851.
117. Ward S, Chapin RE, Harris MW, et al. The general and reproductive toxicity of the photovoltaic material copper gallium diselene (CGS). Toxicologist 1994; 14(1):266.
118. Mast T, Dill J, Greenspan B, et al. The developmental toxicity of inhaled gallium arsenide in rodents. Teratology 1991; 43(5):455–456.
119. Ferm VH, Carpenter SJ. Teratogenic and embryopathic effects of indium, gallium, and germanium. Toxicol Appl Pharmacol 1970; 16(1):166–170.
120. Formigli L, Scelsi R, Poggi P, et al. Thallium-induced testicular toxicity in the rat. Environ Res 1986; 40(2):531–539.

121. Gregotti C, Di Nucci A, Costa LG, et al. Effects of thallium on primary cultures of testicular cells. J Toxicol Environ Health 1992; 36(1):59–69.
122. Gibson JE, Becker BA. Placental transfer, embryotoxicity, and teratogenicity of thallium sulfate in normal and potassium-deficient rats. Toxicol Appl Pharmacol 1970; 16(1):120–132.
123. Hoffman RS. Thallium poisoning during pregnancy: a case report and comprehensive literature review. J Toxicol 2000; 38(7):767–775.
124. Hoyes KP, Johnson C, Johnston RE, et al. Testicular toxicity of the transferrin binding radionuclide 114mIn in adult and neonatal rats. Reprod Toxicol 1995; 9(3):297–305.
125. Hoyes KP, Sharma HL, Jackson H, et al. Spermatogenic and mutagenic damage after paternal exposure to systemic indium-114m. Radiat Res 1994; 139(2):185–193.
126. Rao DV, Sastry KS, Grimmond HE, et al. Cytotoxicity of some indium radiopharmaceuticals in mouse testes. J Nucl Med 1988; 29(3):375–384.
127. Ungvary G, Szakmary E, Tatrai E, et al. Embryotoxic and teratogenic effects of indium chloride in rats and rabbits. J Toxicol Environ Health A 2000; 59(1):27–42.
128. Nakajima M, Takahashi H, Sasaki M, et al. Developmental toxicity of indium chloride by intravenous or oral administration in rats. Teratog Carcinog Mutagen 1998; 18(5):231–238.
129. Ungvary G, Tatrai E, Szakmary E, et al. The effect of prenatal indium chloride exposure on chondrogenic ossification. J Toxicol Environ Health A 2001; 62(5):387–396.
130. Nakajima M, Takahashi H, Nakazawa K, et al. Fetal cartilage malformation by intravenous administration of indium trichloride to pregnant rats. Reprod Toxicol 2007; 24(3–4):409–413.
131. Nakajima M, Takahashi H, Sasaki M, et al. Comparative developmental toxicity study of indium in rats and mice. Teratog Carcinog Mutagen 2000; 20(4):219–227.
132. Chapin RE, Harris MW, Hunter ES III, et al. The reproductive and developmental toxicity of indium in the Swiss mouse. Fundam Appl Toxicol 1995; 27(1):140–148.
133. Baltrukiewicz Z, Marciniak M, Urbaniak B. Distribution and retention of indium113m in the organism of pregnant rats. Acta Physiol Pol 1976; 27(2):191–197.
134. Nakajima M, Usami M, Nakazawa K, et al. Developmental toxicity of indium: embryotoxicity and teratogenicity in experimental animals. Congenit Anom 2008; 48(4):145–150.
135. Llobet JM, Colomina MT, Sirvent JJ, et al. Reproductive toxicology of aluminum in male mice. Fundam Appl Toxicol 1995; 25(1):45–51.
136. Bataineh H, Al-Hamood MH, Elbetieha AM. Assessment of aggression, sexual behavior and fertility in adult male rat following long-term ingestion of four industrial metals salts. Hum Exp Toxicol 1998; 17(10):570–576.
137. Yousef MI, El-Morsy AM, Hassan MS. Aluminium-induced deterioration in reproductive performance and seminal plasma biochemistry of male rabbits: protective role of ascorbic acid. Toxicology 2005; 215(1–2):97–107.
138. Domingo JL, Paternain JL, Llobet JM, et al. The effects of aluminium ingestion on reproduction and postnatal survival in rats. Life Sci 1987; 41(9):1127–1131.
139. Domingo JL, Paternain JL, Llobet JM, et al. Effects of oral aluminum administration on perinatal and postnatal development in rats. Res Commun Chem Pathol Pharmacol 1987; 57(1):129–132.
140. Golub M, Domingo JL. What we know and what we need to know about developmental aluminum toxicity. J Toxicol Environ Health 1996; 48:585–597.
141. Bishop N, McGraw M, Ward B. Aluminium in infant formulas. Lancet 1989; 1(8637):565.
142. Arfsten DP, Still KR, Ritchie GD. A review of the effects of uranium and depleted uranium exposure on reproduction and fetal development. Toxicol Ind Health 2001; 17(5–10):180–191.
143. Domingo JL. Reproductive and developmental toxicity of natural and depleted uranium: a review. Reprod Toxicol 2001; 15(6):603–609.
144. Squibb KS, McDiarmid MA. Depleted uranium exposure and health effects in Gulf War veterans. Philos Trans R Soc Lond 2006; 361(1468):639–648.
145. McDiarmid MA, Engelhardt SM, Dorsey CD, et al. Surveillance results of depleted uranium-exposed Gulf War I veterans: sixteen years of follow-up. J Toxicol Environ Health A 2009; 72(1):14–29.

146. Arfsten DP, Bekkedal M, Wilfong ER, et al. Study of the reproductive effects in rats surgically implanted with depleted uranium for up to 90 days. J Toxicol Environ Health A 2005; 68(11–12):967–997.
147. Arfsten DP, Schaeffer DJ, Johnson EW, et al. Evaluation of the effect of implanted depleted uranium on male reproductive success, sperm concentration, and sperm velocity. Environ Res 2006; 100(2):205–215.
148. Raymond-Whish S, Mayer LP, O'Neal T, et al. Drinking water with uranium below the U.S. EPA water standard causes estrogen receptor-dependent responses in female mice. Environ Health Perspect 2007; 115(12):1711–1716.
149. Choe SY, Kim SJ, Kim HG, et al. Evaluation of estrogenicity of major heavy metals. Sci Total Environ 2003; 312(1–3):15–21.
150. Darbre PD. Metalloestrogens: an emerging class of inorganic xenoestrogens with potential to add to the oestrogenic burden of the human breast. J Appl Toxicol 2006; 26(3):191–197.
151. Stoica A, Pentecost E, Martin MB. Effects of arsenite on estrogen receptor-alpha expression and activity in MCF-7 breast cancer cells. Endocrinology 2000; 141(10):3595–3602.
152. Stoica A, Pentecost E, Martin MB. Effects of selenite on estrogen receptor-alpha expression and activity in MCF-7 breast cancer cells. Journal of cellular biochemistry 2000; 79(2):282–292.
153. Zhang X, Wang Y, Zhao Y, et al. Experimental study on the estrogen-like effect of mercuric chloride. Biometals 2008; 21(2):143–150.

20

Omics in reproductive and developmental toxicology

Susan Sumner, Rodney Snyder, Jason Burgess, Timothy Fennell, and Rochelle W. Tyl

In recent years, ground-breaking research in genomic applications in the area of reproductive and developmental toxicology has been successful in linking specific genes to effects induced by drugs or chemicals. Gene expression profiling has demonstrated the ability to provide mechanistic insight into cellular mechanisms of drug and chemical-induced effects, but cannot reveal insights into mechanisms that are not regulated at the gene level. Incorporation of proteomics into the study of reproduction and development provides the means to investigate protein level interactions, including posttranslation modifications that are known to be involved in the generation of a number of fetal malformations and are functions that cannot be evaluated through measurement of mRNA or through DNA sequencing. In recent years, the application of proteomics in the study of reproduction and development has rapidly increased. Applications of metabolomics in reproduction and development are at their infancy, but clearly, this technology will bring huge advantages in the development of non-invasive markers to stage disease, monitor treatments, and link exposure with health outcomes. The goal of this chapter is to summarize recent advances in genomic and proteomic applications in reproductive and developmental toxicology and to provide a review and examples of metabolomic applications for the study of reproductive and developmental toxicology.

GENOMICS

Genomics involves the comprehensive study of the genotype (DNA) or transcriptome (mRNA) of cells and tissues. This field was established in the 1970s when sequencing tools, together with what are now referred to as bioinformatics approaches, were used to determine the genome of a virus. Since that time, the genomes of many species have been sequenced, and most modern genomics studies in toxicological research use microarray technologies to investigate functional genomics: expression of a wide array of genes under a variety of normal physiological conditions or environmental stressors. Technologies used in the toxicological community to study the genome have been recently summarized with a perspective regarding the strengths of advances in new technologies (1); a significant portion of studies involve the use of microarray technologies to study amplifications, deletions, epigenetic change, mRNA translation, and the control of mRNA.

Many scientists have questioned whether genomics, and particularly the use of microarray technologies, has lived up to the original promise to define the genomic signature of cells and tissues and provide mechanistic insights into the response of the genome to environmental stressors. While the original concept that genomics would provide an easy, rapid, and less expensive method for screening drug candidates and accelerate drug development has not been fully realized, the reality is that genomics technologies have and will continue to be successfully applied in combination with more traditional approaches to advance our scientific understanding of the influence of environmental factors (e.g., nutrition, chemicals, in utero environment) on reproduction and development. A key factor in making such research endeavors successful is ensuring that the link is made between alterations in gene patterns and the dose- and time to response for the development of adverse response (i.e., the phenotypic anchor). In the absence of phenotypic anchors (e.g., an accepted endpoint that is used to define a developmental phase, a disease state, or a malformation), gene expression data can only inform us about the phenomenon of up- or downregulation. The use of phenotypic anchors enables a pattern of gene expression to be linked to an effect, and the use of dose- and time-to-response studies enables us to determine trends in the data at low doses, or at early times, that have a causal relation to effects generated at higher doses or following longer term exposures.

In this decade, the field of developmental and reproductive toxicology began a transformation to integrate new molecular biology, imaging, and genomic technologies. In 2002, the National Institute of Environmental Health Sciences, the National Institutes of Health Office of Rare Disease, the U.S. Environmental Protection Agency, and the American Chemistry Council held a conference aimed at planning the future for the integration of emerging technologies in developmental toxicology research (2). This conference provided a framework for government agencies to begin the needed investment in scientific approaches that have helped to establish the framework of a systems biology approach for developmental and reproductive biology. Since that time, the application of genomics tools in the study of reproductive and developmental biology has grown rapidly.

In developmental and reproductive toxicological research, genomics technologies have been used to assess samples collected from dose- and time-to-response studies conducted in in vivo or in vitro model systems, with chemicals that have potential for endocrine disruption often being a focus of these investigations. Examples of early studies include the investigation of (*i*) the influence of specific chemicals on sensitive mammalian developmental processes (3); (*ii*) the use of gene expression data derived from high-dose and low-dose studies to develop dose-response significance (4); (*iii*) the use of gene expression to evaluate how estrogen agonists or estrogen antagonists affect embryonic and fetal development of the rat testis and epididymis (5); (*iv*) chemical-induced expression of genes during the development of the craniofacies (6); (*v*) teratogen-induced gene alterations in the embryo (7,8); (*vi*) identification of specific genes involved in the development of the uterus and ovary following estrogen exposure (4,9,10); (*vii*) how insulin exposure during the preimplantation stage alters the expression of imprinted genes and affects fetal development (11); (*viii*) the expression and methylation levels of the growth-related imprinted genes H19 and Igf2 in fetuses of diabetic mice (12); (*ix*) how exposure to 2,3,7,8-tetrachlorodibenzo alters the methylation status of imprinting genes (13); and (*x*) the use of pathway analysis to provide insights into the molecular mechanisms of estrogens (14).

Highlighted example: A study was conducted to investigate the effects of 17-α-ethynyl estradiol (EE) given at 0.001 to 10 mg/kg/day for 4 days [postnatal day (PND) 20–23, by subcutaneous injection] to 20-day-old Sprague–Dawley rats (15). These investigators report the modification of the expression level of over 300 genes in the uterus and ovaries. Of those genes, 88 were shown to exhibit a significant dose-dependent relationship. Further studies (5) were conducted to evaluate the dose dependency of the gene pattern that is responsive to EE, genistein (Ges), and bisphenol A (BPA) during fetal development of the rat testis and epididymis. These investigators demonstrated that gene expression profiles of target tissues were modified in a dose-responsive manner, even when morphological changes were not observed, and that 50 genes were expressed in a common direction across the three chemicals, pointing to cellular pathways that are affected by estrogen exposure, and the potential use of this methodology to define the shape of the dose-response curve at concentrations below the effect level. Recent investigations (16) of EE-induced gene expression in Ishikawa cells were conducted, and results of this in vitro study were compared with the response to EE in the rat uterus. Seventy-one genes were regulated in a similar manner in vivo and in vitro, suggesting that gene expression profiling in in vitro models is a viable tool to predict in vivo estrogenic activities.

PROTEOMICS

While gene expression profiling has demonstrated the ability to provide mechanistic insight into cellular mechanisms of drug and chemical-induced effects, proteomics provides advantages in areas not directly obtainable through gene studies. For example, proteins and posttranslational modifications are known to be directly involved in the generation of some malformations (17), functions that cannot be evaluated through DNA sequencing, or measurement of mRNA levels. In the last five years, the application of proteomics in toxicology studies has rapidly grown (18), as has the field of toxicogenomics that involves the integration of conventional toxicological examinations with gene, protein, or metabolite expression profiles (19).

Proteomics involves the use of molecular and analytical methods to capture a broad spectrum of proteins that are associated with the proteome. The proteome is the set of proteins, including their modifications (such as ubiquitination, methylation, acetylation, oxidation, nitrosylation, phosphorylation), that comprises a system (e.g., cell, tissue) at a given time. Because specific cell types can have different gene activations, specific cell types can also have distinct proteomes, which can be perturbed by genetic or environmental stressors. In many cases, the protein may become active only after a modification event; thus, the proteomic approach will involve study designs that enable the identification of a protein that correlates with a phenotypic anchor and an evaluation of protein structure for posttranslational modification.

There is a wide variety of methods to study the proteome (20–25). In general, the early proteomics methods included the use of antibodies raised to a specific protein (or a modified form) and the use of ELISA assays for protein quantification. Because of the complexity of the number of proteins in a given tissue or cell type, gel electrophoresis methods are commonly used to separate proteins (or denatured proteins), based on mass, isoelectric point (the pH at which the surface is not charged), and/or electrophoretic mobility (related to the length or folding of the protein). Communications between proteins (such as cell signaling) are being studied with traditional methods of yeast two-hybrid analysis, new methods in protein microarrays, and immunoaffinity chromatography.

Mass spectrometry (MS) based methodologies are used in combination with a variety of sample preparation methods to enable separation of proteins and molecular weight determinations. These include the use of liquid chromatography (LC), coupled with MS to conduct shotgun proteomics (detection of a broad spectrum of proteins), the use of digestion of proteins prior to MS analysis (termed bottom up proteomics) to detect selected proteins, or the use of ion trapping and mass measurement (termed top-down proteomics). The mass or the mass spectra are then compared with information contained within a sequence database to predict structure.

Proteomics in reproductive and developmental biology is a rapidly growing area, which has focused largely on the identification of the proteome of reproductive fluids or the identification of noninvasive biomarkers that can characterize the status of male fertility or female pregnancy-related pathologies. These studies hold promise for gaining a mechanistic understanding of the etiology of the pathology for the use in the development of interventional strategies and treatments. A summary of the applications of mass spectrometry-based proteomics in the characterization of the proteome of reproductive fluids has recently been published (26).

Proteomics analysis of amniotic fluid and maternal serum has been used to reveal markers of intrauterine infection or premature rupture of membranes (27–36). Research on the processes in mammalian preimplantation and embryonic development (37), and in the study of the cellular communication between the germinal vesicle oocytes and surrounding cumulus calls (38) has also included the use of proteomics methods. The proteome of the chicken egg (39–41), rat embryo proper, and yolk sac membrane (42), and the proteome of semen have been investigated (43). In addition, the identification of biomarkers that correlate with fertility rates (44) and interindividual variations in seminal plasma of fertile men (45) have been investigated using proteomics. Proteomics has been included in the study of bisphenol A–induced alterations in the proteome of prenatally exposed mice (46), and to evaluate antiandrogenic effects of drugs and chemicals for use in Hershberger Assay (47).

Highlighted example: Surface-enhanced laser desorption/ionization time-of-flight mass spectrometry was used to analyze amniotic fluid collected from women with clinical presentation of chronic hypertension or preeclampsia (48). Analysis of samples from women that did not present with these findings (normotensive controls) were also conducted. Signals were revealed that distinguished hypertensive or pre-eclamptic samples from the noromotensive controls. Subsequent isolation of associated signals using high-performance LC, followed by gel electrophoresis and in-gel tryptic digestion, enabled the identification of proapolipoprotein A-I and a functionally obscure peptide, SBBI42. This study demonstrated the promise of proteomics methods to develop markers to detect, stage, and monitor for hypertensive disorders of pregnancy.

METABOLOMICS
Metabolomics is a term used to describe the study of the low molecular weight (e. g., <2000 molecular weight) component of cells, tissues, and biological fluids. It is generally presented (49) that the mammalian metabolome consists of less than 5000 metabolites that can be related to the approximately 115 known specific mammalian metabolic pathways; a small universe compared with the estimated 100,000 proteins or 20,000–40,000 genes. In addition to the endogenous metabolome, there are considerable additional metabolites (e.g., derived from chemicals, foods, or drugs) that must be considered in the analysis of metabolomics data.

While the study of the metabolome is technically challenging, it is believed that the metabolome has considerable qualitative conservation across species, which affords the opportunity to make cross species comparisons for the development of translational markers or for the extrapolation of risks. Unlike genomics or proteomics, metabolomics tools are readily applied to the investigation of the metabolome of biological fluids. Most metabolomics studies involve making comparisons of metabolic profiles from samples obtained under normal physiological conditions (or nondisease states), with samples obtained from specific disease states or specified environmental exposures (e.g., drugs, toxicants, stress, nutrition). This technology is ideally suited for the development of noninvasive markers for the early detection, diagnosis, and treatment of disease, and providing mechanistic insights for the development of new interventional strategies.

Metabolomic (also termed metabonomic) analysis of toxicity to the adult animal species has been applied many times to examine the effects of drug or environmental chemicals on target organs in a pattern recognition/classification approach. The work of the Consortium for Metabonomic Toxicology was recently summarized (50). This consortium used metabolomics of urine and serum to construct a predictive system for liver and kidney toxicity, and through identification of metabolites that were most significant for prediction, were able to determine the endogenous biochemicals and their related pathways that were responsible for the associations. In addition, metabolomics investigations have been conducted that demonstrate the utility of this methodology for developing noninvasive markers for disease detection. Some recent examples include the use of metabolomics to investigate central nervous diseases (51,52) to develop biomarkers of prostate cancer (53), to determine population diversity in metabolic phenotype and the correlation of specific markers with diet and blood pressure (54), to study mechanisms by which fat impacts pancreatic disease (55), and to evaluate glucose disorders (reviewed in Ref. 56). Metabolomics is believed to hold huge promise for providing mechanistic insights for linking exposure to disease, extrapolating across species, and environmental monitoring (57,58; summarized in Ref. 59,60).

Metabolomics investigations have most commonly employed nuclear magnetic resonance (NMR) and chromatography-coupled MS methods, while additional spectroscopic methods (e.g., IR, Raman, ECD) are not as widely used but have demonstrated power in revealing metabolomics-based biomarkers. Each of these analytical methods has distinct advantages (e.g., in resolution or sensitivity) and inherent disadvantages (e.g., in throughput, longitudinal stability, or complexity of operations). When selecting a metabolomics method, investigators consider if the question is best answered by the analysis of (*i*) polar or nonpolar components; (*ii*) noninvasive biological fluids (e.g., urine and blood) or tissue extracts; (*iii*) the need for the nonselective capture of a broad range of low molecular weight analytes; or (*iv*) the need for selected metabolite/ pathway monitoring. In addition, some methods are more longitudinally stable, and thus consideration of the duration of the study period (e.g., longitudinal or cross-sectional study designs) in regard to the selected method is critical. Additional, essential components of a metabolomics analysis includes the use of empirical and standards libraries, species- and tissue-specific metabolite databases, and implementing tools for data reduction and pathway mapping.

For hypothesis-driven metabolomic studies, where the investigator seeks to understand a specific pathway, "targeted profiling" for specific analytes is often employed. This is most commonly accomplished through the use of gas chromatography (GC) MS or LC MS methods, where standards are used to develop specific chromatographic methods for the separation of compounds in tissues or biological fluids and detection of the signal for comparison with known standards. These types of methods have been used for decades for the selected monitoring of organic acids, amino acids and, more recently, using tandem mass spectrometry for analysis of free and total carnitine in human plasma (61). NMR methods also afford the selection of targeted analytes, and quantitative targeted profiling via NMR has gained considerable traction in the metabolomics community (62). This method, most commonly applied to urine, is superior to MS-based methods in terms of longitudinal stability, rapid acquisition, and rapid spectral interpretation, with available quantitative libraries for over 300 metabolites important to the study of the metabolome. The MS-based targeted profiling methods rely on targeted synthetic standards, and thus are powerful in providing sensitivity and resolution for these targeted analytes.

Many investigators prefer the use of broader spectrum profiling methodologies for nonhypothesis driven research. Most of these studies are conducted with an attempt to "move away from the lamp post" to discover novel markers and glean new mechanistic insights. Metabolomics via NMR has been the most common approach published for the nonbiased detection of a wide range of analytes and is amenable to the analysis of many biological fluids as well as tissue extracts. The NMR data acquisition approach is the same as for targeted profiling. However, the data analysis approach involves the use of fixed bin sizes that are created by integration of a fixed spectral width (typically 0.04 ppm) across the NMR spectrum (50,63). The NMR platform for metabolomics is high throughput, reproducible between laboratories, stable over long-term studies, and has advanced software for spectral deconvolution, metabolite libraries, quantitative libraries, and metabolite identification. The most widely used broad-spectrum MS methods involves time-of-flight (tof) MS. Time of flight uses an electric field to accelerate ions to the same potential, and the ions advance to the detector through a "flight" tube, where the time to reach the detector is related to the mass of the ion. Using tof for broad-spectrum analysis increases mass accuracy and mass resolution, as well as improves sensitivity, acquisition, and dynamic range when profiling over a broad molecular weight range. LC and ultra performance LC (UPLC), coupled with tof (e.g., UPLC-tof-MS), is ideal for the analysis of a broad spectrum of polar metabolites such as amino acids, aromatic amino acids, organics, and sugars. Incorporation of ion mobilization with LC-tof-MS methods improves the ability to separate compounds of similar masses. GC coupled with tof (i.e., GC-tof-MS) is ideal for the analysis of a broad spectrum of nonpolar metabolites such as lipids, steroids, terpenes, and flavonoids, and the GCxGC-tof-MS systems provide a higher resolution chromatographic system by employing a repeated reinjection of effluent from the first column to the second column.

Any one of these methods enables the measurement of hundreds of metabolites that map to biochemical pathways and assign a fraction of the metabolome. Nonetheless, many spectroscopic methods, individually, have been proven successful in providing patterns of metabolites in tissues and biological fluids to differentiate "groups" based on how the biochemical profile is influenced by exposure (e.g., chemical, drug, or diet), gender, age, genetic background, or clinically defined illness. Through assignment of signals and linking metabolites to known biochemical pathways, investigators have developed mechanistic insights that show promise for the utilization of this technology to stage disease, monitor therapeutic treatments, or develop or improve intervention strategies.

While genomics and proteomic applications in reproductive and developmental toxicology have rapidly increased in the literature in recent years, reproductive toxicology studies that incorporate metabolomics are just beginning to emerge in the literature. Recent studies have demonstrated the use of metabolomics to distinguish between plasma samples from healthy women and those from women with preeclampsia (64). Researchers are using metabolomics to provide markers for the early detection of epithelial ovarian cancer (EOC), a promise demonstrated through a pattern recognition study of serum from healthy women and women with EOC (65). Metabolomics has widely been used in the study of cancer; one example revealed differences in the metabolome of ovarian carcinomas and borderline tumors (66). Researchers have also demonstrated that metabolomics could be used to predict the viability of an embryo to result in a positive pregnancy outcome at a significantly greater rate than using traditional morphology measurements (67; http://www.reuters.com/article/pressRelease/idUS131968+07-Jul-2008+PRN20080707). The use of metabolomics in the study of male fertility is growing, with recent advances that associate alterations in the sperm metabolome with diabetic status (68).

The use of metabolomics in the study of the consequences of environmental stressors in reproduction and development has also grown. Significant metabolic perturbations in the early life stages of Chinook salmon exposed to pesticides have been demonstrated using metabolomics (69). A study (70) of pregnant rats on feed restriction, compared with pregnant rats receiving feed ad libitum, with measurement of glucose and lipid levels in the offspring for up to six months after birth, demonstrated an alteration in the metabolic syndrome phenotype.

Highlighted example: The use of metabolomics for improving our ability to draw correlations between early life exposures and reproductive and/or developmental outcomes was recently evaluated (71–73). To demonstrate the use of metabolomics in studies of reproductive health and developmental outcomes, pregnant rats were orally dosed with butylbenzylphthalate (BBP) daily during gestation [gestation day (gd) 14–22], with collection of urine from the dam on gd 18, and following lactation and weaning (pnd 21), and from the male and female pups several days after weaning (pnd 25). Traditional phenotypic anchors (e.g., nipple retention, anogenital distance, hormones) were measured. BBP doses were selected that are known to induce effects (750 mg/kg/day from gd 14 to 21) and that had not previously been shown to induce effects (25 mg/kg/day from gd 14 to 21). Quantitative NMR-based metabolomics analysis (which excluded the measurement of BBP and BBP-derived metabolites) was performed on the urine samples collected from the dam and from the male and female pups. As expected, analysis of dam urine collected on gd 18 (during the time of exposure) showed significant differences in the concentrations of many endogenous metabolites for dams that received the high dose of BBP compared with the control group (data not shown). At the low-dose no-effect level, the endogenous metabolites in urine from dams on gd 18 were also significantly different from controls. Principal component analysis (PCA), applied to the metabolic profiles of urine that was collected three weeks (pnd 21) after the gestational exposure had ended, was used to differentiate dams that had previously experienced (during gestation) a low-dose no-effect level BBP exposure from dams that had vehicle during gestational exposure or had pregnancy complications from high-dose BBP exposures (Fig. 1). Compounds associated with tryptophan and niacinamide metabolism (e.g., quinolate, trigonelline, indole-3-acetate, and nicotinate) were decreased in the pnd 21 high-dose group relative to the control group. The high-dose, pnd 21 dam urine was lower in o-phosphoethanolamine (glycerophospholipid metabolism and glycine serine, threonine metabolism) and lower in protocatechuate (phenylalanine and tyrosine, tryptophan metabolism) relative to the control group.

PCA analysis of urine collected from male and female pups on pnd 23 (23 days after delivery from the dams that had received gestational exposure) demonstrated that metabolomics could be used to differentiate pups based on prior exposure (Fig. 2, male

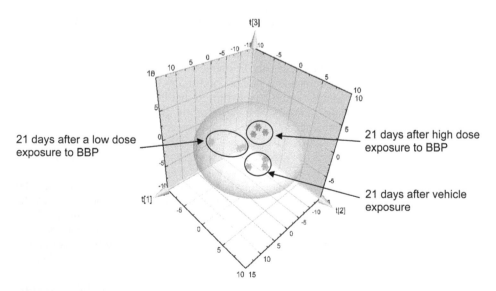

FIGURE 1 *Maternal urine*: PCA analysis of pnd 21 urine could be used to differentiate dams that received vehicle, low-dose BBP, or high-dose BBP exposures daily during gestation (gd 14–21). *Abbreviation*: PCA, principal component analysis.

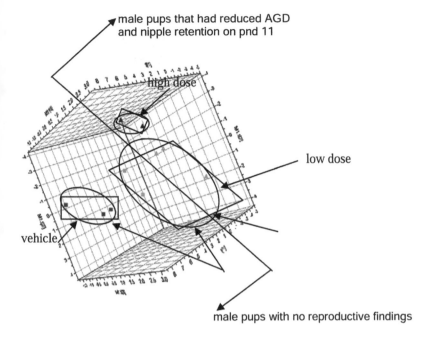

male pups that had reduced AGD and nipple retention on pnd 11

high dose

low dose

vehicle

male pups with no reproductive findings

FIGURE 2 PCA of pup urine collected three weeks after birth to BBP-exposed dams was used to differentiate pups with effects from pups with no effects and was used to differentiate pups at the low dose that had no effects measured from the vehicle control. *Abbreviation*: PCA, principal component analysis.

pups) and to differentiate pups based on traditional phenotypic findings, while the measurement of hormones was insufficient for making this differentiation (72). Male pups born to dams with a low-dose gestational exposure, and that had no measurable reproductive endpoint at any study time-point, were best classified by metabolites that map to the glycine, serine, and threonine pathway, or to alanine metabolism. Male pups born to dams that had a high-dose gestational exposure were best classified by compounds involved in the Krebs cycle, alanine, and glycine, serine, threonine, and pyrimidine, purine metabolism. All pups that had measurable reproductive endpoints at any study timepoint also had elevated 2-oxoglutarate (relative to the time-matched control). For both the low- and high-dose groups, male pups showed decreases in metabolites related to tryptophan and nicotinate metabolism, tyrosine metabolism, purine metabolism. This study demonstrated the use of metabolomics in the classification of prior in utero exposure and the promise of using this technology to gain mechanistic insights to link exposure and disease outcomes.

In a second study, male and female CD rats were fed AIN-76A (phytoestrogen free) or AIN-76A diet with genistein added (250 or 500 ppm) through acclimation and breeding. Female rats were maintained on these diets through gestation, lactation, and weaning (pnd 21), and their offspring were maintained on the mothers' diet following weaning. Metabolomics of endogenous compounds in urine could differentiate dams that received each of the three diets (Fig. 3) and could differentiate the pups based on gender, genistein level in the diet, and prepubertal (Fig. 4), or postpubertal age (data not shown). Metabolites most significant to classification of dams receiving the highest level of genistein in diet included gluconate, carnosine, histidine, phenylacetate, ethanol, and

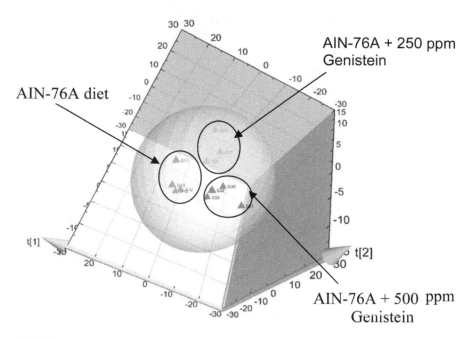

FIGURE 3 Multivariate analysis of urine from dams (pnd 21) fed differing concentrations of genistein in diet prior to mating and through weaning: phytoestrogen-free diet (AIN-76A), or AIN-76A + 250 ppm genistein, AIN-76A + 500 ppm genistein.

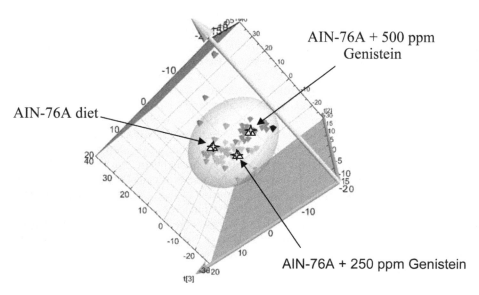

FIGURE 4 Multivariate analysis of urine from pups (pnd 56) that were fed differing concentrations of genistein in diet from birth to dams with corresponding diet: phytoestrogen-free diet (AIN-76A, or AIN-76A + 250 ppm genistein, AIN-76A + 500 ppm genistein. ⟨⟩ Represents the mean of each of the three groups.

anserine. Alanine, pyroglutamate, glutamate, citrate, cysteine, and oxoglutarate were important in classification of rats receiving the phytoestrogen-free diet. Metabolites most significant to classification of pups receiving the highest level of genistein in diet included carnosine, histidine, phenylacetate, isoeuguenol, and phenylalanine. N-acetylaspartate, glutamate, succinate, and N,N-dimethylglucine were important in classification of pups receiving the phytoestrogen-free diet. This pilot investigation revealed differences in how the level of genistein in diet impacted the biochemical profiles of the dams and offspring.

CONCLUSIONS

There has been a critical need in the study of reproduction and development for new molecular methods that can assess prior in utero exposure and that can be used to unravel the mechanisms underlying disease or dysfunction. New approaches are needed to reveal biomarkers that can assess disease or dysfunction early on, for early detection, diagnosis, and intervention. Traditional markers for the assessment of reproductive disease or dysfunction are not early, but are typically markers of the specific reproductive effect, such as traditional measurements of nipple retention, reduced anogenital distance, or andrology assessment. There is a need for noninvasive markers for the early detection and diagnosis of disease and dysfunction, and that can be used to gain mechanistic insights regarding alterations in biochemical processes that have taken place. While the measurement of hormones (such as gonadal testosterone and estradiol, hypothalamic gonadotrophin–releasing hormones, anterior pituitary hormone, follicle stimulating hormone, and luteinizing hormone) provides valuable information, it is known that normal cyclical fluctuations in hormonal responses often compromise study interpretations and can arise from interindividual fluctuations in cyclical production and release. Study designs that include hormone measurements often require a large number of samples taken over many timepoints. It is likely the case that hormonal changes that occur early in life have a consequence on the overall biochemistry of the developing offspring that is longer lived than the hormonal fluctuation. Thus, the incorporation of -omics technologies in the study of reproduction and development provides a means to detect markers of the consequences of hormonal imbalances that occur as a result of an environmental insult (be it a chemical, drug, stressor, etc.).

Genomic, proteomic, and metabolomics approaches presented herein provide compelling examples of the promise of these technologies in providing early markers and noninvasive markers for the detection of disease or dysfunction. In addition, both genomic and metabolomics technologies have been demonstrated in their use in providing a classification approach to the assessment of prior in utero exposure. These advances have and will continue to enhance our understanding of the processes that are involved in normal or abnormal development of the offspring and the consequences of environmental insult.

REFERENCES

1. Gant TW. Novel and future applications of microarrays in toxicological research. Expert Opin Drug Metab Toxicol 2007; 3 (4):599–608.
2. Mirkes P, McClure ME, Heindel JJ, et al. Developmental toxicology in the 21st century: multidisciplinary approaches using model organisms and genomics. Birth Defects Res A Clin Mol Teratol 2003; 67(1):21–34.
3. Clausen I, Kietz S, Fischer B. Lineage-specific effects of polychlorinated biphenyls (PCB) on gene expression in the rabbit blastocyst. Reprod Toxicol 2005; 20(1):47–56.
4. Daston GP, Naciff JM. Gene expression changes related to growth and differentiation in the fetal and juvenile reproductive system of the female rat: evaluation of microarray results. Reprod Toxicol 2005; 19(3):381–394.

5. Naciff JM, Hess KA, Overmann GJ, et al. Gene expression changes induced in the testis by transplacental exposure to high and low doses of 17{alpha}-ethynyl estradiol, genistein, or bisphenol A. Toxicol Sci 2005; 86(2):396–416.
6. Gelineau-van Waes J, Bennett GD, Finnell RH. Phenytoin-induced alterations in craniofacial gene expression. Teratology 1999; 59(1):23–34.
7. Mikheeva S, Barrier M, Little SA, et al. Alterations in gene expression induced in day-9 mouse embryos exposed to hyperthermia (HS) or 4-hydroperoxycyclophosphamide (4CP): analysis using cDNA microarrays. Toxicol Sci 2004; 79(2): 345–359.
8. Kultima K, Nystrom AM, Scholz B, et al. Valproic acid teratogenicity: a toxicogenomics approach. Environ Health Perspect 2004; 112(12):1225–1235.
9. Moggs JG, Tinwell H, Spurway T, et al. Phenotypic anchoring of gene expression changes during estrogen-induced uterine growth. Environ Health Perspect 2004; 112(16): 1589–1606.
10. Naciff JM, Daston GP. Toxicogenomic approach to endocrine disrupters: identification of a transcript profile characteristic of chemicals with estrogenic activity. Toxicol Pathol 2004; 32(suppl 2):59–70.
11. Shao WJ, Tao LY, Xie JY, et al. Exposure of preimplantation embryos to insulin alters expression of imprinted genes. Comp Med 2007; 57(5):482–486.
12. Shao WJ, Tao LY, Gao C, et al. Alterations in methylation and expression levels of imprinted genes H19 and Igf2 in the fetuses of diabetic mice. Comp Med 2008; 58(4): 341–346.
13. Wu Q, Ohsako S, Ishimura R, et al. Exposure of mouse preimplantation embryos to 2,3,7,8-tetrachlorodibenzo-p-dioxin (TCDD) alters the methylation status of imprinted genes H19 and Igf2. Biol Reprod 2004; 70(6):1790–1797.
14. Currie RA, Orphanides G, Moggs JG. Mapping molecular responses to xenoestrogens through Gene Ontology and pathway analysis of toxicogenomic data. Reprod Toxicol 2005; 20(3):433–440.
15. Naciff JM, Overmann GJ, Torontali SM, et al. Gene expression profile induced by 17 alpha-ethynyl estradiol in the prepubertal female reproductive system of the rat. Toxicol Sci 2003; 72(2):314–330.
16. Naciff JM, Khambatta ZS, Thomason RG, et al. The genomic response of a human uterine endometrial adenocarcinoma cell line to 17alpha-ethynyl estradiol. Toxicol Sci 2009; 107(1):40–55.
17. Barrier M, Mirkes PE. Proteomics in developmental toxicology. Reprod Toxicol 2005; 19(3): 291–304.
18. Wetmore BA, Merrick BA. Toxicoproteomics: proteomics applied to toxicology and pathology. Toxicol Pathol 2004; 32(6):619–642.
19. Heijne WH, Kienhuis AS, van Ommen B, et al. Systems toxicology: applications of toxicogenomics, transcriptomics, proteomics and metabolomics in toxicology. Expert Rev Proteomics 2005; 2(5):767–780.
20. Issaq H, Veenstra T. Two-dimensional polyacrylamide gel electrophoresis (2D-PAGE): advances and perspectives. Biotechniques 2008; 44(5):697–698.
21. Paoletti AC, Washburn MP. Quantitation in proteomic experiments utilizing mass spectrometry. Biotechnol Genet Eng Rev 2006; 22:1–19.
22. America AH, Cordewener JH. Comparative LC-MS: a landscape of peaks and valleys. Proteomics 2008; 8(4):731–749.
23. Simpson RJ, Bernhard OK, Greening DW, et al. Proteomics-driven cancer biomarker discovery: looking to the future. Curr Opin Chem Biol. 2008; 12(1):72–77.
24. Deutsch EW, Lam H, Aebersold R. Data analysis and bioinformatics tools for tandem mass spectrometry in proteomics. Physiol Genomics 2008; 33(1):18–25.
25. Matt P, Fu Z, Fu Q, et al. Biomarker discovery: proteome fractionation and separation in biological samples. Physiol Genomics 2008; 33(1):12–17.
26. Kolialexi A, Mavrou A, Spyrou G, et al. Mass spectrometry-based proteomics in reproductive medicine. Mass Spectrom Rev 2008; 27(6): 624–634.
27. Buhimschi IA, Buhimschi CS, Christner R, et al. Proteomic profiling and intra-amniotic infection. JAMA 2004; 292:2338.
28. Buhimschi IA, Buhimschi CS, Christner R, et al. Proteomics technology for the accurate diagnosis of inflammation in twin pregnancies. BJOG 2005; 112:250–255.

29. Buhimschi IA, Buhimschi CS, Weiner CP, et al. Proteomic but not enzyme-linked immunosorbent assay technology detects amniotic fluid monomeric calgranulins from their complexed calprotectin form. Clin Diagn Lab Immunol 2005; 12:837–844.
30. Buhimschi IA, Christner R, Buhimschi CS. Proteomic biomarker analysis of amniotic fluid for identification of intra-amniotic inflammation. BJOG 2005; 112:173–181.
31. Gravett MG, Novy MJ, Rosenfeld RG, et al. Diagnosis of intra-amniotic infection by proteomic profiling and identification of novel biomarkers. JAMA 2004; 292:462–469.
32. Klein LL, Freitag BC, Gibbs RS, et al. Detection of intra-amniotic infection in a rabbit model by proteomics-based amniotic fluid analysis. Am J Obstet Gynecol 2005; 193: 1302–1306.
33. Nilsson S, Ramstrom M, Palmblad M, et al. Explorative study of the protein composition of amniotic fluid by LC electrospray ionization Fourier transform ion cyclotron resonance mass spectrometry. J Proteome Res 2004; 3:884–889.
34. Ruetschi U, Rosen A, Karlsson G, et al. Proteomic analysis using protein chips to detect biomarkers in cervical and amniotic fluid in women with intra-amniotic inflammation. J Proteome Res 2005; 4:2236–2242.
35. Thadikkaran L, Crettaz D, Siegenthaler MA, et al. The role of proteomics in the assessment of premature rupture of fetal membranes. Clin Chim Acta 2005; 360(1–2): 27–36.
36. Vuadens F, Benay C, Crettaz D, et al. Identification of biologic markers of the premature rupture of fetal membranes: proteomic approach. Proteomics 2003; 3:1521–1525.
37. Katz-Jaffe MG, Linck DW, Schoolcraft WB, et al. A proteomic analysis of mammalian preimplantation embryonic development. Reproduction 2005; 130(6):899–905.
38. Memili E, Peddinti D, Shack LA, et al. Bovine germinal vesicle oocyte and cumulus cell proteomics. Reproduction 2007; 133(6):1107–1120.
39. Mann K. Proteomic analysis of the chicken egg vitelline membrane. Proteomics 2008; 8(11): 2322–2332.
40. Mann K, Mann M. The chicken egg yolk plasma and granule proteomes. Proteomics 2008; 8(1):178–191.
41. Mann K. The chicken egg white proteome. Proteomics 2007; 7(19):3558–3568.
42. Usami M, Mitsunaga K, Nakazawa K. Two-dimensional electrophoresis of protein from cultured postimplantation rat embryos for developmental toxicity studies. Toxicol In Vitro 2007; 21(3):521–526.
43. Li LW, Fan LQ, Zhu WB, et al. Establishment of a high-resolution 2-D reference map of human spermatozoal proteins from 12 fertile sperm-bank donors. Asian J Androl 2007; 9(3): 321–329.
44. Peddinti D, Nanduri B, Kaya A, et al. Comprehensive proteomic analysis of bovine spermatozoa of varying fertility rates and identification of biomarkers associated with fertility. BMC Syst Biol 2008; 22(2):19.
45. Yamakawa K, Yoshida K, Nishikawa H, et al. Comparative analysis of interindividual variations in the seminal plasma proteome of fertile men with identification of potential markers for azoospermia in infertile patients. J Androl 2007; 28(6):858–865.
46. Yang M, Lee HS, Pyo MY. Proteomic biomarkers for prenatal bisphenol A-exposure in mouse immune organs. Environ Mol Mutagen 2008; 49(5):368–373.
47. Tinwell H, Friry-Santini C, Rouquié D, et al. Evaluation of the antiandrogenic effects of flutamide, DDE, and linuron in the weanling rat assay using organ weight, histopathological, and proteomic approaches. Toxicol Sci 2007; 100(1):54–65.
48. Park JS, Oh KJ, Norwitz ER, et al. Identification of proteomic biomarkers of preeclampsia in amniotic fluid using SELDI-TOF mass spectrometry. Reprod Sci 2008; 5(5):457–468.
49. Dettmer K, Hammock BD. Metabolomics—a new exciting field within the "omics" sciences. Environ Health Perspect 2004; 112(7):A396–A397.
50. Lindon JC, Holmes E, Nicholson JK. Metabonomics in pharmaceutical R&D. FEBS J 2007; 274(5):1140–1151.
51. Kaddurah-Daouk R, McEvoy J, Baillie RA, et al. Metabolomic mapping of atypical antipsychotic effects in schizophrenia. Mol Psychiatry 2007; 12(10):934–945.
52. Kaddurah-Daouk R, Krishnan KR. Metabolomics: a global biochemical approach to the study of central nervous system diseases. Neuropsychopharmacology 2009; 34(1):173–186.
53. Sreekumar A, Poisson LM, Rajendiran TM, et al. Metabolomic profiles delineate potential role for sarcosine in prostate cancer progression. Nature 2009; 12,457 (7231):910–914.

54. Holmes E, Loo RL, Stamler J, et al. Human metabolic phenotype diversity and its association with diet and blood pressure. Nature 2008; 453:(7193):396–400.
55. Zyromski NJ, Mathur A, Gowda GA, et al. Nuclear magnetic resonance spectroscopy-based metabolomics of the fatty pancreas: implicating fat in pancreatic pathology. Pancreatology 2009; 19, 9(4):410–419.
56. Sébédio JL, Pujos-Guillot E, Ferrara M. Metabolomics in evaluation of glucose disorders. Curr Opin Clin Nutr Metab Care 2009;12(4):412–418.
57. Weis BK, Balshaw D, Barr JR, et al. Personalized exposure assessment: promising approaches for human environmental health research. Environ Health Perspect 2005; 113(7):840–848.
58. Sumner SCJ, Fennell TR. Biomarkers, omics, and species comparisons. Hum Ecol Risk Assess: Int J 2007; 13(1):111–119.
59. Viant MR. Recent developments in environmental metabolomics. Mol Biosyst 2008; 10: 980–986.
60. Sumner S, Burgess J, Snyder R, et al. A Non-invasive Marker of Isoniazid Induced Liver Injury. Metabolomics 2010; DOI:10.1007.
61. Stevens RD, Hillman SL, Worthy S, et al. Assay for free and total carnitine in human plasma using tandem mass spectrometry. Clin Chem 2000; 46(5):727–729.
62. Weljie AM, Newton J, Mercier PM. Targeted profiling: quantitative analysis of 1H-NMR metabolomics data. Anal Chem 2006; 78(13):4430–4442.
63. Lindon JC, Nicholson JK, Holmes E. The Handbook of Metabonomics and Metabolomics. Elsevier, 2007, ISBN 0444528415, 9780444528414.
64. Kenny LC, Dunn WB, Ellis DI, et al. Novel biomarkers for pre-eclampsia detected using metabolomics and machine learning. Metabolomics 2005; 1:227–234.
65. Odunsi K, Wollman RM, Ambrosone CB, et al. Detection of epithelial ovarian cancer using 1H-NMR-based metabonomics. Int J Cancer 2005; 113 (5):782–788.
66. Denkert C, Budczies J, Kind T, et al. Mass spectrometry-based metabolic profiling reveals different metabolite patterns in invasive ovarian carcinomas and ovarian borderline tumors. Cancer Res 2006; 66(22):10795–10804.
67. Vergouw CG, Botros LL, Roos P, et al. Metabolomic profiling by near-infrared s pectroscopy as a tool to assess embryo viability: a novel, non-invasive method for embryo selection. Human Reprod 2008; 23(7):1499–1504.
68. Mallidis C, Green BD, Rogers D, et al. Metabolic profile changes in the testes of mice with streptozotocin-induced type 1 diabetes mellitus. Int J Androl 2007; 30:9999.
69. Viant MR. Recent developments in environmental metabolomics. Mol Biosyst 2008; 4(10): 980–986.
70. Desai M, Gayle D, Babu J, et al. The timing of nutrient restriction during rat pregnancy/lactation alters metabolic syndrome phenotype. Am J Obstet Gynecol 2007; 196(6):555. e1–e7.
71. Sumner S, Snyder R, Burgess J, et al. Metabolomics: application to the study of phthalates in reproduction and development. Toxicologist 2008; 64:11.
72. Sumner S, Snyder R, Burgess J, et al. Metabolomics in the study of reproduction and development. Toxicologist 2009; 1414:218.
73. Sumner S, Snyder R, Burgess J, et al. Metabolomics in the Assessment of in utero Exposure: application to the study of phthalates. J Applied Toxicol 2009; 29(8):703–714.

Index

Note: Page numbers followed by *f* and *t* indicate figures and tables, respectively.

Milton Keynes UK
Ingram Content Group UK Ltd.
UKHW020321111024
449327UK00040B/1478

9 780367 383602